Graduate Texts in Mathematics 243

Graduate Texts in Mathematics

(continued after index)

Ross Geoghegan

Topological Methods in Group Theory

Springer

Ross Geoghegan
Department of Mathematical Sciences
Binghamton University (SUNY)
Binghamton
NY 13902-6000
USA
ross@math.binghamton.edu

ISBN 978-1-4419-2564-0   e-ISBN 978-0-387-74714-2

Mathematics Subject Classification (2000): 20-xx 54xx 57-xx 53-xx

To Suzanne, Niall and Michael

Preface

This book is about the interplay between algebraic topology and the theory of infinite discrete groups. I have written it for three kinds of readers. First, it is for graduate students who have had an introductory course in algebraic topology and who need bridges from common knowledge to the current research literature in geometric and homological group theory. Secondly, I am writing for group theorists who would like to know more about the topological side of their subject but who have been too long away from topology. Thirdly, I hope the book will be useful to manifold topologists, both high- and low-dimensional, as a reference source for basic material on proper homotopy and locally finite homology.

To keep the length reasonable and the focus clear, I assume that the reader knows or can easily learn the necessary algebra, but wants to see the topology done in detail. Scattered through the book are sections entitled "Review of ..." in which I give statements, without proofs, of most of the algebraic theorems used. Occasionally the algebraic references are more conveniently included in the course of a topological discussion. All of this algebra is standard, and can be found in many textbooks. It is a mixture of homological algebra, combinatorial group theory, a little category theory, and a little module theory. I give references.

As for topology, I assume only that the reader has or can easily reacquire knowledge of elementary general topology. Nearly all of what I use is summarized in the opening section. A prior course on fundamental group and singular homology is desirable, but not absolutely essential if the reader is willing to take a very small number of theorems in Chap. 2 on faith (or, with a different philosophy, as axioms). But this is not an elementary book. My maxim has been: "Start far back but go fast."

In my choice of topological material, I have tried to minimize the overlap with related books such as [29], [49], [106], [83], [110], [14] and [24]. There is some overlap of technique with [91], mainly in the content of my Chap. 11, but the point of that book is different, as it is pitched towards problems in geometric topology.

The book is divided into six Parts. Parts I and III could be the basis for a useful course in algebraic topology (which might also include Sects. 16.1-16.4). I have divided this material up, and placed it, with group theory in mind. Part II is about finiteness properties of groups, including both the theory and some key examples. This is a topic that does not involve asymptotic or end-theoretic invariants. By contrast, Parts IV and V are mostly concerned with such matters – topological invariants of a group which can be seen "at infinity." Part VI consists of essays on three important topics related to, but not central to, the thrust of the book.

The modern study of infinite groups brings several areas of mathematics into contact with group theory. Standing out among these are: Riemannian geometry, synthetic versions of non-positive sectional curvature (e.g., hyperbolic groups, CAT(0) spaces), homological algebra, probability theory, coarse geometry, and topology. My main goal is to help the reader with the last of these.

In more detail, I distinguish between topological methods (the subject of this book) and metric methods. The latter include some topics touched on here in so far as they provide enriching examples (e.g., quasi-isometric invariants, CAT(0) geometry, hyperbolic groups), and important methods not discussed here at all (e.g., train-tracks in the study of individual automorphisms of free groups, as well as, more broadly, the interplay between group theory and the geometry of surfaces.) Some of these omitted topics are covered in recent books such as [48], [134], [127], [5] and [24].

I am indebted to many people for encouragement and support during a project which took far too long to complete. Outstanding among these are Craig Guilbault, Peter Hilton, Tom Klein, John Meier and Michael Mihalik. The late Karl Gruenberg suggested that there is a need for this kind of book, and I kept in mind his guidelines. Many others helped as well – too many to list; among those whose suggestions are incorporated in the text are: David Benson, Robert Bieri, Matthew Brin, Ken Brown, Kai-Uwe Bux, Dan Farley, Wolfgang Kappe, Peter Kropholler, Francisco Fernandez Lasheras, Gerald Marchesi, Holgar Meinert, Boris Okun, Martin Roller, Ralph Strebel, Gadde Swarup, Kevin Whyte, and David Wright.

I have included Source Notes after some of the sections. I would like to make clear that these constitute merely a subjective choice, mostly papers which originally dealt with some of the less well-known topics. Other papers and books are listed in the Source Notes because I judge they would be useful for further reading. I have made no attempt to give the kind of bibliography which would be appropriate in an authoritative survey. Indeed, I have omitted attribution for material that I consider to be well-known, or "folklore," or (and this applies to quite a few items in the book) ways of looking at things which emerge naturally from my approach, but which others might consider to be "folklore".

Lurking in the background throughout this book is what might be called the "shape-theoretic point of view." This could be summarized as the transfer of the ideas of Borsuk's shape theory of compact metric spaces (later enriched by the formalism of Grothendieck's "pro-categories") to the proper homotopy theory of ends of open manifolds and locally compact polyhedra, and then, in the case of universal covers of compact polyhedra, to group theory. I originally set out this program, in a sense the outline of this book, in [68]. The formative ideas for this developed as a result of extensive conversations with, and collaboration with, David A. Edwards in my mathematical youth. Though those conversations did not involve group theory, in some sense this book is an outgrowth of them, and I am happy to acknowledge his influence.

Springer editor Mark Spencer was ever supportive, especially when I made the decision, at a late stage, to reorganize the book into more and shorter chapters (eighteen instead of seven). Comments by the anonymous referees were also helpful.

I had the benefit of the TeX expertise of Marge Pratt; besides her ever patient and thoughtful consideration, she typed the book superbly. I am also grateful for technical assistance given me by my mathematical colleagues Collin Bleak, Keith Jones and Erik K. Pedersen, and by Frank Ganz and Felix Portnoy at Springer.

Finally, the encouragement to finish given me by my wife Suzanne and my sons Niall and Michael was a spur which in the end I could not resist.

Binghamton University (SUNY Binghamton),

May 2007

Ross Geoghegan

Notes to the Reader

1. **Shorter courses:** Within this book there are two natural shorter courses. Both begin with the first four sections of Chap. 1 on the elementary topology of CW complexes. Then one can proceed in either of two ways:

- *The homotopical course*: Chaps. 3, 4, 5 (omitting 5.4), 6, 7, 9, 10, 16 and 17.
- *The homological course*: Chaps. 2, 5, 8, 11, 12, 13, 14 and 15.

2. **Notation:** If the group G acts on the space X on the left, the set of orbits is denoted by $G \backslash X$; similarly, a right action gives X/G. But if R is a commutative ring and (M, N) is a pair of R-modules, I always write M/N for the quotient module. And if A is a subspace of the space X, the quotient space is denoted by X/A.

The term "ring" without further qualification means a commutative ring with $1 \neq 0$.

I draw attention to the notation $X \overset{c}{-} A$ where A is a subcomplex of the CW complex X. This is the "CW complement", namely the largest subcomplex of X whose 0-skeleton consists of the vertices of X which are not in A. If one wants to stay in the world of CW complexes one must use this as "complement" since in general the ordinary complement is not a subcomplex.

The notations $A := B$ and $B =: A$ both mean that A (a new symbol) is defined to be equal to B (something already known).

As usual, the non-word "iff" is short for "if and only if."

3. **Categories:** I assume an elementary knowledge of categories and functors. I sometimes refer to well-known categories by their objects (the word is given a capital opening letter). Thus Groups refers to the category of groups and homomorphisms. Similarly: Sets, Pointed Sets, Spaces, Pointed Spaces, Homotopy (spaces and homotopy classes of maps), Pointed Homotopy, and R-modules. When there might be ambiguity I name the objects and the morphisms (e.g., the category of oriented CW complexes of locally finite type and CW-proper homotopy classes of CW-proper maps).

4. **Website:** I plan to collect corrections, updates etc. at the Internet website

math.binghamton.edu/ross/tmgt

I also invite supplementary essays or comments which readers feel would be helpful, especially to students. Such contributions, as well as corrections and errata, should be sent to me at the web address

ross@math.binghamton.edu

Contents

PART I: ALGEBRAIC TOPOLOGY FOR GROUP THEORY

We have gathered into Part I some topics in algebraic and geometric topology which are useful in understanding groups from a topological point of view: CW complexes, cellular homology, fundamental group, basic homotopy theory, and the most elementary ideas about manifolds and piecewise linear methods. While starting almost from the beginning (though some prior acquaintance with singular homology is desirable) we give a detailed treatment of cellular homology. In effect we present this theory twice, a formal version derived from singular theory and a geometrical version in terms of incidence numbers and mapping degrees. It is this latter version which exhibits "what is really going on": experienced topologists know it (or intuit it) but it is rarely written down in the detail given here.

This is followed by a discussion of the fundamental group and covering spaces, done in a combinatorial way appropriate for working with CW complexes. In particular, our approach makes the Seifert-Van Kampen Theorem almost a tautology.

We discuss some elementary topics in homotopy theory which are useful for group theory. Chief among these are: ways to alter a CW complex without changing its homotopy type (e.g. by homotoping attaching maps, by cell trading etc.), and an elementary proof of the Hurewicz Theorem based on Hurewicz's original proof.*

We end by explaining the elementary geometric topology of simplicial complexes and of topological and piecewise linear manifolds.

---

* Modern proofs usually involve more sophisticated methods.

1

CW Complexes and Homotopy

CW complexes are topological spaces equipped with a partitioning into compact pieces called "cells." They are particularly suitable for group theory: a presentation of a group can be interpreted as a recipe for building a two-dimensional CW complex (Example 1.2.17), and we will see in later chapters that CW complexes exhibit many group theoretic properties geometrically.

Beginners in algebraic topology are usually introduced first to simplicial complexes. A simplicial complex is (or can be interpreted as) an especially nice kind of CW complex. In the long run, however, it is often unnatural to be confined to the world of simplicial complexes, in particular because they often have an inconveniently large number of cells. For this reason, we concentrate on CW complexes from the start. Simplicial complexes are treated in Chap. 5.

1.1 Review of general topology

We review, without proof, most of the general topology we will need. This section can be used for reference or as a quick refresher course. Proofs of all our statements can be found in [51] or [123], or, in the case of statements about k-spaces, in [148].

A *topology* on a set X is a set, $\mathcal{T}$, of subsets of X closed under finite intersection and arbitrary union, and satisfying: $\emptyset \in \mathcal{T}$, $X \in \mathcal{T}$. The pair $(X, \mathcal{T})$ is a *topological space* (or just *space*). Usually we suppress $\mathcal{T}$, saying "X is a space" etc. The members of $\mathcal{T}$ are *open* sets. If every subset of X is open, $\mathcal{T}$ is the *discrete* topology on X. The subset $F \subset X$ is *closed* if the complementary subset[1] $X - F$ is open. For $A \subset X$, the *interior* of A in X, $\mathrm{int}_X A$, is the union of all subsets of A which are open in X; the *closure* of

---

[1] Throughout this book we denote the complement of A in X by $X - A$. More often, we will need $X \stackrel{c}{-} A$, the CW complement of A in X (where X is a CW complex and A is a subcomplex). This is defined in Sect. 1.5.

A in X, $\mathrm{cl}_X A$, is the intersection of all closed subsets of X which contain A; the *frontier* of A in X, $\mathrm{fr}_X A$, is $(\mathrm{cl}_X A) \cap (\mathrm{cl}_X(X - A))$. The frontier is often called the "boundary" but we will save that word for other uses. The subscript X in int_X, cl_X, and fr_X is often suppressed. The subset A is *dense* in X if $\mathrm{cl}_X A = X$; A is *nowhere dense* in X if $\mathrm{int}_X(\mathrm{cl}_X A) = \emptyset$.

If $\mathcal{S}$ is a set of subsets of X and if $\mathcal{T}(\mathcal{S})$ is the topology on X consisting of all unions of finite intersections of members of $\mathcal{S}$, together with X and $\emptyset$, $\mathcal{S}$ is called a *subbasis* for $\mathcal{T}(\mathcal{S})$.

A *neighborhood* of $x \in X$ [resp. of $A \subset X$] is a set $N \subset X$ such that for some open subset U of X, $x \in U \subset N$, [resp. $A \subset U \subset N$].

For $A \subset X$, $\{U \cap A \mid U \in \mathcal{T}\}$ is a topology on A, called the *inherited* topology; A, endowed with this topology, is a *subspace* of X. A *pair* of spaces (X, A) consists of a space X and a subspace A. Similarly, if $B \subset A \subset X$, (X, A, B) is a *triple* of spaces.

A function $f : X \to Y$, where X and Y are spaces, is *continuous* if whenever U is open in Y, $f^{-1}(U)$ is open in X. A continuous function is also called a *map*. A *map of pairs* $f : (X, A) \to (Y, B)$ is a map $f : X \to Y$ such that $f(A) \subset B$. If (X, A) is a pair of spaces, there is an *inclusion* map $i : A \to X$, $a \mapsto a$; another useful notation for the inclusion map is $A \hookrightarrow X$. In the special case where $A = X$, i is called the *identity* map, denoted $\mathrm{id}_X :$ $X \to X$. The composition $X \xrightarrow{f} Y \xrightarrow{g} Z$ of maps f and g is a map, denoted $g \circ f : X \to Z$. If $A \subset X$, and $f : X \to Y$ is a map, $f \mid A : A \to Y$ is the composition $A \hookrightarrow X \xrightarrow{f} Y$; $f \mid A$ is the *restriction of f to A*.

A function $f : X \to Y$ is *closed* [resp. *open*] if it maps closed [resp. open] sets onto closed [resp. open] sets.

A *homeomorphism* $f : X \to Y$ is a map for which there exists an *inverse*, namely a map $f^{-1} : Y \to X$ such that $f^{-1} \circ f = \mathrm{id}_X$ and $f \circ f^{-1} = \mathrm{id}_Y$. A *topological property* is a property preserved by homeomorphisms. If there exists a homeomorphism $X \to Y$ then X and Y are *homeomorphic*. Obviously a homeomorphism is a continuous open bijection and any function with these properties is a homeomorphism.

If $f : X \to Y$ is a map, and $f(X) \subset V \subset Y$, there is an *induced map* $X \to V$, $x \mapsto f(x)$; this induced map is only formally different from f insofar as its target is V rather than Y. More rigorously, the induced map is the unique map making the following diagram commute

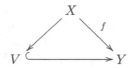

This induced map $X \to V$ is sometimes called the *corestriction* of f to V. The map $f : X \to Y$ is an *embedding* if the induced map $X \to f(X)$ is a homeomorphism; in particular, if A is a subspace of X, $A \hookrightarrow X$ is an

embedding. The embedding $f : X \to Y$ is a *closed embedding* if $f(X)$ is closed in Y; f is an *open embedding* if $f(X)$ is open in Y.

Let X_1 and X_2 be spaces. Their *product* $X_1 \times X_2$ is the set of ordered pairs (x_1, x_2) such that $x_1 \in X_1$ and $x_2 \in X_2$, endowed with the *product topology*; namely, $U \subset X_1 \times X_2$ is open if for each $(x_1, x_2) \in U$ there are open sets U_1 in X_1 and U_2 in X_2 such that $(x_1, x_2) \in U_1 \times U_2 \subset U$. There are *projection maps* $X_1 \xleftarrow{p_1} X_1 \times X_2 \xrightarrow{p_2} X_2$, $(x_1, x_2) \xmapsto{p_i} x_i$; the product topology is the smallest topology making the functions p_1 and p_2 continuous. All finite products, $\prod_{i=1}^{n} X_i$, are defined similarly, and if each $X_i = X$, we use the alternative notation X^n; X^1 and X are identical. There is a convenient way of checking the continuity of functions into products: $f : Z \longrightarrow \prod_{i=1}^{n} X_i$ is continuous if $p_j \circ f : Z \to X_j$ is continuous for each j, where $p_j(x_1, \ldots, x_n) = x_j$. In addition to being continuous, each projection $\prod_{i=1}^{n} X_i \to X_j$ is surjective and maps open sets to open sets. The product $f_1 \times f_2 : X_1 \times X_2 \to Y_1 \times Y_2$ of maps is a map.

In the case of an *arbitrary product*, $\prod_{\alpha \in \mathcal{A}} X_\alpha$, of spaces X_α, a subset U is open if for each point (x_α) there is a finite subset $\mathcal{B} \subset \mathcal{A}$, and for each $\alpha \in \mathcal{B}$ an open neighborhood U_α of x_α in X_α such that $(x_\alpha) \in \prod_{\alpha \in \mathcal{A}} U_\alpha \subset U$ where $U_\alpha = X_\alpha$ whenever $\alpha \notin \mathcal{B}$. This is the smallest topology with respect to which all projections are continuous. As in the finite case, continuity of maps into an arbitrary product can be checked by checking it coordinatewise, and the product of maps is a map.

When we discuss k-spaces, below, it will be necessary to revisit the subject of products.

One interesting space is $\mathbb{R}$, the real numbers with the *usual* topology: $U \subset \mathbb{R}$ is open if for each $x \in U$ there is an "open" interval (a, b) such that $x \in (a, b) \subset U$. "Open" intervals are open in the usual topology! Our definition of X^n defines in particular *Euclidean n-space* $\mathbb{R}^n$. Many of the spaces of interest are, or are homeomorphic to, subspaces of $\mathbb{R}^n$ or are quotients of topological sums of such spaces (these terms are defined below).

Some particularly useful subspaces of $\mathbb{R}^n$ are

$\mathbb{N}$ = the non-negative integers;

$$B^n = \{x \in \mathbb{R}^n \mid |x| \le 1\} = \text{the } n\text{-ball where } |x|^2 = \sum_{i=1}^{n} x_i^2;$$

$S^{n-1} = \operatorname{fr}_{\mathbb{R}^n} B^n = \text{the } (n-1)\text{-} sphere;$

$I = [0,1] \subset \mathbb{R};$

$\mathbb{R}_+^n = \{x \in \mathbb{R}^n \mid x_n \ge 0\};$

$\mathbb{R}_-^n = \{x \in \mathbb{R}^n \mid x_n \le 0\}.$

Addition and scalar multiplication in $\mathbb{R}^n$ are continuous. $\mathbb{R}^0$ and B^0 are other notations for $\{0\}$, and, although $-1 \notin \mathbb{N}$, it is convenient to define $S^{-1} = \emptyset$. An n-ball [resp. an $(n-1)$-sphere] is a space homeomorphic to B^n [resp. S^{n-1}].

It is often useful not to distinguish between $(x_1, \ldots, x_n) \in \mathbb{R}^n$ and $(x_1, \ldots, x_n, 0) \in \mathbb{R}^{n+1}$, that is, to identify $\mathbb{R}^n$ with its image under that closed embedding $\mathbb{R}^n \to \mathbb{R}^{n+1}$. This is implied when we write $S^{n-1} \subset S^n$, $B^n \subset B^{n+1}$, etc.

Let X be a space, Y a set, and $p : X \to Y$ a surjection. The *quotient topology* on Y with respect to p is defined by: U is open in Y iff $p^{-1}(U)$ is open in X. Typically, Y will be the set of equivalence classes in X with respect to some given equivalence relation, $\sim$, on X, and $p(x)$ will be the equivalence class of x; Y is the *quotient space* of X by $\sim$, and Y is sometimes denoted by $X/\sim$. More generally, given spaces X and Y, a surjection $p : X \to Y$ is a *quotient map* if the topology on Y is the quotient topology with respect to p (i.e., $U \subset Y$ is open if $p^{-1}(U)$ is open in X). Obviously, quotient maps are continuous, but they do not always map open sets to open sets. The subset $A \subset X$ is *saturated* with respect to p if $A = p^{-1}(p(A))$; if A is saturated and open, $p(A)$ is open. There is a convenient way of checking the existence and continuity of functions out of quotient spaces: in the following diagram, where $f : X \to Z$ is a given map, and $p : X \to X/\sim$ is a quotient map, there exists a function f' making the diagram commute iff f takes entire equivalence classes in X to points of Z. Moreover, if the function f' exists it is unique and continuous.

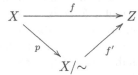

If $A \subset X$, and the equivalence classes under $\sim$ are A and the sets $\{x\}$ for $x \in X - A$ then $X/\sim$ is also written X/A.

If $p : X \to Y$ is a quotient map and $B \subset Y$, one sometimes wishes to claim that $p| : p^{-1}(B) \to B$ is a quotient map. This is not always true, but it is true if B is an open subset or a closed subset of Y.

A closely related notion is that of "weak topology". Let X be a set, and let $\{A_\alpha \mid \alpha \in \mathcal{A}\}$ be a family of subsets of X, such that each A_α has a topology. The family of spaces $\{A_\alpha \mid \alpha \in \mathcal{A}\}$ is *suitable for defining a weak topology on X* if (i) $X = \bigcup_\alpha A_\alpha$, (ii) for all $\alpha, \beta \in \mathcal{A}$, $A_\alpha \cap A_\beta$ inherits the same topology from A_α as from A_β, and (iii) either (a) $A_\alpha \cap A_\beta$ is closed both in A_α and in A_β for all $\alpha, \beta \in \mathcal{A}$, or (b) $A_\alpha \cap A_\beta$ is open both in A_α and in A_β for all $\alpha, \beta \in \mathcal{A}$. The *weak topology* on X with respect to $\{A_\alpha \mid \alpha \in \mathcal{A}\}$ is $\{U \subset X \mid U \cap A_\alpha$ is open in A_α for all $\alpha \in \mathcal{A}\}$. This topology has some useful properties: (i) $S \subset X$ is closed [resp. open] if $S \cap A_\alpha$ is closed [resp. open] in A_α for all α; (ii) each A_α inherits its original topology from the weak topology on X; (iii) there is a convenient way of checking the continuity of functions out of weak topologies: a function $f : X \to Z$ is continuous iff for each α, $f \mid A_\alpha : A_\alpha \to Z$ is continuous; (iv) if (a) in the above definition of suitability holds then each A_α is closed in X, while if (b) holds each A_α is open in X. By custom, if one asserts that X has the weak topology with respect to $\{A_\alpha\}$, it is tacitly assumed (or must be checked) that $\{A_\alpha\}$ is suitable for defining a weak topology on X.

Given a family of spaces $\{X_\alpha \mid \alpha \in \mathcal{A}\}$ (which might not be pairwise disjoint), their *topological sum* is the space $\coprod_{\alpha \in \mathcal{A}} X_\alpha$ whose underlying set is $\bigcup\{X_\alpha \times \{\alpha\} \mid \alpha \in \mathcal{A}\}$ and whose topology is generated by (i.e., is the smallest topology containing) $\{U \times \{\alpha\} \mid U$ is open in $X_\alpha\}$. There are closed-and-open embeddings $i_\beta : X_\beta \to \coprod_{\alpha \in \mathcal{A}} X_\alpha$, $x \mapsto (x, \beta)$, for each $\beta \in \mathcal{A}$. The point is: $i_\beta(X_\beta) \cap i_\alpha(X_\alpha) = \emptyset$ when $\beta \neq \alpha$. In practice, the inclusions i_β are often suppressed, and one writes X_β for $X_\beta \times \{\beta\}$ when confusion is unlikely. In the case of finitely many spaces $X_1, \ldots, X_n$ their topological sum is also written as $\coprod_{i=1}^n X_i$ or $X_1 \amalg \ldots \amalg X_n$. The previously mentioned weak topology on X with respect to $\{A_\alpha \mid \alpha \in \mathcal{A}\}$ is the same as the quotient topology obtained from $p : \coprod_{\alpha \in \mathcal{A}} A_\alpha \to X$, $(a, \alpha) \mapsto a$.

A family $\{A_\alpha \mid \alpha \in \mathcal{A}\}$ of subsets of a space X is *locally finite* if for each $x \in X$ there is a neighborhood N of x such that $N \cap A_\alpha = \emptyset$ for all but finitely many $\alpha \in \mathcal{A}$.

It is often the case that one has: spaces X and Y, a family $\{A_\alpha \mid \alpha \in \mathcal{A}\}$ of subsets of X whose union is X, and a function $f : X \to Y$ such that $f \mid A_\alpha : A_\alpha \to Y$ is continuous for each α, where A_α has the inherited topology. One wishes to conclude that f is continuous. This is true if every A_α is open in X. It is also true if every A_α is closed in X and $\{A_\alpha\}$ is a locally finite family of subsets of X.

An *open cover* of the space X is a family of open subsets of X whose union is X. A space X is *compact* if every open cover of X has a finite subcover. If $f : X \to Y$ is a continuous surjection and if X is compact, then Y is compact. Products and closed subsets of compact spaces are compact. In particular, closed and bounded subsets of $\mathbb{R}^n$ are compact, since, by the well-known Heine-Borel Theorem, I is compact. The space X is *locally compact* if every point of X has a compact neighborhood. The space $\mathbb{R}$ is locally compact but not compact.

The space X is *Hausdorff* if whenever $x \neq y$ are points of X, x and y have disjoint neighborhoods. A compact subset of a Hausdorff space is closed. In particular, one-point subsets are closed. If X is compact and Y is Hausdorff, any continuous bijection $X \to Y$ is a homeomorphism and any continuous surjection $X \to Y$ is a quotient map. Products, topological sums, and subspaces of Hausdorff spaces are Hausdorff. The space $\mathbb{R}$ is clearly Hausdorff, hence all subsets of $\mathbb{R}^n$ are Hausdorff. In general, a quotient space of a Hausdorff space need not be Hausdorff; example: $X = \mathbb{R}$, $x \sim y$ if either x and y are rational or $x = y$.

A Hausdorff space X is a *k-space* (or *compactly generated space*) if X has the weak topology with respect to its compact subsets; for example, locally compact Hausdorff spaces are k-spaces. (The Hausdorff condition ensures that the compact subsets form a family suitable for defining a weak topology.) One might wish that a category whose objects are k-spaces should be closed under the operation of taking finite products, but this is not always the case. However, there is a canonical method of "correcting" the situation: for any Hausdorff space X one defines kX to be the set X equipped with the weak topology with respect to the compact subspaces of X. Indeed, this k defines a functor from Spaces to k-Spaces, with kf defined in the obvious way for every map f. In the category of k-spaces one redefines "product" by declaring the product of X and Y to be $k(X \times Y)$. This new kind of product has all the properties of "product" given above, provided one consistently replaces any space occurring in the discussion by its image under k. The main class of spaces appearing in this book is the class of CW complexes, whose definition and general topology are discussed in detail in the next section; CW complexes are k-spaces. It should be noted that throughout this book, when we discuss the product of two CW complexes, the topology on their product is always to be understood in this modified sense.

A *path*[2] in the space X is a map $\omega : I \to X$; its *initial point* is $\omega(0)$ and its *final point* is $\omega(1)$. A *path in X from $x \in X$ to $y \in X$* is a path whose initial point is x and whose final point is y. Points $x, y \in X$ *are in the same path component* if there is a path in X from x to y. This generates an equivalence relation on X; an equivalence class, with the topology inherited from X, is called a *path component* of X. The space X is *path connected* if X has exactly one path component. The empty space, $\emptyset$, is considered to have no path components, hence it is not path connected. The set of path components of X is denoted by $\pi_0(X)$. If $x \in X$ the notation $\pi_0(X, x)$ is used for the pointed set whose base point is the path component of x.

---

[2] Sometimes it is convenient to replace I by some other closed interval $[a, b]$.

A space X is *connected* if whenever $X = U \cup V$ with U and V disjoint and open in X then either $U = \emptyset$ or $V = \emptyset$. A *component* of X is a maximal connected subspace. Components are pairwise disjoint, and X is the union of its components. The empty space is connected and has one component.

X is *locally path connected* if it satisfies any of three equivalent conditions: (i) for every $x \in X$ and every neighborhood U of x there is a path connected neighborhood V of x such that $V \subset U$; (ii) for every $x \in X$ and every neighborhood U of x there is a neighborhood V of x lying in a path component of U; (iii) each path component of each open subset of X is an open subset of X. Each path component of a space X lies in a component of X. For non-empty locally path connected spaces, the components and the path components coincide. In particular, this applies to non-empty open subsets of $\mathbb{R}^n$ or B^n or S^n or $\mathbb{R}^n_+$ or I.

A *metric space* is a pair (X, d) where X is a set and $d : X \times X \to \mathbb{R}$ is a function (called a *metric*) satisfying (i) $d(x, y) \geq 0$ and $d(x, y) = 0$ iff $x = y$; (ii) $d(x, y) = d(y, x)$; and the *triangle inequality* (iii) $d(x, z) \leq d(x, y) + d(y, z)$. For $x \in X$ and $r > 0$ the *open ball of radius r about x* is

$$B_r(x) = \{y \in X \mid d(x, y) < r\}.$$

The *closed ball of radius r about x* is

$$\bar{B}_r(x) = \{y \in X \mid d(x, y) \leq r\}$$

The *diameter* of $A \subset X$ is diam $A = \sup\{d(a, a') \mid a, a' \in A\}$. The metric d *induces* a topology $\mathcal{T}_d$ on X: $U \in \mathcal{T}_d$ iff for every $x \in U$ there is some $r > 0$ such that $B_r(x) \subset U$. Two metrics on X which induce the same topology on X are *topologically equivalent*. A topological space $(X, \mathcal{T})$ is *metrizable* if there exists a metric d on X which induces $\mathcal{T}$. Metrizable spaces are Hausdorff. Countable products of metrizable spaces are metrizable. The *Euclidean metric* on $\mathbb{R}^n$ is given by $d(x, y) = |x - y|$.

A family $\{U_\alpha\}$ of neighborhoods of the point x in the space X is a *basis for the neighborhoods of x* if every neighborhood of x has some U_α as a subset. The space X is *first countable* if there is a countable basis for the neighborhoods of each $x \in X$. Every metrizable space is first countable, so the absence of this property is often a quick way to show non-metrizability. Every first countable Hausdorff space, hence every metrizable space, is a k-space.

If X and Y are spaces, we denote by $C(X, Y)$ the set of all maps $X \to Y$ endowed with the *compact-open* topology: for each compact subset K of X and each open subset U of Y let $\langle K, U \rangle$ denote the set of all $f \in C(X, Y)$ such that $f(K) \subset U$; the family of all such sets $\langle K, U \rangle$ is a subbasis for this topology. An important feature is that the *exponential correspondence* $C(X \times Y, Z) \to C(X, C(Y, Z))$, $f \mapsto \hat{f}$ where $\hat{f}(x)(y) = f(x, y)$, is a well-defined bijection when Y is locally compact and Hausdorff, and also when all three spaces are k-spaces, with "product" understood in that sense. The map $\hat{f}$ is called the *adjoint* of the map f and vice versa. When X is compact and

(Y, d) is a metric space the compact-open topology is induced by the metric $\rho(f, g) = \sup\{d(f(x), g(x)) \mid x \in X\}$.

1.2 CW complexes

We begin by describing how to build a new space Y from a given space A by "gluing n-balls to A along their boundaries."

Let $\mathcal{A}$ be an indexing set, let $n \geq 0$, and let $B^n(\mathcal{A}) = \coprod_{\alpha \in \mathcal{A}} B_\alpha^n$ where each $B_\alpha^n = B^n$, i.e., the topological sum of copies of B^n indexed by $\mathcal{A}$. Let $S^{n-1}(\mathcal{A}) = \coprod_{\alpha \in \mathcal{A}} S_\alpha^{n-1}$ where $S_\alpha^{n-1} = S^{n-1}$. Let $f : S^{n-1}(\mathcal{A}) \to A$ be a map. Let $\sim$ be the equivalence relation on $A \coprod B^n(\mathcal{A})$ generated by: $x \sim f(x)$ whenever $x \in S^{n-1}(\mathcal{A})$. Then the quotient space $Y := (A \coprod B^n(\mathcal{A}))/\sim$ is *the space obtained by attaching $B^n(\mathcal{A})$ to A using f.* See Fig. 1.1.

Proposition 1.2.1. *Let $q : A \coprod B^n(\mathcal{A}) \to Y$ be the quotient map taking each point to its equivalence class. Then $q \mid A : A \to Y$ is a closed embedding, and $q \mid B^n(\mathcal{A}) - S^{n-1}(\mathcal{A})$ is an open embedding.*

Proof. Equivalence classes in $A \coprod B^n(\mathcal{A})$ have the form $\{a\} \cup f^{-1}(a)$ with $a \in A$, or the form $\{z\}$ with $z \in B^n(\mathcal{A}) - S^{n-1}(\mathcal{A})$. Thus $q \mid A$ and $q \mid B^n(\mathcal{A}) - S^{n-1}(\mathcal{A})$ are injective maps. If C is a closed subset of A, $q^{-1}q(C) = f^{-1}(C) \cup C$ which is closed in $A \coprod B^n(\mathcal{A})$, hence $q(C)$ is closed in Y. Thus $q \mid A$ is a closed embedding. If U is an open subset of $B^n(\mathcal{A}) - S^{n-1}(\mathcal{A})$, $q^{-1}q(U) = U$, hence $q(U)$ is open in Y. Thus $q \mid B^n(\mathcal{A}) - S^{n-1}(\mathcal{A})$ is an open embedding. $\qquad\square$

There is some terminology and notation to go with this construction. In view of 1.2.1, it is customary to identify $a \in A$ with $q(a) \in Y$ and hence to regard A as a closed subset of Y. Let $e_\alpha^n = q(B_\alpha^n)$. The sets e_α^n are called the *n-cells* of the pair (Y, A). Let $\overset{\circ}{e}_\alpha^n = e_\alpha^n - A$. By 1.2.1, $\overset{\circ}{e}_\alpha^n$ is open in Y. Let[3] $\overset{\bullet}{e}_\alpha^n = e_\alpha^n \cap A$. The map $q_\alpha := q \mid B_\alpha^n : (B_\alpha^n, S_\alpha^{n-1}) \to (e_\alpha^n, \overset{\bullet}{e}_\alpha^n)$ is called *the characteristic map* for the cell e_α^n. The map $f_\alpha := f \mid S_\alpha^{n-1} : S_\alpha^{n-1} \to A$ is called *the attaching map* for the cell e_α^n. The map f itself is *the simultaneous attaching map.*

Before proceeding, consider some simple cases: (i) If Y is obtained from A by attaching a single 0-cell then Y is $A \coprod \{p\}$ where p is a point and $p \notin A$; this is because B^0 is a point and S^{-1} is empty. (ii) If Y is obtained from A by attaching a 1-cell then the image of the attaching map $f : S^0 \to A$ might meet two path components of A, in which case they would become part of a

---

[3] $\overset{\bullet}{e}_\alpha^n$ is sometimes called the "boundary" of e_α^n, and $\overset{\circ}{e}_\alpha^n$ is the "interior" of e_α^n. Distinguish this use of "interior" from how the word is used in general topology (Sect. 1.1) and in discussing manifolds (Sect. 5.1).

single path component of Y; or this image might lie in one path component of A, in which case it might consist of two points (an embedded 1-cell) or one point (a 1-cell which is not homeomorphic to B^1). (iii) When $n \geq 2$ the image of the attaching map $f : S^{n-1} \to A$ lies in a single path component of A.

Proposition 1.2.2. *If A is Hausdorff, Y is Hausdorff. Hence $e_\alpha^n = \mathrm{cl}_Y \, \overset{\circ}{e}{}_\alpha^n$.*

Proof. Let $y_1 \neq y_2 \in Y$. We seek saturated disjoint open subsets $U_1, U_2 \subset B^n(\mathcal{A}) \coprod A$ whose images contain y_1 and y_2 respectively. There are three cases: (i) $q^{-1}(y_i) = \{z_i\}$ where $z_i \in B^n(\mathcal{A}) - S^{n-1}(\mathcal{A})$ for $i = 1, 2$; (ii) $q^{-1}(y_1) = \{z_1\}$ where $z_1 \in B^n(\mathcal{A}) - S^{n-1}(\mathcal{A})$ and $q^{-1}(y_2) = \{a_2\} \cup f^{-1}(a_2)$ where $a_2 \in A$; (iii) $q^{-1}(y_i) = \{a_i\} \cup f^{-1}(a_i)$ where $a_i \in A$ for $i = 1, 2$. In Case (i), pick U_1 and U_2 to be disjoint open subsets of $B^n(\mathcal{A}) - S^{n-1}(\mathcal{A})$ containing z_1 and z_2: these clearly exist and are saturated. In Case (ii) let $z_1 \in B_\alpha^n - S_\alpha^{n-1}$ where $\alpha \in \mathcal{A}$. There is clearly an open set U, containing z_1, whose closure lies in $B_\alpha^n - S_\alpha^{n-1}$. Then U and the complement of $cl\ U$ are the required saturated sets. In Case (iii), use the fact that A is Hausdorff to pick disjoint open subsets W_1 and W_2 of A containing a_1 and a_2. Then $f^{-1}(W_1)$ and $f^{-1}(W_2)$ are disjoint open subsets of $S^{n-1}(\mathcal{A})$. There exist disjoint open subsets V_1 and V_2 of $B^n(\mathcal{A})$ such that $V_i \cap S^{n-1}(\mathcal{A}) = f^{-1}(W_i)$ (exercise!). The sets $V_1 \cup W_1$ and $V_2 \cup W_2$ are the required saturated sets. So Y is Hausdorff. It follows that e_α^n, being compact, is closed in Y. So $\mathrm{cl}_Y \overset{\circ}{e}{}_\alpha^n \subset e_\alpha^n$. Since $q^{-1}(\mathrm{cl}_Y \overset{\circ}{e}{}_\alpha^n) \cap B_\alpha^n$ is closed in B_α^n and contains $B_\alpha^n - S_\alpha^{n-1}$, $B_\alpha^n \subset q^{-1}(\mathrm{cl}_Y \overset{\circ}{e}{}_\alpha^n)$. So $e_\alpha^n = q(B_\alpha^n) \subset q(q^{-1}(\mathrm{cl}_Y \overset{\circ}{e}{}_\alpha^n)) = \mathrm{cl}_Y \overset{\circ}{e}{}_\alpha^n$. $\square$

We have described the space Y in terms of $(A, \mathcal{A}, n, f : S^{n-1}(\mathcal{A}) \to A)$. Note that there is a bijection between $\mathcal{A}$ and the set of path components of $Y - A$. In practice, we might not be dealing with the pair (Y, A) but with a pair "equivalent" to it: we now say what this means.

Let (Y, A) be a pair, and let $\{\overset{\circ}{e}_\alpha \mid \alpha \in \mathcal{A}\}$ be the set of path components of $Y - A$. Let $n \in \mathbb{N}$. Y *is obtained from A by attaching n-cells* if there exists a quotient map $p : A \coprod B^n(\mathcal{A}) \to Y$ such that $p(S^{n-1}(\mathcal{A})) \subset A$, $p \mid A : A \hookrightarrow Y$, and p maps $B^n(\mathcal{A}) - S^{n-1}(\mathcal{A})$ homeomorphically onto $Y - A$. This implies that A is closed in Y, and that each $\overset{\circ}{e}_\alpha$ is open in Y. See Fig. 1.1.

Proposition 1.2.3. *Let $f = p \mid S^{n-1}(\mathcal{A}) : S^{n-1}(\mathcal{A}) \to A$. Let X be the space obtained by attaching $B^n(\mathcal{A})$ to A using f. Let $q : A \coprod B^n(\mathcal{A}) \to X$ be the quotient map. There is a homeomorphism $h : X \to Y$ such that $h \circ q = p$.*

Proof. Consider the following commutative diagram:

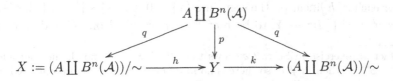

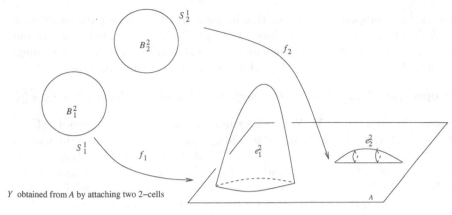

Y obtained from A by attaching two 2–cells

Fig. 1.1.

By definition of p and q, functions h and k exist as indicated. Since p and q are quotient maps, h and k are continuous. By uniqueness of maps induced on quotient spaces, $k \circ h = \text{id}$. Similarly $h \circ k = \text{id}$. $\qquad\square$

Let Y be obtained from the Hausdorff space A by attaching n-cells. Then $Y - A$ is a topological sum[4] of copies of $B^n - S^{n-1}$. By 1.2.2 and 1.2.3, Y is Hausdorff. Let $p : A \coprod B^n(\mathcal{A}) \to Y$ be as above. Denote $p(B_\alpha^n)$ by e_α^n; by 1.2.2 and 1.2.3, e_α^n is the closure in Y of a path component of $Y - A$. The sets e_α^n are called n-*cells* of (Y, A). Write $\overset{\circ}{e}{}_\alpha^n = e_\alpha^n - A$. As before, $\overset{\circ}{e}{}_\alpha^n$ is open in Y. Write $\overset{\bullet}{e}{}_\alpha^n = e_\alpha^n \cap A$. The map $p_\alpha := p \mid B_\alpha^n : (B_\alpha^n, S_\alpha^{n-1}) \to (e_\alpha^n, \overset{\bullet}{e}{}_\alpha^n)$ is *a characteristic map* for the cell e_α^n. The map $f_\alpha := p \mid S_\alpha^{n-1} : S_\alpha^{n-1} \to A$ is *an attaching map* for e_α^n. $p \mid S^{n-1}(\mathcal{A}) : S^{n-1}(\mathcal{A}) \to A$ is *a simultaneous attaching map*. We emphasize the indefinite article in the latter definitions. They depend on p, not simply on (Y, A), as we now show.

Example 1.2.4. Let $(Y, A) = (B^2, S^1)$. Here Y is obtained from A by attaching 2-cells, since we can take $p : S^1 \coprod B^2 \to Y$ to be the identity on B^2 and the inclusion on S^1. The corresponding attaching map $f : S^1 \to S^1$ for e^2 is the identity map. But we could instead take $p' : S^1 \coprod B^2 \to Y$ to be $r : (x_1, x_2) \mapsto (-x_1, x_2)$ on B^2 and to be the inclusion on S^1. Then the attaching map $f' : S^1 \to S^1$ would be $r \mid S^1$, which is a homeomorphism but not the identity. Nor is it true that attaching maps are unique up to composition with homeomorphisms as in the case of f and f'. For example, define $h : I \times I \to I$ by $(x, 0) \mapsto 0$, $(x, \frac{1}{3}) \mapsto \frac{1}{3} + \frac{x}{6}$, $(x, \frac{2}{3}) \mapsto \frac{2}{3} - \frac{x}{6}$, $(x, 1) \mapsto 1$, and for each x, h linear on the segments $\{x\} \times [0, \frac{1}{3}]$, $\{x\} \times [\frac{1}{3}, \frac{2}{3}]$, $\{x\} \times [\frac{2}{3}, 1]$. Define $p'' : S^1 \coprod B^2 \to Y$ to be $r \, e^{2\pi i t} \mapsto r \, e^{2\pi i h(r, t)}$ on B^2 and to be the

---

[4] In fact an exercise in Sect. 2.2 shows that $B^n - S^{n-1}$ is not homeomorphic to $B^m - S^{m-1}$ when $m \neq n$, so there is no $m \neq n$ such that Y is also obtained from A by attaching m-cells; however, we will not use this until after that section.

inclusion on S^1. Then the attaching map $f'' : S^1 \to S^1$ is $e^{2\pi it} \mapsto e^{2\pi ih(1,t)}$ $(0 \le t \le 1)$; f'' is constant on $\{e^{2\pi it} \mid \frac{1}{3} \le t \le \frac{2}{3}\}$, hence f'' is not the composition of f and a homeomorphism.

Proposition 1.2.5. *Let A be Hausdorff and let Y be obtained from A by attaching n-cells. Then the space Y has the weak topology with respect to $\{e_\alpha^n \mid \alpha \in \mathcal{A}\} \cup \{A\}$.*

Proof. First, we check that $\{e_\alpha^n \mid \alpha \in \mathcal{A}\} \cup \{A\}$ is suitable for defining a weak topology (see Sect. 1.1). By 1.2.1, A inherits its original topology from Y. When $\alpha \ne \beta$, $e_\alpha^n \cap e_\beta^n = \overset{\bullet}{e}_\alpha^n \cap \overset{\bullet}{e}_\beta^n$ which is compact, hence closed in e_α^n and in e_β^n (by 1.2.2 and 1.2.3). The subspace $e_\alpha^n \cap A = \overset{\bullet}{e}_\alpha^n$ is compact, hence closed in A and in e_α^n. Next, let $C \subset Y$ be such that $C \cap A$ is closed in A and $C \cap e_\alpha^n$ is closed in e_α^n for all α. We have $p^{-1}(C) = \bigcup_\alpha p_\alpha^{-1}(C \cap e_\alpha^n) \cup (C \cap A)$. Each term in this union is a closed subset of a summand of the topological sum

$$A \coprod \left(\coprod_\alpha B_\alpha^n \right),$$

and different terms correspond to different summands. Hence their union is closed, hence $p^{-1}(C)$ is closed, hence C is closed in Y. $\square$

Proposition 1.2.6. *Let A be Hausdorff and let Y be obtained from A by attaching n-cells. Any compact subset of Y lies in the union of A and finitely many cells of (Y, A).*

Proof. Suppose this were false. Then there would be a compact subset C of Y such that $C \cap \overset{\circ}{e}_\alpha^n \ne \emptyset$ for infinitely many values of α. For each such α, pick $x_\alpha \in \overset{\circ}{e}_\alpha^n \cap C$. Let D be the set of these points x_α. For each α, $D \cap e_\alpha$ contains at most one point, so $D \cap e_\alpha$ is closed in e_α. And $D \cap A = \emptyset$. So, by 1.2.5, D is closed in Y. For the same reason, every subset of D is closed in Y. So D inherits the discrete topology from Y. Since $D \subset C$, D is an infinite discrete compact space. Such a space cannot exist, for the one-point sets would form an open cover having no finite subcover. $\square$

Note that, in spite of 1.2.6, a compact set can meet infinitely many cells.

If we are given a pair of spaces (Y, A) how would we recognize that Y is obtained from A by attaching n-cells? Here is a convenient way of recognizing this situation:

Proposition 1.2.7. *Let (Y, A) be a Hausdorff pair,[5] let $\{\overset{\circ}{e}_\alpha \mid \alpha \in \mathcal{A}\}$ be the set of path components of $Y - A$, and let $n \in \mathbb{N}$. Write $e_\alpha = \mathrm{cl}_Y \overset{\circ}{e}_\alpha$. The space Y is obtained from A by attaching n-cells if*

(i) *for each $\alpha \in \mathcal{A}$, there exists a map $p_\alpha : (B^n, S^{n-1}) \to (A \cup \overset{\circ}{e}_\alpha, A)$ such that p_α maps $B^n - S^{n-1}$ homeomorphically onto $\overset{\circ}{e}_\alpha$;*

---

[5] This is just a short way of saying that (Y, A) is a pair of spaces and Y (hence also A) is Hausdorff.

(ii) Y has the weak topology with respect to $\{e_\alpha \mid \alpha \in \mathcal{A}\} \cup \{A\}$.

Proof. If Y is obtained from A by attaching n-cells then (i) is clear and (ii) follows from 1.2.5. For the converse, we first show that if (i) holds, then the family of sets $\{e_\alpha \mid \alpha \in \mathcal{A}\} \cup \{A\}$ is suitable for defining a weak topology on Y. Since Y is Hausdorff, the compact set $p_\alpha(B^n)$ is closed in Y, hence $e_\alpha :=$ $\mathrm{cl}_Y \overset{\circ}{e}_\alpha \subset p_\alpha(B^n)$. If e_α were a proper subset of $p_\alpha(B^n)$, then $p_\alpha^{-1}(e_\alpha)$ would be a proper closed subset of B^n containing $B^n - S^{n-1}$, which is impossible. Hence $e_\alpha = p_\alpha(B^n)$. It follows that each e_α is compact, hence that each $e_\alpha \cap e_\beta$ is closed in e_α and in e_β. The fact that $e_\alpha = p_\alpha(B^n)$, when combined with (i), also gives $e_\alpha \cap A = p_\alpha(S^{n-1})$; this implies that $e_\alpha \cap A$ is compact, and therefore closed, in the Hausdorff spaces e_α and A. So $\{e_\alpha \mid a \in \mathcal{A}\} \cup \{A\}$ is indeed suitable.

Assume (i) and (ii). For each α, we have seen that $e_\alpha = p_\alpha(B^n)$. By (i), $\overset{\circ}{e}_\alpha = p_\alpha(B^n - S^{n-1})$. Hence $e_\alpha - \overset{\circ}{e}_\alpha = p_\alpha(S^{n-1})$ which is closed in e_α. Thus $\overset{\circ}{e}_\alpha$ is open in e_α, and therefore each $\overset{\circ}{e}_\alpha$ is an open subset of Y. In particular, an open subset of $\overset{\circ}{e}_\alpha$ is open in $Y - A$.

Define $p : A \coprod B^n(\mathcal{A}) \to Y$ to agree with p_α on B^n_α and with inclusion on A. p is onto, and $p(S^{n-1}(\mathcal{A})) \subset A$. The restriction $p \mid: B^n(\mathcal{A}) - S^{n-1}(\mathcal{A}) \to Y - A$ is clearly a continuous bijection; to see that it is a homeomorphism, note that any p_α maps an open subset of $B^n - S^{n-1}$ homeomorphically onto an open subset of $\overset{\circ}{e}_\alpha$ (by (i)) and hence onto an open subset of $Y - A$, since $\overset{\circ}{e}_\alpha$ is open in $Y - A$.

It only remains to show that p is a quotient map. Let $C \subset Y$ be such that $p^{-1}(C)$ is closed in $A \coprod B^n(\mathcal{A})$. Note that $p^{-1}(C) \cap A = C \cap A$ is closed in A. For each α, $p_\alpha^{-1}(C) \cap B^n_\alpha = B^n_\alpha \cap p^{-1}(C \cap e_\alpha)$ is closed in B^n_α, hence is compact, hence $p_\alpha^{-1}(C)$ is closed in B^n, hence $p_\alpha^{-1}(C)$ is compact, hence $C \cap e_\alpha := p_\alpha p_\alpha^{-1}(C)$ is compact, hence $C \cap e_\alpha$ is closed in e_α. By (ii), C is closed in Y, so p is a quotient map.   □

In the special case where $\mathcal{A}$ is finite this theorem becomes simpler:

Proposition 1.2.8. *Let (Y, A) be a Hausdorff pair such that $\{\overset{\circ}{e}_a \mid \alpha \in \mathcal{A}\}$, the set of path components of $Y - A$, is finite. Let $n \in \mathbb{N}$. Y is obtained from A by attaching n-cells if A is closed in Y and, for each $\alpha \in \mathcal{A}$, there is a map $p_\alpha : (B^n, S^{n-1}) \to (A \cup \overset{\circ}{e}_\alpha, A)$ such that p_α maps $B^n - S^{n-1}$ homeomorphically onto $\overset{\circ}{e}_\alpha$.*

Proof. "Only if" is clear. To prove "if", note that (i) of 1.2.7 is given. Writing $e_\alpha = \mathrm{cl}_Y \overset{\circ}{e}_\alpha$, it follows, as in 1.2.7, that $\{e_\alpha \mid \alpha \in \mathcal{A}\} \cup \{A\}$ is suitable for defining a weak topology on Y. It only remains to check (ii) of 1.2.7. Let $C \subset Y$ be such that $C \cap e_\alpha$ is closed in e_α for each α, and $C \cap A$ is closed in A. Since A and each e_α are closed in Y, $C \cap A$ and each $C \cap e_\alpha$ are closed in

Y. The set $C = (C \cap A) \cup \left(\bigcup_\alpha C \cap e_\alpha \right)$, which is a finite union of closed sets, so C is closed in Y. Thus (i) and (ii) of 1.2.7 hold. $\qquad\square$

In applying the "if" halves of 1.2.7 and 1.2.8 it is usually easy to check in a given situation that p_α maps $B^n - S^{n-1}$ bijectively (and continuously) onto $\overset{\circ}{e}_\alpha$. In general, a continuous bijection is not a homeomorphism. However, given that our continuous bijection extends to a map $p_\alpha : (B^n, S^{n-1}) \to (A \cup \overset{\circ}{e}_\alpha, A)$, and that Y is Hausdorff, we have:

Proposition 1.2.9. *Let (Y, A), $\{\overset{\circ}{e}_\alpha \mid \alpha \in \mathcal{A}\}$ and $n \in \mathbb{N}$ be as in 1.2.7, and let $p_\alpha : (B^n, S^{n-1}) \to (A \cup \overset{\circ}{e}_\alpha, A)$ be a map such that p_α maps $B^n - S^{n-1}$ bijectively onto $\overset{\circ}{e}_\alpha$. Then p_α maps $B^n - S^{n-1}$ homeomorphically onto $\overset{\circ}{e}_\alpha$.*

Proof. Since B^n is compact, $p_\alpha : B^n \to e_\alpha^n$ is a quotient map. The restriction $p_\alpha \mid: B^n - S^{n-1} \to \overset{\circ}{e}{}_\alpha^{\,n}$ is a bijective quotient map, hence a homeomorphism. $\square$

Example 1.2.10. The one-point space $\{v\}$ is obtained from $\emptyset$ by attaching a 0-cell. The "figure eight", with topology inherited from $\mathbb{R}^2$, is obtained from its center point $\{v\}$ by attaching two 1-cells. It is also obtained from the discrete two-point space by attaching three 1-cells; see Fig. 1.2(a). The torus, with topology inherited from $\mathbb{R}^3$, is obtained from the "figure eight" by attaching a 2-cell; see Fig. 1.2(b). The space Y illustrated in Fig. 1.2(c), with topology inherited from $\mathbb{R}^2$, is not obtained from the indicated subspace A by attaching a 2-cell, even though $Y - A$ is homeomorphic to $B^2 - S^1$, because (i) of 1.2.7 fails. The space illustrated in Fig. 1.2(d), with topology inherited from $\mathbb{R}$, is not obtained from $\emptyset$ by attaching 0-cells, because (ii) of 1.2.7 fails. The space Y illustrated in Fig. 1.2(e), with topology inherited from $\mathbb{R}^2$, is not obtained from A by attaching a 1-cell because (ii) of 1.2.7 fails – this demonstrates the need, in 1.2.8, for A to be closed in Y.

Now we can define the spaces of interest to us.

A CW *complex* consists of a space X and a sequence $\{X^n \mid n \geq 0\}$ of subspaces such that

(i) X^0 is discrete;

(ii) For $n \geq 1$, X^n is obtained from X^{n-1} by attaching n-cells;

(iii) $X = \bigcup_n X^n$;

(iv) X has the weak topology[6] with respect to $\{X^n\}$.

---

[6] By (i)–(iii), $\{X^n\}$ is suitable for defining a weak topology on X.

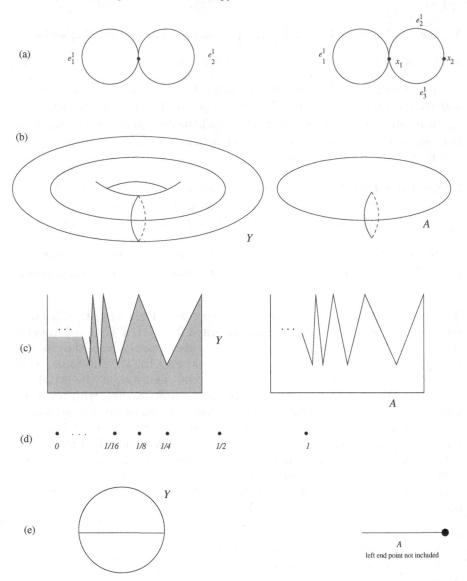

Fig. 1.2.

It is customary to say "let X be a CW complex," etc. but two different CW complexes could have the same underlying space, as in Fig. 1.2(a) for example. For induction arguments it is convenient to write $X^{-1} = \emptyset$.

The subspace X^n is called the *n-skeleton* of the CW complex X. If $X = X^n$ for some n, then the *dimension* of the CW complex is $\min\{n \mid X = X^n\}$. Otherwise the *dimension* is ∞. A 1-dimensional CW complex is also called a graph.[7] The CW complex X is *countable* [resp. has *finite type*] if each X^n is obtained from X^{n-1} by attaching countably many [resp. finitely many] n-cells. The CW complex is *finite* if it has finitely many cells.

Clearly, each X^n is closed in X, and, by 1.2.2, each X^n is Hausdorff. A little more work shows:

Proposition 1.2.11. *Every CW complex is Hausdorff.*

Proof. Let X be a CW complex. Let $x_1 \neq x_2 \in X$. For some n, $x_1 \in X^n$ and $x_2 \in X^n$. By 1.2.2 there are disjoint open sets U_n and V_n of X^n such that $x_1 \in U_n$ and $x_2 \in V_n$. X^{n+1} is obtained from X^n by attaching $(n+1)$-cells. Let $p : X^n \coprod B^{n+1}(\mathcal{A}) \to X^{n+1}$ be a quotient map such that $p(S^n(\mathcal{A})) \subset X^n$, $p \mid X^n$ = inclusion, and p maps $B^{n+1}(\mathcal{A}) - S^n(\mathcal{A})$ homeomorphically onto $X^{n+1} - X^n$. We find disjoint open subsets of X^{n+1}, U_{n+1} and V_{n+1}, such that $U_{n+1} \cap X^n = U_n$ and $V_{n+1} \cap X^n = V_n$ as follows: just as in the proof of 1.2.2, there exist disjoint open sets U'_{n+1} and V'_{n+1} of $B^{n+1}(\mathcal{A})$ such that $U'_{n+1} \cap S^n(\mathcal{A}) = p^{-1}(U_n) \cap S^n(\mathcal{A})$ and $V'_{n+1} \cap S^n(\mathcal{A}) = p^{-1}(V_n) \cap S^n(\mathcal{A})$; the required sets are $U_{n+1} = p(U'_{n+1}) \cup U_n$ and $V_{n+1} = p(V'_{n+1}) \cup V_n$. Proceed by induction to build disjoint open subsets of X^m, U_m and V_m, such that $U_m \cap X^{m-1} = U_{m-1}$ and $V_m \cap X^{m-1} = V_{m-1}$, for each $m \geq n+1$. Then $U = \bigcup_{m=n}^{\infty} U_m$ and $V = \bigcup_{m=n}^{\infty} V_m$ are disjoint open subsets of X, $x_1 \in U$ and $x_2 \in V$. $\square$

The *n-cells* of X are the n-cells of (X^n, X^{n-1}) as previously defined. If e^n is an n-cell of X, $\mathring{e}^n = e^n - X^{n-1}$ and $\dot{e}^n = e^n \cap X^{n-1}$. An n-cell has *characteristic maps* and *attaching maps* as before. A 0-cell is often called a *vertex*.

Proposition 1.2.12. *A CW complex X has the weak topology with respect to its cells.*

Proof. This is proved for n-dimensional CW complexes by induction on n, using 1.2.5. For an arbitrary CW complex X, $U \subset X$ is open iff $U \cap X^n$ is open in X^n for all n, iff $U \cap e^i$ is open in e^i for every i-cell e^i of X such that $i \leq n$ and for every n, iff $U \cap e$ is open in e for every cell e of X. $\square$

Proposition 1.2.13. *A compact subset of a CW complex lies in the union of finitely many cells. In particular, a CW complex is finite iff its underlying space is compact.*

Proof. Suppose this were false. Then there would be a compact subset C of X such that $C \cap \mathring{e}_\alpha \neq \emptyset$ for infinitely many values of α (where $\{e_\alpha \mid \alpha \in \mathcal{A}\}$ is the set of cells of X). For each such α, pick $x_\alpha \in \mathring{e}_\alpha \cap C$. Let D be the set of

---

[7] In discussing graphs it is sometimes sensible to enlarge the definition to include 0-dimensional (i.e., discrete) CW complexes and we will occasionally do this.

these points x_α. By an induction argument similar to that given in the proof of 1.2.6, $D \cap X^n$ is finite for each n. So every subset of D is closed in X. As in the proof of 1.2.6, one concludes that D is an infinite discrete compact space, a contradiction. $\qquad\square$

Here is a way of recognizing that a given space admits the structure of a CW complex.

Proposition 1.2.14. *Let X be a Hausdorff space and let $\{\overset{\circ}{e}_\alpha \mid \alpha \in \mathcal{A}\}$ be a family of subspaces with the following properties:*

(i) $X = \bigcup_\alpha \{\overset{\circ}{e}_\alpha\}$ *and* $\overset{\circ}{e}_\alpha \neq \overset{\circ}{e}_\beta \Rightarrow \overset{\circ}{e}_\alpha \cap \overset{\circ}{e}_\beta = \emptyset$;

(ii) *for each α, there is[8] $n(\alpha) \in \mathbb{N}$ such that $\overset{\circ}{e}_\alpha$ is homeomorphic to the space $B^{n(\alpha)} - S^{n(\alpha)-1}$;*

(iii) *letting $X^k = \cup\{\overset{\circ}{e}_\beta \mid n(\beta) \leq k\}$, there exists for each α (writing $n = n(\alpha)$) a map $p_\alpha : (B^n, S^{n-1}) \to (X^{n-1}\cup\overset{\circ}{e}_\alpha, X^{n-1})$ such that p_α maps $B^n - S^{n-1}$ homeomorphically onto $\overset{\circ}{e}_\alpha$;*

(iv) *(writing $e_\alpha = \text{cl}_X\overset{\circ}{e}_\alpha$, and calling e_α an n-cell of X when $n = n(\alpha)$) each cell of X lies in the union of finitely many members of $\{\overset{\circ}{e}_\alpha\}$;*

(v) *X has the weak topology with respect to the set $\{e_\alpha\}$ of all cells.*

Then $(X, \{X^k\})$ is a CW complex. Conversely, with cell notation as before, any CW complex[9] $(X, \{X^k\})$ possesses Properties (i)–(v).

Proof. Let $A \subset X^0$ and let e_α be a cell of X. Then $A \cap e_\alpha$ is finite by (iv), hence compact, hence closed in e_α. So A is closed in X, hence also in X^0. So X^0 is discrete.

Next, we show that X^n has the weak topology with respect to $\{e_\alpha \mid \dim e_\alpha \leq n\}$ where $\dim e_\alpha = n(\alpha)$. Let $C \subset X^n$ be such that $C \cap e_\alpha$ is closed in e_α whenever $\dim e_\alpha \leq n$. Let $\dim e_\beta > n$. By (iv), there are only finitely many cells $e_{\alpha_1}, \ldots, e_{\alpha_s}$ such that $e_\beta \cap \overset{\circ}{e}_{\alpha_i} \neq \emptyset$. Assume $\dim e_{\alpha_i} \leq n$ if $i \leq r$ and $\dim e_{\alpha_i} > n$ if $i > r$. For $i \leq r$, $e_{\alpha_i} \subset X^n$. We have

$$C \cap e_\beta = C \cap e_\beta \cap X^n \subset \bigcup_{i=1}^{r}(C \cap e_\beta \cap e_{\alpha_i})$$

because

$$e_\beta \cap X^n \subset \bigcup_{i=1}^{s} e_\beta \cap \overset{\circ}{e}_{\alpha_i} \cap X^n \subset \bigcup_{i=1}^{r} e_\beta \cap e_{\alpha_i}.$$

---

[8] We will see in Sect. 2.2 Exercise 3 that this $n(\alpha)$ is unique.

[9] The C in CW stands for "closure finite" which is a name for (iv); the W stands for "weak topology". Note that, by (i)–(iv), $\{e_\alpha\}$ is suitable for defining a weak topology on X. As in the proof of 1.2.7, $e_\alpha = p_\alpha(B^n)$ and is therefore compact.

And $\bigcup_{i=1}^{r}(C \cap e_\beta \cap e_{\alpha_i}) \subset C \cap e_\beta$. It follows, since each $C \cap e_{\alpha_i}$ is compact, that $C \cap e_\beta$ is compact, hence closed in e_β. By (v), C is closed in X, hence in X^n.

By induction, it follows that X^n has the weak topology with respect to $\{e_\alpha \mid \dim e_\alpha = n\} \cup \{X^{n-1}\}$. This and (iii) together imply that X^n is obtained from X^{n-1} by attaching n-cells (by 1.2.7).

Obviously $X = \bigcup_n X^n$ and X has the weak topology with respect to $\{X^n\}$, the latter following from (v) and what has just been established. Thus $(X, \{X^k\})$ is a CW complex.

The converse is clear from the previous propositions. $\qquad\square$

In the finite case we get a simpler criterion:

Proposition 1.2.15. *Let X be a Hausdorff space and let $\{\overset{\circ}{e}_\alpha \mid \alpha \in \mathcal{A}\}$ be a finite family of subspaces which satisfy (i), (ii), and (iii) of 1.2.14. Then $(X, \{X^n\})$ is a finite CW complex. In particular, X is compact.* $\qquad\square$

Example 1.2.16. Let $x \in S^{n-1}$. Let $\overset{\circ}{e}_1 = \{x\}$ and $\overset{\circ}{e}_2 = S^{n-1} - \{x\}$. This gives S^{n-1} the structure of a CW complex with one 0-cell e_1 and one $(n-1)$-cell e_2. Now consider $B^n \supset S^{n-1}$. Let $\overset{\circ}{e}_3 = B^n - S^{n-1}$. This makes B^n into a CW complex with the 0-cell e_1, the $(n-1)$-cell e_2 and an n-cell e_3. Note that all this makes sense when $n = 1$.

Our next example is very important. While the notion of "fundamental group" only appears in Chap. 3, this example shows how to create, for any group G, a CW complex whose fundamental group is isomorphic to G:

Example 1.2.17. Let $W = \{x_\alpha \mid \alpha \in \mathcal{A}\}$ freely generate the free group $F(W)$, abbreviated to F, let $R = \{y_\beta \mid \beta \in \mathcal{B}\}$ be a (possibly empty) set, and let $\rho : R \to F$ be a function. Then $\langle W \mid R, \rho \rangle$ is a *presentation*, and if the group G is isomorphic to the group $F/N(\rho(R))$, where $N(\rho(R))$ is the normal closure of $\rho(R)$ in F, it is called a *presentation of G*. Often, R will be a subset of F and ρ will be the inclusion, in which case we simplify, denoting the presentation of the group $F/N(R)$ by $\langle W \mid R \rangle$. If $R = \emptyset$, so that the group is the free group generated by W, the presentation is often written as $\langle W \mid \rangle$ rather than $\langle W \mid \emptyset \rangle$.

Associated with $\langle W \mid R, \rho \rangle$ is a CW complex X called a *presentation complex*, having one 0-cell e^0, a 1-cell e_α^1 for each $\alpha \in \mathcal{A}$, and a 2-cell e_β^2 for each $\beta \in \mathcal{B}$, as we now describe.[10] There is only one possibility for attaching the 1-cells to form X^1. Choose a characteristic map $h_\alpha : (B^1, S^0) \to (e_\alpha^1, \overset{\bullet}{e}_\alpha^1)$ for each α. We define an attaching map $f_\beta : S^1 \to X^1$ for the 2-cell e_β^2

---

[10] G is *finitely generated* [resp. *finitely presented*] if W [resp. W and R] can be chosen to be finite. If W and R are finite $\langle W \mid R, \rho \rangle$ is a *finite presentation* of G. For more on presentations and presentation complexes see the Appendix to Sect. 3.1.

as follows. The element $\rho(y_\beta)$ is uniquely expressible as a reduced word in the letters $\{x_\alpha, x_\alpha^{-1} \mid \alpha \in \mathcal{A}\}$ consisting of, say, λ letters. If $\lambda = 0$, let f_β map all of S^1 to the 0-cell e^0. If $\lambda > 0$, $\rho(y_\beta) = x_{\alpha_1}^{\epsilon_1} \ldots x_{\alpha_\lambda}^{\epsilon_\lambda}$ where[11] each $\epsilon_j = \pm 1$ $[x_\alpha^1 := x_\alpha]$. Partition S^1 by the λ roots of unity $\{1, \omega, \ldots, \omega^{\lambda-1}\}$ where $\omega = e^{2\pi i/\lambda}$. This defines a CW complex decomposition of S^1 having λ 0-cells $\{\omega^j \mid 1 \leq j \leq \lambda\}$, and λ 1-cells $\{e_j^1\}$ such that $\overset{\circ}{e}_j^1 = \{e^{2\pi i t/\lambda} \mid j - 1 < t < j\}$; see 1.2.15. As a characteristic map $c_j : [-1, 1] \to e_j^1$ choose a homeomorphism which maps 1 to $e^{2\pi i j/\lambda}$. Let $r_j : [-1, 1] \to [-1, 1]$ be the identity if $\epsilon_j = 1$ and be $t \mapsto -t$ if $\epsilon_j = -1$. The required attaching map f_β is the unique map $S^1 \to X^1$ such that[12] for each $1 \leq j \leq \lambda$, $f_\beta \circ c_j = h_{\alpha_j} \circ r_j$.

$$\coprod_{j=1}^{\lambda} B_j^1 \xrightarrow{(c_j)} S^1$$

$$\downarrow \coprod_{j=1}^{\lambda} r_j \qquad \downarrow i_\beta$$

$$\coprod_{j=1}^{\lambda} B_j^1 \xrightarrow{(h_{\alpha_j})} X^1$$

Up to cell-preserving homeomorphism, X does not depend on the choices of characteristic maps h_α, but it does depend on the presentation rather than on the group $F/N(\rho(R))$.

Here are some simple cases of this construction. The one-point space corresponds to the trivial presentation of the trivial group ($W = R = \emptyset$). The "figure eight" CW complex with one 0-cell and two 1-cells corresponds to the presentation $\langle x_1, x_2 \mid \emptyset \rangle$ of the free group of rank 2. The torus CW complex in 1.2.10 corresponds to the presentation $\langle x_1, x_2 \mid x_1 x_2 x_1^{-1} x_2^{-1} \rangle$ of $\mathbb{Z} \times \mathbb{Z}$.

The distinction made at the beginning of 1.2.17 between presentations of the form $\langle W \mid R, \rho \rangle$ and presentations of the form $\langle W \mid R \rangle$ deserves a comment. The point is that we might wish to allow $y_{\beta_1} = y_{\beta_2}$ when $\beta_1 \neq \beta_2$. For example, let $W = \emptyset$ and $\mathcal{B} = \{1, 2\}$. Then $F(W)$ is the trivial group $\{e\}$. Letting $R = \{y_1, y_2\}$, there is only one function $\rho : R \to \{e\}$. $\langle W \mid R, \rho \rangle$ is a presentation of the trivial group. The corresponding CW complex, X, in the spirit of 1.2.17, has one 0-cell, no 1-cells, and two 2-cells (a "bouquet of two 2-spheres"). On the other hand, there is no presentation of the trivial group having the form $\langle W \mid R \rangle$ which yields that particular CW complex X.

---

[11] x_α^1 means x_α.

[12] Intuitively, this says: attach B_β^2 to X^1 by wrapping the jth of the λ intervals of S^1 around the α_j^{th} 1-cell, in the "preferred" direction (as defined by choice of characteristic maps) if $\epsilon_j = 1$, and in the other direction if $\epsilon_j = -1$. We have written this out formally to illustrate how the terminology and the properties of the quotient topology are used.

We turn to some standard methods for manufacturing new CW complexes from old ones.

Proposition 1.2.18. *If $\{X_\alpha \mid \alpha \in \mathcal{A}\}$ is a set of CW complexes then $X := \coprod_\alpha X_\alpha$ is a CW complex with $X^n = \coprod_\alpha X_\alpha^n$.* □

It is sometimes convenient to have cubes rather than balls as the domains of characteristic maps. This is because the product of two cubes is a cube. An example of this is Proposition 1.2.19, below, so we pause to discuss cubes. Let $I^n = \prod_{i=1}^n X_i$ where each $X_i = [-1, 1]$. Then $I^n \subset \mathbb{R}^n$. Let $\overset{\bullet}{I}{}^n = \mathrm{fr}_{\mathbb{R}^n} I^n$. I^n is the *n-cube*. Note that $I^1 \neq I$. The map $I^n \xrightarrow{a_n} B^n$, defined by: $0 \mapsto 0$ and, for $x \neq 0$, $x \longmapsto (t(x)/|x|)x$, where $t(x) = \max\{|x_1|, \ldots, |x_n|\}$, is a continuous bijection. Since I^n is compact and B^n is Hausdorff, a_n is a homeomorphism. a_n takes $\overset{\bullet}{I}{}^n$ to S^{n-1}. WHEN WE USE $(I^n, \overset{\bullet}{I}{}^n)$ AS THE DOMAIN OF A CHARACTERISTIC MAP, WE ALWAYS MEAN TO IDENTIFY $(I^n, \overset{\bullet}{I}{}^n)$ WITH (B^n, S^{n-1}) VIA a_n. All this makes sense for $n = 0$ if we define $I^0 = \{0\}$.

Proposition 1.2.19. *Let X and Y be CW complexes. Then $X \times Y$ (with the product topology understood to be in the sense of k-spaces) is a CW complex with $(X \times Y)^n = \cup\{X^i \times Y^j \mid i + j = n\}$.* □

The special case of 1.2.19 where $Y = I$ (with two 0-cells, 0 and 1, and one 1-cell) is very useful in constructing homotopies: see Sect. 1.3.

Proposition 1.2.20. *Let $(X, \{X^n\})$ be a CW complex, let $\{e_\alpha \mid \alpha \in \mathcal{A}\}$ be the set of cells of X, let $\mathcal{B} \subset \mathcal{A}$, let $A = \cup\{\overset{\circ}{e}_\alpha \mid \alpha \in \mathcal{B}\}$, and let $A^n = A \cap X^n$. If $\mathcal{B}$ is such that $\overset{\bullet}{e}_\alpha \subset A$ whenever $\alpha \in \mathcal{B}$, then $(A, \{A^n\})$ is a CW complex and A is closed in X.*

A CW complex A formed from a CW complex X by the selection of $\mathcal{B} \subset \mathcal{A}$ as in 1.2.20 is called a *subcomplex* of X. A CW *pair* is a pair (X, A) such that X is a CW complex and A is a subcomplex of X. It is usually convenient not to distinguish between the CW pair $(X, \emptyset)$ and the CW complex X.

Proof. We verify the axioms (i)–(iv) in the definition of a CW complex. Axioms (i) and (iii) clearly hold. To verify Axiom (ii), we show by induction that A^n is obtained from A^{n-1} by attaching n-cells and that A^n is closed in X^n. This is clear when $n = 0$; assume it for $(n-1)$. Then A^{n-1} is closed in X^{n-1}. Let $\mathcal{A}_n = \{\alpha \in \mathcal{A} \mid e_\alpha \text{ is an } n\text{-cell of } X\}$. Apply 1.2.7 to (X^n, X^{n-1}) and $\{\overset{\circ}{e}_\alpha^n \mid \alpha \in \mathcal{A}_n\}$ to conclude that for each $\alpha \in \mathcal{A}_n$ there exists $p_\alpha : (B^n, S^{n-1}) \to (X^{n-1} \cup \overset{\circ}{e}_\alpha^n, X^{n-1})$ such that p_α maps $B^n - S^{n-1}$ homeomorphically onto $\overset{\circ}{e}_\alpha$, and that X^n has the weak topology with respect to $\{e_\alpha \mid \alpha \in \mathcal{A}_n\} \cup \{X^{n-1}\}$. Let $\mathcal{B}_n = \mathcal{A}_n \cap \mathcal{B}$. Consider (A^n, A^{n-1}) and

$\{\mathring{e}^{\,n}_\alpha \mid \alpha \in \mathcal{B}_n\}$. Since A^{n-1} is closed in X^{n-1}, and $\mathring{e}^\alpha \subset A \cap X^{n-1} = A^{n-1}$ when $\alpha \in \mathcal{B}_n$, (i) and (ii) of 1.2.7 hold. Hence A^n is obtained from A^{n-1} by attaching n-cells. The subspace X^n has the weak topology with respect to $\{e^n_\alpha \mid \alpha \in \mathcal{A}_n\} \cup \{X^{n-1}\}$. To show that A^n is closed in X^n, we thus observe: $A^n \cap X^{n-1} = A \cap X^{n-1} = A^{n-1}$ is closed in X^{n-1}; when $\alpha \in \mathcal{B}_n$, $A^n \cap e^n_\alpha = e^n_\alpha$; when $\alpha \in \mathcal{A}_n - \mathcal{B}_n$, $A^n \cap e^n_\alpha = A^n \cap \mathring{e}^{\,n}_\alpha = A^{n-1} \cap \mathring{e}^{\,n}_\alpha$ which is a compact subset, hence a closed subset, of e^n_α. This completes the induction, and establishes Axiom (ii). The fact that A^n is closed in X^n obviously implies Axiom (iv). $\qquad\square$

Note that the union of subcomplexes of X is a subcomplex of X. Note also that each X^n is a subcomplex of X.

Proposition 1.2.21. *Each path component of a CW complex X is a subcomplex, an open subset of X, and a closed subset of X. Hence, a non-empty CW complex is connected iff it is path connected.*

Proof. The first part is clear. For the rest, let A be a path component of the CW complex X. Cells are path connected, being the images of balls under maps. Hence, for each cell e of X, either $e \cap A = e$ or $e \cap A = \emptyset$. This proves A is both open and closed in X. $\qquad\square$

Let X be a CW complex, let $\{A_\alpha\}$ be a family of pairwise disjoint subcomplexes, and let $\sim$ be the equivalence relation on X generated by: $x \sim y$ whenever there exists α such that $x \in A_\alpha$ and $y \in A_\alpha$. Let $X/\sim$ be given the quotient topology, with quotient map $p : X \to X/\sim$. Define $(X/\sim)^n = p(X^n)$. It is in fact always the case that $(X/\sim, \{(X/\sim)^n\})$ is a CW complex; the non-obvious part of the proof is the fact that $X/\sim$ is Hausdorff, which is in turn a consequence of the fact that X is a normal space. However, we will only need:

Proposition 1.2.22. *If there exist pairwise disjoint open sets $U_\alpha \subset X$ such that, for each α, $A_\alpha \subset U_\alpha$, then $(X/\sim, \{(X/\sim)^n\})$ is a CW complex. In particular, if A is a subcomplex of X, then $(X/A, \{(X/A)^n\})$ is a CW complex.*

Proof. Apply 1.2.14. The Hausdorff property is clear under these hypotheses. $\square$

With this CW structure, $X/\sim$ is the *quotient complex*.

To end this section, we discuss continuity of functions $X \to Z$ where X is a CW complex and Z is a space. The recognition problem is easy: by 1.2.12 such a function is continuous iff its restriction to each cell is continuous. On the other hand, the building of continuous functions is usually done by induction on skeleta:

Proposition 1.2.23. *Let (X, A) be a CW pair and let the n-cells of X which are not cells of A be indexed by $\mathcal{A}$. Let $\{h_\alpha : B^n_\alpha \to e^n_\alpha \mid \alpha \in \mathcal{A}\}$ be a set of characteristic maps for those cells. Let Z be a space and let $f_{n-1} : (X^{n-1} \cup$*

$A) \to Z$ and $g : B^n(\mathcal{A}) \to Z$ be maps such that $f_{n-1} \circ h_\alpha \mid S_\alpha^{n-1} = g \mid S_\alpha^{n-1}$. Then there is a unique map $f_n : (X^n \cup A) \to Z$ such that $f_n \mid X^{n-1} \cup A = f_{n-1}$ and $f_n \circ h_\alpha = g \mid B_\alpha^n$.

Proof. By 1.2.7, $X^n \cup A$ is obtained from $X^{n-1} \cup A$ by attaching n-cells. The result follows from the properties of the quotient topology stated in Sect. 1.1. □

Remark 1.2.24. A Hausdorff space Z is *paracompact* if every open cover $\mathcal{U}$ of Z has a locally finite open refinement $\mathcal{V}$; this means: $\mathcal{V}$ is a locally finite open cover of Z and every element of $\mathcal{V}$ is a subset of some element of $\mathcal{U}$. It is a fact that every CW complex is paracompact; for a proof see [105]. This arises, for example, in the proof that a fiber bundle whose base space is a CW complex has the homotopy lifting property.

Historical Note: CW complexes were introduced by J.H.C. Whitehead in [154]. In his exposition primacy was given to the "open cells" $\overset{\circ}{e}_\alpha$ rather than the "closed cells" e_α, presumably because each $\overset{\circ}{e}_\alpha$ is homeomorphic to an open ball while e_α need not be homeomorphic to a closed ball. The shift of primacy to "closed cells" has become standard.

Exercises

1. If $X^0 = \emptyset$ then $X = \emptyset$; why?
2. Show that a compact subset of a CW complex lies in a finite subcomplex.
3. In Example 1.2.10(a) and (b) it is asserted that some familiar spaces have particular CW complex structures. Prove that these spaces (endowed with the topology inherited from the Euclidean spaces in which they live) are homeomorphic to the indicated CW complexes.

1.3 Homotopy

Let X and Y be spaces and let $f_0, f_1 : X \to Y$ be maps. One says that f_0 is *homotopic to* f_1, denoted $f_0 \simeq f_1$, if there exists a map $F : X \times I \to Y$ such that for all $x \in X$, $F(x,0) = f_0(x)$ and $F(x,1) = f_1(x)$. The map F is called a *homotopy* from f_0 to f_1. One often writes $F_t : X \to Y$ for the map $x \mapsto F(x,t)$; then $F_0 = f_0$ and $F_1 = f_1$. If F is a homotopy from f_0 to f_1, one writes $F : f_0 \simeq f_1$.

Proposition 1.3.1. *Homotopy is an equivalence relation on the set of maps from X to Y.*

Proof. Given $f : X \to Y$, define $F : f \simeq f$ by $F(x,t) = f(x)$ for all $t \in I$; $F = f \circ$ (projection: $X \times I \to X$), the composition of two maps. So F is a map. Thus reflexivity. Given $F : f \simeq g$, define $F' : g \simeq f$ by $F'(x,t) =$

$F(x, 1-t)$. The function $q : I \to I, t \mapsto 1-t$, is a map. $F' = F \circ (\mathrm{id} \times q)$, so F' is a map. Thus symmetry. Given $F : f \simeq g$ and $G : g \simeq h$, define $H : X \times I \to Y$ by $H(x, t) = F(x, 2t)$ when $0 \le t \le \frac{1}{2}$ and by $H(x, t) = G(x, 2t - 1)$ when $\frac{1}{2} \le t \le 1$. Let $q_1 : [0, \frac{1}{2}] \to I$ be the map $t \mapsto 2t$. Let $q_2 : [\frac{1}{2}, 1] \to I$ be the map $t \mapsto 2t - 1$. On $X \times [0, \frac{1}{2}]$, H agrees with the map $F \circ (\mathrm{id} \times q_1)$; on $X \times [\frac{1}{2}, 1]$, H agrees with the map $G \circ (\mathrm{id} \times q_2)$. Since $X \times [0, \frac{1}{2}]$ and $X \times [\frac{1}{2}, 1]$ are closed in $X \times I$, and the restriction of H to each is a map, H is a map. Thus transitivity. □

A map $f : X \to Y$ is a *homotopy equivalence* if there exists a map $g : Y \to X$ such that $g \circ f \simeq \mathrm{id} : X \to X$ and $f \circ g \simeq \mathrm{id} : Y \to Y$. Such a map g is called a *homotopy inverse* of f. The spaces X and Y have the *same homotopy type* if there exists a homotopy equivalence from X to Y. If g is a homotopy inverse of the homotopy equivalence f, then g is also a homotopy equivalence. A map homotopic to a homotopy equivalence is a homotopy equivalence. Any two homotopy inverses of f are homotopic, and any map homotopic to a homotopy inverse of f is a homotopy inverse of f. The map id_X is a homotopy equivalence. A homeomorphism is a homotopy equivalence.

Definitions parallel to those of the last paragraph hold for maps of pairs. A map $f : (X, A) \to (Y, B)$ is a *homotopy equivalence* if there exists a map $g : (Y, B) \to (X, A)$ such that $g \circ f \simeq \mathrm{id} : (X, A) \to (X, A)$ and $f \circ g \simeq \mathrm{id} : (Y, B) \to (Y, B)$. Such a map g is called a *homotopy inverse* of f. The pairs (X, A) and (Y, B) have the *same homotopy type* if there exists a homotopy equivalence from (X, A) to (Y, B), etc.

Let (X, A) and (Y, B) be pairs of spaces, let $f_0, f_1 : (X, A) \to (Y, B)$ be maps of pairs, and let $X' \subset X$. One says that f_0 is *homotopic to f_1 relative to X'*, denoted $f_0 \simeq f_1$ rel X', if there exists a map $F : (X \times I, A \times I) \to (Y, B)$ such that for all $x \in X$, $F(x, 0) = f_0(x)$ and $F(x, 1) = f_1(x)$, and for all $x' \in X'$ and $t \in I$, $F(x', t) = f_0(x')$. Then F is a *homotopy relative to X'* from f_0 to f_1. By a proof similar to that of 1.3.1 we have:

Proposition 1.3.2. *Homotopy relative to X' is an equivalence relation on the set of maps from (X, A) to (Y, B).* □

It is customary not to distinguish between the space X and the pair $(X, \emptyset)$, so that homotopy between maps of spaces is considered to be a special case of homotopy between maps of pairs.

Proposition 1.3.3. *Let $f_0, f_1 : (X, A) \to (Y, B)$ be homotopic rel X', let $g_0, g_1 : (Y, B) \to (Z, C)$ be homotopic rel Y', where $f_0(X') \subset Y'$. Then $g_0 \circ f_0$ and $g_1 \circ f_1$ are homotopic rel X'.*

Proof. Let $F : f_0 \simeq f_1$ and $G : g_0 \simeq g_1$ be homotopies which behave as required on A, B, C and X'. Let $p : X \times I \to I$ be projection. Let $(F, p) : X \times I \to Y \times I$ denote the function $(x, t) \mapsto (F(x, t), p(x, t)) = (F(x, t), t)$. (F, p) is a map. The required homotopy is $G \circ (F, p) : X \times I \to Z$. □

The space X is *contractible* if it has the same homotopy type as the one-point space. Equivalently, X is contractible if id_X is homotopic to some constant map from X to X.

Proofs of 1.3.4–1.3.7 below are left as exercises:

Proposition 1.3.4. *Any two maps from a space to a contractible space are homotopic.*   □

Proposition 1.3.5. *Any two maps from a contractible space to a path connected space are homotopic.*   □

Proposition 1.3.6. *The product of two contractible spaces is contractible.* □

Example 1.3.7. The spaces $\mathbb{R}$, $\mathbb{R}_+$, and I are contractible. Hence also (by 1.3.6) $\mathbb{R}^n$, $\mathbb{R}_+^n$ and B^n are contractible. If $p \in S^n$, then $S^n - \{p\}$ is homeomorphic to $\mathbb{R}^n$; hence $S^n - \{p\}$ is contractible.

The subspace $A \subset X$ is a *retract* of X if there is a map $r : X \to A$ such that $A \xhookrightarrow{i} X \xrightarrow{r} A$ is the identity. Such a map r is a *retraction of X to A*.

The subspace $A \subset X$ is a *strong deformation retract* of X if there is a retraction $r : X \to A$ and a homotopy $F : X \times I \to X$ relative to A such that $F_0 = \mathrm{id}_X$ and $F_1 = i \circ r$. Such a homotopy F is a *strong deformation retraction of X to A*. A pair (A, A') is a *strong deformation retract* of the pair (X, X') if $(A, A') \subset (X, X')$ and the retraction r and the homotopy F in the previous sentence are maps of pairs.[13]

Proposition 1.3.8. *If A is a strong deformation retract of X then $A \hookrightarrow X$ is a homotopy equivalence.*

Proof. In the notation above, r is a homotopy inverse for inclusion.   □

For building homotopies involving CW complexes the following example is fundamental.

Lemma 1.3.9. *The subspace $(B^n \times \{0\}) \cup (S^{n-1} \times I)$ is a strong deformation retract of $B^n \times I$.*

Proof. One standard proof involves "radial projection" of $B^n \times I$ onto the subspace $(B^n \times \{0\}) \cup (S^{n-1} \times I)$ from the "light source" $(0, \ldots, 0, 2) \in \mathbb{R}^{n+1}$. We give a variation which avoids complicated formulas.

Let $f : B^n \to I$ be the map $f(x) = 0$ if $|x| \leq \frac{1}{2}$ and $f(x) = 2|x| - 1$ if $\frac{1}{2} \leq |x| \leq 1$. Let $Y = \{(x, t) \in B^n \times \mathbb{R} \mid f(x) \leq t \leq 1\}$. Let $Y_0 = \{(x, t) \in Y \mid$

---

[13] If F is not required to be rel A while r remains a retraction, one says that A is a *deformation retract* of X. If this is further weakened by only requiring $r \circ i$ to be homotopic to id_A then A is a *weak deformation retract* of X. Note that A is a weak deformation retract of X iff $i : A \hookrightarrow X$ is a homotopy equivalence. When A is a subcomplex of the CW complex X the three notions of deformation retract coincide; see Exercises 2 and 3 below.

$t = f(x)\}$. There is a strong deformation retraction $F : Y \times I \to Y$ of Y to Y_0, namely $F(x, t, s) = (x, (1 - s)t + sf(x))$. And there is a homeomorphism $h : (B^n \times I, (B^n \times \{0\}) \cup (S^{n-1} \times I)) \to (Y, Y_0)$, namely $h(x, t) = (\frac{1}{2}(t+1)x, t)$. $\square$

We are preparing for the important Theorem 1.3.15. The building blocks of the proof are Lemma 1.3.9 and the following (compare 1.2.23):

Proposition 1.3.10. *Let (X, A) be a CW pair and let the n-cells of X which are not in A be indexed by $\mathcal{A}$. Let $\{h_\alpha : B^n_\alpha \to X \mid \alpha \in \mathcal{A}\}$ be a set of characteristic maps for those n-cells. Let Z be a space and let $F_{n-1} : (X^{n-1} \cup A) \times I \to Z$ and $g : B^n(\mathcal{A}) \times I \to Z$ be maps such that $F_{n-1} \circ (h_\alpha \times \mathrm{id}) \mid S^{n-1}_\alpha \times I = g \mid S^{n-1}_\alpha \times I$. Then there is a unique map $F_n : (X^n \cup A) \times I \to Z$ such that $F_n \mid (X^{n-1} \cup A) \times I = F_{n-1}$ and $F_n \circ (h_\alpha \times \mathrm{id}) = g \mid B^n_\alpha \times I$.*

The proof of 1.3.10 requires a non-obvious lemma from general topology; see [51, p. 262]:

Lemma 1.3.11. *Let Z and W be Hausdorff spaces. If $q : Z \to W$ is a quotient map then $q \times \mathrm{id} : Z \times I \to W \times I$ is a quotient map.* $\square$

Proof (Proof of 1.3.10). By 1.2.7, $X^n \cup A$ is obtained from $X^{n-1} \cup A$ by attaching n-cells. Thus there is a quotient map $p : (X^{n-1} \cup A) \coprod B^n(\mathcal{A}) \to X^n \cup A$ which agrees with inclusion on $X^{n-1} \cup A$ and with h_α on $B^n_\alpha \subset B^n(\mathcal{A})$. By 1.3.11, $p \times \mathrm{id} : ((X^{n-1} \cup A) \times I) \coprod (B^n(\mathcal{A}) \times I) \to (X^n \cup A) \times I$ is a quotient map agreeing with inclusion on the first summand and with $h_\alpha \times \mathrm{id}$ on the $B^n_\alpha \times I$ part of the second summand, for each $\alpha \in \mathcal{A}$. The desired result therefore follows from the properties of the quotient topology stated in Sect. 1.1. $\square$

Proposition 1.3.12. *If Y is obtained from A by attaching n-cells, then $(Y \times \{0\}) \cup (A \times I)$ is a strong deformation retract of $Y \times I$. Hence any map $(Y \times \{0\}) \cup (A \times I) \to Z$ extends to a map $Y \times I \to Z$.*

Proof. By 1.3.9, there is a strong deformation retraction $F : B^n \times I \times I \to B^n \times I$ of $B^n \times I$ to $(B^n \times \{0\}) \cup (S^{n-1} \times I)$. Let $\mathcal{A}$ index the n-cells. Consider the diagram

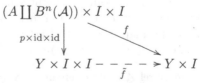

$$(A \coprod B^n(\mathcal{A})) \times I \times I$$

Here, $p : A \coprod B^n(\mathcal{A}) \to Y$ is a quotient map, and f agrees with $A \times I \times I \xrightarrow{\text{projection}} A \times I \xrightarrow{\text{inclusion}} Y \times I$ on $A \times I \times I$ and with $B^n_\alpha \times I \times I \xrightarrow{F} B^n_\alpha \times I \xrightarrow{(p|)\times \mathrm{id}} Y \times I$ on $B^n_\alpha \times I \times I$ for each $\alpha \in \mathcal{A}$.

By 1.3.11, $p \times \mathrm{id} \times \mathrm{id}$ is a quotient map. By the properties of quotient maps stated in Sect. 1.1, the map $\tilde{f}$ exists and is the desired strong deformation retraction. For the last part, any map $g : (Y \times \{0\}) \cup (A \times I) \to Z$ is extended by $g \circ \tilde{f}_1$ to a well-defined map $Y \times I \to Z$. $\qquad\square$

The last sentence of 1.3.12 illustrates the following basic property. A pair of spaces (Y, A) has the *homotopy extension property with respect to a space* Z if every map $(Y \times \{0\}) \cup (A \times I) \to Z$ extends to a map $Y \times I \to Z$. In the presence of this property, a map on A can be extended to X if it is homotopic to a map which can be extended to X.

By 1.2.7 we have:

Proposition 1.3.13. *If A is a subcomplex of X, then $X^n \cup A$ is a subcomplex of X. Moreover $X^n \cup A$ is obtained from $X^{n-1} \cup A$ by attaching n-cells.* $\qquad\square$

Proposition 1.3.14. *Let $Z = \bigcup\limits_{n \geq 1} Z_n$ be a CW complex, where each Z_n is a subcomplex of Z. For all $n > 1$, assume Z_{n-1} is a strong deformation retract of Z_n (in particular $Z_{n-1} \subset Z_n$). Then Z_1 is a strong deformation retract of Z.*

Proof. Let $n \geq 2$. Let $F^{(n)} : Z_n \times [\frac{1}{n}, \frac{1}{n-1}] \to Z_n$ satisfy $F_t^{(n)}(x) = x$ if $t = \frac{1}{n}$ or if $x \in Z_{n-1}$, and $F_{\frac{1}{n-1}}^{(n)}(Z_n) = Z_{n-1}$. Let $\hat{F}_{\frac{1}{n-1}}^{(n)} : Z_n \to Z_{n-1}$ be the map induced by $F_{\frac{1}{n-1}}^{(n)}$. Define $G^{(n)} : Z_n \times I \to Z_n$ to agree with projection $(x, t) \mapsto x$ on $Z_n \times [0, \frac{1}{n}]$, with $F^{(n)}$ on $Z_n \times [\frac{1}{n}, \frac{1}{n-1}]$, with $F^{(n-1)} \circ (\hat{F}_{\frac{1}{n-1}}^{(n)} \times \mathrm{id})$ on $Z_n \times [\frac{1}{n-1}, \frac{1}{n-2}]$, with $F^{(n-2)} \circ (\hat{F}_{\frac{1}{n-2}}^{(n-1)} \times \mathrm{id}) \circ (\hat{F}_{\frac{1}{n-1}}^{(n)} \times \mathrm{id})$ on $Z_n \times [\frac{1}{n-2}, \frac{1}{n-3}], \ldots,$ and with $F^{(2)} \circ (\hat{F}_{\frac{1}{2}}^{(3)} \times \mathrm{id}) \circ \ldots \circ (\hat{F}_{\frac{1}{n-1}}^{(n)} \times \mathrm{id})$ on $Z_n \times [\frac{1}{2}, 1]$. Then $G^{(n)}$ is a strong deformation retraction of Z_n to Z_1, and $G^{(n)}$ agrees with $G^{(n-1)}$ on $Z_{n-1} \times I$. Define $G : Z \times I \to Z$ to agree with $G^{(n)}$ on $Z_n \times I$ for all $n \geq 2$. By 1.2.12, G is continuous. Clearly, G is the required strong deformation retraction of Z to Z_1. $\qquad\square$

Now we are ready for the Homotopy Extension Theorem for CW complexes:

Theorem 1.3.15. (Homotopy Extension Theorem) *If (X, A) is a CW pair, then $(X \times \{0\}) \cup (A \times I)$ is a strong deformation retract of $X \times I$. Hence (X, A) has the homotopy extension property with respect to every space.*

Proof. Let $Z_n = (X \times \{0\}) \cup ((X^n \cup A) \times I)$ and let $Z = X \times I$. Then $Z_{-1} = (X \times \{0\}) \cup (A \times I)$, and $Z = \bigcup\limits_{n \geq -1} Z_n$. By 1.3.13 and 1.2.19, each Z_n is a subcomplex of Z, and of Z_{n+1}. By 1.3.13, $X^n \cup A$ is obtained from $X^{n-1} \cup A$ by attaching n-cells. Hence, by 1.3.12, $((X^n \cup A) \times \{0\}) \cup ((X^{n-1} \cup A) \times I) =$

$(X^n \times \{0\}) \cup ((X^{n-1} \cup A) \times I)$ is a strong deformation retract of $(X^n \cup A) \times I$. It follows immediately that Z_{n-1} is a strong deformation retract of Z_n. By 1.3.14, Z_{-1} is a strong deformation of Z, as claimed. The last part is proved as in the proof of 1.3.12. □

Examination of the proof of 1.3.15 shows a little more which can be useful:

Addendum 1.3.16. *Let (X, A) be a CW pair and let $F : (X \times \{0\}) \cup (A \times I) \to Z$ be a map. The map F extends to a map $\tilde{F} : X \times I \to Z$ such that, for every cell e_α of $X - A$, $\tilde{F}_1(e_\alpha) = F_0(e_\alpha) \cup \tilde{F}(\overset{\bullet}{e}_\alpha \times I)$.* □

The definition of the homotopy extension property has nothing to do with CW complexes. Therefore the last part of 1.3.15 is more powerful than it might appear; it says that the existence of a CW pair structure on the pair of spaces (X, A) ensures that (X, A) has the homotopy extension property with respect to every space.

Corollary 1.3.17. *If (X, A) is a CW pair and Z is a contractible space, every map $A \to Z$ extends to a map $X \to Z$.* □

Exercises

1. Prove 1.3.4–1.3.7.

For the next two exercises (X, A) is a CW pair.

2. Prove that if A is a weak deformation retract of X then A is a deformation retract of X.
3. Prove that if A is a deformation retract of X then A is a strong deformation retract of X.

    Hint for 2. *and* 3.: Apply 1.3.15 appropriately.

1.4 Maps between CW complexes

A map $f : X \to Y$ between CW complexes is *cellular* if $f(X^n) \subset Y^n$ for all $n \geq 0$. A homotopy $F : X \times I \to Y$ is *cellular* if it is a cellular map with respect to the natural CW complex structure on $X \times I$ given by 1.2.19. There are similar definitions for pairs.

The key fact is that arbitrary maps are homotopic (in well controlled ways) to cellular maps. CW complexes are built from euclidean balls of various dimensions, and, although one aims to simplify their homotopy theory by making things formal and combinatorial, there are places in the theory where the geometry of $\mathbb{R}^n$ must appear. This is one such place:

Proposition 1.4.1. *Let $m > n$. Let U be an open subset of $\mathbb{R}^n$ such that the space cl U is compact. Let $f :$ cl $U \to B^m$ be such that $f(\text{fr } U) \subset S^{m-1}$. Then f is homotopic, rel fr U, to a map g such that some point of $B^m - S^{m-1}$ is not in the image of g.*

Proof. There are many proofs in the literature. The most common involve "simplicial approximation" of f (see, for example, [146, 7.6.15],). We give a proof here which assumes instead knowledge of multivariable calculus.

Let $V = f^{-1}(B^m - S^{m-1})$. Define $\epsilon : V \to (0, \infty)$ by $\epsilon(x) = \text{distance}^{14}$ from x to fr V. Using partitions of unity (see [85, pp. 41–45]), there exists a C^1 map $\hat{g} : V \to B^m - S^{m-1}$ such that $|f(x) - \hat{g}(x)| < \epsilon(x)$ for all $x \in V$. Since $|f(x) - \hat{g}(x)| \to 0$ as x approaches fr$_{\mathbb{R}^m} V$, $\hat{g}$ can be extended continuously to $g :$ cl $U \to B^m$ by defining $g = \hat{g}$ on V and $g = f$ on (cl U) $- V$. Then f is homotopic to g by the homotopy

$$H_t(x) = (1 - t)f(x) + tg(x).$$

We have $g(\text{cl } U) \cap (B^m - S^{m-1}) = g(V)$. We will show that $g(V)$ is a proper subset of $B^m - S^{m-1}$.

Identify $\mathbb{R}^n$ with $\mathbb{R}^n \times \{0\} \subset \mathbb{R}^m$. Then $V \subset \mathbb{R}^n \times \{0\} \subset \mathbb{R}^m$ and $g(V) = gp(V)$ where $p : \mathbb{R}^m \to \mathbb{R}^n$ is projection.

An *m-cube C in $\mathbb{R}^m$ of side* $\lambda > 0$ is a product $C = I_1 \times \ldots \times I_m$ of closed intervals each of length λ. The *m-dimensional measure* of C is λ^m. A subset $X \subset \mathbb{R}^m$ has *measure 0* if for every $\epsilon > 0$, X can be covered by a countable set of m-cubes the sum of whose measures is less^{15} than ϵ.

Clearly $V \subset \mathbb{R}^m$ has m-dimensional measure 0. Since $\hat{g}$ is C^1, so also is $\hat{g}p \mid: p^{-1}(V) \to B^m - S^{m-1}$. By the Mean Value Theorem, given any compact subset $K \subset p^{-1}(V)$, there exists $\mu(K) > 0$ such that $|\hat{g}p(x) - \hat{g}p(y)| \le \mu(K)|x - y|$ for all $x, y \in K$; we may take $\mu(K) = \sup\{||D(\hat{g}p)_x|| \mid x \in K\}$. It follows that if $C \subset K$ is an m-cube of side λ, then $gp(C)$ lies in an m-cube C' of side $\sqrt{m}.\mu(K).\lambda$. Thus C' has measure $(\sqrt{m}.\mu(K))^m.(\text{measure of } C)$.

It is a theorem of elementary topology that every open cover of an open subset of $\mathbb{R}^n$ has a countable subcover; see pages 174 and 65 of [51] for a proof. Hence we may write $V = \bigcup_{i=1}^{\infty} V_i$ where each set V_i lies in the interior of a closed m-ball $B_i \subset p^{-1}(V)$. Let $\epsilon > 0$ be given. Each V_i can be covered by countably many m-cubes $C_{i1}, C_{i2}, \ldots$ such that each $C_{ij} \subset \text{int } B_i$, and $\sum_j$ measure $(C_{ij}) < \epsilon$. Hence, by the above discussion, $g(V_i)$ can be covered by countably many cubes C'_{ij} such that $\sum_j$ measure $(C'_{ij}) < \epsilon(\sqrt{m}.\mu(B_i))^m$. So

---

^{14} All distances in euclidean spaces refer to the usual euclidean metric $|x - y| = (\sum(x_i^2 - y_i^2))^{1/2}$.

^{15} Of course, this is m-dimensional Lebesgue measure, but we need almost no measure theory!

$gp(V_i)$ has measure 0. But, clearly, the countable union of sets of measure 0 has measure 0, so $gp(V)(= g(V))$ has measure 0. Clearly, $B^m - S^{m-1}$ does not have measure 0. So $g(V)$ is a proper subset. $\square$

Proposition 1.4.2. Let e_α^n be an n-cell of the CW complex X, and let $z \in \overset{\circ}{e}{}_\alpha^n$. Then $X^n - \overset{\circ}{e}{}_\alpha^n$ is a strong deformation retract of $X^n - \{z\}$.

Proof. Let $h : (B^n, S^{n-1}) \to (e_\alpha^n, \overset{\bullet}{e}{}_\alpha^n)$ be a characteristic map. For any $y \in B^n - S^{n-1}$, the formula $x \mapsto x + (1 - |x|)y$ defines a homeomorphism of B^n fixing each point of S^{n-1} and sending 0 to y. Hence we may assume $h(0) = z$. Let $H : (B^n - \{0\}) \times I \to B^n - \{0\}$ be the homotopy $(x, t) \mapsto (1-t)x + tx/|x|$. Note that H is a strong deformation retraction of $B^n - \{0\}$ onto S^{n-1}. It is an easy exercise to check that a function H' exists making the following diagram commute:

$$
\begin{array}{ccc}
(B^n - \{0\}) \times I & \xrightarrow{\ H\ } & B^n - \{0\} \\
\Big\downarrow{\scriptstyle h|\times \text{ id}} & & \Big\downarrow{\scriptstyle h|} \\
(e_\alpha^n - \{z\}) \times I & \xrightarrow{\ H'\ } & e_\alpha^n - \{z\}
\end{array}
$$

i.e., that if $(h(x_1), t_1) = (h(x_2), t_2)$ then $h(H(x_1, t_1)) = h(H(x_2, t_2))$. Since h is a quotient map, so is $h\,|$, and also $(h\,|) \times$ id by 1.3.11. Thus H' is continuous and is therefore the required strong deformation retraction. $\square$

If X is a CW complex and $S \subset X$, the *carrier* of S is the intersection of all subcomplexes of X which contain S. It is a subcomplex of X, denoted $C(S)$.

Theorem 1.4.3. (Cellular Approximation Theorem) Let $f : X \to Y$ be a map between CW complexes and let A be a subcomplex of X such that $f \mid A : A \to Y$ is cellular. Then f is homotopic, rel A, to a cellular map.

Proof. We will construct $H : X \times I \to Y$ such that $H_0 = f$, $H_t \mid A = f \mid A$ for all t, and $g := H_1$ is cellular. This H is defined by induction on $(X^n \cup A) \times I$.

For each vertex e^0 of $X^0 - A$, there is a unique cell d such that $f(e^0) \in \overset{\circ}{d}$. Define $H \mid \{e^0\} \times I$ to be any path in $C(d)$ from $f(e^0)$ to a vertex of $C(d)$: call that vertex $g(e^0)$. (There is such a path because $C(d)$ is path connected.)

Assume H already defined on $(X^{n-1} \cup A) \times I$ with the desired properties. Extend H to agree with f on $X^n \times \{0\}$. By 1.3.15, H extends to a map $\bar{H} : (X^n \cup A) \times I \to Y$. Let $\bar{g} = \bar{H}_1$. Then $\bar{g} \mid X^{n-1} = g \mid X^{n-1}$ and is cellular, but $\bar{g}$ might not map n-cells of $X - A$ into Y^n. We claim that $\bar{g}$ is homotopic rel $X^{n-1} \cup A$ to a cellular map. For each n-cell e^n of X, the carrier of $\bar{g}(e^n)$, $C(\bar{g}(e^n))$, is a finite subcomplex of Y, by 1.2.13. Let d^m be a top-dimensional cell of $C(\bar{g}(e^n))$. Write $q : e^n \to C(\bar{g}(e^n))$ for the restriction of $\bar{g}$. If $m \leq n$, the next step can be omitted, so suppose $m > n$. Since $q(\overset{\bullet}{e}{}^n) \subset Y^{n-1}$, $q^{-1}(d - \overset{\bullet}{d}) \subset e^n - \overset{\bullet}{e}{}^n$. We want a homotopy of q, rel $\overset{\bullet}{e}{}^n$, to a map which misses a point of $\overset{\circ}{d}{}^{m}$, for then, by 1.4.2, there is a homotopy of q, rel $\overset{\bullet}{e}{}^{n}$, to a map of e^n into $C(\bar{g}(e^n)) - \overset{\circ}{d}$, and after finitely many such homotopies, rel $\overset{\bullet}{e}{}^{n}$, we will obtain a map of e^n into Y^n.

If $q(e^n) \cap \overset{\circ}{d} = \emptyset$, there is nothing to do. Otherwise, let $y \in q(e^n) \cap \overset{\circ}{d}$. $q^{-1}(y)$ is a compact set lying in $\overset{\bullet}{e}{}^{n}$. Let D be a neighborhood of y in $\overset{\circ}{d}$ which is homeomorphic to a closed ball, and let $U = q^{-1}(\text{int } D)$. Then cl U is a compact subset of $\overset{\circ}{e}{}^{n}$ and $q \mid: \text{cl } U \to D$. By 1.4.1, the desired homotopy of q, rel $e^n - U$ (hence rel $\overset{\bullet}{e}{}^{n}$), exists.

Doing this on each n-cell of X, we get the desired homotopy of $\bar{g}$, and, hence, in the obvious way, the desired $H : (X^n \cup A) \times I \to Y$. □

Addendum 1.4.4. *The homotopy $H : X \times I \to Y$ obtained in the proof of 1.4.3 has the property that for each cell e of X, $H(e \times I) \subset C(f(e))$.*

Proof. Assume the induction hypothesis $H(e^m \times I) \subset C(f(e^m))$ for all $m < n$. It is clearly true when $n = 1$. Let $\overset{\bullet}{e}{}^{n} \subset \overset{\circ}{e}{}_{\alpha_1}^{m_1} \cup \ldots \cup \overset{\circ}{e}{}_{\alpha_r}^{m_r}$. Then $\bar{g}(e^n) \subset$

$$\bar{H}(e^n \times I) = H(e^n \times 0) \cup H(\overset{\bullet}{e}{}^{n} \times I) \subset f(e^n) \cup \bigcup_{i=1}^{r} C(f(e_{\alpha_i}^{m_i})) \subset C(f(e^n)).$$

(The last inclusion holds because, if $\overset{\circ}{e}{}^{m} \cap \overset{\bullet}{e}{}^{n} \neq \emptyset$, then $C(e^m) \subset C(e^n)$.) The other half of the homotopy of e^n takes place in $C(\bar{g}(e^n)) \subset C(f(e^n))$. So $H(e^n \times I) \subset C(f(e^n))$. □

Corollary 1.4.5. *Let X be a CW complex; X is path connected iff X^1 is path connected.*

Proof. As in the proof of 1.4.3, there is a path from any point of X to a point of X^1, so X is path connected if X^1 is path connected. For the converse, let $\omega : I \to X$ be a path with $\omega(0), \omega(1) \in X^1$. Without loss of generality assume $\omega(0), \omega(1) \in X^0$. Apply 1.4.3 to produce a cellular path, necessarily in X^1, from $\omega(0)$ to $\omega(1)$. □

Exercise

Give an example in as low a dimension as possible of a cell e of X whose carrier is not covered by the set e.

1.5 Neighborhoods and complements

Let A be a subcomplex of a CW complex X. In general, neither its complement $X - A$ nor the closure $\text{cl}_X(X - A)$ are subcomplexes of X. When we study "end-invariants" of groups, complements will play an important role, and they should be subcomplexes. So we introduce the notion of CW complement, and set out the basic properties. Closely tied to this is our need for a good definition of the "CW neighborhood" of A; this should be a subcomplex of X, a neighborhood of A, and, in a reasonable sense, the tightest subcomplex with these properties.

The subcomplex A is *full* if it is the largest subcomplex of X having A^0 as its 0-skeleton. The full subcomplex of X generated by the vertices of $X^0 - A^0$ is the *CW complement* of A, denoted[16] by $X \stackrel{c}{-} A$.

The *CW neighborhood* of a subcomplex A of X is $N(A) := \bigcup\{C(e) \mid e$ is a cell of X and $C(e) \cap A \neq \emptyset\}$, the union of all cell carriers which meet A. Clearly, $N(A)$ is a subcomplex of X. See Fig. 1.3. If Y is a subcomplex and $A \subset Y$, we write $N_Y(A)$ for the CW neighborhood of A in Y, omitting the subscript when there is no ambiguity. Note that $N_{X^n}(A) \subset N(A) \cap X^n$ but equality does not always hold (see Exercises).

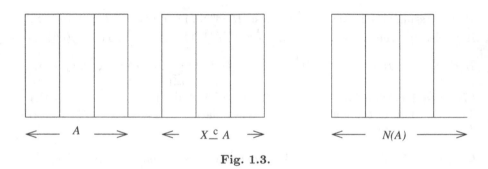

$\longleftarrow \quad A \quad \longrightarrow \qquad \longleftarrow \quad X \stackrel{c}{-} A \quad \longrightarrow \qquad \longleftarrow \quad N(A) \quad \longrightarrow$

Fig. 1.3.

A path $\omega : I \to X$ is *descending* if there is a partition $0 = t_0 < t_1 < \ldots < t_r = 1$ of I such that $\omega([t_{i-1}, t_i)) \subset \overset{\circ}{e}_i$ for some cell e_i, where dim $e_1 > \dim e_2 > \ldots > \dim e_r$. If $r = 1$ there is no dimension condition.

Proposition 1.5.1. *If $x \in \overset{\circ}{e}$ then $C(e) = \{y \in X \mid$ there is a descending path in X from x to $y\}$.*

Proof. By induction on dim e. If dim $e = 0$, the Proposition is trivial. Assume the Proposition for cells of dimension $< n$, and let dim $e = n$. Let $e_1, \ldots, e_k$ be the cells of X such that $\overset{\circ}{e}_i \cap \overset{\bullet}{e} \neq \emptyset$. We have $\overset{\bullet}{e} \subset C(e)$, so each e_i is a cell of $C(e)$, hence $C(e_i)$ is a subcomplex of $C(e)$ for all i. Thus $\overset{\circ}{e} \cup \bigcup_{i=1}^{k} C(e_i) \subset C(e)$.

---

[16] This complement and its symbol $\stackrel{c}{-}$ are non-standard, but are natural in our context. When X is a simplicial complex (see Sect. 5.2) $\mathrm{cl}_X(X - A)$ is indeed a subcomplex of X, but this is not always so when X is a CW complex.

But $\overset{\circ}{e} \cup \bigcup\limits_{i=1}^{k} C(e_i)$ is a subcomplex of X containing e, so $\overset{\circ}{e} \cup \bigcup\limits_{i=1}^{k} C(e_i) = C(e)$. Using induction, it is now an easy exercise to finish the proof. $\qquad\square$

Proposition 1.5.2. *Let A be a subcomplex of X. Then $N(A) = \{z \in X \mid$ for some $y \in A$ and $x \in X$ there are descending paths from x to y and from x to $z\}$.*

Proof. This follows by considering the definition of $N(A)$ in the light of 1.5.1. $\square$

Proposition 1.5.3. *A is full in X iff whenever $x \in X - A$ there is a descending path in X from x to a vertex of $X - A$.*

Proof. Let A be full and let $x \in \overset{\circ}{e} \subset X - A$. Then $C(e) \not\subset A$, so $C(e)^0 \not\subset A^0$. Apply 1.5.1. Conversely, let A not be full. Let B be a subcomplex of X such that $B^0 = A^0$ while $B - A \neq \emptyset$. Let $x \in \overset{\circ}{e} \cap (B - A)$. Then $C(e) \subset B$, so $C(e)^0 \subset B^0$, so, by 1.5.1, every descending path from x to a vertex of X ends in A^0. $\qquad\square$

Note that $N(A)$ can fail to be full even when A is full. For example, let X have two vertices v_0 and v_1 and two 1-cells e_1 and e_2 where $\overset{\bullet}{e}_1 = \{v_0, v_1\}$ and $\overset{\bullet}{e}_2 = \{v_1\}$; and let $A = \{v_0\}$.

Proposition 1.5.4. *Let A be a subcomplex of X and let B be the full subcomplex generated by A^0. Then $B \subset N(A)$ and $X \overset{c}{-} (X \overset{c}{-} A) = B$.*

Proof. Let e be a cell of B. Then $C(e)^0 \subset B^0 = A^0$, so $C(e) \cap A \neq \emptyset$, hence $C(e) \subset N(A)$. The second part is clear. $\qquad\square$

Proposition 1.5.5. *Let A be a subcomplex of X. Then $X = N(A) \cup (X \overset{c}{-} A)$ and $A \cap (X \overset{c}{-} A) = \emptyset$. In particular, $N(A)$ is a neighborhood of A.*

Proof. Let $x \in \overset{\circ}{e} \subset X$. If all descending paths from x to X^0 end in $X^0 - A^0 = (X \overset{c}{-} A)^0$, then $C(e)^0 \subset (X \overset{c}{-} A)^0$ and hence $x \in C(e) \subset X \overset{c}{-} A$. If there exists a descending path from x to a vertex of A, then $C(e) \cap A \neq \emptyset$, by 1.5.1; hence $x \in N(A)$. This proves that $X = N(A) \cup (X \overset{c}{-} A)$. To prove $A \cap (X \overset{c}{-} A) = \emptyset$, suppose the cell e lies in $A \cap (X \overset{c}{-} A)$. Then $C(e)$ is a subcomplex of $X \overset{c}{-} A$, so all its vertices lie in $X^0 - A^0$. On the other hand, $C(e)$ is a subcomplex of A, so all its vertices lie in A^0, a contradiction. The subcomplex $X \overset{c}{-} A$ is closed in X by 1.2.20 and is disjoint from A, so its (set theoretic) complement lies in $N(A)$. Thus $N(A)$ is a neighborhood of A. $\qquad\square$

Exercises

1. Give an example of a finite CW pair (X, A) such that $\mathrm{cl}(X - A)$ is not a sub-complex of X.
2. When (X, A) is a CW pair prove that $(X \overset{c}{-} A)^n = X^n \overset{c}{-} A^n$.
3. Give an example where $N(X \overset{c}{-} A) \cup A \neq X$ (compare with 1.5.5).
4. Give an example where $N_{X^n}(A)$ is a proper subset of $N(A) \cap X^n$.

2

Cellular Homology

The natural homology theory for CW complexes is cellular homology. Cellular homology theory is geometrically appealing at an intuitive level, but some technical subtleties are hidden in the theory and should be exposed to view for a thorough understanding. There are two ways of defining cellular homology: a formal way, given in terms of singular homology (Sects. 2.2, 2.3), and a geometrical way, in terms of incidence numbers and mapping degrees, given in Sect. 2.6 after an extensive discussion of degree and orientation in Sects. 2.4and 2.5. The chapter ends with a presentation of the main points of homology in the cellular context.

2.1 Review of chain complexes

As with our other review sections, this one is intended either for reference or as a quick refresher. Proofs of all our statements can be found in most books on algebraic topology, for example [146, Chap. 4], [77, Chap. 2], or, on homological algebra, for example [83, Chap. 4].

Throughout this book R denotes a commutative ring with an identity element $1 \neq 0$; R-modules are understood to be left modules unless we say otherwise. An important case is $R = \mathbb{Z}$; we use the terms "$\mathbb{Z}$-module" and "abelian group" interchangeably.[1]

A *graded R-module* is a sequence $C := \{C_n\}_{n \in \mathbb{Z}}$ of R-modules. If C and D are graded R-modules, a *(graded) homomorphism of degree d* from C to D is a sequence $f := \{f_n : C_n \to D_{n+d}\}_{n \in \mathbb{Z}}$ of R-module homomorphisms. A *chain complex* over R is a pair (C, ∂) where C is a graded R-module and $\partial : C \to C$ is a homomorphism of degree -1 such that $\partial \circ \partial = 0$. ∂ is the *boundary operator* of C.

---

[1] At various points we will explicitly assume more about R, namely that it is either a principal ideal domain (PID) or a field.

When (C, ∂) is a chain complex, we have $\partial_n : C_n \to C_{n-1}$ for each n. Define $Z_n(C) = \ker \partial_n$, $B_n(C) = \operatorname{im} \partial_{n+1}$, and $H_n(C) = Z_n(C)/B_n(C)$; the last makes sense because $\partial \circ \partial = 0$. Elements of $Z_n(C)$ are *n-cycles*, elements of $B_n(C)$ are *n-boundaries*, and $H_n(C)$ is the *n-dimensional homology* of C. Then $Z(C)$, $B(C)$ and $H(C)$ are graded R-modules.[2]

If (C, ∂) and (D, ∂') are chain complexes over R, a *chain map* is a homomorphism f of degree 0 from C to D such that $\partial' \circ f = f \circ \partial$. If $f, g : C \to D$ are chain maps, a *chain homotopy* from f to g is a homomorphism $h : C \to D$ of degree 1 satisfying $\partial' h + h \partial = f - g$. Since chain maps take cycles to cycles and boundaries to boundaries, it follows that a chain map $f : C \to D$ induces a homomorphism of degree 0, $H(f) : H(C) \to H(D)$. If f is chain homotopic to g (i.e., if there exists h as above), $H(f) = H(g)$. The chain map $f : C \to D$ is a *chain homotopy equivalence* if there is a chain map $g : D \to C$ such that $g \circ f$ and $f \circ g$ are chain homotopic to the respective identity chain maps (of C and of D). The chain map g is a *chain homotopy inverse* for f. If f is a chain homotopy equivalence, $H(f)$ is an isomorphism.

If $C \xrightarrow{f} D \xrightarrow{g} E$ are chain maps, then $H(g \circ f) = H(g) \circ H(f)$. Also, $H(\operatorname{id}_C) = \operatorname{id}_{H(C)}$. Thus, H is a covariant functor from the category of chain complexes and chain homotopy classes of chain maps to the category of graded R-modules and homomorphisms of degree 0.

Morphisms of graded R-modules $C \xrightarrow{f} D \xrightarrow{g} E$ are *exact* at D if $\ker \ g_n = \operatorname{im} f_n$ for all n. A sequence of such morphisms $\ldots \longrightarrow C^{(n+1)} \xrightarrow{f^{(n+1)}} C^{(n)} \xrightarrow{f^{(n)}} C^{(n-1)} \longrightarrow \ldots$ is *exact* if every three-term subsequence of consecutive R-modules is exact at its middle term.

An exact sequence of chain complexes over R of the form $0 \longrightarrow C' \xrightarrow{i} C \xrightarrow{p} C'' \longrightarrow 0$ is called a *short exact sequence*; it induces a "long" exact sequence of R-modules

$$\ldots \longrightarrow H_n(C') \xrightarrow{H(i)} H_n(C) \xrightarrow{H(p)} H_n(C'') \xrightarrow{\partial_*} H_{n-1}(C') \xrightarrow{H(i)} \ldots$$

where $\partial_* \{x''\} = \{i^{-1} \partial p^{-1}(x'')\}$. Surprisingly, ∂_* is well-defined and is therefore a homomorphism; it is called the *connecting homomorphism*. This long exact sequence is *natural* in the sense that if the following diagram of chain maps commutes

$$
\begin{array}{ccccccccc}
0 & \longrightarrow & C' & \longrightarrow & C & \longrightarrow & C'' & \longrightarrow & 0 \\
& & \downarrow{f'} & & \downarrow{f} & & \downarrow{f''} & & \\
0 & \longrightarrow & D' & \longrightarrow & D & \longrightarrow & D'' & \longrightarrow & 0
\end{array}
$$

where the horizontal rows are exact, then the following diagram commutes for all n:

---

[2] Graded $\mathbb{Z}$-modules are often misnamed "graded groups," rather than "graded abelian groups."

$$H_n(C') \longrightarrow H_n(C) \longrightarrow H_n(C'') \xrightarrow{\partial_*} H_{n-1}(C')$$
$$H_n(f') \downarrow \qquad H_n(f) \downarrow \qquad H_n(f'') \downarrow \qquad \qquad \downarrow H_{n-1}(f')$$
$$H_n(D') \longrightarrow H_n(D) \longrightarrow H_n(D'') \xrightarrow{\partial_*} H_{n-1}(D').$$

Exercises

1. Just as with chain complexes, an exact sequence
   $$0 \longrightarrow M' \xrightarrow{\alpha} M \xrightarrow{\beta} M'' \longrightarrow 0$$ of R-modules is called a *short exact sequence*. It *splits* (or is *splittable*) if one of the following three conditions hold. Prove they are equivalent:
   (i) the epimorphism β has a right inverse;
   (ii) the monomorphism α has a left inverse;
   (iii) there is an isomorphism ϕ making the following diagram commute

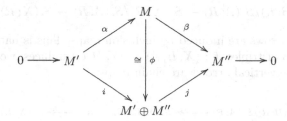

   where $i : x \mapsto (x,0)$ and $j : (x,y) \mapsto y$. (The importance of this is that if the sequence splits, the middle term is isomorphic to the direct sum of the other two.)
2. With $R = \mathbb{Z}$, give an example of an unsplittable short exact sequence of abelian groups.
3. Show that if M'' in Exercise 1 is a free R-module then the sequence splits.

2.2 Review of singular homology

Here we quickly review the basics of singular homology.

The *standard n-simplex* Δ^n is the closed convex hull of the points $p_0, \ldots, p_n$ in $\mathbb{R}^{n+1}$ where p_j has $(j+1)^{\text{th}}$ coordinate 1 and all other coordinates 0. Hence $\Delta^n = \left\{ \sum_{j=0}^{n} t_j p_j \mid 0 \le t_j \le 1 \text{ and } \sum_j t_j = 1 \right\}$. A *singular n-simplex* in the topological space X is a map $\sigma : \Delta^n \to X$. Let R be a ring as in Sect. 2.1. The free R-module generated by the set of all singular n-simplexes is denoted by $S_n(X; R)$ and its elements are called *singular n-chains* (over R). When $n \ge 1$, the ith *face* of the singular n-simplex σ is the composite map $\Delta^{n-1} \xrightarrow{F_i} \Delta^n \xrightarrow{\sigma} X$ where F_i is the affine map which sends

p_j to p_j if $j < i$ and p_j to p_{j+1} if $j \geq i$. A singular 0-simplex has no faces. By definition $S_n(X; R) = 0$ when $n < 0$. When $n > 0$ the *boundary operator* $\partial : S_n(X; R) \to S_{n-1}(X; R)$ is defined by $\partial(\sigma) = \sum_{i=0}^{n}(-1)^i(\sigma \circ F_i)$; when $n \leq 0$, $\partial = 0$. The graded R-module $\{S_n(X; R)\}$ is abbreviated to $S_*(X; R)$. One shows that $\partial \circ \partial = 0$, so $(S_*(X; R), \partial)$ is a chain complex over R called the *singular chain complex* of X. Its *singular homology* modules are denoted[3] by $H_*^{\Delta}(X; R)$. The ring R is the ring of *coefficients*.

A map $f : X \to Y$ induces a chain map $S_*(f) := \{f_{\#} : S_*(X; R) \to S_*(Y; R)\}$ defined by $f_{\#}(\sigma) = f \circ \sigma$. We have[4] $(g \circ f)_{\#} = g_{\#} \circ f_{\#}$, and $\mathrm{id}_{\#} = \mathrm{id}$. We write f_* for $H_n(f) : H_n^{\Delta}(X; R) \to H_n^{\Delta}(Y; R)$, the homomorphism induced by f (or by $f_{\#}$).

For $A \subset X$, the homology of the quotient chain complex $S_*(X; R)/S_*(A; R)$ is denoted by $H_*^{\Delta}(X, A; R)$. When $A = \emptyset$, this is identified with $H_*^{\Delta}(X; R)$. When $A \subset B \subset X$, there is a short exact sequence of chain complexes

$$0 \to S_*(B; R)/S_*(A; R) \to S_*(X; R)/S_*(A; R) \to S_*(X; R)/S_*(B; R) \to 0$$

where the arrows are induced by inclusion maps. This is natural in the sense that a map of triples $f : (X, B, A) \to (Y, D, C)$ induces a commutative diagram whose vertical arrows are chain maps:

$$
\begin{array}{ccccccccc}
0 & \to & S_*(B; R)/S_*(A; R) & \to & S_*(X; R)/S_*(A; R) & \to & S_*(X; R)/S_*(B; R) & \to & 0 \\
 & & \downarrow f_{\#} & & \downarrow f_{\#} & & \downarrow f_{\#} & & \\
0 & \to & S_*(D; R)/S_*(C; R) & \to & S_*(Y; R)/S_*(C; R) & \to & S_*(Y; R)/S_*(D; R) & \to & 0
\end{array}
$$

and hence a natural *long exact sequence of a triple* (X, B, A), namely

$$\cdots \to H_n^{\Delta}(B, A; R) \to H_n^{\Delta}(X, A; R) \to H_n^{\Delta}(X, B; R) \xrightarrow{\partial_*} H_{n-1}^{\Delta}(B, A; R) \to \cdots$$

If $A = \emptyset$, $H_n^{\Delta}(X, \emptyset)$ is identified with $H_n^{\Delta}(X)$ and this becomes the *long exact sequence of a pair* (X, B):

$$\cdots \to H_n^{\Delta}(B; R) \to H_n^{\Delta}(X; R) \to H_n^{\Delta}(X, B; R) \to H_{n-1}^{\Delta}(B; R) \to \cdots$$

Homotopic maps of spaces, or pairs of spaces, or triples of spaces induce chain homotopic chain maps. Thus singular homology is homotopy invariant.

Homology "turns" topological sums into direct sums: i.e., there is an obvious isomorphism $H_n^{\Delta}\left(\coprod_{\alpha \in A} X_\alpha; R\right) \cong \bigoplus_{\alpha \in A} H_n^{\Delta}(X_\alpha; R)$. The homology of the

---

[3] The Δ is for singular homology; we reserve the notation $H_*(X; R)$ for cellular homology.

[4] The reader can formulate what this says about S as a functor.

one-point space $\{p\}$ is given by $H_n^\Delta(\{p\}; R) = 0$ if $n \neq 0$, and $H_0^\Delta(\{p\}; R) \cong R$. It follows that $H_n^\Delta(S^0; R) = 0$ if $n \neq 0$ and $H_0^\Delta(S^0; R) \cong R \oplus R$.

Singular homology satisfies the *excision* property that if $\mathrm{cl}_X U \subset \mathrm{int}_X A$ then the inclusion map $(X - U, A - U) \hookrightarrow (X, A)$ induces an isomorphism of homology modules. Such an inclusion is called an *excision map*. An immediate consequence of this, using homotopy invariance, is that if $U \subset V \subset A$ with U as above and if the pair $(X - V, A - V)$ is a strong deformation retract of the pair $(X - U, A - U)$, then the excision map $(X - V, A - V) \hookrightarrow (X, A)$ also induces an isomorphism of homology modules.

When $X = U \cup V$, where U and V are open subsets, there is a short exact sequence of chain complexes

$$0 \longrightarrow S_*(U \cap V; R) \overset{i}{\longrightarrow} S_*(U; R) \oplus S_*(V; R) \overset{j}{\longrightarrow} S_*(X : R) \longrightarrow 0$$

with $i(c) = (c, -c)$ and $j(c, d) = c + d$. This induces a *Mayer-Vietoris sequence* which is natural in the obvious sense.

$$\cdots \to H_n^\Delta(U \cap V; R) \to H_n^\Delta(U; R) \oplus H_n^\Delta(V; R) \to H_n^\Delta(X; R) \to H_{n-1}^\Delta(U \cap V; R) \to \cdots$$

The following proposition is an easy consequence of the theory we have just outlined:

Proposition 2.2.1. *In the long exact sequence of the pair* (B^n, S^{n-1}), $\partial_* :$ $H_k^\Delta(B^n, S^{n-1}; R) \to H_{k-1}^\Delta(S^{n-1}; R)$ *is an isomorphism for* $2 \leq k \leq n$. *For* $n \geq 0$, $H_n^\Delta(B^n, S^{n-1}; R)$ *is isomorphic to* R, *and for* $k \neq n$ $H_k^\Delta(B^n, S^{n-1}; R)$ *is trivial.* □

Historical Note: Early definitions of homology were applicable only to simplicial complexes (Sect. 5.2). In the period 1925–35 extensions of the theory to large classes of topological spaces appeared: to metric spaces by Vietoris and to compact Hausdorff spaces by Čech (see Sect. 17.7 for the latter). These agree on the overlap - compact metrizable spaces. Singular homology, applicable to all spaces, was introduced in 1944 by Eilenberg. With proper precision of wording all these theories agree on finite CW complexes.

Exercises

1. Prove that S^n is not contractible.
2. Prove that if $m \neq n$, S^m and S^n do not have the same homotopy type.
3. Prove that if $m \neq n$, $\mathbb{R}^m$ and $\mathbb{R}^n$ are not homeomorphic.
4. Prove that if Y is obtained from A by attaching n-cells, and if $m \neq n$ then Y is not obtained from A by attaching m-cells. Hence the dimension of a cell in a CW complex is a topological property.

2.3 Cellular homology: the abstract theory

The homology theory to be used in this book is cellular homology. We give two equivalent treatments. The first, in this section, is the traditional approach which is efficient but is hard to grasp intuitively, even with a good prior understanding of singular homology theory. The second treatment in Sect. 2.4–2.6 is more intuitive. A full understanding involves both approaches.

When X is a CW complex the homology modules $H_*^\Delta(X; R)$ can be computed from a chain complex much smaller than $S_*(X; R)$. Define $C_n(X; R) = H_n^\Delta(X^n, X^{n-1}; R)$. Call its elements *cellular chains in X* (over R).

Proposition 2.3.1. *Let the set $\mathcal{A}$ index the n-cells of X. Then $C_n(X; R)$ is isomorphic to $\bigoplus_{\alpha \in \mathcal{A}} R$ and also to $\bigoplus_{\alpha} H_n^\Delta(e_\alpha^n, \overset{\bullet}{e}_\alpha^n)$.*

Proof. By excision $H_n^\Delta(X^n, X^{n-1}; R)$ is isomorphic to $H_n^\Delta(B^n(\mathcal{A}), S^{n-1}(\mathcal{A}); R)$, hence to $\bigoplus_{\alpha \in \mathcal{A}} H_n^\Delta(B^n, S^{n-1})$, hence, by 2.2.1, to $\bigoplus_{\alpha \in \mathcal{A}} R$. The second statement also follows by excision. $\qquad\square$

For reference we note that excision also gives:

Proposition 2.3.2. *For all k, $H_k^\Delta(X^n, X^{n-1}; R)$ is isomorphic to $H_k^\Delta(B^n(\mathcal{A}), S^{n-1}(\mathcal{A}); R)$, and these are zero when $k \neq n$.* $\qquad\square$

Define $\partial : C_n(X; R) \to C_{n-1}(X; R)$ to be the connecting homomorphism $\partial : H_n^\Delta(X^n, X^{n-1}; R) \to H_{n-1}^\Delta(X^{n-1}, X^{n-2}; R)$ in the long exact sequence of the triple (X^n, X^{n-1}, X^{n-2}).

Proposition 2.3.3. *$(C_*(X; R), \partial)$ is a chain complex; i.e., $\partial \circ \partial = 0$.*

Proof. If we consider the long exact sequences of the triple (X^{n+1}, X^n, X^{n-1}) and of the pairs (X^{n+1}, X^n) and (X^n, X^{n-1}), we obtain a commutative diagram

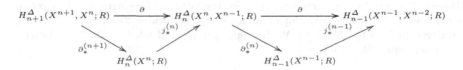

where $j^{(n)} : (X^n, \emptyset) \to (X^n, X^{n-1})$ denotes inclusion. By exactness we have $\partial_*^{(n)} \circ j_*^{(n)} = 0$ so $\partial \circ \partial = 0$. $\qquad\square$

The chain complex $(C_*(X; R), \partial)$ is the *cellular chain complex* of X (over R). We write $Z_*(X; R)$ and $B_*(X; R)$ for the cycles and boundaries respectively.

If X and Y are CW complexes and $f : X \to Y$ is a cellular map, f induces a homomorphism $f_\# : C_n(X; R) \to C_n(Y; R)$ in the obvious way. The proof of the next proposition is an exercise:

Proposition 2.3.4. $\{f_\# : C_*(X;R) \to C_*(Y;R)\}$ *is a chain map. If* $g : Y \to Z$ *is another cellular map then* $(g \circ f)_\# = g_\# \circ f_\# : C_*(X;R) \to C_*(Z;R)$. *For all* X, $\mathrm{id}_\# = \mathrm{id} : C_n(X;R) \to C_n(X;R)$. $\qquad\square$

Thus $C_*(\cdot;R)$ is a functor from the category of CW complexes and cellular maps to the category of chain complexes over R.

The *cellular homology* of X (over R) is the homology, $H_*(X;R)$, of the cellular chain complex $C_*(X;R)$. The fundamental fact is that it is naturally isomorphic to singular homology:

Theorem 2.3.5. *There is a natural isomorphism* $\psi : H_n(X;R) \to H_n^\Delta(X;R)$.

Remark 2.3.6. Here, as usual, "natural" means that if $f : X \to Y$ is a cellular map then $f_* \circ \psi = \psi \circ f_*$.

For the proof of 2.3.5 we need a lemma about the singular homology of CW complexes. The *support* of a singular n-chain z in X is the union of the (finitely many) compact sets $\sigma_\alpha(\Delta^n)$ such that the singular n-simplex σ_α occurs in z with non-zero coefficient. Thus every singular n-chain has compact support. So if X is a CW complex, 1.2.13 implies that the support of the n-chain z lies in X^m for some m. This implies:

Lemma 2.3.7. *If* $\{z\} \in H_n^\Delta(X, X^k)$ *then, for some* m, $\{z\}$ *is in the image of* $i_* : H_n^\Delta(X^m, X^k) \to H_n^\Delta(X, X^k)$. $\qquad\square$

Proof (of Theorem 2.3.5). Consider the following diagram:

$$H_{n+1}^\Delta(X^{n+1}, X^n; R) \xrightarrow{\ \partial_*^{(n+1)}\ } H_n^\Delta(X^n; R) \xrightarrow{\ j_*^{(n)}\ } H_n^\Delta(X^n, X^{n-1}; R) = C_n(X;R)$$

$$\Big\downarrow i_*^{(n)} \qquad\qquad \nearrow \phi$$

$$H_n^\Delta(X^{n+1}; R)$$

$$\Big\downarrow$$

$$0 = H_n^\Delta(X^{n+1}, X^n; R)$$

The homomorphism ϕ is not well defined but the relation $j_*^{(n)}(i_*^{(n)})^{-1}$ has its image in $Z_n(X;R)$ and the ambiguity in defining ϕ is clearly resolved by factoring out $B_n(X;R)$. Thus the relation ϕ defines a homomorphism $\bar\phi : H_n^\Delta(X^{n+1};R) \to H_n(X;R)$ which is easily seen to be an isomorphism.

We claim that the inclusion map $i : X^{n+1} \hookrightarrow X$ induces an isomorphism on n-dimensional singular homology. By 2.2.1 and 2.3.2, $H_k^\Delta(X^{j+1}, X^j; R) = 0$ when $k \le j$. It follows from this and 2.3.7, together with the long exact sequence of a triple, that $H_k^\Delta(X, X^{n+1}; R) = 0$ when $k \le n+1$ and hence, by the long exact sequence of the pair (X, X^{n+1}), that i induces an isomorphism as claimed. The composition of isomorphisms

$$H_n(X; R) \xrightarrow{\bar{\phi}^{-1}} H_n^{\Delta}(X^{n+1}; R) \longrightarrow H_n^{\Delta}(X; R)$$

is natural. $\qquad\square$

We have now seen that singular homology can be computed from the cellular chain complex. For this to be useful we need a better understanding of the homomorphisms ∂ and $f_{\#}$. We do this in two stages, first algebraically in this section, then geometrically in Sect. 2.5.

In 2.3.1 we saw that $C_n(X; R)$ can be viewed as the free R-module generated by the set of n-cells of X, but this is rather vague. We should pick generators for $C_n(X; R)$ in some geometrically meaningful way, reflecting the fact that

$$C_n(X; R) \cong H_n(X^n, X^{n-1}; R) \cong \bigoplus_{\alpha \in \mathcal{A}} H_n^{\Delta}(e_{\alpha}^n, \overset{\bullet}{e}{}_{\alpha}^n).$$

To begin, we choose, once and for all, a generator $\lambda_n \in H_n^{\Delta}(B^n, S^{n-1}; R)$ which is isomorphic to R by 2.2.1. Later (after 2.5.17) we will be more specific about our choice, but here we will only use the fact that we have made a choice. Next, we choose $h_{\alpha} : (B^n, S^{n-1}) \to (e_{\alpha}^n, \overset{\bullet}{e}{}_{\alpha}^n)$, a characteristic map for each n-cell e_{α}^n. It follows from the proof of 2.3.1 that the homomorphism

$$\bigoplus_{\alpha} H_n^{\Delta}(e_{\alpha}^n, \overset{\bullet}{e}{}_{\alpha}^n; R) \to H_n^{\Delta}(X^n, X^{n-1}; R) =: C_n(X; R)$$

induced by the inclusion maps $i_{\alpha} : e_{\alpha}^n \hookrightarrow X^n$ is an isomorphism. Moreover, by excision $h_{\alpha *} : H_n^{\Delta}(B^n, S^{n-1}; R) \to H_n^{\Delta}(e_{\alpha}^n, \overset{\bullet}{e}{}_{\alpha}^n; R)$ is an isomorphism. Indeed, only the homotopy class of the characteristic map h_{α} matters. Writing $\bar{h}_{\alpha} = i_{\alpha} \circ h_{\alpha} : (B^n, S^{n-1}) \to (X^n, X^{n-1})$, we can sharpen 2.3.1 to read:

Proposition 2.3.8. *The R-module $C_n(X; R)$ is freely generated by the set* $\{\bar{h}_{\alpha *}(\lambda_n) \mid \alpha \in \mathcal{A}\}$. $\qquad\square$

Let e_{β}^{n-1} be a cell of X. To understand the boundary homomorphism $\partial : C_n(X; R) \to C_{n-1}(X; R)$ we must understand the coefficient of $\bar{h}_{\beta *}(\lambda_{n-1})$ in $\partial \bar{h}_{\alpha *}(\lambda_n)$. That coefficient is the image of $1 \in R$ under the following R-module homomorphism from R to R (omitting explicit mention of R in the homology modules):

$$
\begin{array}{ccccc}
R \xrightarrow{\cong} H_n^{\Delta}(B^n, S^{n-1}) & \xrightarrow[\cong]{h_{\alpha *}} & H_n^{\Delta}(e_{\alpha}^n, \overset{\bullet}{e}{}_{\alpha}^n) & \xrightarrow{i_{\alpha *}} & H_n^{\Delta}(X^n, X^{n-1}) \\
1 \longmapsto \lambda_n & & & & \downarrow{\partial_*} \\
& & & & H_{n-1}^{\Delta}(X^{n-1}, X^{n-2}) \\
& & & & \cong \uparrow \\
R \xleftarrow{\cong} H_{n-1}^{\Delta}(B^{n-1}, S^{n-2}) & \xrightarrow[\cong]{h_{\beta *}} & H_{n-1}^{\Delta}(e_{\beta}^{n-1}, \overset{\bullet}{e}{}_{\beta}^{n-1}) & \xleftarrow{\text{projection}} & \bigoplus_{\gamma} H_{n-1}^{\Delta}(e_{\gamma}^{n-1}, \overset{\bullet}{e}{}_{\gamma}^{n-1}) \\
1 \longleftarrow\!\mid \lambda_{n-1} & & & &
\end{array}
$$

Similarly, if $f : X \to Y$ is cellular and if $\mathcal{A}$ and $\mathcal{B}$ index the n-cells of X and Y respectively, we understand $f_\# : C_n(X; R) \to C_n(Y; R)$ by understanding the coefficient of $\bar{\bar{h}}_{\beta*}(\lambda_n)$ in the chain $f_\#(\bar{h}_{\alpha*}(\lambda_n))$, where we also choose a characteristic map $\tilde{h}_\beta : (B^n, S^{n-1}) \to (\tilde{e}^n_\beta; \overset{\bullet}{\tilde{e}}\,^n_\beta)$ for each n-cell $\tilde{e}^n_\beta$ of Y. That coefficient is the image of $1 \in R$ under the following R-module homomorphism from R to R:

$$R \xrightarrow{\cong} H_n^\Delta(B^n, S^{n-1}) \xrightarrow{h_{\alpha*}} H_n^\Delta(e^n_\alpha, \overset{\bullet}{e}\,^n_\alpha) \xrightarrow{i_{\alpha*}} H_n^\Delta(X^n, X^{n-1}) \xrightarrow{f_\#} H_n^\Delta(Y^n, Y^{n-1})$$

$$1 \longmapsto \lambda_n$$

$$R \xleftarrow{\cong} H_n^\Delta(B^n, S^{n-1}) \xrightarrow[\cong]{\bar{h}_{\beta*}} H_n^\Delta(\tilde{e}^n_\beta, \overset{\bullet}{\tilde{e}}\,^n_\beta) \xleftarrow{\text{projection}} \bigoplus_\gamma H_n^\Delta(\tilde{e}^n_\gamma, \overset{\bullet}{\tilde{e}}\,^n_\gamma)$$

$$1 \longmapsfrom \lambda_n$$

To understand cellular homology geometrically we must understand these complicated homomorphisms $R \to R$. We tackle this in the next three sections.

2.4 The degree of a map from a sphere to itself

If f and g are maps from S^n to S^n it is important to know when they are homotopic. This turns out to be a simple matter: they are homotopic if and only if they have the same "degree". The degree of f, deg (f), is an integer which counts algebraically the number of times that f wraps S^n around itself. The main issues, all treated in this section, are:

1. The definition of $\deg(f)$.
2. If $f \simeq g$ then $\deg(f) = \deg(g)$.
3. For every $d \in \mathbb{Z}$ and every $n \geq 1$ there exists f with deg $(f) = d$.
4. If $\deg(f) = \deg(g)$ then $f \simeq g$ (except when $n = 0$).
5. Product Theorem for degree.

First, we dispose of the case $n = 0$. There are four maps $S^0 \to S^0$. We define the *degree* of the identity map to be 1, the *degree* of the map which permutes the two points of S^0 to be -1, and the *degree* of both constant maps to be 0. Obviously, 1. and 2. hold while 3. does not.

For the rest of this section we assume $n \geq 1$. The abelian group $H_n^\Delta(S^n; \mathbb{Z})$ is infinite cyclic. The *degree* of f is the integer $\deg(f)$ such that $f_*(\kappa) = \deg(f)\kappa$ where κ is a generator (it doesn't matter which generator) of $H_n^\Delta(S^n; \mathbb{Z})$. By homotopy invariance of singular homology we have:

Proposition 2.4.1. *Homotopic maps $S^n \to S^n$ have the same degree.* $\qquad\square$

To exhibit maps of every degree we start with the case $n = 1$, the circle. For $d \in \mathbb{Z}$ define $f_{1,d} : S^1 \to S^1$ by $e^{2\pi i t} \mapsto e^{2\pi i d t}$, a map with "winding number" d.

Proposition 2.4.2. *The map $f_{1,d}$ has degree d.*

Proof. This is an exercise in applying the definitions given in Sect. 2.2. Apply $f_{1,d}$ to the singular 1-simplex $(1 - t)p_0 + tp_1 \mapsto e^{2\pi i t}$. $\qquad\qquad\square$

Whenever $g : S^n \to S^n$ is a map, its *suspension* $\Sigma g : S^{n+1} \to S^{n+1}$ is illustrated in Fig. 2.1; a formula is:

$$(\Sigma g)(x, t) = \begin{cases} (\sqrt{1 - t^2}\, g(\frac{1}{\sqrt{1 - t^2}} x), t) & \text{if } -1 < t < 1 \\ (0, \ldots, 0, 1) & \text{if } t = 1 \\ (0, \ldots, 0, -1) & \text{if } t = -1 \end{cases}$$

In the exercises, the reader is asked to check that Σg is continuous, and that $g_1 \simeq g_2$ implies $\Sigma g_1 \simeq \Sigma g_2$. Define $\Sigma^r g = \Sigma(\Sigma^{r-1} g)$ by induction. $\Sigma^r g$ is the *r-fold suspension* of g. It is convenient to define $\Sigma^0 g$ to be g. For each $d \in \mathbb{Z}$, and each $n \geq 2$, define $f_{n,d} : S^n \to S^n$ to be $\Sigma^{n-1} f_{1,d}$. The map $f_{n,0}$ maps S^n onto a proper subset of S^n, half of a great $(n-1)$-sphere. By considering $f_{2,d} := \Sigma f_{1,d} : S^2 \to S^2$, and then generalizing, we may think of $f_{n,d}$ as wrapping S^n $|d|$ times around itself, in a "positive" sense if $d > 0$ and a "negative" sense if $d < 0$. This is made precise in 2.4.4.

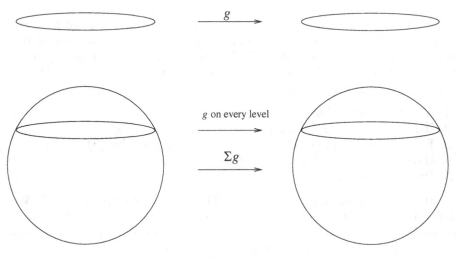

Fig. 2.1.

Proposition 2.4.3. *For $n \geq 1$, $f_{n,1}$ is the identity map of S^n; $f_{n,-1}$ is the map $(x_1, x_2, x_3, \ldots, x_{n+1}) \mapsto (x_1, -x_2, x_3, \ldots, x_{n+1})$; $f_{n,0}$ is homotopic to a constant map.*

Proof. The statements about $f_{n,1}$ and $f_{n,-1}$ are obvious. For the statement about $f_{n,0}$, one proves by induction on n that P_n does not lie in the image of $f_{n,0}$, where $P_n = (-1, 0, \ldots, 0) \in S^n \subset \mathbb{R}^{n+1}$. This is clear for $n = 1$; assume it for n; if $f_{n+1,0}(x, t) = P_{n+1}$, then $t = 0$, and $f_{n,0}(x) = P_n$, a contradiction. By 1.3.7, $S^n - \{P_n\}$ is contractible. So, by 1.3.4, $f_{n,0}$ is homotopic (in $S^n - \{P_n\}$, hence in S^n) to a constant map. $\qquad\square$

Proposition 2.4.4. $\deg(f_{n,d}) = d$.

Proof. The case $n = 1$ is 2.4.2, so we may assume $n \geq 2$. Let $e_{\pm}^n = S^n \cap \mathbb{R}_{\pm}^{n+1}$. Then $S^n = e_+^n \cup e_-^n$ and $S^{n-1} = e_+^n \cap e_-^n$. The spaces $e_{\pm}^n$ are homeomorphic to B^n, hence they are contractible. The Mayer-Vietoris sequence in singular homology gives (omitting the coefficient ring $\mathbb{Z}$):

$$H_n^{\Delta}(e_+^n) \oplus H_n^{\Delta}(e_-^n) \to H_n^{\Delta}(S^n) \xrightarrow{\partial_n} H_{n-1}^{\Delta}(S^{n-1}) \to H_{n-1}^{\Delta}(e_+^n) \oplus H_{n-1}^{\Delta}(e_-^n)$$

implying that, for $n \geq 2$, ∂_n is an isomorphism (between infinite cyclic groups). Naturality of the Mayer-Vietoris sequence implies that for any map $f : S^{n-1} \to S^{n-1}$ the following diagram commutes:

$$\begin{array}{ccc} H_n^{\Delta}(S^n) & \xrightarrow{(\Sigma f)_*} & H_n^{\Delta}(S^n) \\ \downarrow{\partial_n} & & \downarrow{\partial_n} \\ H_{n-1}^{\Delta}(S^{n-1}) & \xrightarrow{f_*} & H_{n-1}^{\Delta}(S^{n-1}) \end{array}$$

This implies that $\deg(f) = \deg(\Sigma f)$. Since $f_{n,d} = \Sigma f_{n-1,d}$ and $\deg(f_{1,d}) = d$, this completes the argument. $\qquad\square$

We turn to Item 4 on our initial list.

Theorem 2.4.5. (Brouwer-Hopf Theorem) *Two maps $S^n \to S^n$ are homotopic iff they have the same degree.*

In view of the previous propositions the only part of this theorem not yet proved[5] is that for $n \geq 1$ every map $f : S^n \to S^n$ of degree d is homotopic to $f_{n,d}$. We begin with the case $n = 1$.

A map $p : E \to B$ is a *covering projection* if for every $b \in B$ there is an open neighborhood U of b such that $p^{-1}(U)$ can be written as the union of pairwise disjoint open subsets of E each mapped homeomorphically by p onto U. Sets $U \subset B$ with this property are said to be *evenly covered* by p. The space B is the *base space* and E is the *covering space*. For example, the map $\exp : \mathbb{R} \to S^1$, $t \mapsto e^{2\pi i t}$, is a covering projection.

The following two theorems about covering spaces are fundamental. The proofs are elementary, and are to be found in numerous books on algebraic topology, for example, [74, Chap. 5].

---

[5] A complete proof of the Brouwer-Hopf Theorem is included here because it is omitted from many books on algebraic topology. The reader may prefer to skip it and go to Theorem 2.4.19.

Theorem 2.4.6. (Homotopy Lifting Property) *Let* $p : E \to B$ *be a covering projection, let* $F : Y \times I \to B$ *and* $f : Y \times \{0\} \to E$ *be maps such that* $p \circ f = F \mid Y \times \{0\}$. *Then there exists* $\bar{F} : Y \times I \to E$ *making the following diagram commute:*

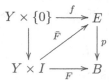

$\square$

Theorem 2.4.7. (Unique Path Lifting) *Let* $p : E \to B$ *be a covering projection. If* $\omega_i : I \to E$ *are paths, for* $i = 1, 2$, *such that* $p \circ \omega_1 = p \circ \omega_2$ *and* $\omega_1(0) = \omega_2(0)$, *then* $\omega_1 = \omega_2$. $\square$

Consider a map $f : S^1 \to S^1$, and the covering projection $\exp : \mathbb{R} \to S^1$. Let $t_0 \in \mathbb{R}$ be such that $f(1) = e^{2\pi i t_0}$ (where $1 \in S^1 \subset \mathbb{C}$). By 2.4.6, the path $\omega : I \to S^1$, $t \mapsto f(e^{2\pi i t})$, lifts through p to a path $\tilde{\omega} : I \to \mathbb{R}$ such that $\tilde{\omega}(0) = t_0$.

Proposition 2.4.8. *The number* $\tilde{\omega}(1) - \tilde{\omega}(0)$ *is an integer and is independent of the choice of* $t_0 \in \exp^{-1}(f(1))$.

Proof. Let $\tilde{\omega}(1) = t_1$. $\omega(0) = \omega(1)$, so $e^{2\pi i t_0} = e^{2\pi i t_1}$, hence $t_0 - t_1 \in \mathbb{Z}$. For $a \in \mathbb{R}$, let $T_a : \mathbb{R} \to \mathbb{R}$ be the translation homeomorphism $x \mapsto x + a$. If $\tilde{\omega}$ is replaced by $\bar{\omega}$ where $\bar{\omega}(0) = t_0'$ and $\exp \circ \bar{\omega} = \omega$, then $T_{(t_0' - t_0)} \circ \tilde{\omega} = \bar{\omega}$, by 2.4.7. Hence $t_1' := \bar{\omega}(1) = t_1 + t_0' - t_0$. Hence $t_1' - t_0' = t_1 - t_0$. $\square$

The integer $\tilde{\omega}(1) - \tilde{\omega}(0)$ will be denoted by $\delta(f)$.

Proposition 2.4.9. *Two maps* $f, g : S^1 \to S^1$ *are homotopic iff* $\delta(f) = \delta(g)$.

Proof. Let $F : S^1 \times I \to S^1$ be a homotopy. Define $G : I \times I \to S^1$ by $G(t, s) = F(e^{2\pi i t}, s)$. Let $\omega(t) = G(0, t)$ and let $\tilde{\omega} : I \to \mathbb{R}$ be such that $\omega = \exp \circ \tilde{\omega}$. By 2.4.6, there is a map $\tilde{G} : I \times I \to \mathbb{R}$ such that $\tilde{G}_0 = \tilde{\omega}$. Note that $\delta(F_s) = \tilde{G}_s(1) - \tilde{G}_s(0)$, a formula continuous in s and taking values in the discrete subspace $\mathbb{Z} \subset \mathbb{R}$, hence constant. Thus $\delta(f) = \delta(g)$.

Conversely, let $\delta(f) = \delta(g) = d$. Let $\omega, \tau : I \to S^1$ be defined by $\omega(t) = f(e^{2\pi i t})$ and $\tau(t) = g(e^{2\pi i t})$. Let $\tilde{\omega}$ and $\tilde{\tau}$ be lifts of ω and τ such that $d = \tilde{\omega}(1) - \tilde{\omega}(0) = \tilde{\tau}(1) - \tilde{\tau}(0)$. Let $\beta : I \to \mathbb{R}$ be a path from $\tilde{\omega}(0)$ to $\tilde{\tau}(0)$. Then $s \mapsto \beta(s) + d$ is a path from $\tilde{\omega}(1)$ to $\tilde{\tau}(1)$. By 1.3.17, there is a map $\tilde{H} : I \times I \to \mathbb{R}$ such that $\tilde{H}(t, 0) = \tilde{\omega}(t)$, $\tilde{H}(t, 1) = \tilde{\tau}(t)$, $\tilde{H}(0, s) = \beta(s)$, and $\tilde{H}(1, s) = \beta(s) + d$. Clearly, there is a function H making the following diagram commute:

$$
\begin{array}{ccc}
I \times I & \xrightarrow{\tilde{H}} & \mathbb{R} \\
{\scriptstyle(\exp\,|)\times \mathrm{id}}\downarrow & & \downarrow{\scriptstyle\exp} \\
S^1 \times I & \xrightarrow{H} & S^1
\end{array}
$$

By 1.3.11, H is continuous. One checks that $H_0 = f$ and $H_1 = g$.   □

Proposition 2.4.10. $\delta(f_{1,d}) = d$.

Proof. Let $d \in \mathbb{Z} \subset \mathbb{R}$. The map $\tilde{\omega} : I \to \mathbb{R}$, $t \mapsto td$ makes the following diagram commute:

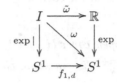

where ω is defined by the diagram. Thus $\delta(f_{1,d}) = \tilde{\omega}(1) - \tilde{\omega}(0) = d$.   □

This completes the proof of 2.4.5 when $n = 1$. Now we assume $n \geq 2$.

We need some preliminaries (2.4.11–2.4.18).

Let K and K' be two CW complex structures on the same underlying space X. The complex K' is a *subdivision* of K if for each cell e_β of K' there is a cell d_α of K such that $\overset{\circ}{e}_\beta \subset \overset{\circ}{d}_\alpha$.

Proposition 2.4.11. *Let L be a subcomplex of K and let K' be a subdivision of K. Then there is a subcomplex L' of K' which is a subdivision of L.*

Proof. Let L' consist of all cells e of K' such that $e \subset L$. We show that L' is a subcomplex and that L' is a subdivision of L.

Let e be a cell of L'. By 1.2.13 there are only finitely many cells of K', $e_1, \ldots, e_r$ such that $\overset{\bullet}{e} \cap \overset{\circ}{e}_i \neq \emptyset$. Since $e \subset L$, $\overset{\bullet}{e} \subset L$, so $\overset{\circ}{e}_i \cap L \neq \emptyset$ for all i. Let d_i be a cell of K such that $\overset{\circ}{e}_i \subset \overset{\circ}{d}_i$. Then for $1 \leq i \leq r$, $\overset{\circ}{d}_i \cap L \neq \emptyset$, hence d_i is a cell of L, hence $e_i \subset L$, hence e_i is a cell of L', hence $\overset{\bullet}{e} \subset L'$. So L' is a subcomplex of L.

Given a cell e of L', there is a cell d of K such that $\overset{\circ}{e} \subset \overset{\circ}{d}$. Since $e \subset L$, $\overset{\circ}{d} \cap L \neq \emptyset$, so $d \subset L$. Thus d is a cell of L. Finally, if $x \in L$, there are cells e_x of K' and d_x of L such that $x \in \overset{\circ}{e}_x \subset \overset{\circ}{d}_x$. So $\overset{\circ}{e}_x \subset L$, so $e_x \subset L$, so x lies in a cell of L'. Thus L' is a subdivision of L.   □

Let $I^1 := [-1, 1]$ have the CW complex structure consisting of two vertices at -1 and $+1$, and one 1-cell. For $k \geq 1$, the kth *barycentric subdivision* of I^1 is the CW complex I^1_k whose vertices are the points $i/2^{k-1}$, where $2^{k-1} \leq i \leq 2^{k-1}$, and whose 1-cells are the closed intervals bounded by adjacent vertices. The kth *barycentric subdivision* of I^{n+1} is the $(n+1)$-fold product CW complex $I^{n+1}_k := I^1_k \times \ldots \times I^1_k$. This is obviously a subdivision of I^{n+1}. By 2.4.11, there is a subcomplex $\overset{\bullet}{I}{}^{n+1}_k$ of I^{n+1}_k subdividing $\overset{\bullet}{I}{}^{n+1}$.

Proposition 2.4.12. *For each open cover $\mathcal{U}$ of I^{n+1} there exists k_0 such that for every $k \geq k_0$ every cell of I^{n+1}_k lies in some element of $\mathcal{U}$.*

Proof. Let $x \in U^{(x)} \subset I^{n+1}$, where $U^{(x)} \in \mathcal{U}$. There are intervals $V_1^{(x)}, \ldots, V_{n+1}^{(x)}$ which are open sets in I^1 such that $x \in V_1^{(x)} \times \ldots \times V_{n+1}^{(x)} =: V^{(x)} \subset U^{(x)}$. $\{V_x \mid x \in X\}$ has a finite subcover $V^{(x_1)}, \ldots, V^{(x_r)}$ since I^{n+1} is compact. Clearly, there is a subdivision $I_{k_i}^1$ of I^1 every cell of which lies in some $V_i^{(x_j)}$. Let $k_0 = \max\{k_1, \ldots, k_{n+1}\}$. Then for each $k \geq k_0$ every cell of I_k^{n+1} lies in some $V^{(x_j)} \subset U^{(x_j)}$. $\square$

Corollary 2.4.13. *For each open cover $\mathcal{U}$ of $\overset{\bullet}{I}{}^{n+1}$ there exists k_0 such that for every $k \geq k_0$ each cell of $\overset{\bullet}{I}{}_k^{n+1}$ lies in some element of $\mathcal{U}$.*

Proof. For each $U \in \mathcal{U}$ pick U' an open subset of I^{n+1} such that $U' \cap \overset{\bullet}{I}{}^{n+1} = U$. Let $\mathcal{U}'$ consist of all the sets U' and the set $I^{n+1} - \overset{\bullet}{I}{}^{n+1}$. Apply 2.4.11 and 2.4.12 to $\mathcal{U}'$. $\square$

Proposition 2.4.14. *Let X be an n-dimensional CW complex, let e^n be an n-cell of X and let $U \subset \overset{\circ}{e}{}^n$ be a neighborhood of $z \in \overset{\circ}{e}{}^n$. There is a homotopy $H : X \times I \to X$ such that $H_0 = \text{id}$, $H_t = \text{id}$ on $X - \overset{\circ}{e}{}^n$ for all t, and $H_1(e^n - U) \subset \overset{\circ}{e}{}^n$.*

Proof. As in the proof of 1.4.2, pick a characteristic map $h : (B^n, S^{n-1}) \to (e^n, \overset{\bullet}{e}{}^n)$ such that $h(0) = z$. $h^{-1}(U)$ is a neighborhood of 0 in B^n, so there is a $\epsilon > 0$ such that $h(B_\epsilon) \subset U$, where $B_\epsilon = \{x \in B^n \mid |x| \leq \epsilon\}$. By 1.4.2, there is a homotopy $H : (X - \text{int } h(B_\epsilon)) \times I \to X$ such that $H_0 = \text{id}$, $H_t = \text{id}$ on $X - \overset{\circ}{e}{}^n$ for all t, and $H_1(e^n - \text{int } h(B_\epsilon^n)) \subset \overset{\bullet}{e}{}^n$. Extend H_0 to be the identity on all of $X \times \{0\}$. By 1.2.8, X is obtained from $X - \text{int } h(B_\epsilon)$ by attaching an n-cell. So, by 1.3.12, H further extends to map $X \times I$ to X as required. $\square$

Proposition 2.4.15. *Let $x_1, \ldots, x_m$, $y_1, \ldots, y_m$ be distinct points of S^n, where $n \geq 2$. There is a homeomorphism $k : S^n \to S^n$, which is homotopic to id, such that $k(x_i) = y_i$ for each i.*

Proof. For any $z \in S^n$, $S^n - \{z\}$ is homeomorphic to $\mathbb{R}^n$. By choosing z different from each x_i and each y_i, we may work in $\mathbb{R}^n$. Let $M > 0$ be such that every x_i and every y_i lies in $(-M, M)^n$. We will show that there is a homeomorphism h of $[-M, M]^n$ which fixes every point of the frontier, and takes x_i to y_i for all i. Clearly this will be enough.

The proof requires two elementary lemmas whose proofs are left to the reader.

Lemma 2.4.16. *Let N be a compact convex neighborhood of a point $x \in \mathbb{R}^n$ and let $y \in \text{int } N$. There is a homotopy $H : N \times I \to N$ such that $H_0 = \text{id}$, $H_t = \text{id}$ on $\text{fr } N$ for all t, H_t is a homeomorphism for all t, and $H_1(x) = y$.* $\square$

Lemma 2.4.17. *If $n \geq 3$, there exist compact convex pairwise disjoint neighborhoods M_i of x_i and N_i of y_i, and points $x_i' \in \text{int } M_i$ and $y_i' \in \text{int } N_i$ such that no four points in $\{x_1', \ldots, x_m', y_1', \ldots, y_m'\}$ are coplanar.* $\square$

We now complete the proof of 2.4.15 for $n \geq 3$. By 2.4.16 and 2.4.17, we may assume that the line segments $L_i := [x_i, y_i]$ in $\mathbb{R}^n$ are pairwise disjoint. Since each L_i is compact and convex, $C_i := \{x \in \mathbb{R}^n \mid |x - u| \leq \epsilon$ for some $u \in L_i\}$ is a compact convex neighborhood of every point of L_i, and if ϵ is small enough, the sets C_i are pairwise disjoint and lie in $(-M, M)^n$. Applying 2.4.16 separately to each C_i to move x_i to y_i, and extending the resulting homotopy to be the identity on $\mathbb{R}^n - \bigcup_{i=1}^m C_i$, we obtain the desired homeomorphism k.
The case $n = 2$ is left as an exercise. $\square$

We remark that 2.4.15 and 2.4.17 are false when $n = 1$. The fundamental geometric Proposition 2.4.15 will be used in the proof of Theorem 2.4.5. It is also the core of the proof that higher homotopy groups are abelian (compare 4.4.5).

For $n \geq 2$, give S^n the CW structure consisting of one 0-cell e^0, lying in $S^{n-1} \subset S^n$, one $(n-1)$-cell $e^{n-1} = S^{n-1}$, and two n-cells $e_{\pm}^n = S^n \cap \mathbb{R}_{\pm}^{n+1}$. A map $f : S^n \to S^n$ *preserves hemispheres* if f is cellular with respect to the given CW structure, $f(e_+^n) \subset e_+^n$, and $f(e_-^n) \subset e_-^n$.

Proposition 2.4.18. *For $n \geq 2$, every map $S^n \to S^n$ is homotopic to a map which preserves hemispheres.*

Proof. Recall that we may identify $\overset{\bullet}{I}{}^{n+1}$ with S^n via the homeomorphism $a_{n+1} |: x \mapsto x/|x|$. By the *upper* [resp. *lower*] *open hemisphere* we mean the set of points whose last coordinate is positive [resp. negative]; their closures are *closed hemispheres*. We are given a map $f_1 : \overset{\bullet}{I}{}^{n+1} \to \overset{\bullet}{I}{}^{n+1}$. Using the Cellular Approximation Theorem with respect to various CW complex structures, we will obtain a map homotopic to f_1 which preserves hemispheres.
Define

$$U_+ := \{(x_1, \ldots, x_{n+1}) \in \overset{\bullet}{I}{}^{n+1} \mid x_{n+1} > -\tfrac{1}{2}\}$$

$$U_- := \{(x_1, \ldots, x_{n+1}) \in \overset{\bullet}{I}{}^{n+1} \mid x_{n+1} < \tfrac{1}{2}\}.$$

By 2.4.13 there exists k such that every cell of the subdivision $\overset{\bullet}{I}{}_k^{n+1}$ lies in $f_1^{-1}(U_+)$ or $f_1^{-1}(U_-)$.
Let $\phi : I^1 \to I^1$ be the piecewise linear map which fixes 0, 1 and -1, which takes $\pm\tfrac{1}{2}$ to 0, and which is affine on $[-1, -\tfrac{1}{2}]$, $[-\tfrac{1}{2}, \tfrac{1}{2}]$, $[\tfrac{1}{2}, 1]$. The homotopy $G : \overset{\bullet}{I}{}_k^{n+1} \times I \to \overset{\bullet}{I}{}_k^{n+1}$ which takes $(x_1, \ldots, x_{n+1}, t)$ to $(x_1, \ldots, x_n, (1-t)x_{n+1} + t\phi(x_{n+1}))$ has the property that $G_0 = \text{id}$, $G_1(U_+)$ lies in the upper closed hemisphere, and $G_1(U_-)$ lies in the lower closed hemisphere. Let $f_2 = G_1 \circ f_1$.

Then f_2 takes each n-cell of $\overset{\bullet}{I}{}^{n+1}_k$ into either the upper or the lower closed hemisphere.

We now regard f_2 as a map $\overset{\bullet}{I}{}^{n+1}_k \to S^n$. Here, S^n has one vertex, one $(n-1)$-cell (which is S^{n-1}), and the two closed hemispheres as n-cells. By 1.4.3 and 1.4.4, f_2 is homotopic to a cellular map f_3 which maps the $(n-1)$-skeleton of $\overset{\bullet}{I}{}^{n+1}_k$ into S^{n-1} and which maps each n-cell of $\overset{\bullet}{I}{}^{n+1}_k$ into one of the closed hemispheres.

Let $e^n_1, \ldots, e^n_m$ be the n-cells of $\overset{\bullet}{I}{}^{n+1}_k$ which are not mapped into S^{n-1} by f_3. Pick $x_i \in \overset{\circ}{e}{}^n_i$ such that $f_3(x_i) \notin S^{n-1}$. Pick y_i in the upper [resp. lower] open hemisphere of $\overset{\bullet}{I}{}^{n+1}$ iff $f_3(x_i)$ is in the upper [resp. lower] open hemisphere of S^n, and such that $y_1, \ldots, y_m$ are distinct. By 2.4.15, there is a homeomorphism $k : \overset{\bullet}{I}{}^{n+1} \to \overset{\bullet}{I}{}^{n+1}$, homotopic to id, taking x_i to y_i for all i. Let $V_i \subset \overset{\circ}{e}{}^n_i$ be an open neighborhood of x_i such that $k(V_i)$ is entirely in the open hemisphere of $\overset{\bullet}{I}{}^{n+1}$ containing y_i. See Fig. 2.2. By 2.4.14, f_3 is homotopic to a map $f_4 : \overset{\bullet}{I}{}^{n+1} \to S^n$ such that, for $1 \le i \le m$, $f_4(e^n_i - V_i) = f_4(\overset{\bullet}{e}{}^n_i) \subset S^{n-1}$, $f_4(e^n_i) = f_3(e^n_i)$, while $f_4 = f_3$ outside $\bigcup_{i=1}^{m} e^n_i$. Since k is homotopic to id, so is k^{-1}, hence f_4 is homotopic to $f_5 := f_4 \circ k^{-1}$. If y_i is in the upper [resp. lower] open hemisphere, then f_5 maps the neighborhood $k(V_i)$ of y_i into the upper [resp. lower] open hemisphere of S^n. f_5 maps $\overset{\bullet}{I}{}^{n+1} - \bigcup_{i=1}^{m} k(V_i)$ into S^{n-1}. Thus f_5 preserves hemispheres. $\qquad\square$

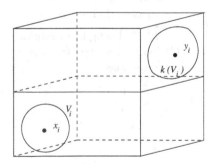
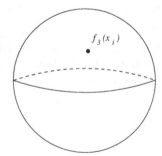

Fig. 2.2.

Proof (of Theorem 2.4.5 (concluded)). Let $f : S^n \to S^n$ be a map. By 2.4.18, it may be assumed that f preserves hemispheres. Let $g : S^{n-1} \to S^{n-1}$ be the restriction of f. Then for each $x \in S^n$, $(\Sigma g)(x)$ and $f(x)$ are not diametrically opposite points of S^n. Thus the line segment in $\mathbb{R}^{n+1}$, $\{\omega_t(x) := (1-t)\Sigma g(x) +$

$tf(x) \mid 0 \leq t \leq 1\}$, does not contain 0. The map $(x,t) \to \omega_t(x)/|\omega_t(x)|$ is a homotopy from Σg to f. By induction, g is homotopic to $\Sigma^{n-2}h$ where $h : S^1 \to S^1$ is a map. By 2.4.9 and 2.4.10, h is homotopic to $f_{1,d}$ for some d. Thus $f \simeq \Sigma^{n-1}h \simeq \Sigma^{n-1}f_{1,d} =: f_{n,d}$. $\qquad\square$

From the definition of degree one deduces:

Theorem 2.4.19. (Product Theorem for Degree) *Let* $f, g : S^n \to S^n$ *be maps,* $n \geq 0$. *Then* $\deg(g \circ f) = \deg(g) \cdot \deg(f)$. $\qquad\square$

Corollary 2.4.20. *A homotopy equivalence* $f : S^n \to S^n$ *has degree* ± 1. $\quad\square$

Together with 2.4.4 and 2.4.5, this corollary implies that a homeomorphism $h : S^n \to S^n$ is homotopic either to $\mathrm{id}_{S^n} = f_{n,1}$ or to the homeomorphism $f_{n,-1}$. If the former, h is *orientation preserving*; if the latter h is *orientation reversing*. (For this to make sense when $n = 0$, define $f_{0,-1}$ to be the map which interchanges the two points of S^0.)

There is another important theorem concerning degree, the Sum Theorem. For this we need the notion of the "wedge" of a family of spaces. If X is a space and $x \in X$, we write (X, x) for the pair $(X, \{x\})$, we call (X, x) a *pointed space*[6], and we call x the *base point*. The *wedge* $\bigvee_\alpha X_\alpha$ of a family $\{(X_\alpha, x_\alpha)\}$ of pointed spaces is the quotient space $\coprod_{\alpha \in \mathcal{A}} X_\alpha/\sim$ where $\sim$ identifies all the base points x_α. The wedge is also called the *one-point union*. The point of $\bigvee_\alpha X_\alpha$ corresponding to the equivalence class of $\{x_\alpha\}$ is the *wedge point*. If each X_α is a CW complex and if for all α the point x_α is a vertex of X_α, then $\bigvee_{\alpha \in \mathcal{A}} X_\alpha$ acquires a quotient CW complex structure as in 1.2.22. The wedge point is a vertex. When each X_α is an n-sphere, $\bigvee_\alpha X_\alpha$ is also called a *bouquet* of n-spheres.

Let the base point v of S^n be $(1, 0, \ldots, 0)$. Then $v \in S^0 \subset S^1 \subset \cdots$.

Consider $\bigvee_{\alpha \in \mathcal{A}} S^n_\alpha$ where (S^n_α, v_α) is a copy of (S^n, v) for each α. There are canonical maps $S^n_\beta \underset{i_\beta}{\overset{q_\beta}{\rightleftarrows}} \bigvee_{\alpha \in \mathcal{A}} S^n_\alpha$ for each $\beta \in \mathcal{A}$; i_β is the inclusion of S^n_β as the β-summand of the wedge; q_β is the identity of S^n_β and maps all other summands to $v_\beta \in S^n_\beta$.

Exercises

1. Fill in the details in the proof of 2.4.2.

---

[6] By a *pointed CW complex* we mean a pointed space (X, v) where X is a CW complex and v is a vertex.

2. In the definition of "suspension" prove that Σg is continuous when g is; and that Σg_1 and Σg_2 are homotopic when g_1 and g_2 are.
3. Prove that homeomorphic CW complexes have the same dimension.
4. Describe a CW complex structure for $\mathbb{R}^n$.
5. (for those who know category theory) Show that "wedge" is the category theoretical coproduct in the category Pointed Spaces. What is the category theoretical coproduct in the category Spaces?

2.5 Orientation and incidence number

Let e_α^n be an n-cell of a CW complex X. Recall that a characteristic map for e_α^n is a map of pairs $h : (B^n, S^{n-1}) \to (e_\alpha^n, \overset{\bullet}{e}_\alpha^n)$ which maps $B^n - S^{n-1}$ homeomorphically onto $\overset{\circ}{e}_\alpha^n = e_\alpha^n - \overset{\bullet}{e}_\alpha^n$. Two characteristic maps for e_α^n are *equivalent* if they are homotopic (as maps of pairs). When $n > 0$, an equivalence class of characteristic maps is called an *orientation* of e_α^n. An *orientation* of a 0-cell is an assignment of 1 or -1 to that cell.

Our first task is to prove (2.5.2) that each cell of X admits exactly two orientations. This is trivial when $n = 0$. From here until after the proof of 2.5.1, we assume $n \geq 1$.

A characteristic map $h : (B^n, S^{n-1}) \to (e_\alpha^n, \overset{\bullet}{e}_\alpha^n)$ induces a map $h' : B^n/S^{n-1} \to e_\alpha^n/\overset{\bullet}{e}_\alpha^n$. Now $e_\alpha^n/\overset{\bullet}{e}_\alpha^n$ is obviously homeomorphic to $X^n/(X^n - \overset{\circ}{e}_\alpha^n)$. By 1.2.22, the latter is Hausdorff. The function h' is a continuous bijection (see Sect. 1.1). Since B^n is compact, so is B^n/S^{n-1}. We conclude that h' is a homeomorphism.

A special case of this occurs when S^n has a CW complex structure consisting of one 0-cell and one n-cell (see 1.2.16). Thus B^n/S^{n-1} is homeomorphic to S^n. WE PICK, ONCE AND FOR ALL, A HOMEOMORPHISM $k_n : B^n/S^{n-1} \to S^n$ so that we may pass back and forth unambiguously between these two standard spaces. Later in this section (see Convention 2.5.16) we will be more specific about this choice.

If h_1 and h_2 are two characteristic maps for e_α^n, where $n \geq 1$, we have a homeomorphism $k_n \circ (h_2')^{-1} \circ h_1' \circ k_n^{-1} : S^n \to S^n$. By 2.4.20, its degree is ± 1. Thus it is homotopic either to id_{S^n} or to the canonical homeomorphism $f_{n,-1}$ of degree -1.

Theorem 2.5.1. *Let $n \geq 1$, and let e_α^n be a cell of X. Let h_1, h_2 be characteristic maps of e_α^n. Then h_1 and h_2 are equivalent iff $k_n \circ (h_2')^{-1} \circ h_1' \circ k_n^{-1}$ has degree 1.*

The proof is given after that of Lemma 2.5.5.

Corollary 2.5.2. *Each cell of a CW complex admits exactly two orientations. If $n \geq 1$ and if $h : (B^n, S^{n-1}) \to (e_\alpha^n, \overset{\bullet}{e}_\alpha^n)$ determines an orientation of e_α^n, then $h \circ r$ determines the other orientation, where $r : (B^n, S^{n-1}) \to (B^n, S^{n-1})$ is the map $(x_1, \ldots, x_n) \mapsto (-x_1, x_2, x_3, \ldots, x_n)$.*

Proof. First, we observe that $r \mid: S^{n-1} \to S^{n-1}$ is not homotopic to id. For $n = 1$, this is obvious. For $n > 1$, consider the homeomorphism $t_{12} : S^{n-1} \to S^{n-1}$ $(x_1, x_2, x_3, \ldots, x_n) \mapsto (x_2, x_1, x_3, \ldots, x_n)$. By 2.4.3, $r \mid = t_{12}^{-1} \circ f_{n-1,-1} \circ t_{12}$. If $r \mid$ were homotopic to id, then $f_{n-1,-1}$ would be homotopic to id (conjugating the homotopy by t_{12}), which is false by 2.4.4.

It follows that $r : (B^n, S^{n-1}) \to (B^n, S^{n-1})$ is not equivalent to (i.e., pairwise homotopic to) the identity map. Clearly $(h \circ r)' : B^n/S^{n-1} \to e_\alpha^n/\overset{\bullet}{e}{}_\alpha^n$ is the composition $h' \circ r'$. So $k_n \circ (h')^{-1} \circ (h \circ r)' \circ k_n^{-1} = k_n \circ r' \circ k_n^{-1}$, and this map does not have degree 1. Thus h and $h \circ r$ determine different orientations of e_α^n. By 2.5.1, every orientation is given by one of these. $\qquad \square$

The proof of 2.5.1 depends on some lemmas.

Lemma 2.5.3. *Let $n \geq 1$. Let $a = (0, \ldots, 0, 1) \in S^n$ and $b = (0, \ldots, 0, -1) \in S^n$. Let e_+^n and e_-^n be the closed upper and lower hemispheres, respectively. Let $f, g : S^n \to S^n$ be maps such that $a \notin f(e_-^n) \cup g(e_-^n)$ and $b \notin f(e_+^n) \cup g(e_+^n)$. If f and g have the same degree, then $f \mid \simeq g \mid : S^{n-1} \to S^n - \{a, b\}$.*

Proof. We consider f and g as maps $(S^n, S^{n-1}) \to (S^n, S^n - \{a, b\})$. As in the first part of the proof of 2.4.18, f and g are (pairwise) homotopic to maps f' and g' which preserve hemispheres, hence, as in the proof of 2.4.5, to suspensions $\Sigma f''$ and $\Sigma g''$, where $f'', g'' : S^{n-1} \to S^{n-1}$. These homotopies restrict to homotopies $S^{n-1} \times I \to S^n - \{a, b\}$. So $f \mid \simeq$ inclusion $\circ f''$ and $g \mid \simeq$ inclusion $\circ g''$ (as maps $S^{n-1} \to S^n - \{a, b\}$). Now degree$(f'') = $ degree$(\Sigma f'') = $ degree$(f) = $ degree$(g) = $ degree$(\Sigma g'') = $ degree(g''). So $f'' \simeq g''$, by 2.4.5, hence $f \mid \simeq g \mid$. Note that, while 2.4.18 is false for $n = 1$, the part of the proof used here works for $n = 1$. $\qquad \square$

Corollary 2.5.4. *Let $f, g : (B^n, S^{n-1}) \to (e_\alpha^n, \overset{\bullet}{e}{}_\alpha^n)$ be characteristic maps inducing $f', g' : B^n/S^{n-1} \to e_\alpha^n/\overset{\bullet}{e}{}_\alpha^n$. For $t \in (0, 1)$, let A_t be the annulus $\{x \in B^n \mid t < |x| < 1\}$ and let $S_t = \{x \in B^n \mid |x| = t\}$. Let $z \in \overset{\circ}{e}{}_\alpha^n$. If there exists t such that $f(A_t) \cup g(A_t) \subset \overset{\circ}{e}{}_\alpha^n - \{z\}$, and if $f' \simeq g'$, then $f \mid \simeq g \mid : S_t \to e_\alpha^n - \{z\}$.*

Proof. Since B^n/S^{n-1} and $e_\alpha^n/\overset{\bullet}{e}{}_\alpha^n$ are homeomorphic to S^n, the result follows from 2.5.3. In fact, $f \mid \simeq g \mid$ as maps from S_t to $\overset{\circ}{e}{}_\alpha^n - \{z\}$. $\qquad \square$

Lemma 2.5.5. *Let $z \in \overset{\circ}{e}{}_\alpha^n$, let f and g be two maps $(S^{n-1} \times I, S^{n-1} \times \{1\}) \to (e_\alpha^n - \{z\}, \overset{\bullet}{e}{}_\alpha^n)$, and let $H : S^{n-1} \times \{0\} \times I \to e_\alpha^n - \{z\}$ be a homotopy from $f \mid S^{n-1} \times \{0\}$ to $g \mid S^{n-1} \times \{0\}$. Then there is a homotopy $K : (S^{n-1} \times I \times I, S^{n-1} \times \{1\} \times I) \to (e_\alpha^n - \{z\}, \overset{\bullet}{e}{}_\alpha^n)$ from f to g extending H.*

Proof. The map $(S^{n-1} \times \{0\} \times I) \cup (S^{n-1} \times I \times \{0, 1\}) \to e_\alpha^n - \{z\}$, which agrees with H on $S^{n-1} \times \{0\} \times I$, with f on $S^{n-1} \times I \times \{0\}$ and with g on $S^{n-1} \times I \times \{1\}$, extends, by 1.3.15, to a map $L : S^{n-1} \times I \times I \to e_\alpha^n - \{z\}$.

The set $\overset{\circ}{e}{}^{\,n}_{\alpha}-(\text{image of } L) =: U$ is a neighborhood of z in $\overset{\circ}{e}{}^{\,n}_{\alpha}$. By 2.4.14, there is a homotopy $D : X \times I \to X$ such that $D_0 = \text{id}$, $D_t = \text{id}$ on $X - \overset{\circ}{e}{}^{\,n}_{\alpha}$ for all t, and $D_1(e^n_{\alpha} - U) \subset \overset{\bullet}{e}{}^{\,n}$. Define $M : S^{n-1} \times I \times I \to e^n_{\alpha} - \{z\}$ by $M(x,s,t) = L(x,s,2t)$ if $0 \le t \le \frac{1}{2}$ and by $M(x,s,t) = D_{2t-1}(L(x,s,1))$ if $\frac{1}{2} \le t \le 1$. It is a matter of planar geometry to construct a homeomorphism $h : I \times I \to I \times I$ agreeing with id on $I \times \{0\}$, taking $(0,t)$ to $(0,2t)$ and $(1,t)$ to $(1,2t)$ when $0 \le t \le \frac{1}{2}$, and taking the rest of $\text{fr}_{\mathbb{R}^2}(I \times I)$ into $I \times \{1\}$. The required K is $M \circ h^{-1}$. $\square$

Proof (of 2.5.1). "Only if" is obvious. We prove "if". By the hypothesis and 2.4.5 $h'_1 \simeq h'_2 : B^n/S^{n-1} \to e^n_{\alpha}/\overset{\circ}{e}{}^{\,n}_{\alpha}$. Let $z \in \overset{\circ}{e}{}^{\,n}_{\alpha}$. There exists $t \in (0,1)$ such that A_t is mapped into $\overset{\circ}{e}{}^{\,n}_{\alpha} - \{z\}$ by h_1 and by h_2. By 2.5.4, there exists $H : S_t \times I \to e^n_{\alpha} - \{z\}$ such that $H_0 = h_1 \mid S_t$ and $H_1 = h_2 \mid S_t$. Therefore, by 2.5.5 and 1.3.17, $h_1 \simeq h_2$. $\square$

An *oriented CW complex* is a CW complex X together with a choice of orientation on each cell. If A is a subcomplex of X, it is understood to be given the *inherited orientation* (each cell oriented as in X): we call (X, A) an *oriented CW pair*. A quotient complex $X/\sim$, as in Sect. 1.2, is given the *quotient orientation*: the vertices $\{A_{\alpha}\}$ are oriented by $+1$; all other cells receive their orientation via the quotient map $X \to X/\sim$.

Let X be an oriented CW complex, e^n_{α} an n-cell and e^{n-1}_{β} an $(n-1)$-cell. First, assume $n \ge 2$. Choose characteristic maps $h_{\alpha} : (B^n, S^{n-1}) \to (e^n_{\alpha}, \overset{\bullet}{e}{}^{\,n}_{\alpha})$ and $h_{\beta} : (B^{n-1}, S^{n-2}) \to (e^{n-1}_{\beta}, \overset{\bullet}{e}{}^{\,n-1}_{\beta})$ representing the orientations. Consider the commutative diagram:

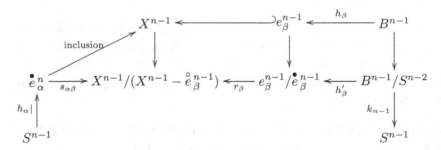

Here r_{β} is induced by inclusion and is clearly a homeomorphism. Unmarked arrows are quotient maps. $s_{\alpha\beta}$ is the indicated composition, and k_{n-1} is as before. Since h'_{β} is a homeomorphism, we obtain a map

$$k_{n-1} \circ (h'_{\beta})^{-1} \circ r_{\beta}^{-1} \circ s_{\alpha\beta} \circ h_{\alpha} \mid : S^{n-1} \longrightarrow S^{n-1}$$

whose degree, denoted by $[e^n_{\alpha} : e^{n-1}_{\beta}]$, is the *incidence number* of the oriented cells e^n_{α} and e^{n-1}_{β}. One should think of it as the (algebraic) number of times e^n_{α} is attached to e^{n-1}_{β}. By 2.5.1 we have:

Proposition 2.5.6. *For $n \geq 2$, the definition of $[e_\alpha^n : e_\beta^{n-1}]$ depends on the orientations but not on the specific maps h_α and h_β.* $\square$

Note that the definition of $[e_\alpha^n : e_\beta^{n-1}]$ also depends on k_{n-1}. Indeed, the homeomorphisms $\{k_n \mid n \geq 1\}$ have been chosen once and for all. We will say more about how we want them to have been chosen later (see 2.5.16).

We now extend the definition of incidence number to the case $n = 1$. Given an oriented 1-cell e_α^1 and a 0-cell e_β^0 oriented by $\epsilon \in \{\pm 1\}$, choose a characteristic map $h_\alpha : (B^1, S^0) \to (e_\alpha^1, \overset{\bullet}{e}{}_\alpha^1)$ representing the orientation. Define the *incidence number* $[e_\alpha^1 : e_\beta^0]$ by:

$$[e_\alpha^1 : e_\beta^0] = \begin{cases} \epsilon & \text{if } h_\alpha(1) = e_\beta^0 \text{ and } h_\alpha(-1) \neq e_\beta^0 \\ -\epsilon & \text{if } h_\alpha(1) \neq e_\beta^0 \text{ and } h_\alpha(-1) = e_\beta^0 \\ 0 & \text{if } h_\alpha(1) = h_\alpha(-1) = e_\beta^0 \\ 0 & \text{if } h_\alpha(1) \neq e_\beta^0 \text{ and } h_\alpha(-1) \neq e_\beta^0. \end{cases}$$

Clearly we have:

Proposition 2.5.7. *The definition of $[e_\alpha^1 : e_\beta^0]$ depends only on the orientations.* $\square$

Proposition 2.5.8. *Let e_α^n and e_β^{n-1} be (oriented) cells, $n \geq 1$. If e_β^{n-1} is not a subset of e_α^n, then $[e_\alpha^n : e_\beta^{n-1}] = 0$.*

Proof. When $n \geq 2$, the map $S^{n-1} \to S^{n-1}$ whose degree defines $[e_\alpha^n : e_\beta^{n-1}]$ is not surjective, so, by 1.3.4 and 1.3.7, it is homotopic to a constant map. By 2.4.3 its degree is 0. When $n = 1$, the Proposition is obvious. $\square$

We say e_β is a *face* of e_α if $e_\beta \subset e_\alpha$.

Now we turn to numbers associated with maps. Let X and Y be oriented CW complexes, and let $f : X \to Y$ be a map such that, for some n, $f(X^n) \subset Y^n$ and $f(X^{n-1}) \subset Y^{n-1}$; of course cellular maps have this property for all n. For each e_α^n in X and $\tilde{e}_\beta^n$ in Y we wish to define an integer $[e_\alpha^n : \tilde{e}_\beta^n : f]$ measuring the (algebraic) number of times f maps e_α^n onto $\tilde{e}_\beta^n$. First, let $n \geq 1$ and consider the commutative diagram:

Here $\tilde{e}_\beta^n$ is a cell of Y; $\tilde{h}_\beta$ is a characteristic map defining the orientation; $f'_{\alpha\beta}$ is induced by f, and is well defined since $f(X^n, X^{n-1}) \subset (Y^n, Y^{n-1})$; r_β is a homeomorphism as before. Let $[e_\alpha^n : \tilde{e}_\beta^n : f]$ be the degree of the map

$$k_n \circ (\tilde{h}'_\beta)^{-1} \circ r_\beta^{-1} \circ f'_{\alpha\beta} \circ h'_\alpha \circ k_n^{-1} : S^n \to S^n.$$

For the case $n = 0$, let $e_\alpha^0 \subset X$ have orientation ϵ and let $\tilde{e}_\beta^0 \subset Y$ have orientation $\tilde{\epsilon}$. Define $[e_\alpha^0 : \tilde{e}_\beta^0 : f] = \epsilon\tilde{\epsilon}$ if $f(e_\alpha^0) = \tilde{e}_\beta^0$, and $[e_\alpha^0 : \tilde{e}_\beta^0 : f] = 0$ otherwise. From 2.5.1 we conclude:

Proposition 2.5.9. *The definition of $[e_\alpha^n : \tilde{e}_\beta^n : f]$ depends only on the orientations.*   □

As with 2.5.8 we have:

Proposition 2.5.10. *Let e_α^n and $\tilde{e}_\beta^n$ be oriented n-cells of X and Y respectively. If $\tilde{e}_\beta^n$ is not a subset of $f(e_\alpha^n)$ then $[e_\alpha^n : \tilde{e}_\beta^n : f] = 0$.*   □

The number $[e_\alpha^n : \tilde{e}_\beta^n : f]$ is called the *mapping degree* of f with respect to e_α^n and $\tilde{e}_\beta^n$.

Next, we discuss product orientations. Let X and Y be oriented CW complexes. We define the *product orientation* on the CW complex $X \times Y$ by specifying an orientation for each $e_\alpha^m \times \tilde{e}_\beta^n$ where e_α^m, $\tilde{e}_\beta^n$ are cells of X and Y respectively: (i) if $m > 0$ and $n > 0$, and if $h_\alpha : (I^m, \overset{\bullet}{I}{}^m) \to (e_\alpha^m, \overset{\bullet}{e}{}_\alpha^m)$ and $\tilde{h}_\beta : (I^n, \overset{\bullet}{I}{}^n) \to (\tilde{e}_\beta^n, \overset{\bullet}{\tilde{e}}{}_\beta^n)$ are characteristic maps defining the separate orientations, $e_\alpha^m \times \tilde{e}_\beta^n$ is to be oriented by $h_\alpha \times \tilde{h}_\beta$. (ii) If $m > 0$ and $n = 0$, so that $\tilde{e}_\beta^0 = \{y_\beta\}$, if h_α orients e_α^m, and if 1 orients $\tilde{e}_\beta^0$, then $e_\alpha^m \times \tilde{e}_\beta^0$ is to be oriented by the map $I^m \to e_\alpha^m \times \tilde{e}_\beta^0$, $x \mapsto (h_\alpha(x), y_\beta)$; if -1 orients $\tilde{e}_\beta^0$, the opposite orientation is to be used. (iii) If $m = 0$ and $n > 0$, the rule of orientation is similar to that given in (ii). (iv) If $m = 0 = n$, with e_α^0 oriented by ϵ, and $\tilde{e}_\beta^0$ oriented by $\tilde{\epsilon}$, then $e_\alpha^0 \times \tilde{e}_\beta^0$ is to be oriented by $\epsilon\tilde{\epsilon}$.

We can use this to define a preferred orientation on the CW complex I^n as follows: when $n = 0$, the *canonical orientation* is 1; when $n = 1$, the *canonical orientation* on I^1 (with two vertices and one 1-cell) is: 1 on each vertex, and (the equivalence class of) id : $I^1 \to I^1$ on the 1-cell. The *canonical orientation* on I^n is then defined inductively to be the product orientation on $I^{n-1} \times I^1$. We leave the proofs of the next two statements to the reader:

Proposition 2.5.11. *If I^n is factorized as $I^r \times I^s$, and each factor carries the canonical orientation, then the product orientation on $I^r \times I^s$ coincides with the canonical orientation on I^n.*   □

Proposition 2.5.12. *Let $r \geq 1$. Let e be the r-cell of I^n obtained by constraining the $i_j{}^{th}$ coordinate to be $\epsilon_j \in \{1, -1\}$, for $1 \leq j \leq n - r$. Let the remaining r coordinates be indexed by $p_1 < p_2 < \ldots < p_r$. The canonical* *orientation on e is given by the characteristic map $(I^r, \overset{\bullet}{I}{}^r) \to (e, \overset{\bullet}{e})$ taking $(x_1, \ldots, x_r)$ to $(y_1, \ldots, y_n)$ where $y_{i_j} = \epsilon_j$ for $1 \le j \le n - r$ and $y_{p_j} = x_j$ for $1 \le j \le r$.* $\square$

Proposition 2.5.13. *Let $r_i : \overset{\bullet}{I}{}^n \to \overset{\bullet}{I}{}^n$ be the homeomorphism $(x_1, \ldots, x_n) \mapsto (x_1, \ldots, x_{i-1}, -x_i, x_{i+1}, \ldots, x_n)$. Then r_i is orientation reversing.*

Proof. If $n = 1$, this is obvious. Next, let $n = 2$ and $i = 1$, let M_t be the matrix $\begin{bmatrix} \cos(t\pi) & \sin(t\pi) \\ \sin(t\pi) & -\cos(t\pi) \end{bmatrix}$, and let $a_n : I^n \to B^n$ be the homeomorphism given in Sect. 1.2. Then $a_2^{-1} M a_2 \mid \overset{\bullet}{I}{}^2$ gives a homotopy from $a_2^{-1} f_{1,-1} a_2$ to r_1. For all other cases, embed this matrix in the appropriate $n \times n$ matrix. (For another proof, adapt the first paragraph of the proof of 2.5.2.) $\square$

Now we discuss the incidence numbers of I^n with its $(n-1)$-dimensional faces. For $\epsilon = \pm 1$, let $F_{i,\epsilon}$ be $\{(x_1, \ldots, x_n) \in I^n \mid x_i = \epsilon\}$. Each $F_{i,\epsilon}$ is an $(n-1)$-cell of I^n. All cells are canonically oriented.

Proposition 2.5.14. $[I^n : F_{i,1}] = -[I^n : F_{i,-1}]$.

Proof. This is clear when $n = 1$. Assume $n \ge 2$. Recall our permanent choice of homeomorphism, a_n, for identifying I^n with B^n; see Sect. 1.2. Consider the following commutative diagram:

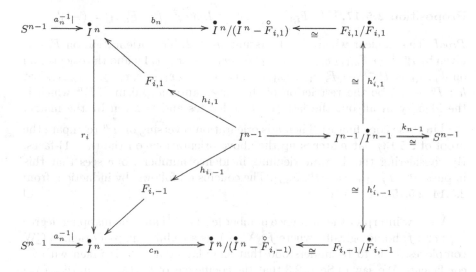

Here, $h_{i,\epsilon} : I^{n-1} \to F_{i,\epsilon}$ is the canonical characteristic map $(x_1, \ldots, x_{n-1}) \mapsto (x_1, \ldots, x_{i-1}, \epsilon, x_{i+1}, \ldots, x_{n-1})$ of $F_{i,\epsilon}$, given by 2.5.12, and r_i is as in 2.5.13. All other arrows are the obvious quotients or inclusions or are induced by such. By 2.5.13, r_i is orientation reversing.

The required incidence numbers are the degrees of f and g where f [resp. g] is the composition of maps leading from the top left [resp. bottom left] copy of S^{n-1} to the right copy of S^{n-1}. Note that $f = g \circ (a_n \circ r_i \circ a_n^{-1})$. By 2.5.13, r_i has degree -1. So, by 2.4.19, $\deg(f) = -\deg(g)$. $\square$

Proposition 2.5.15. *The maps b_n and c_n in the above diagram are homotopy equivalences. Hence the incidence numbers in 2.5.14 are ± 1.*

Proof. Let $e^n(t) = \{x \in S^n \mid t \le x_{n+1} \le 1\}$. The space $e^n(0)$ is the upper closed hemisphere, while $e^n(1) = \{$north pole$\}$. The reader can easily construct a homotopy $H : S^n \times I \to S^n$ such that $H_0 = \mathrm{id}$, $H_t(e^n(0)) = e^n(t)$, and H_t maps $S^n - e^n(0)$ bijectively onto $S^n - e^n(t)$, for all t. The map H_1 is a quotient map (see Sect. 1.1), and since $H_1 \simeq \mathrm{id}$, H_1 is a homotopy equivalence. Clearly this implies that b_n and c_n are homotopy equivalences. The second part then follows from 2.4.20. $\square$

Proposition 2.5.14 makes a relative statement about incidence numbers. But even though we know, by 2.5.15, that these incidence numbers are ± 1, we cannot say which is which (when $n \ge 2$) until we specify the homeomorphism k_{n-1} which "was chosen once and for all" earlier in this section. In fact we are only making our choice now:

Convention 2.5.16. *For $n \ge 2$, the canonical homeomorphism $k_{n-1} : B^{n-1}/S^{n-2} \to S^{n-1}$ is to be chosen to make $[I^n : F_{n,1}] = (-1)^{n+1}$.*

Proposition 2.5.17. $[I^n : F_{i,1}] = (-1)^{i+1}$. *Hence*[7] $[I^n : F_{i,-1}] = (-1)^i$.

Proof. This is clear when $n = 1$. Assume $n \ge 2$. The orientation on $F_{n,1}$ is given by $I^{n-1} \to F_{n,1}, (x_1, \cdots, x_{n-1}) \mapsto (x_1, \cdots, x_{n-1}, 1)$; and the orientation on $F_{n-1,1}$ is $I^{n-1} \to F_{n-1,1}, (x_1, \cdots, x_{n-1}) \mapsto (x_1, \cdots, x_{n-2}, 1, x_{n-1})$. Let $h : I^n \to I^n$ be the restriction of the linear automorphism of $\mathbb{R}^n$ which is the identity in all but the last two coordinates and is given by the matrix $\left[\begin{smallmatrix} 0 & 1 \\ 1 & 0 \end{smallmatrix}\right]$ in those coordinates. Then h is orientation reversing on $\overset{\bullet}{I}{}^n$ (compare the proof of 2.5.13) but matches up the above orientations on the $(n-1)$-faces. By considering the diagram defining incidence numbers, one sees that this implies $[I^n : F_{n-1,1}] = -[I^n, F_{n,1}]$. The conclusion follows, by induction, from 2.5.14, 2.5.15 and 2.5.16. $\square$

We now interpret the incidence number $[e_\alpha^n : e_\beta^{n-1}]$ and the mapping degree $[e_\alpha^n : \tilde{e}_\gamma^n : f]$ homologically, where $f : X \to Y$ is a cellular map of oriented CW complexes. It is to be understood that singular homology is taken with $\mathbb{Z}$-coefficients. We saw in Sect. 2.3 that the coefficient of $\bar{h}_{\beta*}(\lambda_{n-1})$ in $\partial \bar{h}_{\alpha*}(\lambda_n)$ is the image of 1 under a certain homomorphism $\mathbb{Z} \to \mathbb{Z}$. We wish to claim

---

[7] The choice in 2.5.16 which leads to the incidence numbers stated in 2.5.17 is not a universal convention. For example, choices used in [42] lead to $[I^n : F_{i,1}] = (-1)^{n-i}$.

that this integer is precisely the incidence number $[e_\alpha^n : e_\beta^{n-1}]$. However, for this to be true some further once-and-for-all choices must be made correctly.

By an easy exercise in singular homology using excision as outlined in Sect. 2.2, there are canonical isomorphisms

$$H_n^\Delta(B^n, S^{n-1}) \xrightarrow{\cong} H_n^\Delta(B^n/S^{n-1}, v) \xleftarrow{\cong} H_n^\Delta(B^n/S^{n-1}) \xrightarrow{k_{n*}} H_n^\Delta(S^n)$$

when $n \geq 1$. Let $\kappa_n \in H_n^\Delta(S^n)$ be the image of the generator $\lambda_n \in H_n^\Delta(B^n, S^{n-1})$ under this composition of isomorphisms. We also have a homomorphism $\partial_* : H_n^\Delta(B^n, S^{n-1}) \to H_{n-1}^\Delta(S^{n-1})$ which is an isomorphism for $n \geq 2$ and a monomorphism when $n = 1$; see 2.2.1. Indeed, the image of λ_1 under $\partial_* : H_1^\Delta(B^1, S^0) \to H_0^\Delta(S^0) \cong \mathbb{Z} \oplus \mathbb{Z}$ is the homology class of the singular 0-cycle $\pm(\{1\} - \{-1\})$, with the $+$ or $-$ depending on which generator of $H_1^\Delta(B^1, S^0)$ has been chosen as λ_1.

Now we make our choices. Let $\kappa_0 \in H_0^\Delta(S^0)$ be the homology class of $\{1\} - \{-1\}$. Define the generators $\lambda_n \in H_n^\Delta(B^n, S^{n-1})$ and $\kappa_n \in H_n^\Delta(S^n)$ by

$$H_0^\Delta(S^0) \longleftarrow H_1^\Delta(B^1, S^0) \xrightarrow{\cong} H_1^\Delta(S^1) \xleftarrow{\cong} H_2^\Delta(B^2, S^1) \xrightarrow{\cong} H_2^\Delta(S^2) \longleftarrow \cdots$$

$$\kappa_0 \longleftarrow\!\!\shortmid \lambda_1 \longmapsto \kappa_1 \longleftarrow\!\!\shortmid \lambda_2 \longmapsto \kappa_2 \longleftarrow\!\!\shortmid \cdots$$

This specifies λ_n, $n \geq 1$, for the first time. We also define $\lambda_0 \in H_0^\Delta(B^0)$ to be the homology class of the singular 0-cycle $\{0\}$. With these choices we get:

Proposition 2.5.18. *The coefficient of $\bar{h}_{\beta*}(\lambda_{n-1})$ in $\partial \bar{h}_{\alpha*}(\lambda_n)$ is the incidence number $[e_\alpha^n : e_\beta^{n-1}]$.*

Proof. We leave the case $n = 1$ as an exercise. For $n \geq 2$ the claim can be read off from the following commutative diagram of singular homology groups (Δ's omitted):

The point here is that, by our choices, $\lambda_n \mapsto \kappa_{n-1}$ on the left and $\lambda_{n-1} \mapsto \kappa_{n-1}$ on the right. $\qquad\square$

Similarly, one proves (exercise):

Proposition 2.5.19. *The coefficient of* $\bar{h}_{\beta*}(\lambda_n)$ *in* $f_\#(\bar{h}_{\alpha*}(\lambda_n))$ *is the mapping degree* $[e_\alpha^n : \tilde{e}_\beta^n : f]$. □

Exercises

1. Let X and Y be oriented CW complexes. Give $X \times Y$ the product orientation. Prove
   (a) $[e_\alpha^n \times \tilde{e}_\gamma^m : e_\beta^{n-1} \times \tilde{e}_\gamma^m] = [e_\alpha^n : e_\beta^{n-1}]$
   (b) $[e_\alpha^n \times \tilde{e}_\gamma^m : e_\alpha^n \times \tilde{e}_\delta^{m-1}] = (-1)^n [\tilde{e}_\gamma^m : \tilde{e}_\delta^{m-1}]$
2. Prove 2.5.19.
3. Prove the $n = 1$ case of 2.5.18.

2.6 The geometric cellular chain complex

In Sect. 2.3 we defined $C_n(X; R)$ to be $H_n^\Delta(X^n, X^{n-1}; R)$ and we called its elements "cellular chains." There is a more geometrical way of defining cellular chains which (this author believes) gives more insight into what homology is about; we have been preparing for this in the last two sections. Temporarily we will call the new chains "geometric cellular chains" but we will drop this distinction (and the "geom" superscript) after this section.

For X an oriented CW complex and $n \geq 0$ define $C_n^{\text{geom}}(X; R)$ to be the free R-module generated by the set of (already oriented) n-cells[8] of X. For $n < 0$, define $C_n^{\text{geom}}(X; R)$ to be the trivial R-module 0. Elements of $C_n^{\text{geom}}(X; R)$ are called *cellular n-chains* in X (over R).

There is a canonical ring homomorphism $\mathbb{Z} \to R$ taking $1 \in \mathbb{Z}$ to $1 \in R$. If $n \in \mathbb{Z}$ we also denote its image in R by n. This applies, in particular, to incidence numbers and mapping degrees in what follows. In the context of R-coefficients they are interpreted as elements of R.

For $n \geq 1$, define $\partial : C_n^{\text{geom}}(X; R) \to C_{n-1}^{\text{geom}}(X; R)$ by

$$\partial(e_\alpha^n) = \sum_\beta [e_\alpha^n : e_\beta^{n-1}] e_\beta^{n-1}$$

---

[8] Because of potential confusion, we spell this out. Each n-cell e_α^n of X has been equipped with an orientation at the start. When $n \geq 1$, this orientation is the homotopy class $[h]$ of some characteristic map $h : (B^n, S^{n-1}) \to (e_\alpha^n, \overset{\bullet}{e}{}_\alpha^n)$. The ordered pair $(e_\alpha^n, [h])$ is a generator of $C_n(X; R)$. If $[h \circ r]$ is the other orientation of e_α^n (see 2.5.2), the ordered pair $(e_\alpha^n, [h \circ r])$ is not in $C_n(X; R)$. Similar remarks apply to the case $n = 0$. Thus, although there is a bijection between the n-cells of X and the generators of $C_n(X; R)$, it can be misleading to think of $C_n(X; R)$ as simply the free R-module generated by the set of n-cells.

where α and β index the n-cells and $(n-1)$ cells, respectively. This is a finite sum. For $n \leq 0$, define $\partial = 0$.

Define $\phi : C_n^{\text{geom}}(X; R) \to C_n(X; R)$ to be the canonical extension of the function $e_\alpha^n \mapsto \bar{h}_{\alpha_*}(\lambda_n)$. By 2.3.8 ϕ is an isomorphism. The discussion at the end of the last section then shows that the following diagram commutes:

$$
\begin{array}{ccc}
C_n^{\text{geom}}(X; R) & \xrightarrow{\phi} & C_n(X; R) \\
\downarrow{\partial} & & \downarrow{\partial} \\
C_{n-1}^{\text{geom}}(X; R) & \xrightarrow{\phi} & C_{n-1}(X; R)
\end{array}
$$

Thus $(C_n^{\text{geom}}(X; R), \partial)$ is a chain complex and is canonically chain isomorphic to $(C_n(X; R), \partial)$ as defined in Sect. 2.3.

If $f : X \to Y$ is a cellular map between oriented CW complexes, define $f_\# : C_n^{\text{geom}}(X; R) \to C_n^{\text{geom}}(Y; R)$ by

$$
f_\#(e_\alpha^n) = \sum_\beta [e_\alpha^n : \tilde{e}_\beta^n : f]\tilde{e}_\beta^n.
$$

By 2.5.19 the following diagram commutes

$$
\begin{array}{ccc}
C_n^{\text{geom}}(X; R) & \xrightarrow{\phi} & C_n(X; R) \\
\downarrow{f_\#} & & \downarrow{f_\#} \\
C_n^{\text{geom}}(Y; R) & \xrightarrow{\phi} & C_n(Y; R)
\end{array}
$$

Thus $f_\#$ is a chain map.

Theorem 2.6.1. (Sum Theorem for Degree) *Let $n \geq 1$ and let*

$$
S^n \xrightarrow{F} \bigvee_{\alpha \in \mathcal{A}} S_\alpha^n \xrightarrow{G} S^n
$$

be maps. The degree of $G \circ F$ is $\sum_{\alpha \in \mathcal{A}} \deg (G \circ i_\alpha \circ q_\alpha \circ F)$.

Proof. We use the CW structures with one vertex and one n-cell for each sphere. We have $\deg(G \circ i_\alpha) = [S_\alpha^n : S^n : G]$ and $\deg(q_\alpha \circ F) = [S^n : S_\alpha^n : F]$. By 2.4.19

$$
\sum_\alpha \deg(G \circ i_\alpha \circ q_\alpha \circ F) = \sum_\alpha [S_\alpha^n : S^n : G][S^n : S_\alpha^n : F].
$$

At the level of cellular chains we have:

$$
(G \circ F)_\#(S^n) = \sum_\alpha [S_\alpha^n : S^n : G][S^n : S_\alpha^n : F]S^n.
$$

Thus $\deg(G \circ F)$ is as claimed. $\qquad\square$

In the exercises, the reader is asked to provide direct proofs of: $\partial \circ \partial = 0$, $f_\# \circ \phi = \phi \circ f_\#$ and $(g \circ f)_\# = g_\# \circ f_\#$ using the definitions of this section as well as basic facts about degree (Sect. 2.4) and singular homology (Sect. 2.2). These are instructive in that they show why it is not easy, perhaps impossible, to give a self-contained treatment of cellular homology from the geometric viewpoint – one which completely avoids singular homology.

We have now reached our goal of defining cellular homology in terms of oriented cells, incidence numbers and mapping degrees. In the rest of the chapter we establish the basic properties of homology in these terms AND WE DROP THE DISTINCTION BETWEEN $C_n^{\mathrm{geom}}(X; R)$ AND $C_n(X; R)$.

Historical Note: Cellular homology as described in Sect. 2.3 was introduced by N. Steenrod in lectures and was popularized as an appendix to [Milnor-2] which circulated widely as a preprint for many years before it was published as a book. The geometric treatment preferred here in Sect. 2.6 (and throughout this book) is folklore.

Exercises

1. Prove: $\partial(I^m \times I^n) = (\partial I^m) \times I^n + (-1)^m (I^m \times \partial I^n)$.

Prove the following from the definitions given in this section, without using the equivalent formulation in Sect. 2.3:

2. The composition $C_{n+1}(X; \mathbb{Z}) \xrightarrow{\partial} C_n(X; \mathbb{Z}) \xrightarrow{\partial} C_{n-1}(X; \mathbb{Z})$ is zero, where X is an oriented CW complex.
3. If $f : X \to Y$ is a cellular map between oriented CW complexes, then $\partial \circ f_\# = f_\# \circ \partial$.
4. If $X \xrightarrow{f} Y \xrightarrow{g} Z$ are cellular maps between oriented CW complexes, then $(g \circ f)_\# = g_\# \circ f_\#$ and $\mathrm{id}_\# = \mathrm{id}$. *Hint*: Use 2.6.1.
5. If $f : X \to Y$ is homotopic to a constant map then $f_\# : H_n(X; R) \to H_n(Y; R)$ is the zero homomorphism when $n > 0$.

2.7 Some properties of cellular homology

We begin with some properties whose proofs are exercises:

Proposition 2.7.1. *If X is path connected, $H_0(X; R) \cong R$.*   □

Proposition 2.7.2. *If $\{X_\alpha \mid \alpha \in \mathcal{A}\}$ is the set of path components of X, then each X_α is a subcomplex of X, $X = \coprod_\alpha X_\alpha$, and $H_n(X; R)$ is isomorphic to $\oplus_\alpha H_n(X_\alpha; R)$ for all n. In particular, $H_0(X; R)$ is a free R-module whose rank is the cardinal number of $\mathcal{A}$.*   □

Corollary 2.7.3. *The homology of the one-point CW complex, $\{v\}$, is as follows: $H_0(\{v\}; R) \cong R$; $H_n(\{v\}; R) = 0$ for $n > 0$.* $\square$

Proposition 2.7.4. *Let $\langle W \mid R, \rho \rangle$ be a presentation of the group $G := F/N(\rho(R))$, and let X be the 2-dimensional CW complex constructed in Example 1.2.17. Then, for any choice of orientations on X, $H_1(X; \mathbb{Z})$ is isomorphic to the abelianization of G, $G/[G, G]$.* $\square$

The last proposition shows how incidence number is related to the notion of exponent sum in a group relation.

Proposition 2.7.5. *If X has dimension d, then $H_n(X; R) = 0$ for all $n > d$.* $\square$

Proposition 2.7.6. *If $i : X^{n+1} \hookrightarrow X$, then $i_* : H_j(X^{n+1}; R) \to H_j(X; R)$ is an isomorphism for $j \leq n$ and an epimorphism for $j = n + 1$.* $\square$

Proposition 2.7.7. *If X has finite type and R is a PID then $H_n(X; R)$ is finitely generated for each n.*

Proof. Since R is a PID the modules of cycles are (free and) finitely generated. $\square$

Now we turn to homotopy invariance; it is particularly easy to understand in the geometric cellular chain complex.

Let $f, g : X \to Y$ be cellular maps between oriented CW complexes. We wish to show that if $f \simeq g$ then $f_* = g_* : H_n(X; R) \to H_n(Y; R)$. By 1.4.3, there is a cellular homotopy $F : X \times I \to Y$ from f to g. We regard I as a CW complex having two vertices and one 1-cell; it carries the *canonical orientation*: 1 on each vertex, and (the equivalence class of) $(I^1, \overset{\bullet}{I}{}^1) \to (I, \{0, 1\})$, $t \mapsto \frac{1}{2}(t + 1)$. We give $X \times I$ the product orientation.

By considering the definition of incidence number together with 2.5.8 and 2.5.17, the reader can check:

Proposition 2.7.8. *If e_α^{n-1} and e_β^{n-1} are $(n-1)$-cells of X, then we have $[e_\alpha^{n-1} \times I : e_\beta^{n-1} \times \{j\}] = (-1)^{n+j} \delta_{\alpha\beta}$. If e_γ^{n-2} is an $(n-2)$-cell of X, then $[e_\alpha^{n-1} \times I : e_\gamma^{n-2} \times I] = [e_\alpha^{n-1} : e_\gamma^{n-2}]$.* $\square$

A chain $c \in C_n(X; R)$ has the form $\sum\limits_{\alpha \in \mathcal{A}} m_\alpha e_\alpha^n$ where $\mathcal{A}$ indexes the n-cells of X, and only finitely many m_α's are non-zero. Define $c \times I = \sum\limits_{\alpha \in \mathcal{A}} m_\alpha (e_\alpha^n \times I) \in C_{n+1}(X \times I; R)$.

Corollary 2.7.9. $\partial(c \times I) = (-1)^{n+1}(c \times \{0\} - c \times \{1\}) + (\partial c) \times I.$ $\square$

Recall that $Z_n(Y) = \ker(\partial : C_n(Y; R) \to C_{n-1}(Y; R))$ is the R-module of n-cycles, $B_n(Y; R) = \text{image}(\partial : C_{n+1}(Y; R) \to C_n(Y; R))$ is the R-module of n-boundaries, and $H_n(Y; R) = Z_n(Y; R)/B_n(Y; R)$. Cycles $z, z' \in Z_n(Y; R)$ represent the same element of $H_n(Y; R)$ iff $z - z' \in B_n(Y; R)$, in which case z and z' are *homologous*.

Theorem 2.7.10. (Homotopy Invariance) *If $f, g : X \to Y$ are homotopic cellular maps, then $f_* = g_* : H_n(X; R) \to H_n(Y; R)$ for all n.*

Proof. Let $z \in Z_n(X; R)$ represent $\{z\} \in H_n(X; R)$. Then $f_*(\{z\}) = \{f_\#(z)\}$ and $g_*(\{z\}) = \{g_\#(z)\}$. We must show that $f_\#(z)$ and $g_\#(z)$ are homologous; i.e., $f_\#(z) - g_\#(z) \in B_n(Y; R)$. By 1.4.3 there is a cellular homotopy $F : X \times I \to Y$ from f to g.

$$\partial(F_\#(z \times I)) = F_\#(\partial(z \times I))$$
$$= (-1)^{n+1}(F_\#(z \times \{0\}) - F_\#(z \times \{1\})) \quad \text{by 2.7.9}$$
$$= (-1)^{n+1}(f_\#(z) - g_\#(z)).$$

Thus $f_\#(z) - g_\#(z) = \partial((-1)^{n+1}F_\#(z \times I)) \in B_n(Y; R)$. $\qquad\square$

By 1.4.3, every map is homotopic to a cellular map. If $f : X \to Y$ is a map between oriented CW complexes, we can define $f_* : H_n(X; R) \to H_n(Y; R)$ by $f_* = g_*$ where g is any cellular map homotopic to f. This is well defined by 2.7.10. In fact, we can regard $H_*(\,\cdot\,; R)$ as a covariant functor from the category of oriented CW complexes and homotopy classes of maps to the category of graded R-modules and homomorphisms of degree 0.

Corollary 2.7.11. *If X and Y are oriented CW complexes and $f : X \to Y$ is a homotopy equivalence, then $f_* : H_n(X; R) \to H_n(Y; R)$ is an isomorphism for all n.* $\qquad\square$

Corollary 2.7.12. *If X_1 and X_2 represent the same CW complex X oriented in two different ways, then the identity map $\text{id} : X_1 \to X_2$ induces an isomorphism $H_*(X_1) \to H_*(X_2)$.* $\qquad\square$

In this sense, cellular homology is independent of the choices of orientation on cells. So from now on we will mention an orientation only when discussing chains.

Corollary 2.7.13. *If the space X admits two CW complex structures X_1 and X_2, these yield isomorphic homology groups.*

Proof. The identity map id_X is a homotopy equivalence $X_1 \to X_2$. $\qquad\square$

Our proof of Theorem 2.7.10 was direct. We showed that for any cycle z, $f_\#(z)$ and $g_\#(z)$ differ by the boundary $\partial(F_\#(z \times I))$. When we come to the corresponding theorems for cohomology in Sect. 12.1, it will be helpful also to have a slightly stronger statement:

Proposition 2.7.14. *Let* $F : X \times I \to Y$ *be a cellular homotopy from* f *to* g*. Define* $D_n : C_n(X) \to C_{n+1}(Y)$ *by* $D_n(e_\alpha^n) = (-1)^{n+1} F_\#(e_\alpha^n \times I)$*. Then* $\{D_n\}$ *is a chain homotopy between* $f_\#$ *and* $g_\#$*.*

Proof. We use the proof of 2.7.10 and 2.7.9 to get:

$$\partial D_n(e_\alpha^n) = (-1)^{n+1} \partial F_\#(e_\alpha^n \times I)$$
$$= (-1)^{n+1} F_\# \partial(e_\alpha^n \times I)$$
$$= F_\#(e_\alpha^n \times \{0\} - e_\alpha^n \times \{1\}) + (-1)^{n+1} F_\#((\partial e_\alpha^n) \times I)$$
$$= F_\#(e_\alpha^n \times \{0\} - e_\alpha^n \times \{1\}) - D_{n-1}(\partial e_\alpha^n).$$

So $\partial D_n(e_\alpha^n) + D_{n-1}\partial(e_\alpha^n) = f_\#(e_\alpha^n) - g_\#(e_\alpha^n)$. $\qquad\square$

We end with a generalization of homotopy equivalence which will be needed later. A cellular map $f : (X, A) \to (Y, B)$ between CW pairs[9] is an *n-equivalence* if there is a cellular map $g : (Y, B) \to (X, A)$ such that $g \circ f \mid: (X^{n-1}, A^{n-1}) \to (X, A)$ is homotopic to $(X^{n-1}, A^{n-1}) \hookrightarrow (X, A)$ and $f \circ g \mid: (Y^{n-1}, B^{n-1}) \to (Y, B)$ is homotopic to $(Y^{n-1}, B^{n-1}) \hookrightarrow (Y, B)$. The map g is an *n-inverse* for f.

It is straightforward to prove:

Proposition 2.7.15. *An n-equivalence induces isomorphisms on homology modules in dimensions* $\leq n - 1$. $\qquad\square$

Exercise

Prove all the unproved statements in this section.

2.8 Further properties of cellular homology

Recall from Sect. 2.1 that whenever

$$0 \longrightarrow C' \overset{\alpha}{\longrightarrow} C \overset{\beta}{\longrightarrow} C'' \longrightarrow 0$$

is a short exact sequence of chain complexes, there is a *connecting homomorphism* $\partial_* : H_n(C'') \to H_{n-1}(C')$ for each n, giving rise to an exact sequence

$$\dots \longrightarrow H_n(C') \overset{\alpha_*}{\longrightarrow} H_n(C) \overset{\beta_*}{\longrightarrow} H_n(C'') \overset{\partial_*}{\longrightarrow} H_{n-1}(C') \overset{\alpha_*}{\longrightarrow} \dots .$$

This purely algebraic fact can be used in several ways. Throughout this section, let (X, A) be an oriented CW pair. There is a short exact sequence of chain complexes

---

[9] Some authors have a slightly different definition of *n*-equivalence.

$$0 \longrightarrow C_*(A; R) \xrightarrow{i_\#} C_*(X; R) \xrightarrow{p} C_*(X; R)/C_*(A; R) \longrightarrow 0$$

where the boundary homomorphism on the quotient chain complex $\{C_n(X; R)/C_n(A; R)\}$ is induced by the boundary on $C_n(X; R)$. The alternative notation $C_*(X, A; R)$ is also used for this chain complex. The nth *relative cellular homology module of* (X, A), denoted $H_n(X, A; R)$, is the nth homology module of the chain complex $C_*(X; R)/C_*(A; R)$. There results the *homology exact sequence* of (X, A):

Proposition 2.8.1. *The following sequence is exact:*

$$\ldots \longrightarrow H_n(A; R) \xrightarrow{i_*} H_n(X; R) \xrightarrow{p_*} H_n(X, A; R) \xrightarrow{\partial_*} H_{n-1}(A; R) \longrightarrow \ldots \ .$$

$\square$

This is only formal until we interpret $H_n(X, A; R)$ geometrically. We may do so in two ways.

First, let $Z_n(X, A; R) = \{c \in C_n(X; R) \mid \partial c \in C_{n-1}(A; R)\}$ and let $B_n(X, A; R) = \{c \in C_n(X; R) \mid c$ is homologous to an element of $C_n(A; R)\}$. The elements of $Z_n(X, A; R)$ and of $B_n(X, A; R)$ are called *relative n-cycles* and *relative n-boundaries* respectively. Clearly, $B_n(X, A; R) \subset Z_n(X, A; R)$.

Proposition 2.8.2. $H_n(X, A; R) \cong Z_n(X, A; R)/B_n(X, A; R)$.

Proof. Consider the commutative diagram

$$
\begin{array}{ccc}
C_n(X; R) & \xrightarrow{\ \partial\ } & C_{n-1}(X; R) \\
\downarrow{\scriptstyle p} & & \downarrow{\scriptstyle p} \\
C_n(X; R)/C_n(A; R) & \xrightarrow{\ \bar\partial\ } & C_{n-1}(X; R)/C_{n-1}(A; R)
\end{array}
$$

Here $\bar\partial$ is the boundary homomorphism. Note that $Z_n(X, A; R) = p^{-1}(\ker \bar\partial)$, and $B_{n-1}(X, A; R) = p^{-1}(\text{image } \bar\partial)$. By the Noether Isomorphism Theorem (i.e., $(M/P)/(N/P) \cong M/N$) the result follows. $\square$

The second interpretation of $H_n(X, A; R)$ is in terms of the oriented quotient CW complex X/A:

Proposition 2.8.3. $H_n(X/A; R) \cong \begin{cases} H_n(X, A; R) & \text{if } n \neq 0 \\ \\ H_0(X, A; R) \oplus R & \text{if } n = 0. \end{cases}$

Proof. There is a unique homomorphism r making the following diagram commute:

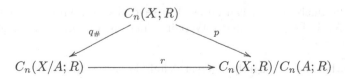

Here, $q : X \to X/A$ is the quotient map, and X/A has the quotient orientation. Since p and $q_\#$ are chain maps, and $q_\#$ is onto for each n, r is a chain map. Clearly, r induces a bijection of generators when $n > 0$. When $n = 0$, ker r is isomorphic to R, generated by the (oriented) vertex $\{A\}$ of X/A; the other generators are mapped bijectively to the generators of $C_0(X; R)/C_0(A; R)$. The result follows. $\qquad\square$

Homology of CW pairs works like homology of CW complexes. There is a pairwise Cellular Approximation Theorem:

Proposition 2.8.4. *A map* $f : (X, A) \to (Y, B)$ *between CW pairs which is already cellular on the subcomplex Z of X is pairwise homotopic, rel Z, to a cellular map.*

Proof. The proof of 1.4.3 achieves this. $\qquad\square$

A cellular map $f : (X, A) \to (Y, B)$ of oriented CW pairs induces a chain map $f_\# : C_n(X; R)/C_n(A; R) \to C_n(Y; R)/C_n(B; R)$ for each n, and hence $f_* : H_n(X, A; R) \to H_n(Y, B; R)$. Theorem 2.7.10 has an analog (with similar proof: use 2.8.2):

Theorem 2.8.5. (Homotopy Invariance) *If $f, g : (X, A) \to (Y, B)$ are pairwise homotopic cellular maps, then $f_* = g_* : H_n(X, A; R) \to H_n(Y, B; R)$ for all n.* $\qquad\square$

The connecting homomorphism ∂_* satisfies the following "naturality" property for algebraic reasons (see Sect. 2.1):

Proposition 2.8.6. *Let $f : (X, A) \to (Y, B)$ be a cellular map of oriented CW pairs. Then, for all n, the following diagram commutes:*

$$
\begin{array}{ccc}
H_n(X, A; R) & \xrightarrow{\ \partial_*\ } & H_{n-1}(A; R) \\
\downarrow{\scriptstyle f_*} & & \downarrow{\scriptstyle (f|)_*} \\
H_n(Y, B; R) & \xrightarrow{\ \partial_*\ } & H_{n-1}(B; R) \quad .
\end{array}
$$

$\qquad\square$

Thus, we have a commutative diagram

$$
\begin{array}{ccccccccc}
\cdots \longrightarrow & H_n(A; R) & \xrightarrow{\ i_*\ } & H_n(X; R) & \xrightarrow{\ j_*\ } & H_n(X, A; R) & \xrightarrow{\ \partial_*\ } & H_{n-1}(A; R) & \longrightarrow \cdots \\
& \downarrow{\scriptstyle (f|)_*} & & \downarrow{\scriptstyle f_*} & & \downarrow{\scriptstyle f_*} & & \downarrow{\scriptstyle (f|)_*} & \\
\cdots \longrightarrow & H_n(B; R) & \xrightarrow{\ i_*\ } & H_n(Y; R) & \xrightarrow{\ j_*\ } & H_n(Y, B; R) & \xrightarrow{\ \partial_*\ } & H_{n-1}(B; R) & \longrightarrow \cdots
\end{array}
$$

whose horizontal rows are exact.

Next, we turn to the excision property. Let (X, A) and (Y, B) be oriented CW pairs where X is a subcomplex of Y and A is a subcomplex of B. The inclusion $i : (X, A) \to (Y, B)$ is an excision map iff $X - A = Y - B$. This could be taken as the definition of excision in our context; it is compatible with the definition in Sect. 2.2.

Theorem 2.8.7. *Excision maps induce isomorphisms on relative homology groups.*

Proof. Let i be an excision map as above. Then $Y = X \cup B$ and $A = X \cap B$. By elementary algebra

$$C_n(X; R)/C_n(A; R) \cong [C_n(X; R) + C_n(B; R)]/C_n(B; R) = C_n(Y; R)/C_n(B; R)$$

and this canonical isomorphism commutes with ∂'s. $\qquad\square$

The same algebraic trick used to derive the homology sequence of the oriented CW pair (X, A) can also be used to derive an exact sequence expressing the homology of a CW complex in terms of the homology of two subcomplexes which cover it. To see this, let X be an oriented CW complex having two subcomplexes A and B such that $X = A \cup B$; then $A \cap B$ is also a subcomplex. There is a short exact sequence of chain complexes

$$0 \longrightarrow C_*(A \cap B; R) \xrightarrow{i} C_*(A; R) \oplus C_*(B; R) \xrightarrow{j} C_*(A \cup B; R) \longrightarrow 0$$

where $i(c) = (c, -c)$ and $j(c, d) = c + d$. [Here, of course, we consider $C_n(A; R) \subset C_n(A \cup B; R)$, etc. The direct sum of two chain complexes is defined in the obvious way, and $H_n(C \oplus C') \cong H_n(C) \oplus H_n(C')$.] The general procedure for generating homology exact sequences can be applied here, this time yielding the *Mayer-Vietoris sequence* of the complexes A and B:

$$\ldots \longrightarrow H_n(A \cap B; R) \xrightarrow{i_*} H_n(A; R) \oplus H_n(B; R) \xrightarrow{j_*} H_n(X; R) \xrightarrow{\partial_*} H_{n-1}(A \cap B; R) \longrightarrow \ldots.$$

This sequence is "natural" in the sense that if $f : X \to X'$ is cellular, with $f(A) \subset A'$ and $f(B) \subset B'$, and if $X = A \cup B$ and $X' = A' \cup B'$, then the Mayer-Vietoris sequence of A and B maps to that of A' and B' by homomorphisms induced by f, so as to give a commutative diagram.

In closing, we remark that cellular homology can be axiomatized. Suppose that to each oriented CW pair (X, A) and $n \geq 0$ we associate an R-module $h_n(X, A)$ and a homomorphism[10] $\partial_* : h_{n+1}(X, A) \to h_n(A, \emptyset)$, and to each map of CW pairs $f : (X, A) \to (Y, B)$ and $n \geq 0$ we associate a homomorphism $f_* : h_n(X, A) \to h_n(Y, B)$ satisfying:

---

[10] Convention: write $h_n(A)$ for $h_n(A, \emptyset)$

(i) $\mathrm{id}_* = \mathrm{id}$ and $(g \circ f)_* = g_* \circ f_*$;

(ii) $(f|)_* \partial_* = \partial_* f_* : h_{n+1}(X, A) \to h_n(B)$;

(iii) the sequence

$$\cdots \longrightarrow h_n(A) \xrightarrow{i_*} h_n(X) \xrightarrow{j_*} h_n(X, A) \xrightarrow{\partial_*} h_{n-1}(A) \longrightarrow \cdots$$

is exact;[11]

(iv) whenever $f, g : (X, A) \to (Y, B)$ are homotopic, then $f_* = g_* : h_n(X, A) \to h_n(Y, B)$ for all $n \geq 0$;

(v) excision maps $i : (X, A) \to (Y, B)$ induce isomorphisms $i_* : h_n(X, A) \to h_n(Y, B)$ for all $n \geq 0$;

(vi)
$$h_n(\{v\}) \cong \begin{cases} 0 & \text{if } n \neq 0 \\ R & \text{if } n = 0 \end{cases}$$

where $\{v\}$ is the one-point CW complex;

and, finally, suppose $\phi_0 : h_0(\{v\}) \to H_0(\{v\}; R)$ is an isomorphism (see 2.7.1). Then for every finite CW pair (X, A) and every $n \geq 0$ there is an isomorphism $\phi_{(X,A)} : h_n(X, A) \to H_n(X, A; R)$ such that $\phi_{(\{v\}, \emptyset)} = \phi_0$ and all diagrams of the following kinds commute:

$$
\begin{array}{ccc}
h_n(X, A) & \xrightarrow{f_*} & h_n(Y, B) \\
\downarrow{\phi_{(X,A)}} & & \downarrow{\phi_{(Y,B)}} \\
H_n(X, A; R) & \xrightarrow{f_*} & H_n(Y, B; R)
\end{array}
\qquad
\begin{array}{ccc}
h_{n+1}(X, A) & \xrightarrow{\partial_*} & h_n(A) \\
\downarrow{\phi_{(X,A)}} & & \downarrow{\phi_A} \\
H_{n+1}(X, A; R) & \xrightarrow{\partial_*} & H_n(A; R).
\end{array}
$$

Properties (i)–(vi) are the *Eilenberg-Steenrod axioms* for finite CW pairs. This Uniqueness Theorem, which we shall not prove (see for example [61]) implies that any "homology theory" h_* satisfying these axioms – e.g., singular homology – agrees with cellular homology on oriented finite CW pairs. We have seen in this chapter that cellular homology $H_*(\cdot ; R)$, with the connecting homomorphism ∂_*, satisfies them.

Exercises

1. Let (X, A, B) be a triple of CW complexes. Establish an exact sequence of the form

$$\cdots \longrightarrow H_n(A, B; R) \longrightarrow H_n(X, B; R) \longrightarrow H_n(X, A; R)$$

$$\longrightarrow H_{n-1}(A, B; R) \longrightarrow \cdots$$

2. Discover a Mayer-Vietoris sequence for the triple (X, A, B) in the spirit of Exercise 1.

---

[11] Here $(A, \emptyset) \xrightarrow{i} (X, \emptyset) \xrightarrow{j} (X, A)$ and $h_{-1}(A) = \{0\}$.

2.9 Reduced homology

An *augmented CW complex* is a CW complex X together with an additional cell of dimension -1 called the *empty cell*. If X is a CW complex, the *augmentation* of X is the augmented CW complex so obtained. There is only one empty cell; all augmented CW complexes have at least that cell in common. The empty cell is considered to admit a unique orientation. Thus if X is oriented, its augmentation is oriented too.

The cellular chain complex of the augmentation of X is traditionally called the *augmented cellular chain complex* of X and will be denoted by $(\tilde{C}_n(X;R), \tilde{\partial})$. Here, $\tilde{C}_n(X;R) = C_n(X;R)$ for all $n \neq -1$, $\tilde{C}_{-1}(X;R) = R$ (think of the free R-module generated by the empty cell), $\tilde{\partial} = \partial : C_n(X;R) \to C_{n-1}(X;R)$ for all $n \neq 0$, and[12] $\tilde{\partial}(e_\alpha^0) = \epsilon_\alpha \in \{\pm 1\} \subset R$, where ϵ_α is the chosen orientation[13] of e_α^0. Clearly $\tilde{\partial} \circ \tilde{\partial} = 0$, so $(\tilde{C}_*(X;R), \tilde{\partial})$ is a chain complex. Its homology modules, $\tilde{H}_n(X;R)$, are the *reduced cellular homology modules* of X.

Proposition 2.9.1. $\tilde{H}_n(\emptyset; R) = 0$ if $n \neq -1$, and $\tilde{H}_{-1}(\emptyset; R) \cong R$. When $X \neq \emptyset$, $\tilde{H}_n(X;R) \cong H_n(X;R)$ if $n \neq 0$, and there is a short exact sequence

$$0 \longrightarrow \tilde{H}_0(X;R) \xrightarrow{q_*} H_0(X;R) \longrightarrow R \longrightarrow 0$$

where q_* is induced by the chain map $q : \tilde{C}_*(X;R) \to C_*(X;R)$ which is the identity in dimensions $\neq -1$ and is zero in dimension -1. $\qquad\square$

Combining this with 2.7.1 and 2.7.2 gives:

Corollary 2.9.2. $\tilde{H}_0(X;R)$ is a free R-module with rank α, where $\alpha + 1$ is the (cardinal) number of path components of X. $\qquad\square$

Corollary 2.9.3. If X is path connected, $\tilde{H}_n(X;R) \cong H_n(X;R)$ for all $n \neq 0$ and $\tilde{H}_0(X;R) = 0$. In particular, the reduced homology of a contractible space is trivial. $\qquad\square$

A cellular map $f : X \to Y$ between oriented CW complexes induces $f_\# : \tilde{C}_*(X;R) \to \tilde{C}_*(Y;R)$; $f_\#$ is defined as in Sect. 2.7 when $n \geq 0$, as $\mathrm{id} : R \to R$ when $n = -1$, and trivially when $n < -1$.

Proposition 2.9.4. $f_\# : \tilde{C}_*(X;R) \to \tilde{C}_*(Y;R)$ is a chain map.

---

[12] This homomorphism $C_0(X;R) \to R$ is called the *augmentation* and is usually denoted by ϵ.

[13] Unless there is a reason to do otherwise, one should give the orientation $+1$ to each vertex. In that case the statement $\tilde{\partial}(e_\alpha^0) = 1$ can be interpreted as saying that the empty cell is the boundary of each vertex.

Proof. By 2.3.4, it is only necessary to check the commutativity of

$$
\begin{array}{ccc}
\tilde{C}_0(X;R) & \xrightarrow{\ \tilde{\partial}\ } & R \\
\downarrow{f_\#} & & \downarrow{id} \\
\tilde{C}_0(Y;R) & \xrightarrow{\ \tilde{\partial}\ } & R.
\end{array}
$$

If $f(e_\alpha^0) = \tilde{e}_\beta^0$, where e_α^0 is a cell of X oriented by ϵ_α and $\tilde{e}_\beta^0$ is a cell of Y oriented by $\tilde{\epsilon}_\beta$, then $\tilde{\partial}(e_\alpha^0) = \epsilon_\alpha$, while $\tilde{\partial}f_\#(e_\alpha^0) = \epsilon_\alpha\tilde{\epsilon}_\beta\tilde{\epsilon}_\beta = \epsilon_\alpha$. $\quad\square$

Thus, a cellular map $f : X \to Y$ induces a homomorphism $f_* : \tilde{H}_*(X;R) \to \tilde{H}_*(Y;R)$. Clearly this satisfies $\mathrm{id}_* = \mathrm{id}$ and $(g \circ f)_* = g_* \circ f_*$.

Proposition 2.9.5. *If $f, g : X \to Y$ are homotopic cellular maps between non-empty oriented CW complexes, then $f_* = g_* : \tilde{H}_n(X;R) \to \tilde{H}_n(Y;R)$ for all n.* $\quad\square$

If A is a subcomplex of the oriented CW complex X, the chain map $q : \tilde{C}_*(X;R) \to C_*(X;R)$ defined in 2.9.1 clearly induces an isomorphism of quotient chain complexes $\tilde{C}_*(X;R)/\tilde{C}_*(A;R) \to C_*(X;R)/C_*(A;R)$. Thus there is no need to define relative reduced homology modules. The short exact sequence of chain complexes:

$$
0 \longrightarrow \tilde{C}_*(A;R) \longrightarrow \tilde{C}_*(X;R) \longrightarrow C_*(X;R)/C_*(A;R) \longrightarrow 0
$$

gives the *reduced homology exact sequence* of (X, A):

$$
\cdots \longrightarrow \tilde{H}_n(A;R) \longrightarrow \tilde{H}_n(X;R) \longrightarrow H_n(X, A;R) \longrightarrow \tilde{H}_{n-1}(A;R) \longrightarrow \cdots
$$

This sequence is "natural" with respect to cellular maps $(X, A) \to (Y, B)$, just as in Sect. 2.8.

We get a more elegant version of 2.8.3:

Proposition 2.9.6. $H_n(X, A; R) \cong \tilde{H}_n(X/A; R)$ *for all n.* $\quad\square$

Finally, if $X = A \cup B$ where A and B are two subcomplexes we get a short exact sequence of chain complexes

$$
0 \longrightarrow \tilde{C}_*(A \cap B;R) \longrightarrow \tilde{C}_*(A;R) \oplus \tilde{C}_*(B;R) \longrightarrow \tilde{C}_*(A \cup B;R) \longrightarrow 0
$$

giving a *reduced Mayer-Vietoris sequence*

$$
\cdots \longrightarrow \tilde{H}_n(A \cap B;R) \longrightarrow \tilde{H}_n(A;R) \oplus \tilde{H}_n(B;R) \longrightarrow \tilde{H}_n(X;R) \longrightarrow \tilde{H}_{n-1}(A \cap B;R) \longrightarrow \cdots
$$

as in Sect. 2.8, which is again "natural" with respect to maps $f : X \to X'$ taking A to A' and B to B'.

We repeat that reduced homology of non-empty CW complexes only differs from homology in dimension 0; for the empty CW complex the difference occurs in dimension -1.

A closing word on terminology. If $\tilde{\partial}_0 : \tilde{C}_0(X; R) \to \tilde{C}_{-1}(X; R) := R$ is the 0th boundary homomorphism in the augmented chain complex, it is easy to check that the reduced homology modules $\tilde{H}_*(X; R)$ are also the homology modules of the following *reduced cellular chain complex*:

$$\ldots \longrightarrow C_2(X; R) \longrightarrow C_1(X; R) \xrightarrow{\partial} \ker \tilde{\partial}_0 \longrightarrow 0 \longrightarrow \ldots$$

This explains the name "reduced homology."

Exercise

Let X be a CW complex and let Y be obtained from X by attaching an n-cell to X^{n-1} using the attaching map $f : S^{n-1} \to X^{n-1}$. Write i for $X^{n-1} \hookrightarrow X$ and $\bar{f} = i \circ f : S^{n-1} \to X$. The following formulas express the changes which occur in homology when an n-cell is attached (coefficients omitted):

(i) $\tilde{H}_k(Y) \cong \tilde{H}_k(X)$ when $k \neq n$ or $n - 1$;

(ii) $0 \to \text{image}(\bar{f}_*) \to \tilde{H}_{n-1}(X) \to \tilde{H}_{n-1}(Y) \to 0$ is exact;

(iii) $0 \to \tilde{H}_n(X) \to \tilde{H}_n(Y) \to \ker(\bar{f}_*) \to 0$ is exact

where $\bar{f}_* : H_{n-1}(S^{n-1}) \to H_{n-1}(X)$. Describe the unmarked arrows and prove (i)–(iii). Establish analogous formulas when simultaneous attaching (in the sense of Sect. 1.2) of a set of n-cells occurs. Deduce that finite generation of homology in each dimension is preserved when finitely many new cells are attached in this manner.

3

Fundamental Group and Tietze Transformations

In §3.1 we define and study the "combinatorial" fundamental group of a CW complex and we show how to read off a presentation from the 2-skeleton. As an application, we prove a version of the Seifert-Van Kampen Theorem. We end by showing that the abelianization of the fundamental group is the first homology group. In §3.2 we give a combinatorial treatment of covering spaces suitable for later group theoretic discussions, and we introduce Cayley graphs in this context. In the remaining sections we relate this to the fundamental group as it is usually defined in topology. It is good to understand both points of view. The final two theorems of the chapter, 3.4.9 and 3.4.10, are extremely useful in group theory: they give precise information about $p^{-1}(A)$ where $p : \bar{X} \to X$ is a covering projection and A is a subcomplex of X.

3.1 Combinatorial fundamental group, Tietze transformations, Van Kampen Theorem

Let X be a CW complex. A *non-degenerate edge* of X is an oriented 1-cell e_β^1 of X. If $h_\beta : (B^1, S^0) \to (e_\beta^1, \overset{\bullet}{e}{}_\beta^1)$ is a characteristic map representing the given orientation, then $h_\beta(-1)$ and $h_\beta(1)$ are the *initial point* and the *final point* of e_β^1, respectively; they are vertices of X, and might not be distinct. A *degenerate edge* is a vertex v of X. Its *initial* and *final* points are also v. An *edge* of X is a non-degenerate edge or a degenerate edge.

An *edge path* in X is a finite non-empty sequence of edges $\tau := (\tau_1, \ldots, \tau_k)$ such that the final point of τ_i is the initial point of τ_{i+1} for all i. The initial point of τ_1 is the *initial point* of the edge path, and the final point of τ_k is the *final point* of the edge path.

Edges have *inverses*: if τ_i is non-degenerate, τ_i^{-1} denotes the same 1-cell with the opposite orientation; if τ_i is degenerate, we define $\tau_i^{-1} = \tau_i$. Edge paths have *inverses*: define $(\tau_1, \ldots, \tau_k)^{-1}$ to be the edge path $(\tau_k^{-1}, \ldots, \tau_1^{-1})$.

Certain edge paths can be multiplied. If the final point of τ is also the initial point of σ the *product* edge path is $\tau.\sigma := (\tau_1, \ldots, \tau_k, \sigma_1, \ldots, \sigma_m)$. Whenever a triple product is defined it is obviously associative.

Any edge path has a unique "reduction." Intuitively this is obtained by removing, step by step, a degenerate edge or an adjacent pair τ_i, τ_i^{-1} or τ_i^{-1}, τ_i of edges (unless this leads to the "empty edge path," in which case the reduction is (v) where v is the initial point of the given edge path). To say this in a correct way, choose an orientation for each 1-cell of X, and let F be the free group generated by the set of (oriented) 1-cells. Define a function ϕ from the set of edges of X to F as follows: ϕ sends all degenerate edges to $1 \in F$; if τ_i is a non-degenerate edge whose orientation agrees with the chosen orientation (so that τ_i is a generator of F), ϕ sends τ_i to τ_i and τ_i^{-1} to τ_i^{-1} (the two meanings of τ_i^{-1} are clear). Extend ϕ to a function Φ from the set of edge paths to F, sending $\tau := (\tau_1, \ldots, \tau_k)$ to $\Phi(\tau) := \phi(\tau_1)\phi(\tau_2) \ldots \phi(\tau_k) \in F$. Call τ *reduced* if either $\Phi(\tau)$ is a k-letter word in F, or $k = 1$ and τ_1 is degenerate. By convention $1 \in F$ is a 0-letter word. The *reduction* of an edge path σ is the unique reduced edge path τ such that $\Phi(\sigma) = \Phi(\tau)$. Clearly, the definition of reduction is independent of the orientations we chose for the 1-cells of X in defining F.

An *edge loop* at v is an edge path whose initial and final points are v. A *cyclic edge loop* is an equivalence class of edge loops under cyclic permutation.

We consider an oriented 2-cell e_γ^2. In Sect. 2.6 we defined the "homological boundary" $\partial e_\gamma^2 = \Sigma [e_\gamma^2 : e_\alpha^1] e_\alpha^1$; we saw in 2.5.8 that the only possible non-zero terms in this sum involve 1-cells contained in e_γ^2, and the order in which these 1-cells are considered is irrelevant. Now we define the "homotopical boundary" Δe_γ^2 to be the reduced cyclic edge loop obtained by taking those same 1-cells with their incidence numbers, but in strict cyclic order as one goes around the "edge" of e_γ^2 in the positive direction. This is made precise in the following paragraphs.

Let $h_\gamma : (B^2, S^1) \to (e_\gamma^2, \overset{\bullet}{e}_\gamma^2)$ be a characteristic map representing the chosen orientation on e_γ^2, and let $f = h_\gamma| : S^1 \to X^1$ be the corresponding attaching map. First, suppose $f^{-1}(X^0) \neq \emptyset$. Each path component of $S^1 - f^{-1}(X^0)$ is an open interval mapped by f into some $\overset{\circ}{e}_\beta^1$. By elementary analysis there are countably many such open intervals: label them $\overset{\circ}{I}_j$, $j \geq 1$. Let $e_{\beta_j}^1$ be the unique 1-cell such that $f(\overset{\circ}{I}_j) \subset \overset{\circ}{e}_{\beta_j}^1$.

Proposition 3.1.1. f *maps* $\overset{\circ}{I}_j$ *onto* $\overset{\circ}{e}_{\beta_j}^1$ *for at most finitely many* j.

Proof. Let the 1-cells of X be $\{e_\alpha^1 \mid \alpha \in \mathcal{A}\}$. Pick $z_\alpha \in \overset{\circ}{e}_\alpha^1$. There is an open cover of X^1 consisting of $X^1 - \{z_\alpha \mid \alpha \in \mathcal{A}\} =: U$ and each $\overset{\circ}{e}_\alpha^1$. The open cover $\{f^{-1}(U)\} \cup \{f^{-1}(\overset{\circ}{e}_\alpha^1) \mid \alpha \in \mathcal{A}\}$ of S^1 has a finite subcover $\mathcal{U} = \{f^{-1}(U), f^{-1}(\overset{\circ}{e}_{\alpha_1}^1), \ldots, f^{-1}(\overset{\circ}{e}_{\alpha_n}^1)\}$. If the proposition were false, at least one of the sets $f^{-1}(\overset{\circ}{e}{}^{1}_{\alpha_i})$ could be written as $\bigcup\limits_{k=1}^{\infty} \overset{\circ}{I}_{j_k}$ with all j_k's distinct and no

$\overset{\circ}{I}_{j_k}$ covered by any member of $\mathcal{U}$ other than $f^{-1}(\overset{\circ}{e}{}^{1}_{\alpha_i})$; this is because each $\overset{\circ}{I}_{j_k}$
meets $f^{-1}(z_{\alpha_i})$. Replacing $f^{-1}(\overset{\circ}{e}{}^{1}_{\alpha_i})$, in $\mathcal{U}$, by all the intervals $\overset{\circ}{I}_{j_k}$, we would
have an open cover of S^1 having no finite subcover, a contradiction since S^1
is compact. $\qquad\qquad\qquad\qquad\qquad\qquad\qquad\qquad\qquad\qquad\qquad$ □

For convenience, label the intervals $\overset{\circ}{I}_j$ so that $\overset{\circ}{I}_1,\dots,\overset{\circ}{I}_k$ are those mentioned in 3.1.1 (mapped onto 1-cells). Arrange further that $\overset{\circ}{I}_1,\dots,\overset{\circ}{I}_k$ are
labeled in cyclic order with respect to positive rotation on S^1. First, assume
there is at least one such interval (i.e., $k \geq 1$). For each $1 \leq j \leq k$, we define
a *local incidence number* $i_j = \pm 1$ of e^2_γ with $e^1_{\beta_j}$ along $\overset{\circ}{I}_j$ as the degree of the
self-map of S^1 indicated in the following commutative diagram:

$$
\begin{array}{ccc}
\mathrm{cl}\,\overset{\circ}{I}_j & \xrightarrow{\ f| \ } & e^1_{\beta_j} \\
\Big\downarrow{\scriptstyle p_j} & & \Big\downarrow{\scriptstyle q_j} \\
S^1 & \xrightarrow{\ r_j \ } & e^1_{\beta_j}/\overset{\bullet}{e}{}^1_{\beta_j} \xleftarrow[\cong]{\ h'_{\beta_j}\ } B^1/S^0 \xrightarrow[\cong]{\ k_1\ } S^1.
\end{array}
$$

Here, p_j is the restriction to $\mathrm{cl}\,\overset{\circ}{I}_j$ of a degree 1 map $S^1 \to S^1$ which is constant
on $S^1 - \overset{\circ}{I}_j$ and is injective on $\overset{\circ}{I}_j$; q_j is the quotient map; r_j is the unique map
making the left square commute; h'_{β_j} and k_1 are as in Sect. 2.5. Clearly, $f|$
is homotopic, rel fr $\overset{\circ}{I}_j$, to a map which takes $\overset{\circ}{I}_j$ homeomorphically onto $\overset{\circ}{e}{}^1_{\beta_j}$,
hence r_j is homotopic to a homeomorphism. By 2.4.20, $i_j = \pm 1$, as claimed.

The *homotopical boundary* of the oriented 2-cell e^2_γ is the reduced cyclic
edge loop Δe^2_γ represented by the reduction of the edge loop $(\tau^{i_1}_{\beta_1},\dots,\tau^{i_k}_{\beta_k})$,
where τ_{β_j} is the edge $e^1_{\beta_j}$ with the preferred orientation.

Proposition 3.1.2. $\left(\tau^{i_1}_{\beta_1},\dots,\tau^{i_k}_{\beta_k}\right)$ *is indeed an edge loop, and is unique up
to cyclic permutation.*

Proof. Consider[1] $\overset{\circ}{I}_j$ where $j > k$. Since $f(\overset{\circ}{I}_j)$ is a proper subset of $\overset{\circ}{e}{}^1_{\beta_j}$, $f(\mathrm{fr}\,I_j)$
is a single point of $\overset{\bullet}{e}{}^1_{\beta_j}$. Thus there is a map $g : S^1 \to X^1$, agreeing with f on
$\overset{\circ}{I}_j$ when $j \leq k$, such that $g\left(S^1 - \left(\bigcup\limits_{j=1}^{k}\overset{\circ}{I}_j\right)\right) \subset X^0$. Indeed, adjusting g on

---

[1] It is not to be thought that the labels $> k$ occur in cyclic order: there might be
infinitely many intervals $\overset{\circ}{I}_j$, perhaps infinitely many in each quadrant of S^1.

each $\overset{\circ}{I}_j$ $(1 \leq j \leq k)$ by a homotopy rel fr $\overset{\circ}{I}_j$, we may assume that g maps $\overset{\circ}{I}_j$ homeomorphically onto $\overset{\circ}{e}_{\beta_j}$ when $1 \leq j \leq k$. It is then obvious that we have a cyclic edge loop as claimed.   □

There remain two cases. If $f^{-1}(X^0) = \emptyset$, then there is a single 1-cell e_β^1 such that $f(S^1) \subset \overset{\circ}{e}_\beta^1$, and we define Δe_γ^2 to be the degenerate cyclic edge loop represented solely by the initial point of e_β^1. If $f^{-1}(X^0) \neq \emptyset$ but $k = 0$, there is a single vertex v common to all the 1-cells which meet $f(S^1)$, and we define Δe_γ^2 to be the degenerate cyclic edge loop (v).

Clearly, a change of orientation on the cell e_γ^2 causes an inversion of Δe_γ^2. That Δe_γ^2 depends only on the orientation rather than a specific characteristic map h_γ is an exercise.

We can now define equivalence of edge paths. Let σ and τ be edge paths in the oriented CW complex X having the same initial point and the same final point. Write $\sigma \simeq \tau$ iff either τ is the reduction of σ, or there is an (oriented) 2-cell e_γ^2 such that $\sigma.\tau^{-1} = \lambda.\mu_1.\nu.\nu^{-1}.\mu_2.\lambda^{-1}$ where $\mu_1.\mu_2$ is an edge loop representing the cyclic edge loop Δe_γ^2; see Fig. 3.1. We will call either of these an *elementary equivalence* between σ and τ. The relation $\simeq$ generates an equivalence relation on the set of edge paths in X, which we call *equivalence*.

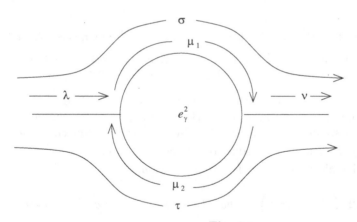

Fig. 3.1.

The next three propositions are left as exercises:

Proposition 3.1.3. *Products of equivalent edge paths are equivalent. Inverses of equivalent edge paths are equivalent.*   □

Proposition 3.1.4. *Let the edge path τ have initial point v_1 and final point v_2. Then τ is equivalent to both $(v_1).\tau$ and $\tau.(v_2)$.*   □

Proposition 3.1.5. *If τ is an edge path with initial point v_1 and final point v_2, then $\tau.\tau^{-1}$ is equivalent to (v_1) and $\tau^{-1}.\tau$ is equivalent to (v_2).*   □

Let v be a vertex of X. Our discussion shows that the set of equivalence classes of edge loops at v is a group whose multiplication is induced by the product operation, whose identity is the equivalence class of v, and whose inversion is induced by the inverse operation. This group is the *fundamental group of X based at v*. It is denoted by $\pi_1(X, v)$.

Proposition 3.1.6. *The definition of equivalence of edge paths is independent of the orientation chosen for X. In particular, $\pi_1(X, v)$ does not depend on the orientation of X.*   □

Proposition 3.1.6 clarifies the role of the chosen orientation of X in this section. We must choose an orientation in order to define equivalence of edge loops at v, but the definition of equivalence turns out to be independent of this choice.

Proposition 3.1.7. *$\pi_1(X, v)$ depends only on the 2-skeleton of the path component of X containing v.*   □

Theorem 3.1.8. *Let X be an oriented CW complex having only one vertex, v. Let F be the free group generated by the set W of (oriented) 1-cells of X, let R be the set of (oriented) 2-cells of X, for each $e_\gamma^2 \in R$ let $\tau := \tau(\gamma)$ be an edge loop representing Δe_γ^2, and let $\rho : R \to F$ be the function[2] $e_\gamma^2 \mapsto \Phi(\tau(\gamma))$. Then $\langle W \mid R, \rho \rangle$ is a presentation of $\pi_1(X, v)$.*

Proof. Let $G = \langle W \mid R, \rho \rangle$. It is enough to define an epimorphism $\Psi : G \to \pi_1(X, v)$ and a function $\Phi' : \pi_1(X, v) \to G$ such that $\Phi' \circ \Psi = \mathrm{id}_G$.

Each edge of X is an edge-loop at v. Define $\psi : W \to \pi_1(X, v)$ to take $\sigma \in W$ to the equivalence class of the edge-loop (σ); ψ extends uniquely to an epimorphism $\Psi' : F \to \pi_1(X, v)$. Fix $e_\gamma^2 \in R$. Let $\tau := \tau(\gamma) = (\tau_1^{i_1}, \ldots, \tau_n^{i_n})$, where each $\tau_j \in W$ and $i_j = \pm 1$. Using the notation introduced earlier in this section, $\Phi(\tau) = \phi(\tau_1)^{i_1} \ldots \phi(\tau_n)^{i_n} \in F$. If the reduced form of this element of F is $\sigma_1^{j_1} \ldots \sigma_k^{j_k}$ (each $\sigma_m \in W$, each $j_m = \pm 1$), then $\Psi'(\Phi(\tau))$ is the equivalence class of the edge loop $(\sigma_1^{j_1}, \ldots, \sigma_k^{j_k})$. This in turn is the reduction of the edge loop $(\tau_1^{i_1}, \ldots, \tau_n^{i_n})$, which is equivalent to the edge loop (v). So $\Psi'(\Phi(\tau)) = 1 \in \pi_1(X, v)$. Thus there is an induced epimorphism $\Psi : G \to \pi_1(X, v)$.

To define Φ', consider $\Phi : \{\text{edge loops in } X \text{ at } v\} \to F$; Φ maps an edge loop and its reduction to the same element of F. If σ and η are two edge loops such that $\sigma.\eta^{-1} = \lambda.\mu_1.\nu.\nu^{-1}.\mu_2.\lambda^{-1}$, as above, where $\mu_1.\mu_2$ represents Δe_γ^2, then $\Phi(\sigma)\Phi(\eta)^{-1} = \Phi(\lambda)\Phi(\mu_1.\mu_2)\Phi(\lambda)^{-1}$ which is conjugate to $\Phi(\tau)$, since τ and $\mu_1.\mu_2$ differ by a cyclic permutation. Thus Φ induces a function $\Phi' : \pi_1(X, v) \to G$ which is clearly a left inverse for Ψ.   □

---

[2] Recall $\Phi : \{\text{edge paths}\} \to F$, above.

Thus, Example 1.2.17 gives a procedure for constructing a 2-dimensional CW complex having just one vertex, and any prescribed group as fundamental group.

Corollary 3.1.9. *Let X be an oriented 1-dimensional CW complex having only one vertex v. Then the (oriented) 1-cells are edge loops whose equivalence classes freely generate $\pi_1(X, v)$.* □

Corollary 3.1.10. *Let X be a CW complex having one vertex v and no 1-cells. Then $\pi_1(X, v) = \{1\}$.* □

If A is a subcomplex of X, with $v \in A$, the inclusion map $i : (A, v) \to (X, v)$ induces an obvious homomorphism $i_\# : \pi_1(A, v) \to \pi_1(X, v)$. The example $X = B^2$, with cells $\{v\}$, S^1 and B^2, and $A = S^1$, shows that ker $i_\#$ need not be trivial (by 3.1.8 and 3.1.9). Indeed, this homomorphism $i_\#$ is a special case of "the homomorphism induced by a map," but that definition is best left until Sect. 3.3.

There is a sense in which $\pi_1(X, v)$ is independent of the base vertex v, provided that X is path connected. Let τ be an edge path in X with initial point v and final point v'; τ induces a function $h_\tau : \pi_1(X, v) \to \pi_1(X, v')$, $[\sigma] \mapsto [\tau^{-1}.\sigma.\tau]$, where $[\sigma] \in \pi_1(X, v)$ denotes the equivalence class of the edge loop σ at v. Clearly, we have

Proposition 3.1.11. *h_τ is well defined and is a group isomorphism whose inverse is $h_{\tau^{-1}}$.* □

The CW complex X is *simply connected* if X is path connected and $\pi_1(X, v)$ is trivial for some (equivalently, any) vertex v of X.

A *tree* is a non-empty CW complex, T, of dimension ≤ 1 in which, given vertices v_1 and v_2 of T, there is exactly one reduced edge path in T with initial vertex v_1 and final vertex v_2.

Proposition 3.1.12. *Every tree is simply connected. Every simply connected CW complex of dimension ≤ 1 is a tree. Every tree is contractible.*

Proof. A tree T is clearly simply connected, for if $(\tau_1, \ldots, \tau_n)$ is a reduced edge loop in T at the vertex v then $n = 1$. Conversely, let X be simply connected and of dimension ≤ 1 and let v be a vertex. In the absence of 2-cells, the equivalence relation on edge loops at v defining $\pi_1(X, v)$ boils down to reduction: no two distinct reduced edge loops are equivalent. So there is only one reduced edge loop. It follows that there is only one reduced edge path from v to each vertex of X.

The fact that trees are contractible follows easily from 1.3.14. □

We say that T *is a tree in* X if T is a tree and is a subcomplex of X; T is a *maximal tree* if there is no larger subcomplex of X which is a tree. If X is path connected it follows that a maximal tree in X contains X^0 and, conversely, any tree in X containing X^0 is maximal.

Proposition 3.1.13. *Let (X, A) be a CW pair with X and A path connected. Let T_A be a maximal tree in A. There is a maximal tree, T_X, in X such that $T_X \cap A = T_A$.*

Proof. Let $\mathcal{T}$ denote the set of trees T in X such that $T \cap A = T_A$. This set $\mathcal{T}$ is non-empty and is partially ordered by inclusion. Let $\{T_i\}$ be a linearly ordered subset of $\mathcal{T}$ and let $\tilde{T} = \bigcup_i T_i$. Clearly, $\tilde{T} \in \mathcal{T}$. By Zorn's Lemma, $\mathcal{T}$ contains a maximal element, T_X, as required. $\qquad\qquad\qquad\qquad$ $\square$

Next, we review the rules for altering a presentation of a group. Let $\langle W \mid R, \rho \rangle$ be a presentation of the group[3] G. A *Tietze transformation of Type I* replaces $\langle W \mid R, \rho \rangle$ by $\langle W \mid R', \rho' \rangle$ where: S is a set, $\psi : S \to N(\rho(R)) \subset F(W)$ is a function, $R' = R \coprod S$, and $\rho' : R' \to F(W)$ agrees with ρ on R and with ψ on S. A *Tietze transformation of Type II* replaces $\langle W \mid R, \rho \rangle$ by $\langle W' \mid R', \rho' \rangle$, where: S is a set, $W' = W \coprod S$, $\psi : S \to F(W) \subset F(W')$ is a function, $R' = R \coprod S$, and $\rho' : R' \to F(W')$ agrees with (inclusion) $\circ \rho$ on R, and[4] with $s \mapsto s.\psi(s)^{-1}$ on S.

Example 3.1.14. Starting with $\langle a \mid a^3 \rangle$, a presentation of the cyclic group of order 3, a Type I transformation might change this to $\langle a \mid a^3, a^6 \rangle$; a Type II transformation might change this latter presentation to $\langle a, b, c, \mid a^3, a^6, ba, ca^2 \rangle$.

The following is well-known (see [107] or [106] for a proof):

Theorem 3.1.15. *Tietze transformations of either type applied to a presentation of G yield another presentation of G. Conversely, if $\langle W_1 \mid R_1, \rho_1 \rangle$ and $\langle W_2 \mid R_2, \rho_2 \rangle$ are presentations of G, then there are presentations and Tietze transformations as follows:*

$$\langle W_1 | R_1, \rho_1 \rangle \xLongrightarrow{\text{Type II}} \langle W | R', \rho' \rangle \xLongrightarrow{\text{Type I}} \langle W | R, \rho \rangle \xLongleftarrow{\text{Type I}} \langle W | R'', \rho'' \rangle \xLongleftarrow{\text{Type II}} \langle W_2 | R_2, \rho_2 \rangle.$$

Moreover, if the given presentations are finite, then the intervening presentations can be chosen to be finite. $\qquad\qquad\qquad\qquad$ $\square$

We are now ready to generalize 3.1.8 by explaining how to read off a presentation of the fundamental group of any path connected CW complex from the 2-skeleton. Note that (by 3.1.13 applied with A a one-vertex subcomplex) such complexes always contain maximal trees.

Theorem 3.1.16. *Let X be an oriented path connected CW complex, let T be a maximal tree in X and let v be a vertex of X. Let F be the free group*

---

[3] Recall the notation for presentations introduced in 1.2.17.

[4] Note that by our conventions $S \subset F(S) \subset F(W')$.

generated by the set W of (oriented) 1-cells of X. Let R be the set of (oriented) 2-cells of X and let S be the set of (oriented) 1-cells of T. For each $e_\gamma^2 \in R$, let $\tau := \tau(\gamma)$ be an edge loop in X representing $\Delta(e_\gamma^2)$. Let $\rho : R \coprod S \to F$ take each $e_\gamma^2 \in R$ to $\Phi(\tau(\gamma))$, and take each $\sigma \in S$ to the one-letter word $\sigma \in F$. Then $\langle W \mid R \coprod S, \rho \rangle$ is a presentation of $\pi_1(X, v)$.

Proof. We claim $\pi_1(X, v) \cong \pi_1(X/T, \bar{v}) \cong \langle W - S \mid R, \bar{\rho} \rangle \cong \langle W \mid R \coprod S, \rho \rangle$, where $\bar{v}$ is the only vertex of X/T, $q : F(W) \to F(W - S)$ is the epimorphism of free groups sending w to w when $w \in W - S$, and w to 1 when $w \in S$, and $\bar{\rho} = q \circ \rho \mid: R \to F(W - S)$.

There is an obvious epimorphism $\alpha : \pi_1(X, v) \to \pi_1(X/T, \bar{v})$. The proof that α is an isomorphism boils down to showing that if Δe_γ^2 is represented by the edge loop $(\tau_1^{i_1}, \ldots, \tau_n^{i_n})$ in X, then the edge loop in X/T obtained by deleting those edges τ_j which lie in T represents $\Delta \bar{e}_\gamma^2$, where $\bar{e}_\gamma^2$ is the 2-cell of X/T corresponding to e_γ^2, which is clear. This is our first isomorphism. The second isomorphism comes from 3.1.8. The third isomorphism comes from Tietze transformations

$$\langle W - S \mid R, \bar{\rho} \rangle \xrightarrow{\text{Type II}} \langle W \mid R \coprod S, \rho' \rangle \xrightarrow{\text{Type I}} \langle W \mid (R \coprod S) \coprod R, \rho'' \rangle \xleftarrow{\text{Type I}} \langle W \mid S \coprod R, \rho \rangle$$

where $\rho' = \bar{\rho}$ on R, $\rho' = \mathrm{id}$ on S, $\rho'' = \rho'$ on $R \coprod S$, and $\rho'' = \rho$ on the second copy of R. $\qquad\square$

Corollary 3.1.17. *Let X be a path connected CW complex. If X^1 is finite, then $\pi_1(X, v)$ is finitely generated. If X^2 is finite, then $\pi_1(X, v)$ is finitely presented.* $\qquad\square$

Another corollary of 3.1.16 expresses the fundamental group of a path connected CW complex in terms of the fundamental groups of path connected subcomplexes:

Theorem 3.1.18. (Seifert-Van Kampen Theorem)[5] *Let the CW complex X have subcomplexes X_1, X_2 and X_0 such that $X = X_1 \cup X_2$ and $X_0 = X_1 \cap X_2$. Assume X_1, X_2 and X_0 are path connected. Let $v \in X_0$ and let $i_1 : X_0 \hookrightarrow X_1$, $i_2 : X_0 \hookrightarrow X_2$, $j_1 : X_1 \hookrightarrow X$, and $j_2 : X_2 \hookrightarrow X$ be the inclusions. Then[6] $(j_{1\#}, j_{2\#}) : \pi_1(X_1, v) * \pi_1(X_2, v) \to \pi_1(X, v)$ is an epimorphism whose kernel is the normal closure of $\{i_{1\#}(g) i_{2\#}(g)^{-1} \mid g \in \pi_1(X_0, v)\}$.*

Proof. By 3.1.13, there is a maximal tree T_X such that $T_{X_i} := X_i \cap T_X$ is a maximal tree in X_i for $i = 0, 1, 2$. Let $\langle W_X \mid R_X \coprod S_X, \rho_X \rangle$ be a presentation of $\pi_1(X, v)$ as in 3.1.16. Using restrictions of the items in this presentation,

---

[5] Theorem 6.2.11 is a useful generalization of this.
[6] Notation: $G * H$ denotes the free product of G and H.

form presentations $\langle W_{X_i} \mid R_{X_i} \coprod S_{X_i}, \rho_{X_i} \rangle$ for $\pi_1(X_i, v)$, $i = 0, 1, 2$. Abusing notation, we have obvious homomorphisms

$$\langle W_{X_0} \mid R_{X_0} \coprod S_{X_0}, \rho_{X_0} \rangle \xrightarrow{\;\;i_{1\#}\;\;} \langle W_{X_1} \mid R_{X_1} \coprod S_{X_1}, \rho_{X_1} \rangle$$

$$i_{2\#} \downarrow \qquad\qquad\qquad\qquad\qquad \downarrow j_{1\#}$$

$$\langle W_{X_2} \mid R_{X_2} \coprod S_{X_2}, \rho_{X_2} \rangle \xrightarrow{\;\;j_{2\#}\;\;} \langle W_X \mid R_X \coprod S_X, \rho_X \rangle.$$

Note that $W_{X_0} \subset W_{X_i} \subset W_X$, $R_{X_0} \subset R_{X_i} \subset R_X$, $S_{X_0} \subset S_{X_i} \subset S_X$, for $i = 1, 2$. FOR THIS PROOF ONLY we introduce a convention. Let $\tilde{W}_{X_0}$, $\tilde{R}_{X_0}$ and $\tilde{S}_{X_0}$ be copies of the sets W_{X_0}, R_{X_0} and S_{X_0} respectively. If $\sigma \in W_{X_0}$ we denote the corresponding element of $\tilde{W}_{X_0}$ by $\tilde{\sigma}$, etc.

The desired result is obtained by applying two Tietze transformations:

$$\langle W_X \mid R_X \coprod S_X, \rho_X \rangle \xrightarrow{\text{Type II}} \langle W_X \coprod \tilde{W}_{X_0} \mid R_X \coprod S_X \coprod \tilde{W}_{X_0}, \rho' \rangle \xrightarrow{\text{Type I}}$$

$$\langle W_X \coprod \tilde{W}_{X_0} \mid R_X \coprod S_X \coprod \tilde{W}_{X_0} \coprod (\tilde{R}_{X_0} \coprod \tilde{S}_{X_0}), \rho'' \rangle$$

where $\rho' = (\text{inclusion}) \circ \rho_X$ on $R_X \coprod S_X$, and, for each $\tilde{\sigma} \in \tilde{W}_{X_0}$, $\rho'(\tilde{\sigma}) = \tilde{\sigma}.\sigma^{-1}$; $\rho'' = \rho'$ on $R_X \coprod S_X \coprod \tilde{W}_{X_0}$, and, for each $\tilde{\tau} \in \tilde{R}_{X_0} \coprod \tilde{S}_{X_0}$, $\rho''(\tilde{\tau}) = \rho_X(\tau) \in F(W_X) \subset F(W_X \coprod \tilde{W}_{X_0})$.

The latter presentation can be rewritten as

$$\langle W_{X_1} \coprod (W_X - W_{X_1}) \coprod \tilde{W}_{X_0} \mid R_{X_1} \coprod (R_X - R_{X_1}) \coprod S_{X_1}$$

$$\coprod (S_X - S_{X_1}) \coprod \tilde{W}_{X_0} \coprod (\tilde{R}_{X_0} \coprod \tilde{S}_{X_0}), \rho'' \rangle.$$

If we identify $(W_X - W_{X_1}) \coprod \tilde{W}_{X_0} = W_{X_2}$, $(R_X - R_{X_1}) \coprod \tilde{R}_{X_0} = R_{X_2}$, $(S_X - S_{X_1}) \coprod \tilde{S}_{X_0} = S_{X_2}$, and $W_{X_0} =$ the appropriate subset of W_{X_1}, then this becomes:

$$\langle W_{X_1} \coprod W_{X_2} \mid R_{X_1} \coprod S_{X_1} \coprod R_{X_2} \coprod S_{X_2} \coprod \tilde{W}_{X_0}, \bar{\rho} \rangle$$

where $\bar{\rho} = (\text{inclusion}) \circ \rho_{X_i}$ on $R_{X_i} \coprod S_{X_i}$ for $i = 1$ and 2, and for each $\tilde{\sigma} \in \tilde{W}_{X_0}$, $\bar{\rho}(\tilde{\sigma}) = \tilde{\sigma}.\sigma^{-1}$. This is clearly a presentation of

$$\langle W_{X_1} \mid R_{X_1} \coprod S_{X_1}, \rho_{X_1} \rangle * \langle W_{X_2} \mid R_{X_2} \coprod S_{X_2}, \rho_{X_2} \rangle / N$$

where N is the normal closure of $\{i_{1\#}(g) i_{2\#}(g)^{-1} \mid g \in \langle W_{X_0} \mid R_{X_0} \coprod S_{X_0}, \rho_{X_0} \rangle \}$. $\square$

Some notation is useful here. If G_1 and G_2 are groups and if S is a subset of the free product $G_1 * G_2$, then we write $\langle G_1, G_2 \mid S \rangle$ for a presentation of $G_1 * G_2 / N(S)$. The notation indicates that generating sets for G_1 and G_2 are on the left of the vertical bar, and relation data for G_1 and G_2 are on its right, together with the additional relations S. In this notation, the conclusion of the Seifert-Van Kampen Theorem 3.1.18 can be rewritten[7]:

---

[7] In the Appendix to this section we discuss this and other variants of "presentation".

$$\pi_1(X, v) \cong \langle \pi_1(X_1, v), \pi_1(X_2, v) \mid i_{1\#}(g) i_{2\#}(g)^{-1}, \ \forall \ g \in \pi_1(X_0, v) \rangle$$

In the special case where $i_{1\#}$ and $i_{2\#}$ are monomorphisms this becomes the free product with amalgamation which is defined in Sect. 6.2.

To end this section, we point out a variation on 2.7.4 showing the relationship between $\pi_1(X, v)$ and $H_1(X; \mathbb{Z})$. Let $\tau := (\tau_1^{i_1}, \ldots, \tau_k^{i_k})$ be an edge loop in X at v where τ_j has the preferred orientation (X is oriented) and $i_j = \pm 1$. Then $\tilde{h}(\tau) := \sum_{j=1}^{k} i_j \tau_j$ is a 1-cycle in X.

Theorem 3.1.19. *This function $\tilde{h}$ induces a homomorphism $h : \pi_1(X, v) \to H_1(X; \mathbb{Z})$ whose kernel is the commutator subgroup of $\pi_1(X, v)$. If X is path connected, h is an epimorphism.*

Proof. Clearly, h is a well defined homomorphism. We may assume that X is path connected: otherwise we could work with the path component containing v. We first deal with the special case in which X has only one vertex. Then the result follows from 3.1.8, since the effect of abelianizing the group $\langle W \mid R, \rho \rangle$ in that theorem is to produce the abelian[8] group $H_1(X; \mathbb{Z})$.

In the general case, pick a maximal tree, T, in X. Consider the diagram:

$$
\begin{array}{ccc}
\pi_1(X, v) & \xrightarrow[\cong]{\alpha} & \pi_1(X/T, \bar{v}) \\
\downarrow{\scriptstyle h} & & \downarrow{\scriptstyle h'} \\
H_1(X; \mathbb{Z}) & \xrightarrow{q_*} & H_1(X/T; \mathbb{Z})
\end{array}
$$

Here, α is the isomorphism discussed in the proof of 3.1.16, h' is the version of h for X/T, and q_* is the homomorphism induced by the (cellular) quotient map $q : X \to X/T$. The diagram clearly commutes. By the special case, h' is an epimorphism whose kernel is the commutator subgroup. It only remains to show that q_* is an isomorphism. The discussion of relative homology in Sect. 2.8 and Sect. 2.9 shows that the following diagram commutes and the top line is exact:

$$
\begin{array}{ccccccc}
H_1(T; \mathbb{Z}) & \longrightarrow & H_1(X; \mathbb{Z}) & \xrightarrow{j_\#} & H_1(X, T; \mathbb{Z}) & \longrightarrow & \tilde{H}_0(T; \mathbb{Z}) \\
& & & \searrow^{q_*} & \uparrow{\scriptstyle \cong} & & \\
& & & & H_1(X/T; \mathbb{Z}) & &
\end{array}
$$

By 3.1.12, $j_\#$ is an isomorphism, hence also q_*. $\qquad\square$

Thus, for path connected X, $H_1(X; \mathbb{Z})$ is the *abelianization* of $\pi_1(X, *)$.

The homological, or abelianized, version of the Seifert-van Kampen theorem is:

---

[8] The free abelian group generated by W is $Z_1(X; \mathbb{Z})$, and the abelianization of Δe_γ^2 is ∂e_γ^2, so the result is $Z_1(X; \mathbb{Z})/B_1(X; \mathbb{Z})$.

Proposition 3.1.20. *With hypotheses as in 3.1.18, $j_{1*} + j_{2*} : H_1(X_1; \mathbb{Z}) \oplus H_1(X_2; \mathbb{Z}) \to H_1(X; \mathbb{Z})$ is an epimorphism whose kernel is $\{i_{1*}(z) - i_{2*}(z) \mid z \in H_1(X_0; \mathbb{Z})\}$.*

Proof (First Proof). Apply 3.1.18 and 3.1.19.   □

Proof (Second Proof). The Mayer-Vietoris sequence (Sect. 2.8) gives an exact sequence

$$H_1(X_0; \mathbb{Z}) \xrightarrow{(i_{1*}, -i_{2*})} H_1(X_1; \mathbb{Z}) \oplus H_1(X_2; \mathbb{Z}) \xrightarrow{j_{1*} + j_{2*}} H_1(X; \mathbb{Z}) \xrightarrow{\partial_*}$$

$$H_0(X_0; \mathbb{Z}) \xrightarrow{(i_{1*}, -i_{2*})} H_0(X_1; \mathbb{Z}) \oplus H_0(X_2; \mathbb{Z})$$

The rightmost arrow is a monomorphism, by 2.7.2 and the definition of i_{1*}. Hence $\partial_* = 0$. The result follows by exactness.   □

Appendix: Presentations

We defined a "presentation of G" in 1.2.17. However, the term is also used in subtly different (though roughly equivalent) ways.

Given G, a subset $S \subset G$ and a set T of products of elements of $S \cup S^{-1}$ each of which is obviously trivial in G, one may read that $\langle S \mid T \rangle$ is a *presentation* of G. Strictly speaking this is nothing new since $T \subset F(S)$. But it can cause confusion. For example, one says that

$$\left\langle \begin{bmatrix} 0 & -1 \\ 1 & 0 \end{bmatrix}, \begin{bmatrix} 0 & -1 \\ 1 & 1 \end{bmatrix} \ \middle| \ \begin{bmatrix} 0 & -1 \\ 1 & 0 \end{bmatrix}^4, \begin{bmatrix} 0 & -1 \\ 1 & 0 \end{bmatrix}^2 \begin{bmatrix} 0 & -1 \\ 1 & 1 \end{bmatrix}^{-3} \right\rangle$$

is a presentation of $SL_2(\mathbb{Z})$. This really means that there is an isomorphism[9] $\langle a, b \mid a^4, a^2 b^{-3} \rangle \to SL_2(\mathbb{Z})$ taking a to $\begin{bmatrix} 0 & -1 \\ 1 & 0 \end{bmatrix}$ and b to $\begin{bmatrix} 0 & -1 \\ 1 & 1 \end{bmatrix}$.

Another way of writing "presentations" is this: Let G be a group, W a set disjoint from G, and R a subset of $F(G \cup W)$. Then elements of $F(G \cup W)$ are "words" in the "alphabet" $G \cup W$. By saying $\langle G, W \mid R \rangle$ "is a presentation of the group H" one means: letting $\langle S \mid T' \rangle$ be a presentation of G (i.e., $T \subset G$ as above), $\langle S \cup W \mid T' \cup R' \rangle$ is a presentation of H, where $T' \cup R' \subset F(S \cup W)$ with T' mapping bijectively onto $T \subset G$ and R' mapping bijectively onto R. We give two examples:

Example 3.1.21. If G is a group with subgroups A and B and if $\phi : A \to B$ is an isomorphism, the HNN extension of G by ϕ (see Sect. 6.2) has a presentation $\langle G, t \mid t^{-1} a t \phi(a)^{-1} \ \forall \ a \in A \rangle$. Here $W = \{t\}$, abbreviated to t, and $R = \{t^{-1} a t \phi(a)^{-1} \mid a \in A\}$.

---

[9] There is obviously such an epimorphism; that it is an isomorphism requires proof.

An obvious variant replaces G by more than one group:

Example 3.1.22. If G_1, G_2 and A are groups and $\phi_i : A \rightarrowtail G_i$ are monomorphisms, the free product with amalgamation of G_1 and G_2 across A (see Sect. 6.2) has a presentation $\langle G_1, G_2 \mid \phi_1(a)\phi_2(a)^{-1} \ \forall \ a \in A \rangle$.

Many authors prefer to write relations as equations. For example, they would write the presentation given above for $SL_2(\mathbb{Z})$ as $\langle a, b \mid a^4 = 1, a^2 = b^3 \rangle$. Our previous presentation $\langle a \mid a^3 \rangle$ for $\mathbb{Z}/3\mathbb{Z}$ would instead be written $\langle a \mid a^3 = 1 \rangle$. And the presentations in the last two examples would be written

$$\langle G, t \mid t^{-1}at = \phi(a) \ \forall a \in A \rangle$$

and

$$\langle G_1, G_2 \mid \phi_1(a) = \phi_2(a) \ \forall a \in A \rangle.$$

Exercises

1. Prove 3.1.3–3.1.5.
2. Prove that our definition of Δe_γ^2 only depends on the orientation and not on the particular characteristic map h_γ. (Material in Sect. 3.4 may be helpful here.)
3. Describe the abelianizations of the groups in Examples 3.1.21 and 3.1.22 in the common language of abelian groups.
4. Prove that there are only countably many isomorphism classes of finitely presented groups.[10] We will see explicit examples of finitely generated groups which are not finitely presented in Sect. 8.3.
5. Show using Tietze transformations that the following presentations define the same group: (i) $\langle a, b, c \mid ab = ba, c^2 = b \rangle$; (ii) $\langle x, y \mid t^{-1}xt = y, t^{-1}yt = x \rangle$. (See Exercise 3 of Sect. 4.3 for another way of looking at this question.)

3.2 Combinatorial description of covering spaces

Let Y be a CW complex. An *automorphism* of Y is a homeomorphism $h : Y \to Y$ such that whenever e is a cell of Y so is $h(e)$. The group of all automorphisms of Y will be denoted by Aut Y: the multiplication is composition, $(h' \circ h)(y) = h'(h(y))$.

Let G be a group. A *G-CW complex* is a CW complex Y together with a homomorphism $\alpha : G \to$ Aut Y. We write $g.y$ for $\alpha(g)(y)$. The homomorphism α is also called a *cell-permuting left action* of G on Y. The *stabilizer* of $A \subset Y$ is $G_A = \{g \in G \mid g(A) = A\}$. The action is *free*, and Y is a *free G-CW complex*, if the stabilizer of each cell is trivial. The action is *rigid*, and Y is a *rigid G-CW complex*, if the stabilizer of each cell, e, acts trivially on e. There

---

[10] The interest here is that there are uncountably many isomorphism classes of finitely generated groups – see [106, p. 188].

is a sense (explained at the end of Sect. 4.1) in which "rigid G-CW complex" rather than "arbitrary G-CW complex" is the natural equivariant analog of "CW complex".[11]

An action of G defines an equivalence relation on Y: $y_1 = y_2$ mod G iff there is some $g \in G$ such that $g.y_1 = y_2$. The set of equivalence classes (also called *orbits*) is denoted by $G\backslash Y$: it is given the quotient topology.[12]

Proposition 3.2.1. *Let Y be a free G-CW complex. The quotient map $q :$ $Y \to G\backslash Y =: X$ is a covering projection, and X admits the structure of a CW complex whose cells are $\{q(e) \mid e$ is a cell of $Y\}$.*

Proof. Let $X^n = \cup\{q(e) \mid e$ is a cell of Y having dimension $\leq n\}$. X^0 is discrete. $X = \bigcup_n X^n$. If $\mathcal{A}$ indexes the n-cells of Y, the G-action on Y induces a G-action on $\mathcal{A}$. Write $\mathcal{B} = G\backslash\mathcal{A}$. Consider the following diagram:

$$
\begin{array}{ccc}
B^n(\mathcal{A}) \coprod Y^{n-1} & \xrightarrow{\;p\;} & Y^n \\
\downarrow{r} & & \downarrow{q|} \\
B^n(\mathcal{B}) \coprod X^{n-1} & \xrightarrow{\;p'\;} & X^n
\end{array}
$$

Here, $B^n(\mathcal{A})$ is regarded as a G-space, where $g \in G$ maps B^n_α to $B^n_{g.\alpha}$ by the "identity". Thus $B^n(\mathcal{A}) \coprod Y^{n-1}$ is a G-CW complex. The map p is the quotient map described in 1.2.1; it can be chosen to satisfy $g.p(x) = p(g.x)$ for all $x \in B^n(\mathcal{A}) \coprod Y^{n-1}$ by defining p to agree with a characteristic map h_α on B^n_α for one α in each G-orbit of $\mathcal{A}$, and then defining p on the other B^n_α's using the group action. The map r agrees with q on Y^{n-1} and maps B^n_α homeomorphically onto B^n_β, where $\beta \in \mathcal{B}$ is the orbit of α. Since r is a quotient map (see Sect. 1.1), there is a map p' making the diagram commute. Moreover, $p' \mid X^{n-1} = $ inclusion, $p'(S^{n-1}(\mathcal{B})) \subset X^{n-1}$, and p' maps $B^n(\mathcal{B}) - S^{n-1}(\mathcal{B})$ homeomorphically onto $X^n - X^{n-1}$. The map p' is a quotient map, since the other three maps in the diagram are quotient maps. So X^n is obtained from X^{n-1} by attaching n-cells. Since q is a quotient map and Y is a CW complex, X has the weak topology with respect to $\{X^n\}$. Thus X is a CW complex.

Next we describe the inductive step in the proof that q is a covering projection. Suppose U_{n-1} is an open subset of X^{n-1} having the property that $q^{-1}(U_{n-1}) = q^{-1}(U_{n-1}) \cap Y^{n-1} = \cup\{g.V_{n-1} \mid g \in G\}$, where V_{n-1} is open in Y^{n-1}, and $g_1.V_{n-1} \cap g_2.V_{n-1} = \emptyset$ whenever $g_1 \neq g_2$. Then certainly U_{n-1} is evenly covered by $q \mid Y^{n-1}$. Let $W_{n-1} = p^{-1}q^{-1}(U_{n-1}) \cap B^n(\mathcal{A})$, let $S_{n-1} = p^{-1}(V_{n-1}) \cap B^n(\mathcal{A})$, and let $T_{n-1} = (p')^{-1}(U_{n-1}) \cap B^n(\mathcal{B})$. Then

---

[11] Looking ahead, and using terminology from Sect. 5.3, we find a useful way of rigidifying: when Y is a regular G-CW complex, its "first derived" $|\mathrm{sd}\, Y|$ is a rigid G-CW complex.

[12] This action α is a left action of G on Y, hence the notation $G\backslash Y$; we reserve Y/G for right actions.

T_{n-1} is evenly covered by W_{n-1} (via r) and $r \mid: S_{n-1} \to T_{n-1}$ is a homeomorphism. The sets S_{n-1} and W_{n-1} are open in $S^{n-1}(\mathcal{A})$, while T_{n-1} is open in $S^{n-1}(\mathcal{B})$. Choose S_n open in $B^n(\mathcal{A})$ so that $S_n \cap S^{n-1}(\mathcal{A}) = S_{n-1}$. Write $W_n = \bigcup\{gS_n \mid g \in G\}$. Then W_n is open in $B^n(\mathcal{A})$ and $T_n = r(W_n) = r(S_n)$ is open in $B^n(\mathcal{B})$, while $T_n \cap S^{n-1}(\mathcal{B}) = T_{n-1}$. We leave it as an exercise to show that S_n can be so chosen that $gS_n \cap S_n = \emptyset$ whenever $g \in G$. Define $V_n = p(S_n \cup V_{n-1})$ and $U_n = p'(T_n \cup U_{n-1})$. Then U_n is open in X^n and is evenly covered by $p(W_n \cup V_{n-1}) = \bigcup\{gV_n \mid g \in G\}$.

To see that q is a covering projection, let $x \in X$. There is a unique m and a unique cell e_δ^m such that $x \in \overset{\circ}{e}{}_\delta^{\,m}$. We have $q^{-1}(\overset{\circ}{e}{}_\delta^{\,m}) = \cup\{g.\overset{\circ}{e}{}_{\gamma_\delta}^{\,m} \mid g \in G\}$ where $e_{\gamma_\delta}^m$ is an arbitrarily chosen m-cell over e_δ^m. Clearly, $g_1.\overset{\circ}{e}{}_\gamma^{\,m} \cap g_2.\overset{\circ}{e}{}_\gamma^{\,m} = \emptyset$ whenever $g_1 \neq g_2$. Write $U_m = \overset{\circ}{e}{}_\delta^{\,m}$. By induction, choose $U_m \subset U_{m+1} \subset \cdots$ as above. Let $U = \bigcup_{k \geq m} U_k$. Then U is an open subset of X evenly covered by q. $\qquad\square$

Neither part of 3.2.1 need hold if the action is not free: for example, take $G = \mathbb{Z}_2$, $Y = [-1, 1]$ with 0-cells at ± 1 and one 1-cell, and let the non-trivial element of G act on Y by $t \mapsto -t$. However, we have:

Proposition 3.2.2. *Let Y be a rigid G-CW complex and let $q : Y \to G\backslash Y$ be the quotient map. Then $G\backslash Y$ admits a CW complex structure whose cells are $\{q(e) \mid e$ is a cell of $Y\}$.*

Proof. Similar to the first part of the proof[13] of 3.2.1. $\qquad\square$

The map q in 3.2.2 is not, in general, a covering projection (unlike the situation in 3.2.1): for example, take $G = \mathbb{Z}_2$, $Y = [-1, 1]$ with the CW structure consisting of 0-cells at -1, 0 and 1, and 1-cells $[-1, 0]$ and $[0, 1]$; let the non-trivial element of G act on Y by $t \mapsto -t$.

We will sometimes refer to the n-cells of X (in 3.2.1 and 3.2.2) as the *n-cells of Y mod G*. We say Y is *finite mod G* if $G\backslash Y$ is a finite CW complex.[14]

Our next task is to show that when Y is a simply connected free G-CW complex, and v is a vertex of $X := G\backslash Y$, then $\pi_1(X, v) \cong G$. For this we must define lifts of edge paths in X to edge paths in Y. If τ_1 is an edge of X with initial point $v = q(\tilde{v})$, there is a unique edge $\tilde{\tau}_1$ of Y with initial point $\tilde{v}$ which maps to τ_1, by 2.4.6 and 2.4.7. Call $\tilde{\tau}_i$ the *lift of τ_i at $\tilde{v}$*. By induction we define the *lift* of an edge path $\tau := (\tau_1, \ldots, \tau_k)$ to be the unique edge path $\tilde{\tau} := (\tilde{\tau}_1, \ldots, \tilde{\tau}_k)$ with initial point $\tilde{v}$ such that $q(\tilde{\tau}) = \tau$.

Pick a vertex $\tilde{v} \in Y$ with $q(\tilde{v}) = v$. Assuming Y simply connected, define $\chi : G \to \pi_1(X, v)$ as follows: choose an edge path $\tilde{\tau}$ in Y from $\tilde{v}$ to $g.\tilde{v}$, and let $\chi(g)$ be the element of $\pi_1(X, v)$ represented by the edge loop $q(\tilde{\tau})$. Path connectedness ensures that $\tilde{\tau}$ exists.

---

[13] The reader should consider why the argument would fail without rigidity.

[14] Alternatively, one says that G acts *cocompactly* on Y.

Proposition 3.2.3. χ *is well defined and is an isomorphism.*

Proof. We first show that χ is well defined. Let $\tilde{\sigma}$ and $\tilde{\tau}$ be edge paths from $\tilde{v}$ to $g.\tilde{v}$. Since Y is simply connected, $\tilde{\sigma}$ and $\tilde{\tau}$ are equivalent. (To see this [using $\simeq$ for equivalence] note that $\sigma \simeq \sigma.(g.\tilde{v}) \simeq \sigma.(\sigma^{-1}.\tau) = (\sigma.\sigma^{-1}).\tau \simeq \tau$; the first and last equivalences are elementary, the other comes from 3.1.3.) Each elementary equivalence on the way from $\tilde{\sigma}$ to $\tilde{\tau}$ induces an elementary equivalence on the way from $q(\tilde{\sigma})$ to $q(\tilde{\tau})$. Hence $[q(\tilde{\sigma})] = [q(\tilde{\tau})]$.

χ is obviously a homomorphism. Let $\chi(g) = 1$. If $\tilde{\tau}$ is an edge path in Y from $\tilde{v}$ to $g.\tilde{v}$, the corresponding edge loop τ at v is equivalent to the degenerate edge loop v. If two loops at v differ by an elementary equivalence, their lifts at $\tilde{v}$ have the same final point. Hence this also holds if they are merely equivalent. It follows that $g.\tilde{v} = \tilde{v}$, hence $g = 1$. So χ is a monomorphism. But χ is clearly onto, for given an edge loop τ at v in X, let $g.\tilde{v}$ be the final point of $\tilde{\tau}$. Then $\chi(g)$ is the element of $\pi_1(X, v)$ represented by τ. $\square$

We immediately conclude, using 3.1.17:

Theorem 3.2.4. *Let Y be a simply connected free G-CW complex such that there are only finitely many 1-cells [resp. 1-cells and 2-cells] mod G. Then G is finitely generated [resp. finitely presented].* $\square$

We have seen in 3.2.3 that the quotient of a simply connected free G-CW complex has fundamental group G. Conversely, given a path connected CW complex X, we now show how to construct a simply connected free $\pi_1(X, v)$-CW complex $\tilde{X}$ having quotient X. This is the "universal cover" construction. It is common to define $\tilde{X}$ as a quotient space of a function space – an efficient but non-constructive procedure. We prefer to construct $\tilde{X}$ as a CW complex, skeleton by skeleton. We shall see in Chaps. 13, 14 and 16 that even the 1-skeleton and 2-skeleton of $\tilde{X}$, as constructed here, exhibit interesting "end" invariants of the group $\pi_1(X, v)$, so the work involved in the construction will be worthwhile.

Let X be a path connected CW complex. Choose an orientation for X, a maximal tree $T \subset X$, and a vertex $v \in X$ as base point. Write $\pi = \pi_1(X, v)$. Give π the discrete topology.

Let $\tilde{X}^0 = \pi \times X^0$, let $p_0 : \tilde{X}^0 \to X^0$ be projection on the X^0-factor, and let π act on $\tilde{X}^0$ by $\bar{g}.(g, v_\alpha) = (\bar{g}g, v_\alpha)$. This π-action is free and p_0 is its quotient map; p_0 is a covering projection. Pick as base vertex $\tilde{v} := (1, v) \in \pi \times X^0 = \tilde{X}^0$.

Next, we define the 1-skeleton $\tilde{X}^1$. Part of the 1-skeleton is $\pi \times T$, but we will attach more 1-cells. Let the (already oriented) 1-cells of X which are not in T be $\{e_\beta^1 \mid \beta \in \mathcal{B}\}$. For each $\beta \in \mathcal{B}$, let $g_\beta \in \pi$ be the element represented by the edge loop $\lambda.e_\beta^1.\mu^{-1}$, where λ and μ are the unique reduced edge paths in T from v to the initial and final points of e_β^1; see the proof of 3.1.12. By 3.1.16, the elements g_β generate π. Pick a characteristic map $h_\beta : (B^1, S^0) \to (e_\beta^1, \overset{\bullet}{e}_\beta^1)$ representing the chosen orientation; let $f_\beta : S^0 \to X^0$ be the corresponding attaching map. For each $\beta \in \mathcal{B}$ and each $g \in \pi$ attach a 1-cell $e_{\beta,g}^1$ to $\pi \times T$

by the attaching map $f_{\beta,g} : S^0 \to \tilde{X}^0 \subset \pi \times T$, $-1 \mapsto (g, h_\beta(-1))$ and $1 \mapsto$

$(gg_\beta, h_\beta(1))$. The resulting CW complex is $\tilde{X}^1 = \left((\pi \times T) \amalg \coprod_{\beta,g} B^1_{\beta,g} \right) /{\sim}$,

where $\sim$ is defined in the obvious way by the maps $f_{\beta,g}$. Note that $\pi \times T$ is a

subcomplex of $\tilde{X}^1$. Now, $X^1 = \left(T \amalg \coprod_\beta B^1_\beta \right) /{\sim}$ where $\sim$ is defined by the

maps f_β. The map $(\pi \times T) \amalg \coprod_{\beta,g} B^1_{\beta,g} \to T \amalg \coprod_\beta B^1_\beta$ which is "projection onto T"

on $\pi \times T$ and is "identity": $B^1_{\beta,g} \to B^1_\beta$ on $B^1_{\beta,g}$ induces a map $p_1 : \tilde{X}^1 \to X^1$

extending p_0. For each $\bar{g} \in \pi$, the self-homeomorphism $\tilde{d}_{\bar{g}}$ of $(\pi \times T) \amalg \coprod_{\beta,g} B^1_{\beta,g}$

which is $(g, u) \mapsto (\bar{g}g, u)$ on $\pi \times T$ and is "identity": $B^1_{\beta,g} \to B^1_{\beta,\bar{g}g}$ on $B^1_{\beta,g}$

induces a self-homeomorphism $d_{\bar{g}}$ of $\tilde{X}^1$. Clearly, $p_1 \circ d_{\bar{g}} = p_1$. Moreover, the

homomorphism $\bar{g} \mapsto d_{\bar{g}}$ makes $\tilde{X}^1$ into a free π-CW complex containing the

previously defined π-CW complex $\tilde{X}^0$ as a π-subcomplex. By 3.2.1, p_1 is a

covering projection, and the cells of X^1 are the p_1-images of the cells of $\tilde{X}^1$.

It is easy to check that $\tilde{X}^1$ is path connected.

   We remark that if X has just one vertex, then $T = \{v\}$ and our construc-

tion of $\tilde{X}^1$ is called the *Cayley graph*[15] of π with respect to the generators

$\{g_\beta\}$: a vertex for each element of π, and an edge joining g to gg_β for each

$g \in \pi$ and each $(g, \beta) \in G \times \mathcal{B}$.

Proposition 3.2.5. *Let τ be an edge loop at $v \in X$, and let $\tilde{\tau}$ be the lift of*
τ with initial point $(g, v) \in \tilde{X}^1$. The final point of $\tilde{\tau}$ is $(g\bar{g}, v)$ where $\bar{g}$ is the
element of π represented by τ. In particular, either every lift of τ is an edge
loop, or none is.

Proof. Let $(g, u) \in \pi \times X^0 = (\tilde{X}^1)^0$. A non-degenerate edge τ_i in T from u to
w lifts to an edge in $\tilde{X}^1$ from (g, u) to (g, w). If τ_β is the edge e^1_β ($\beta \in \mathcal{B}$) with
the preferred orientation, having initial point u and final point w, τ_β lifts to
an edge of $\tilde{X}^1$ with initial point (g, u) and final point (gg_β, w); τ_β^{-1} lifts to an
edge of $\tilde{X}^1$ with initial point (g, w) and final point (gg_β^{-1}, u). Applying these
remarks inductively to $\tau := (\tau_1^{i_1}, \ldots, \tau_k^{i_k})$, we see that $\tilde{\tau}$ has initial point (g, v)
and final point $(g\bar{g}, v)$ as claimed.[16]   □

---

[15] The Cayley graph of a group with respect to a finite set of generators is an
   important construction in group theory. We will see that the number of ends of
   this graph is a quasi-isometry invariant (Sect. 18.2) and gives information about
   the structure of the group (Sect. 13.5). It is the basis for the "word metric" on
   the group (Sect. 9.1) and its geometry determines whether or not the group is
   "hyperbolic." Some examples are discussed in the Appendix.

[16] To simplify notation, some details are omitted here: when τ_i is in T the π-
   coordinate is unchanged; when $\tau_i = \tau_\beta^{i_\beta}$ the π-coordinate is right multiplied by

Let $h_\gamma : (B^2, S^1) \to (e_\gamma^2, \dot{e}_\gamma^2)$ be a characteristic map representing the given orientation of e_γ^2. Let $f_\gamma : S^1 \to X^1$ be the corresponding attaching map. By 3.2.5, any representative edge loop μ_γ of the cyclic edge loop Δe_γ^2 lifts to an edge loop in $\tilde{X}^1$. From this we deduce:

Proposition 3.2.6. *There are maps $\tilde{f}_\gamma : S^1 \to \tilde{X}^1$ such that $p_1 \circ \tilde{f}_\gamma = f_\gamma$. If $\tilde{f}_\gamma$ is one such, then the others are $d_g \circ \tilde{f}_\gamma$ where $g \in \pi$.*

Proof. Let $\mu_\gamma = (\tau_1^{i_1}, \ldots, \tau_m^{i_m})$ where τ_j has the chosen orientation, and $i_j = \pm 1$. Let K be the CW complex structure on S^1 having vertices at the mth roots of unity (compare 1.2.17). By 1.4.2 and 3.1.1, f_γ is homotopic to a map $f' : S^1 \to X^1$ such that the restriction of f' maps the interior of the jth 1-cell of K homeomorphically onto $\overset{\circ}{\tau}_j$ with orientation indicated by i_j; or else $m = 1$, τ_1 is degenerate and f' is constant. By 3.2.5, f' lifts. By 2.4.7, if $\tilde{f}'$ is one lift, then the others are $d_g \circ \tilde{f}'$ where $g \in \pi$. By 2.4.6, the same is true of $\tilde{f}_\gamma$. $\qquad\square$

We now define $\tilde{X}^2$ and $p_2 : \tilde{X}^2 \to X^2$. For each 2-cell e_γ^2 of X choose h_γ, as above; choose a lift $\tilde{f}_\gamma$ of f_γ, and, for each $g \in \pi$, attach a 2-cell $e_{\gamma,g}^2$ to $\tilde{X}^1$ by the attaching map $d_g \circ \tilde{f}_\gamma$. The resulting CW complex is $\tilde{X}^2 = \left(\tilde{X}^1 \amalg \coprod_{\gamma,g} B_{\gamma,g}^2 \right) / \sim$ where $\sim$ is defined by the maps $d_g \circ \tilde{f}_\gamma$. Then $X^2 = \left(X^1 \amalg \coprod_\gamma B_\gamma^2 \right) / \sim$, where $\sim$ comes from the maps f_γ. Just as before, the map $\tilde{X}_1 \amalg \coprod_{\gamma,g} B_{\gamma,g}^2 \to X_1 \amalg \coprod_\gamma B_\gamma^2$ which is p_1 on $\tilde{X}_1$ and is "identity": $B_{\gamma,g}^2 \to B_\gamma^2$ on $B_{\gamma,g}^2$ induces a map $p_2 : \tilde{X}^2 \to X^2$ extending p_1. And, just as before, the free π-action on $\tilde{X}^1$ extends to make $\tilde{X}^2$ into a free π-CW complex for which p_2 is the quotient map. By 3.2.1, p_2 is a covering projection, and the cells of X^2 are the p_2-images of the cells of $\tilde{X}^2$.

Theorem 3.2.7. *$\tilde{X}^2$ is simply connected.*

Proof. Let $\tilde{\tau}$ be an edge loop in $\tilde{X}^2$ at $\tilde{v}$. Then $\tau := p_2(\tilde{\tau})$ is an edge loop in X at v. By 3.2.5, τ represents $1 \in \pi_1(X, v)$. Recall that equivalence of edge loops is defined in terms of elementary equivalences each of which is either a reduction or a formal move across a 2-cell (see Fig. 3.1). We say that τ is of distance $\leq n$ from the trivial edge loop, (v), if it is possible to pass from τ to (v) by n elementary equivalences. We prove, by induction on n, that if τ is of distance $\leq n$ from (v) then $\tilde{\tau}$ is of distance $\leq n$ from $(\tilde{v})$. If $n = 0$, $\tau = (v)$ and $\tilde{\tau} = (\tilde{v})$. The induction is completed by observing that if τ differs from σ by

---

$g_\beta^{i_\beta}$. The resulting $\bar{g}$ is indeed the element of π represented by τ, as we saw in the proof of 3.1.16.

one elementary equivalence, then the same is true of the difference between $\tilde{\tau}$ and $\tilde{\sigma}$. Indeed, when the difference is a reduction, this is clear. When the difference is an elementary equivalence across a 2-cell e_γ^2, this follows from the fact that if μ is an edge loop representing Δe_γ^2 then for each $g \in \pi$, some lift of μ represents $\Delta e_{\gamma,g}^2$. $\qquad\qquad\qquad\square$

A *covering transformation* (or *deck transformation*) of a covering projection $p : E \to B$ is a homeomorphism $d : E \to E$ such that $p \circ d = p$. The covering transformations form a group of homeomorphisms (with composition as the group multiplication).

In the present case, $p_2 : \tilde{X}^2 \to X^2$ is a covering projection, and the elements g of π give rise to covering transformations $d_g : t \mapsto g.t$. Since p_2 is the quotient map of the π-action, these are the only covering transformations. Note that when $g \neq \bar{g} \in \pi$, $d_g \neq d_{\bar{g}}$.

Here is a general property of covering projections, which we are about to use (compare 3.3.4):

Proposition 3.2.8. *Let $p : E \to B$ be a covering projection, let $n \geq 2$, and let $g : S^n \to B$ be a map. There is a map $\tilde{g} : S^n \to E$ such that $p \circ \tilde{g} = g$ (call such $\tilde{g}$ a "lift" of g) and any other lift of g has the form $d \circ \tilde{g}$ where d is a covering transformation of p.* $\qquad\qquad\qquad\square$

Now we are ready to define $p : \tilde{X} \to X$ extending p_2. By induction, assume that for some $n \geq 2$, a free π-CW complex $\tilde{X}^n$ has been defined whose quotient map is $p_n : \tilde{X}^n \to X^n$. As above, we denote by $d_g : \tilde{X}^n \to \tilde{X}^n$ the covering transformation corresponding to $g \in \pi$. By 3.2.1, the cells of X^n are the p_n-images of the cells of $\tilde{X}^n$. Choose a characteristic map $h_\delta : (B^{n+1}, S^n) \to (e_\delta^{n+1}, \overset{\bullet}{e}_\delta^{n+1})$ representing the given orientation of each $(n + 1)$-cell, e_δ^{n+1}, of X. Let $f_\delta : S^n \to X^n$ be the corresponding attaching map. By 3.2.8, there is a lift $\tilde{f}_\delta : S^n \to \tilde{X}^n$, and all lifts have the form $d_g \circ \tilde{f}_\delta$ where $g \in \pi$. Attach an $(n+1)$-cell $e_{\delta,g}^{n+1}$ to $\tilde{X}^n$ by the attaching map $d_g \circ \tilde{f}_\delta$, for each $g \in \pi$. The resulting CW complex is $\tilde{X}^{n+1} = \left(\tilde{X}^n \amalg \coprod_{\delta,g} B_{\delta,g}^{n+1} \right) / \sim$ where $\sim$ is defined by the maps $d_g \circ \tilde{f}_\delta$. $X^{n+1} = \left(X^n \amalg \coprod_{\delta} B_\delta^{n+1} \right) / \sim$ where $\sim$ comes from the maps f_δ. We define $p_{n+1} : \tilde{X}^{n+1} \to X^{n+1}$ extending p_n, and we extend the free π-action on $\tilde{X}^n$ to a free π-action on $\tilde{X}^{n+1}$ just as before. This is easily seen to complete the induction. Let $\tilde{X} = \bigcup_n \tilde{X}^n$. Define $p : \tilde{X} \to X$ by $p \mid \tilde{X}^n = p_n$.

By 3.1.7 and 3.2.7, $\tilde{X}$ is simply connected. Summarizing:

Proposition 3.2.9. *Given a path connected oriented CW complex X, a vertex $v \in X$, and a maximal tree T in X, the above construction yields a simply connected free $\pi_1(X, v)$-CW complex $\tilde{X}$ and a covering projection $p : \tilde{X} \to X$*

which is the quotient map of the $\pi_1(X, v)$-*action. Moreover, the cells of* X *are the* p-*images of the cells of* $\tilde{X}$.   □

We will review in Sect. 3.3 the well-known fact that this action can be defined in a purely topological manner.

Remark on Notation. For $n \geq 2$, $(\tilde{X})^n = (X^n)^\sim$. But for $n = 0$ or 1, these can be different; for example, consider $X = B^2$. In ambiguous cases $\tilde{X}^n$ will always mean $(\tilde{X})^n$.

Propositions 3.2.9 and 3.1.8 imply:

Corollary 3.2.10. *For any group* G, *there exists a simply connected free* G-*CW complex.*   □

We will recall in Sect. 3.3 that the simply connected covering space $\tilde{X}$ is, in a certain sense, unique and is a covering space of all other path connected covering spaces of X. Anticipating that, we call $\tilde{X}$ the *universal cover* of X (remembering that our particular construction of $\tilde{X}$ appears to depend on many choices).

Knowing $\tilde{X}$, we can easily construct a path connected covering space of X with any subgroup H of $\pi_1(X, v)$ as fundamental group. Let $\bar{X}(H) = H \backslash \tilde{X}$. Consider the following diagram:

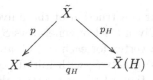

Here, p_H is the quotient map. The construction of $\tilde{X}$ gave that space a natural base point $\tilde{v} = (1, v)$. We give $\bar{X}(H)$ the base point $\bar{v} = p_H(\tilde{v})$.

Proposition 3.2.11. *There is a map* q_H *making this diagram commute. Both* p_H *and* q_H *are covering projections.* $\bar{X}(H)$ *admits a CW complex structure whose cells are the* p_H-*images of the cells of* $\tilde{X}$. *The cells of* X *are the* q_H-*images of the cells of* $\bar{X}(H)$. $\pi_1(\bar{X}(H), \bar{v}) \cong H$.

Proof. There is obviously a function q_H making the diagram commute; p_H is a quotient map by definition, so q_H is continuous. It is not hard to show that an open subset of X evenly covered by p is evenly covered by q_H. By 3.2.1, the CW complex structures on $\bar{X}(H)$ and X are as claimed. The isomorphism of $\pi_1(\bar{X}(H), \bar{v})$ and H comes from 3.2.3.   □

Here is a well-known corollary.

Theorem 3.2.12. *Every subgroup of a free group is free.*[17]

---

[17] This, together with Exercise 6, is the Nielsen-Schreier Subgroup Theorem.

Proof. Let F be a free group and let $H \leq F$ be a subgroup. By 3.1.8, there is a 1-dimensional CW complex X (having exactly one vertex v) such that $\pi_1(X, v)$ is isomorphic to F. Form the covering space $\bar{X}(H)$; by 3.2.11, it is a CW complex whose fundamental group is H. Being a covering space of a 1-dimensional complex, $\bar{X}(H)$ is 1-dimensional. By 3.1.16, H is free.   □

If we let $G = \pi_1(X, v)$ and consider the subgroup $H \leq G$, we may ask: what special property does $\bar{X}(H)$ exhibit when the index $[G : H]$ of H in G is finite? To answer this, we observe that the isomorphism χ of 3.2.3 actually defines a bijection between G and the set $p^{-1}(v) \subset \bar{X}$, under which $g \in G$ is mapped to $g.\tilde{v} \in p^{-1}(v)$. The action of H on $p^{-1}(v)$ partitions $p^{-1}(v)$ into equivalence classes in bijective correspondence with the cosets $\{Hg \mid g \in G\}$. These are also in bijective correspondence with $q_H^{-1}(v) \subset \bar{X}(H)$, where $q_H : \bar{X}(H) \to X$ is as in 3.2.11. This proves:

Proposition 3.2.13. *If $[G : H] = n \leq \infty$ then the covering projection $q_H :$ $\bar{X}(H) \to X$ is an n to 1 function. If X is a finite CW complex, then $\bar{X}(H)$ is finite iff H has finite index in G.*   □

Appendix: Cayley graphs

Cayley graphs arose in our construction of the universal cover, but they deserve further discussion. Given a set of generators S for a group G, the associated Cayley graph has a vertex for each $g \in G$, and a non-degenerate edge for each ordered pair $(g, s) \in G \times S$, with initial point g and final point gs. Thus, if we give the orientation $+1$ to each vertex, the Cayley graph, denoted here by $\Gamma(G, S)$, is an oriented graph.

If $1 \in S$ then $\Gamma(G, S)$ includes a loop at each vertex. If $s \in S$ has order 2, then two edges (g, s) and (gs, s) join the vertex g to the vertex gs, one in either direction.[18] The graph $\Gamma(G, S)$ is a G-graph under the action of G on the vertex set (also G) by left translation. In fact, as we have noted, this is a free action with quotient graph $G\backslash\Gamma(G, S)$ having one vertex and an edge for each member of S.

The pictures in Fig. 3.2 are intended to give some insight into Cayley graphs. We name the groups by presentations, but it is the named generators which determine the Cayley graph:

(a) $G = \{1\}$, $S = \emptyset$.
(b) $G = \{1\}$, $S = \{1\}$.
(c) $G = \langle x \mid x^2 \rangle$, $S = \{x\}$.
(d) $G = \langle x \mid x^2 \rangle$, $S = \{1, x\}$.
(e) $G = \langle x \mid x^3 \rangle$, $S = \{x\}$.

---

[18] Some authors disregard orientation in $\Gamma(G, S)$ and include only one edge from g to gs when s has order 2.

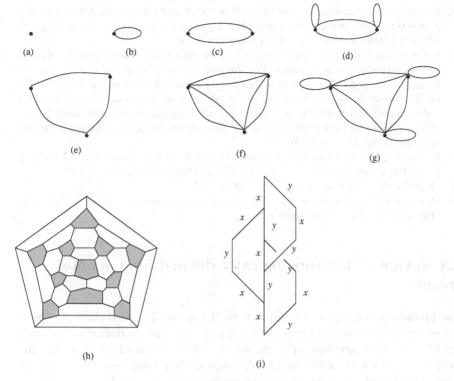

Fig. 3.2.

(f) $G = \langle x \mid x^3 \rangle$, $S = \{x, x^2\}$.

(g) $G = \langle x \mid x^3 \rangle$, $S = \{1, x, x^2\}$.

(h) $G = \langle x, y \mid x^5, y^2, (xy)^3 \rangle$, $S = \{x, y\}$.

(In this picture the sides of the pentagons are x-edges; the others are y-edges and should be double since y has order 2; the reader should compare this picture with the pattern on a soccer ball.)

(i) $G = \langle x, y \mid xyx = yxy \rangle$, $S = \{x, y\}$.

The group in (h) is the alternating group A_5. The group in (i) is the (infinite) three-strand braid group B_3; only a small portion of the Cayley graph is shown since these groups are infinite and so have infinite Cayley graphs.

Exercises

1. Fill in the missing details of the inductive step in the proof of Proposition 3.2.1.

2. Prove that an automorphism of a CW complex maps n-cells onto n-cells, and that the covering projection in 3.2.1 maps the interior of an n-cell homeomorphically onto the interior of an n-cell.

3. Prove: Let τ be an edge loop at the vertex $v \in X$ and let $\bar{\tau}$ be the lift of τ to the 1-skeleton $\bar{X}(H)^1$ with initial point (Hg, v); then the final point of $\bar{\tau}$ is $(Hg\bar{g}, v)$ where $\bar{g}$ is the element of $\pi_1(X, v)$ represented by τ.

4. Let $\sim$ be the equivalence relation on S^n whose equivalence classes consist of pairs of diametrically opposite points. The quotient space $S^n/\!\sim$ is *real n- dimensional projective space* $\mathbb{R}P^n$. This is also the quotient space of the action of $\mathbb{Z}_2 = \langle t \mid t^2 \rangle$ on S^n, where $t.x = -x$. Find a $\mathbb{Z}_2$-CW complex structure on S^n and hence a CW-complex structure on $\mathbb{R}P^n$. ($\mathbb{R}P^2$ is called the *[real] projective plane*).

5. What is the fundamental group of $\mathbb{R}P^n$? Answer this using covering space theory and also using 3.1.8 or 3.1.16.

6. Prove that if F is a free group of rank n and $H \leq F$ is a subgroup of index j then H is a free group of rank $1 + j(n - 1)$. (Compare Theorem 3.2.12.)

7. Describe the universal cover of $(S^1 \vee S^1) \times S^1$.

8. Describe the universal cover of the presentation complex of $\langle x, t \mid t^{-1}xtx^{-2} \rangle$. (For more on this see Example 6.2.10.)

3.3 Review of the topologically defined fundamental group

The fundamental group as defined in Sect. 3.1 is usually called the "edge path group" or "combinatorial fundamental group" because its definition involves the CW complex structure of X rather than just the topology of X. In this section, we review the more usual "topological" definition and its elementary properties. Proofs of all assertions can be found in [74] or [146].

Let (Y, y) be a pointed space. A *loop* in Y at y is a map $\omega : (I^1, \dot{I}^1) \to (Y, y)$; its *inverse* is ω^{-1}, the loop $t \mapsto -t \overset{\omega}{\mapsto} \omega(-t)$. The *product* $\omega_1.\omega_2$ of two loops ω_1 and ω_2 is the loop defined by $t \mapsto 2t + 1 \overset{\omega_1}{\mapsto} \omega_1(2t + 1)$ when $t \leq 0$ and by $t \mapsto 2t - 1 \overset{\omega_2}{\mapsto} \omega_2(2t - 1)$ when $t \geq 0$. Two loops ω_1 and ω_2 (in Y at y) are *homotopic* if the maps of pairs $\omega_1, \omega_2 : (I^1, \dot{I}^1) \to (Y, y)$ are homotopic. The *trivial loop* is the loop taking I^1 to y. Products of homotopic loops are homotopic. Inverses of homotopic loops are homotopic. If ω is a loop, $\omega.\omega^{-1}$ and $\omega^{-1}.\omega$ are homotopic to the trivial loop. If ω_1 is trivial, ω_2 is homotopic both to $\omega_1.\omega_2$ and to $\omega_2.\omega_1$.

Let $\pi_1(Y, y)$ be the set of homotopy classes of loops in Y at y. If $[\omega]$ denotes the homotopy class of the loop ω, then the pairing $([\omega_1], [\omega_2]) \mapsto [\omega_1.\omega_2]$ defines a multiplication on $\pi_1(Y, y)$. This multiplication is associative. If we define $[\omega]^{-1} = [\omega^{-1}]$ and $1 = [\text{trivial loop}]$, $\pi_1(Y, y)$ becomes a group, the *fundamental group of the space Y with base point y*. We will see in the next section that when X is a CW complex the two definitions of fundamental group agree.

If ω is a loop in Y at y and $f : (Y, y) \to (Z, z)$ is a map, then $f \circ \omega$ is a loop in Z at z. If ω_1 and ω_2 are homotopic loops in Y at y, then $f \circ \omega_1$ and $f \circ \omega_2$ are homotopic loops in Z at z.

Let ω be a loop in Y at y, let $f, g : (Y, y) \to (Z, z)$ be maps, let $F : Y \times I \to Z$ be a homotopy such that $F_0 = f$ and $F_1 = g$, and let $\alpha : (I^1, \dot{I}^1) \to (Z, z)$ be the loop $t \mapsto F(y, \frac{1}{2}(t+1))$. Then $f \circ \omega$ is homotopic to $\alpha.(g \circ \omega).\alpha^{-1}$ rel $\dot{I}^1$. Define $f_\# : \pi_1(Y, y) \to \pi_1(Z, z)$ by $f_\#([\omega]) = [f \circ \omega]$. This $f_\#$ is well defined, and if $g : (Y, y) \to (Z, z)$ is such that $f \simeq g$ rel $\{y\}$, then $f_\# = g_\#$. Moreover, $f_\#$ is a homomorphism.

If $(Y, y) \xrightarrow{f} (Z, z) \xrightarrow{g} (W, w)$ are maps, $(g \circ f)_\# = g_\# \circ f_\#$, and $(\text{id}_Y)_\# = \text{id}$. Thus π_1 is a covariant functor from the category of pointed spaces and homotopy classes rel base point to the category of groups and homomorphisms. Nevertheless, in most cases of interest, the homotopy invariance can be expressed without restriction to base point preserving homotopies. In order to say this precisely, we define a base point $y \in Y$ to be *good* if the pair $(Y, \{y\})$ has the homotopy extension property with respect to any space. We need not dwell on the pathological situations under which a base point might fail to be good. The important case for us is an immediate consequence of 1.3.15, namely:

Proposition 3.3.1. *If X is a CW complex and v is a vertex of X, then v is a good base point for X.*   □

The desired homotopy invariance theorem is:

Proposition 3.3.2. *Let $f : (Y, y) \to (Z, z)$ be a map such that $f : Y \to Z$ is a homotopy equivalence, and let z be a good base point. Then $f_\# : \pi_1(Y, y) \to \pi_1(Z, z)$ is an isomorphism.*[19]

Proof. Let $g : Z \to Y$ be a homotopy inverse for f. Let α be a path in Y from $g(z)$ to y. Define $H : (Z \times \{0\}) \cup (\{z\} \times I) \to Y$ by $H(x, 0) = g(x)$, for $z \in Z$, and $H(z, t) = \alpha(t)$. H extends to a map $\bar{H} : Z \times I \to Y$. Let $\bar{g} = \bar{H}(\cdot, 1)$. Then $\bar{g}$ is also a homotopy inverse for f, and $\bar{g}(z) = y$. By our previous remarks, $(\bar{g} \circ f)_\#([\omega]) = [\beta].[\omega].[\beta]^{-1}$ and $(f \circ \bar{g})_\#([\sigma]) = [\gamma].[\sigma].[\gamma]^{-1}$ for suitable $[\beta] \in \pi_1(Y, y)$ and $[\gamma] \in \pi_1(Z, z)$. Thus $f_\#$ is an isomorphism.   □

Let $p : \tilde{Y} \to Y$ be a covering projection, where $\tilde{Y}$ is path connected. Let $\tilde{y} \in \tilde{Y}$ and let $y = p(\tilde{y})$. Let G be the group of covering transformations. Assuming $\pi_1(\tilde{Y}, \tilde{y})$ trivial, define $\chi' : G \to \pi_1(Y, y)$ as follows: choose a path $\tilde{\omega}$ in $\tilde{Y}$ from $\tilde{y}$ to $g.\tilde{y}$, and let $\chi'(g)$ be the element of $\pi_1(Y, y)$ represented by the loop $p \circ \tilde{\omega}$. Path connectedness ensures that $\tilde{\omega}$ exists, and the triviality of $\pi_1(\tilde{Y}, \tilde{y})$ ensures that $\chi'(g)$ is well defined. By analogy with 3.2.3 we have:

Proposition 3.3.3. *χ' is an isomorphism. Its inverse is explicitly described as follows: let $[\omega] \in \pi_1(Y, y)$, let $\tilde{z} \in \tilde{Y}$, and let $\tilde{\tau}$ be a path in $\tilde{Y}$ from $\tilde{y}$ to $\tilde{z}$; $(\chi')^{-1}([\omega])(\tilde{z}) = \widetilde{(\tau^{-1}\omega\tau)}(1)$ where $\widetilde{(\tau^{-1}\omega\tau)}$ is the unique lift of $\tau^{-1}\omega\tau$ with initial point $\tilde{z}$.*   □

---

[19] For variations on 3.3.2, see Exercise 2, and Remark 4.1.6.

The explicit description of $(\chi')^{-1}$ in 3.3.3 is important, for it shows how $\pi_1(Y, y)$ acts on $\tilde{Y}$ on the left. We have already seen this action in the CW complex context in 3.2.9. There, the action appeared to depend on the CW complex structure, on orientations of the cells, and on a maximal tree. Now we see that the action depends on none of these. However, it does depend on the choice of base point $\tilde{y} \in Y$ such that $p(\tilde{y}) = y$. (For more on this, see Exercise 1.)

Finally, we recall the important "lifting criterion" for covering spaces, a special case of which we met in 3.2.8:

Proposition 3.3.4. *[Lifting Criterion] Let $p : (E, e) \to (B, b)$ be a covering projection, let Y be a path connected, locally path connected space, and let $y \in Y$. A map $f : (Y, y) \to (B, b)$ lifts to a map $\bar{f} : (Y, y) \to (E, e)$ iff $f_\#(\pi_1(Y, y)) \le p_\#(\pi_1(E, e))$. If $\bar{f}$ exists, it is unique.* $\qquad\square$

We have reviewed the abstract part of covering space theory (here and in 2.4.6 and 2.4.7). The other part of the theory consists of existence and classification theorems. Since we have built the universal cover of a CW complex in Sect. 3.2, we will get that part of the theory (for CW complexes) at essentially no cost in effort. It will be given in the next section.

Exercises

1. If $p : \tilde{Y} \to Y$ is a covering projection, the group of covering transformations G (being a group of homeomorphisms) acts on the left on $\tilde{Y}$; this is independent of the choice of $\tilde{y}$. But the isomorphism χ' depends on $\tilde{y}$, so the corresponding left action of $\pi_1(Y, y)$ on $\tilde{Y}$, given by 3.3.3, depends on $\tilde{y}$. For $[\omega] \in \pi_1(Y, y)$ and $\tilde{z} \in p^{-1}(y)$ define $\tilde{z}.[\omega] = \tilde{\omega}(1)$ where $\tilde{\omega}(0) = \tilde{z}$ and $\tilde{\omega}$ covers ω. Prove this defines a right action of $\pi_1(Y, y)$ on $p^{-1}(y)$ which is independent of $\tilde{y}$. Write down explicitly the corresponding right action of G on $p^{-1}(y)$.
2. Prove that a 2-equivalence induces an isomorphism of fundamental groups (this generalizes 3.3.2).
3. Let $f, g : (X, x) \to (Y, y)$ be pointed maps and let $H : f \simeq g$ be a homotopy between f and g. Let ω be the loop $t \mapsto H(x, t)$. Prove that $f_\#$ and $g_\# : \pi_1(X, x) \to \pi_1(Y, y)$ are conjugate homomorphisms, more precisely that $f_\#([\tau]) = [\omega]g_\#([\tau])[\omega]^{-1}$.

3.4 Equivalence of the two definitions of the fundamental group of a CW complex

Now we will show that our two definitions of fundamental group agree. Let X be an oriented CW complex and let v be a vertex of X. Just until after Theorem 3.4.1, we will denote by $\pi_1^{edge}(X, v)$ the "edge path" fundamental group defined in Sect. 3.1, and by $\pi_1^{top}(X, v)$ the "topological" fundamental group defined in Sect. 3.3. If $\omega_1, \omega_2 : I^1 \to X$ are maps such that $\omega_1(1) = \omega_2(-1)$, their *product* $\omega_1.\omega_2 : I^1 \to X$ is defined by the same formula used for loops, namely, $t \mapsto \omega_1(2t+1)$ when $t \leq 0$, and $t \mapsto \omega_2(2t-1)$ when $t \geq 0$. We call $\omega_1, \omega_2, \omega_1.\omega_2$, etc., "paths" even though the domain is I^1 rather than I. A characteristic map $h : (I^1, \overset{\bullet}{I}{}^1) \to (e^1_\alpha, \overset{\bullet}{e}{}^1_\alpha)$ of a 1-cell e^1_α defines a path $h : I^1 \to X$. For each non-degenerate edge τ_i (= oriented 1-cell) in X, pick a characteristic map h_{τ_i} for the underlying 1-cell representing the orientation of τ_i; regard h_{τ_i} as a path in X. For each degenerate edge τ_i, let h_{τ_i} be the constant path at the point τ_i. With each edge loop $\tau := (\tau_1, \ldots, \tau_k)$ at v, associate the product path $h_\tau := (\ldots((h_{\tau_1}.h_{\tau_2}).h_{\tau_3}).\ldots).h_{\tau_k}$. Thus h_τ is a loop at v, a *parametrization of* τ.

Theorem 3.4.1. *This association induces an isomorphism* $\alpha : \pi_1^{edge}(X, v) \to \pi_1^{top}(X, v)$.

Proof. We claim $\pi_1^{top}(\tilde{X}, v)$ is trivial. Thus the isomorphism χ' of 3.3.3 is defined. By 3.2.3 and 3.2.9, the isomorphism χ is well defined. Let $\alpha = \chi' \circ \chi^{-1}$. Then α is indeed induced by the association $\tau \mapsto h_\tau$.

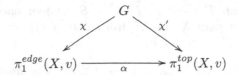

It remains to prove the claim. Let $\omega : (I^1, \overset{\bullet}{I}{}^1) \to (\tilde{X}, \tilde{v})$ be a loop. By 1.4.3, ω is homotopic to a loop in $\tilde{X}^1$. Clearly, any loop in $\tilde{X}^1$ is homotopic to a loop of the form h_τ for some edge loop τ in $\tilde{X}$ at $\tilde{v}$. By 3.2.7, τ can be transformed into the trivial edge loop by elementary equivalences. If σ is a reduction of τ, it is clear that the loops h_σ and h_τ are homotopic. If σ differs from τ by a formal move across a 2-cell, so that (in the notation of Sect. 3.1) $\sigma.\tau^{-1} = \lambda.\mu_1.\nu.\nu^{-1}.\mu_2.\lambda^{-1}$, then $h_\sigma.h_{\tau^{-1}} \simeq h_{\sigma.\tau^{-1}} \simeq h_\lambda.h_{\mu_1.\mu_2}.h_{\lambda^{-1}}$, where $\mu_1.\mu_2$ is an edge loop representing some Δe^2_γ. Careful consideration of the definition of Δe^2_γ will convince the reader that $h_{\mu_1.\mu_2}$ is homotopic to the constant loop at the final point of λ. Hence $h_\sigma \simeq h_\tau$. Finally, note that if τ is the trivial edge loop at $\tilde{v}$, h_τ is the constant loop at $\tilde{v}$. $\qquad\square$

From now on, we will write $\pi_1(X, v)$ for both groups, understanding them to be identified by the isomorphism α of 3.4.1.

Just as with cellular homology, it follows that we may speak of $\pi_1(X, v)$ without reference to a particular CW complex structure on X, and that, by 3.1.16, different CW complex structures lead to different presentations of the same group. To take a simple example, choose for S^n the CW complex structure consisting of one vertex, v, and one n-cell. Then $\pi_1(S^1, v) \cong \mathbb{Z}$ by 3.1.9, and for $n \geq 2$, $\pi_1(S^n, v) = \{1\}$ by 3.1.10. For this, the combinatorial approach is simplest. On the other hand, the topological approach allows a trivial proof that π_1 preserves products:

Proposition 3.4.2. *Let $\{(X_\alpha, x_\alpha)\}_{\alpha \in \mathcal{A}}$ be a family of pointed spaces and let*

$$p_\beta : \prod_\alpha X_\alpha \to X_\beta \text{ be the projection map. Then } p_\# : \pi_1\left(\prod_\alpha X_\alpha, (x_\alpha)\right) \to$$

$$\prod_\alpha \pi_1(X_\alpha, x_\alpha) \text{ is an isomorphism, where } p(x) := (p_\alpha(x)). \qquad \square$$

Hence, writing T^n for the n-fold product of copies of S^1 (T^n is the n-*torus*), we get $\pi_1(T^n, v) \cong \mathbb{Z}^n$.

Of course, we have seen another proof that $\pi_1(S^1, v)$ is $\mathbb{Z}$. Give $\mathbb{R}$ the CW complex structure with vertex set $\mathbb{Z}$ and 1-cells $[n, n+1]$ for each $n \in \mathbb{Z}$. Let $\mathbb{Z}$ act on $\mathbb{R}$ by $n.x = x + n$. Then $\mathbb{R}$ is a simply connected free $\mathbb{Z}$-CW complex whose quotient is homeomorphic to S^1. By 3.2.3, $\pi_1(S^1, v)$ is isomorphic to $\mathbb{Z}$. Similar remarks apply to T^n since its universal cover is $\mathbb{R}^n$.

Since we are concerned with presentations of groups, it is useful to reformulate 3.1.16 topologically. Let (X, v) be a pointed CW complex. By 3.1.16, $\pi_1(X^1, v)$ is a free group. Choose an attaching map $f_\gamma : S^1 \to X^1$ for each 2-cell e_γ^2 of X. Then $f_\gamma \circ \bar{k}_1 : I^1 \to X^1$ is a loop, where $I^1 \xrightarrow{\bar{k}_1} S^1$ is the quotient map $I^1 \longrightarrow I^1/\dot{I}^1 \xrightarrow{k_1} S^1$ chosen once and for all in Sect. 2.5. Choose a path λ_γ in X^1 from v to $f_\gamma(1) := f_\gamma \circ \bar{k}_1(-1)$. Let $g_\gamma = [\lambda_\gamma . (f_\gamma \circ \bar{k}_1) . \lambda_\gamma^{-1}] \in \pi_1(X^1, v)$.

Proposition 3.4.3. *Let $i : (X^1, v) \hookrightarrow (X, v)$. The homomorphism $i_\# : \pi_1(X^1, v) \to \pi_1(X, v)$ is an epimorphism whose kernel is the normal closure of $\{g_\gamma \mid e_\gamma^2 \text{ is a 2-cell of } X\}$.* $\qquad \square$

By 3.3.1 and 3.3.2 we have:

Proposition 3.4.4. *Let $f : X \to Y$ be a homotopy equivalence between CW complexes taking the vertex v to the vertex w. Then $f_\# : \pi_1(X, v) \to \pi_1(Y, w)$ is an isomorphism.* $\qquad \square$

We can now improve the theory of covering spaces of CW complexes begun in Sect. 3.2.

Let X be a path connected CW complex and let v be a vertex of X. Recall from Sect. 3.2 that for each subgroup $H \leq \pi_1(X, v)$ there is a covering projection $q_H : (\bar{X}(H), \bar{v}) \to (X, v)$ such that $\pi_1(\bar{X}(H), \bar{v}) \cong H$. We now have the language to strengthen that statement. The universal cover $\tilde{X}$ is a free left $\pi_1(X, v)$-CW complex, hence also a free left H-CW complex. Moreover, $\tilde{X}$ is simply connected. Therefore there is a canonical isomorphism $\chi : H \to \pi_1(\bar{X}(H), \bar{v})$ defined in Sect. 3.2. On the other hand, q_H induces a homomorphism $q_{H\#} : \pi_1(\bar{X}(H), \bar{v}) \to \pi_1(X, v)$.

Proposition 3.4.5. $q_{H\#} \circ \chi = inclusion : H \hookrightarrow \pi_1(X, v)$. $\qquad \square$

Corollary 3.4.6. $q_{H\#}$ *is a monomorphism whose image is H.* $\qquad \square$

It remains to show that these covering spaces $\bar{X}(H)$ are essentially the only path connected covering spaces of X. For this we need to know that CW complexes are locally path connected. This is easily proved in three steps (the details are an exercise):

Proposition 3.4.7. (i) B^n *is locally path connected;* (ii) *if* Y *is obtained from the locally path connected space* A *by attaching n-cells, then* Y *is locally path connected;* (iii) *every CW complex is locally path connected.* $\square$

Proposition 3.4.8. *Let* $q : (E, e) \to (X, v)$ *be a covering projection, where* E *is path connected and* X *is a CW complex. Let* $H = q_\#(\pi_1(E, e)) \le \pi_1(X, v)$. *Then there is a homeomorphism* $h : (E, e) \to (\bar{X}(H), \bar{v})$ *making the following diagram commute. In particular,* $q_\#$ *is a monomorphism.*

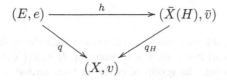

Proof. Apply 3.3.4 to q and to q_H. Uniqueness implies that the resulting lifts are mutually inverse. $\square$

Pointed covering projections $p_1 : (E_1, e_1) \to (X, x)$ and $p_2 : (E_2, e_2) \to (X, x)$, where E_1 and E_2 are path connected, are said to be *equivalent* if there is a homeomorphism $h : (E_1, e_1) \to (E_2, e_2)$ such that $p_1 = h \circ p_2$.

So, up to equivalence, there is a bijection between the path connected pointed covering spaces of the pointed CW complex (X, v) and the subgroups of $\pi_1(X, v)$, such that the fundamental group of the covering space corresponding to H is mapped isomorphically to H. Proposition 3.4.8 also completes the explanation given in Sect. 3.2 for the name "universal cover" – a path connected pointed covering space of (X, v) which covers all path connected pointed covering spaces of (X, v).

We close with two very useful theorems linking topology and group theory, one a special case of the other. Their proofs are left as exercises.

Theorem 3.4.9. *Let* (X, A) *be a pair of path connected CW complexes. Let* $i : (A, v) \hookrightarrow (X, v)$ *be the inclusion of a subcomplex and let* $p : (\tilde{X}, \tilde{v}) \to (X, v)$ *be the universal cover.* (i) *There is a bijection between the set of path components of* $p^{-1}(A)$ *and the set of cosets*

$$\{g.(image\ i_\#) \mid g \in \pi_1(X, v)\}.$$

(ii) *If* $A_{\tilde{v}}$ *denotes the path component of* $p^{-1}(A)$ *containing* $\tilde{v}$, *then* $\pi_1(A_{\tilde{v}}, \tilde{v})$ *is isomorphic to* $\ker i_\#$. $\square$

Theorem 3.4.10. *With notation as in Theorem 3.4.9, let $(\bar{X}(H), \bar{v})$ be the pointed covering space corresponding to H. (i) There is a bijection between the set of path components of $q_H^{-1}(A)$ and the set of double cosets*

$$\{H.g.(image\ i_\#) \mid g \in \pi_1(X, v)\}$$

. (ii) If $A_{\bar{v}}$ denotes the path component of $q_H^{-1}(A)$ containing $\bar{v}$, then $\pi_1(A_{\bar{v}}, \bar{v})$ is isomorphic to $i_\#^{-1}(H)$. □

In summary: (i) $\pi_0(q_H^{-1}(A)) \cong H \backslash \pi_1(X, v)/image\ i_\#$, and (ii) $A_{\bar{v}} = \bar{A}(i_\#^{-1}(H))$.

Exercises

1. Prove 3.4.7.
2. Theorem 3.3.3 establishes an isomorphism between the group of covering transformations of $\tilde{X}$ and $\pi_1(X, v)$. For a subgroup $H \leq \pi_1(X, v)$ establish a similar isomorphism between the group of covering transformations of $\bar{X}(H)$ and the group $N(H)/H$ where $N(H)$ denotes the *normalizer* of H, i.e., the largest subgroup of $\pi_1(X, v)$ in which H is normal. In particular, when H is normal[20] in G the group of covering transformations is isomorphic to G/H.
3. Prove 3.4.9 and 3.4.10.
4. Prove that if G is finitely generated there are only finitely many subgroups of a given finite index in G.
5. Prove that the intersection of finitely many subgroups of finite index in G has finite index in G.
6. Prove that if H has finite index in G the set of subgroups conjugate to H is finite.
7. Describe a CW complex X having one vertex v, one 1-cell and one 2-cell so that $X - \{v\}$ is a path connected space whose fundamental group is not finitely generated.
8. Prove the following transitivity property of regular (= normal) covering projections: given two cells e_1 and e_2 mapped to the same cell of X, there is a covering transformation taking e_1 onto e_2.

---

[20] When H is normal in G the covering projection $q_\# : \bar{X}(H) \to X$ is said to be *regular* or *normal*.

4

Some Techniques in Homotopy Theory

This chapter deals with five topics which help in understanding homotopy type and how to alter a CW complex within its homotopy type; for example, to reduce the number of cells in a dimension of interest. The most important theorems are 4.1.7 and 4.1.8 which are the key ingredients in the Rebuilding Lemma 6.1.4. That in turn tells us much about topological finiteness properties of groups (in Chapter 7). The last topic, the Hurewicz Theorem, is of fundamental importance in algebraic topology.

4.1 Altering a CW complex within its homotopy type

In this section we define "adjunction complexes" and show how to alter them without altering their homotopy types. As a first application we study Tietze transformations of group presentations from the topologist's point of view. If we are given a path connected CW complex whose fundamental group G is known to be finitely generated, we will show how to alter the complex within its homotopy type to a complex whose 1-skeleton is finite, and we will see what can be done to the 2-skeleton when G is finitely presented.

Let (X, A) be a pair of spaces and let $f : A \to Y$ be a map. The *adjunction space* $Y \cup_f X$ is the quotient space $Y \coprod X/\sim$ where $\sim$ is generated by the relation $a \sim f(a)$ for all $a \in A$. If, in addition, (X, A) is a CW pair, Y is a CW complex, and $f : A \to Y$ is a cellular map, then we have:

Proposition 4.1.1. *Let* $q : Y \coprod X \to Y \cup_f X$ *be the quotient map. Then* $Y \cup_f X$ *admits a CW complex structure whose cells are* $\{q(e) \mid e \text{ is a cell of } Y \text{ or a cell of } X \text{ which is not in } A\}$.

Proof. This is similar to the first part of the proof of 3.2.1. The n-skeleton is $Y^n \cup_{f|} X^n$. □

With this CW complex structure, $Y \cup_f X$ is called the *adjunction complex* of f. For example, if K is a CW complex and $f : \coprod_{\alpha \in \mathcal{A}} S_\alpha^{n-1} \to K^{n-1}$ is a simultaneous attaching map for the n-cells, then there is a homeomorphism $K^n \to K^{n-1} \cup_f \left(\coprod_{\alpha \in \mathcal{A}} B_\alpha^n \right)$ which matches cells bijectively.

Another example is the *mapping cylinder* of a cellular map $f : X \to Y$. This is the adjunction complex $Y \cup_{f_0} (X \times I)$ where $f_0 : X \times \{0\} \to Y$ is the map $(x, 0) \mapsto f(x)$. We will denote the mapping cylinder of f by $M(f)$. The map $X \to X \times I$ sending x to $(x, 1)$ induces an embedding $i : X \to M(f)$; the identity map of Y induces an embedding $j : Y \to M(f)$. Both i and j take each cell of the domain homeomorphically onto a cell of $M(f)$; one frequently suppresses i and j, identifying X with $i(X)$ and Y with $j(Y)$, writing $X \subset M(f)$ and $Y \subset M(f)$. The map $X \times I \to Y$, taking (x, t) to $f(x)$, and the identity map on Y, together induce the *collapse* $r : M(f) \to Y$.

The following diagram commutes:

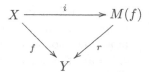

Proposition 4.1.2. *The map r is a homotopy inverse for j, so r is a homotopy equivalence. Indeed there is a strong deformation retraction $D :$ $M(f) \times I \to M(f)$ of $M(f)$ onto Y such that $D_1 = r$.*

Proof. The required D is induced by projection: $Y \times I \to Y$ and the map $X \times I \times I \to X \times I$, $(x, t, s) \mapsto (x, t(1 - s))$. $\qquad\square$

The proof of 4.1.2 gives a "canonical" strong deformation retraction of $M(f)$ onto Y. Thus the same proof gives:

Proposition 4.1.3. *Let $X = A \cup B$ and $X' = A' \cup B'$, where A and B are subcomplexes of X, while A' and B' are subcomplexes of X'. Let $f : X \to X'$ be a cellular map such that $f(A) \subset A'$ and $f(B) \subset B'$. Then there is a strong deformation retraction of $M(f)$ onto X' which restricts to strong deformation retractions of $M(f \mid A)$ onto A', $M(f \mid B)$ onto B', and $M(f \mid A \cap B)$ onto $A' \cap B'$.* $\qquad\square$

Returning to the general adjunction complex $Y \cup_f X$, where (X, A) is a CW pair, Y is a CW complex, and $f : A \to Y$ is a cellular map, we now show that the homotopy type of $Y \cup_f X$ only depends on the homotopy types of (X, A) and Y, and the homotopy class of f. We need some preliminaries (4.1.4 and 4.1.5) which have independent interest.

Let $n \geq 0$ be an integer and let (X, A) be a pair of CW complexes with X non-empty. The pair (X, A) is *n-connected* if for each $0 \leq k \leq n$ every map $(B^k, S^{k-1}) \to (X, A)$ is homotopic rel S^{k-1} to a map whose image lies in A. Thus $(X, \emptyset)$ is never n-connected, and (X, A) is 0-connected iff each path component of X has non-empty intersection with A. For any $x \in X$, $(X, \{x\})$ is 1-connected iff X is simply connected.

Proposition 4.1.4. (Whitehead Theorem)[1] *Let (X, A) be a CW pair. The following are equivalent:*

(i) *A is a strong deformation retract of X;*

(ii) *$A \overset{i}{\hookrightarrow} X$ is a homotopy equivalence;*

(iii) *(X, A) is n-connected for all n.*

Proof. (i) $\Rightarrow$ (ii) is clear; see 1.3.8.

For (ii) $\Rightarrow$ (iii), let $r : X \to A$ be a homotopy inverse for i. Since $r \mid A$ is homotopic to id_A, 1.3.15 implies that r is homotopic to a map $r' : X \to A$ such that $r' \mid A = \mathrm{id}_A$. So we may assume $r \mid A = \mathrm{id}_A$. Let $f : (B^k, S^{k-1}) \to (X, A)$ be a map where $k \leq n$, and let $H : X \times I \to X$ be a homotopy from id_X to $i \circ r$. Define

$$F : (B^k \times I \times \{0\}) \cup (B^k \times \{0, 1\} \times I) \cup (S^{k-1} \times I \times I) \to X$$

by

$$F(x, t, s) = \begin{cases} H(f(x), t) & \text{on } B^k \times I \times \{0\} \\ f(x) & \text{on } B^k \times \{0\} \times I \\ H(irf(x), 1 - s) & \text{on } B^k \times \{1\} \times I \\ H(f(x), (1 - s)t) & \text{on } S^{k-1} \times I \times I. \end{cases}$$

By 1.3.15, F extends to a map $\hat{F} : B^k \times I \times I \to X$. Let $G : B^k \times I \to X$ be the map $G(x, t) = \hat{F}(x, t, 1)$. Then $G(x, 0) = f(x)$, and $G(x, 1) = H(irf(x), 0) = irf(x) \in A$. If $x \in S^{k-1}$, $G(x, t) = H(f(x), 0) = f(x) \in A$. It is easy to see that $((B^k \times \{1\}) \cup (S^{k-1} \times I), S^{k-1} \times \{0\})$ is homeomorphic to (B^k, S^{k-1}) (exercise). It follows that f is homotopic rel S^{k-1} to a map whose image lies in A (exercise).

For (iii) $\Rightarrow$ (i), let $f_0 = \mathrm{id}_X$. The pair (X, A) is 0-connected, so $f_0 \mid: X^0 \cup A \hookrightarrow X$ is homotopic, rel A, to a map into A. By 1.3.15, f_0 is homotopic rel A to a map $f_1 : X \to X$ such that $f_1(X^0 \cup A) \subset A$. The pair (X, A) is 1-connected, so by 1.3.10, $f_1 \mid : X^1 \cup A \to X$ is homotopic, rel $X^0 \cup A$, to a map into A. Again by 1.3.15, f_1 is homotopic, rel $X^0 \cup A$, to a map $f_2 : X \to X$ such that $f_2(X^1 \cup A) \subset A$. Proceeding thus by induction, and observing that the homotopy $f_n \simeq f_{n+1}$ is rel $X^{n-1} \cup A$, we get a well defined limit map $f : X \to X$ such that $f \mid A = \mathrm{id}$ and $f(X) \subset A$. Combining all the homotopies, we can get a homotopy $\mathrm{id}_X \simeq f$. The details of this last step are left as an exercise. $\qquad\square$

The last two propositions lead us to an important theorem on piecing together homotopy equivalences:

Theorem 4.1.5. *Let $X = A \cup B$ and $X' = A' \cup B'$, where A and B are subcomplexes of X, while A' and B' are subcomplexes of X'. Let $f : X \to X'$*

---

[1] This is the version we will need most often. The full Whitehead Theorem is stated in Exercise 1 of Sect. 4.4.

be a cellular map such that $f(A) \subset A'$ *and* $f(B) \subset B'$. *If* $f| : A \to A'$, $f| : B \to B'$, *and* $f| : A \cap B \to A' \cap B'$ *are all homotopy equivalences, then* $f : X \to X'$ *is a homotopy equivalence. Moreover, there is a homotopy inverse* $g : X' \to X$ *for* f, *taking* A' *to* A, B' *to* B, *and* $A' \cap B'$ *to* $A \cap B$, *and homotopies* $g \circ f \simeq \mathrm{id}_X$ *and* $f \circ g \simeq \mathrm{id}_{X'}$ *which restrict to homotopies* $g \circ f \,|\simeq \mathrm{id}_A$, $f \circ g \,|\simeq \mathrm{id}_{A'}$, $g \circ f \,|\simeq \mathrm{id}_B$, $f \circ g \,|\simeq \mathrm{id}_{B'}$, $g \circ f \,|\simeq \mathrm{id}_{A \cap B}$, *and* $f \circ g \,|\simeq \mathrm{id}_{A' \cap B'}$.

Remark 4.1.6. Even the special case in which $B \subset A$ and $B' \subset A'$ is of interest. It says that if $f : (A, B) \to (A', B')$ is a cellular map of pairs such that the induced maps $A \to A'$ and $B \to B'$ are homotopy equivalences then f is a homotopy equivalence of pairs. Note, in particular, what this says when B and B' are single points.

Proof (of 4.1.5). By 4.1.2 f is a homotopy equivalence iff $i : X \hookrightarrow M(f)$ is a homotopy equivalence. Write $f_1 = f \,|: A \to A'$, $f_2 = f \,|: B \to B'$, $f_0 = f \,|: A \cap B \to A' \cap B'$. For $k = 0, 1, 2$, f_k is a homotopy equivalence. So, by 4.1.2, $A \hookrightarrow M(f_1)$, $B \hookrightarrow M(f_2)$ and $A \cap B \hookrightarrow M(f_0)$ are homotopy equivalences. By 4.1.4, the pairs $(M(f_1), A)$, $(M(f_2), B)$ and $(M(f_0), A \cap B)$ are n-connected for all n. The proof of (iii) $\Rightarrow$ (i) in 4.1.4 shows that we can construct a strong deformation retraction of $M(f_0)$ to $A \cap B$ which extends to strong deformation retractions of $M(f_1)$ to A and of $M(f_2)$ to B. So X is a strong deformation retract of $M(f)$. For the second part, combine this with 4.1.3. $\qquad\square$

Now we are ready for the main theorems, 4.1.7 and 4.1.8. Suppose (X, A) and (X', A') are CW pairs, Y and Y' are CW complexes, $f : A \to Y$ and $f' : A' \to Y'$ are cellular maps, $g : (X, A) \to (X', A')$ is a map of pairs, $g : X \to X'$, $g| : A \to A'$ and $k : Y \to Y'$ are homotopy equivalences, and the following diagram commutes:

$$
\begin{array}{ccccc}
X & \longleftarrow A & \xrightarrow{\ f\ } & Y \\
\downarrow{\scriptstyle g} & \quad\downarrow{\scriptstyle g|} & & \downarrow{\scriptstyle k} \\
X' & \longleftarrow A' & \xrightarrow{\ f'\ } & Y'
\end{array}
$$

Theorem 4.1.7. *The induced map* $G : Y \cup_f X \to Y' \cup_{f'} X'$ *is a homotopy equivalence.*

Proof. Consider the commutative diagram

$$
\begin{array}{ccc}
M(f) \cup X & \xrightarrow{\ h\ } & Y \cup_f X \\
\downarrow{\scriptstyle p} & & \downarrow{\scriptstyle G} \\
M(f') \cup X' & \xrightarrow{\ h'\ } & Y' \cup_{f'} X'
\end{array}
$$

Here and in what follows, we write $M(f) \cup X$ for $M(f) \cup_i X$ where $i : A \to M(f)$ is the canonical inclusion – i.e., we literally apply the convention of writing $A \subset M(f)$; p is induced by $(g \mid) \times \mathrm{id} : A \times I \to A' \times I$, $k : Y \to Y'$ and $g : X \to X'$; h is induced by the collapse $r : M(f) \to Y$ and id_X; similarly for h'. By 4.1.5, p is a homotopy equivalence.

To see that h is a homotopy equivalence, consider the commutative diagram

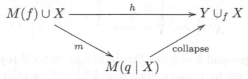

See Fig. 4.1. Here, $q : Y \coprod X \to Y \cup_f X$ is the defining quotient map. The inclusion $m : M(f) \cup X \hookrightarrow M(q \mid X)$ sends X to the copy of X in $M(q \mid X)$; $M(f) \subset M(q \mid X)$ because $q \mid A = \mathrm{inclusion} \circ f$. By 4.1.2, we need only show that m is a homotopy equivalence. By 1.3.15, $(X \times \{1\}) \cup (A \times I)$ is a strong deformation retract of $X \times I$. Any strong deformation retraction of $X \times I$ to $(X \times \{1\}) \cup (A \times I)$ induces a strong deformation retraction of $M(q \mid X)$ to $M(f) \cup X$. $\qquad\square$

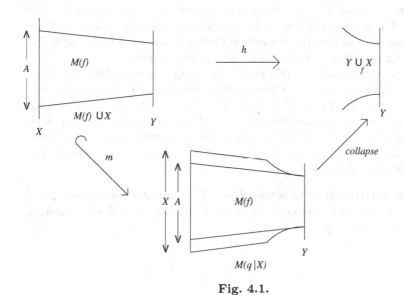

Fig. 4.1.

Theorem 4.1.8. *If the hypotheses of 4.1.7 are weakened from $k \circ f = f' \circ g \mid$ to $k \circ f \simeq f' \circ g \mid$, it is still the case that $Y \cup_f X$ and $Y' \cup_{f'} X'$ have the same homotopy type.*

Proof. Consider the diagram

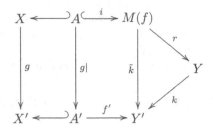

Here r is the collapse and $\tilde{k} = k \circ r$. The right square homotopy commutes. By 1.3.15, $\tilde{k}$ is homotopic to a map $\bar{k}$ such that $\bar{k} \circ i = f' \circ g \mid$. By 4.1.7, $M(f) \cup X$ is homotopy equivalent to $Y' \cup_{f'} X'$. By the proof of 4.1.7, $M(f) \cup X$ is homotopy equivalent to $Y \cup_f X$.   □

Here is a useful application of 4.1.7:

Corollary 4.1.9. *Let X be a CW complex and A a contractible subcomplex. The quotient $q : X \to X/A$ is a homotopy equivalence.*

Proof. Apply 4.1.7 with $X = X'$, $A = A'$, $g = \mathrm{id}$, $Y = A$, $Y' = \{q(A)\}$.   □

Theorems 4.1.7 and 4.1.8 are powerful technical tools. For example, 4.1.8 implies that if Y is obtained from A by attaching n-cells, then the homotopy type of Y only depends on the homotopy classes of the attaching maps. We now discuss an application of this to Tietze transformations.

For each presentation $P = \langle W \mid R, \rho \rangle$ of a group G a procedure was given in Example 1.2.17 for building a presentation complex X_P, having just one vertex v, such that $\pi_1(X_P, v) \cong G$ (by 3.1.8). The 1-cells and 2-cells of X_P are in bijective correspondence with the sets W and R, respectively, in such a way that if $P \Rightarrow P' := \langle W \mid R', \rho' \rangle$ is a Tietze transformation of Type I, then X_P is, in a natural way, a subcomplex of $X_{P'}$, such that all cells of $X_{P'}$ which are not cells of X_P are 2-cells.

Proposition 4.1.10. *Let P' be obtained from P by a Tietze transformation of Type I. Then there is a homotopy equivalence h making the following diagram commute up to homotopy:*

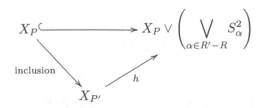

Proof. The attaching maps for the 2-cells of $X_{P'}$ which are not in X_P are homotopic in X_P to constant maps. Apply 4.1.8.   □

Similarly, if $P \Rightarrow P'' = \langle W' \mid R', \rho' \rangle$ is a Tietze transformation of Type II, then X_P is a subcomplex of $X_{P''}$ and we have:

Proposition 4.1.11. *The map $X_P \hookrightarrow X_{P''}$ is a homotopy equivalence.*

Proof. $(X_{P''})^1 = X_P^1 \vee \left(\bigvee_\alpha S_\alpha^1 \right)$, i.e. the wedge of X_P^1 and a bouquet of circles. For each α, the 2-cell e_α^2 of $X_{P''}$ which is not in X_P has a characteristic map of the form $f_\alpha : I^2 \to X_{P''}$ where $f_\alpha(I^1 \times \{-1\}) \subset X_P$, $f_\alpha(\pm 1, t) = f_\alpha(\pm 1, -1)$ for all $t \in I^1$, and $f_\alpha \mid I^1 \times \{1\}$ is a characteristic map for S_α^1. The strong deformation retraction of I^2 onto $I^1 \times \{-1\}$, $(s, t, u) \mapsto (s, (1-u)(1+t) - 1)$ for $0 \leq u \leq 1$, induces a strong deformation retraction of $X_P^1 \cup S_\alpha^1 \cup e_\alpha^2$ onto X_P^1. This can be done simultaneously for all α, giving a strong deformation retraction of $X_{P''}$ onto X_P. $\square$

Proposition 4.1.12. *For $i = 1$ and 2, let $P_i := \langle W_i \mid R_i, \rho_i \rangle$ be presentations of the group G. There are homotopy equivalent CW complexes Y_{P_1} and Y_{P_2} obtained from X_{P_1} and X_{P_2} by attaching 3-cells. Moreover, if P_1 and P_2 are finite presentations, Y_{P_1} and Y_{P_2} can be obtained by attaching finitely many 3-cells.*

Proof. If in the proof of 4.1.10 we attach a 3-cell to $X_P \vee \left(\bigvee_{\alpha \in R'-R} S_\alpha^2 \right)$, for each α, by a homeomorphism $S^2 \to S_\alpha^2$ we obtain a 3-dimensional complex homotopy equivalent to X_P. By 4.1.8, we can attach 3-cells to $X_{P'}$ itself to get a 3-dimensional complex homotopy equivalent to X_P whose 2-skeleton is $X_{P'}$.

Applying this remark in the context of 3.1.15 gives us:

$$P_1 \xoverset{\text{Type II}}{\Longrightarrow} P' \xoverset{\text{Type I}}{\Longrightarrow} P \xoverset{\text{Type I}}{\Longleftarrow} P'' \xoverset{\text{Type II}}{\Longleftarrow} P_2.$$

In terms of associated CW complexes, this gives:

$$X_{P_1} \xoverset{\text{Type II}}{\Longrightarrow} X_{P'} \xoverset{\text{Type I}}{\Longrightarrow} X_P \hookrightarrow X_P \cup \bigcup \{e_\alpha^3 \mid \alpha \in \mathcal{A}\} := Z_1$$

$$X_{P_2} \xoverset{\text{Type II}}{\Longrightarrow} X_{P''} \xoverset{\text{Type I}}{\Longrightarrow} X_P \hookrightarrow X_P \cup \bigcup \{e_\beta^3 \mid \beta \in \mathcal{B}\} := Z_2$$

Indeed, the proof of 3.1.15 shows that $\mathcal{A}$ and $\mathcal{B}$ are in bijective correspondence with the sets $W_1 \coprod R_2$ and $W_2 \coprod R_1$ respectively. The spaces X_{P_1}, $X_{P'}$ and Z_1 are homotopy equivalent. The spaces X_{P_2}, $X_{P''}$ and Z_2 are homotopy equivalent. By 4.1.8, we can attach 3-cells to X_{P_1} and to X_{P_2} to get $Y_{P_1} := X_{P_1} \cup \bigcup \{\tilde{e}_\beta^3 \mid \beta \in \mathcal{B}\}$ and $Y_{P_2} := X_{P_2} \cup \bigcup \{\tilde{e}_\alpha^3 \mid \alpha \in \mathcal{A}\}$, both homotopy equivalent to $X_P \cup \bigcup \{e_\gamma^3 \mid \gamma \in \mathcal{A} \coprod \mathcal{B}\}$. The last sentence of the Proposition is clear. $\square$

Remark 4.1.13. There are examples in [52] of finite CW complexes X and Y each having one vertex, such that $X \vee S^2$ is homotopy equivalent to $Y \vee S^2 \vee S^2$, while there is no CW complex Z such that X is homotopy equivalent to $Z \vee S^2$. This is related to the existence of finitely generated projective $\mathbb{Z}G$-modules which are stably free but not free. (See also [7].) In the proof of 4.1.12, X_P has the homotopy type of both $X_{P'} \vee \left(\bigvee_{\alpha \in \mathcal{A}} S_\alpha^2 \right)$ and $X_{P''} \vee \left(\bigvee_{\beta \in \mathcal{B}} S_\beta^2 \right)$. Dunwoody's examples show that one cannot always "cancel" copies of S^2.

Theorem 4.1.14. *Let X be a path connected CW complex whose fundamental group G is finitely generated. Then:*

  (i) *X is homotopy equivalent to a CW complex having finite 1-skeleton.*

 (ii) *If G is finitely presented, there is a CW complex X', obtained from X by attaching 3-cells, which is homotopy equivalent to a CW complex having finite 2-skeleton.*

(iii) *If G is finitely presented, X is homotopy equivalent to a CW complex Z whose 2-skeleton is the wedge of a finite CW complex and a bouquet of 2-spheres.*

Proof. By 3.1.13, 3.1.12 and 4.1.9, we may assume that X has only one vertex. Write $X_{P_1} = X^2$, and let P_2 be a presentation of G which is finite or finitely generated as appropriate. Write $P_i = \langle W_i \mid R_i \rangle$. Using 4.1.8 as in the proof of 4.1.12, we get $X \cup \bigcup \{3\text{-cells}\}$ homotopy equivalent to a CW complex Y such that $Y^2 = X_{P_2}$. Similarly, we get X homotopy equivalent to a CW complex Z such that $Z^2 = X_{P_2} \cup \bigvee_{\beta \in \mathcal{B}} S_\beta^2$ where $|\mathcal{B}| = |W_2| + |R_1|$. We claim Z^1 is finite; a sketch of the argument follows. Using $\simeq$ for "is homotopy equivalent to", and using the notation of the proof of 4.1.12, we get:

$$X_P \simeq X_{P''} \vee \left(\bigvee_{\beta \in \mathcal{B}} S_\beta^2 \right) \simeq X_{P_2} \vee \left(\bigvee_{\beta \in \mathcal{B}} S_\beta^2 \right).$$

$$X_{P_1} \simeq X_P \cup \bigcup \{e_\alpha^3 \mid \alpha \in \mathcal{A}\} \simeq X_{P_2} \vee \left(\bigvee_{\beta \in \mathcal{B}} S_\beta^2 \right) \cup \bigcup \{\tilde{e}_\alpha^3 \mid \alpha \in \mathcal{A}\}.$$

Hence, by 4.1.8,

$$X \simeq X_{P_2} \vee \left(\bigvee_{\beta \in \mathcal{B}} S_\beta^2 \right) \cup \bigcup \{\tilde{e}_\alpha^3 \mid \alpha \in \mathcal{A}\} \cup \bigcup \{\text{cells of dimension} \geq 3\}.$$

The "cells of dimension ≥ 3" (other than the $\tilde{e}_\alpha^3$'s) are in bijective correspondence with those of X. $\qquad\square$

Example 4.1.15. The "finitely generated" and "finitely presented" parts of 4.1.14 would be more similar if we could say, in the finitely presented case, that X is homotopy equivalent to a CW complex having finite 2-skeleton. However, this is false. For example, let X be an infinite bouquet of 2-spheres. We have $\pi_1(X, v)$ trivial by 3.1.11; however, $H_2(X; \mathbb{Z}_2)$ is an infinite-dimensional $\mathbb{Z}_2$-module (vector space), since every 2-chain is a cycle and none is a boundary. If Y is a finite CW complex, $H_2(Y; \mathbb{Z}_2)$ is a finite-dimensional $\mathbb{Z}_2$-vector space since $C_2(Y; \mathbb{Z}_2)$ is finitely generated, hence also $Z_2(Y; \mathbb{Z}_2)$. Hence, by 2.7.7, X is not homotopy equivalent to a CW complex with finite 2-skeleton.

The proof of 4.1.14 also proves the following, which will be useful.

Addendum 4.1.16. *Let X be a path connected CW complex having m_k k-cells for each $k \geq 0$, where $0 \leq m_k \leq \infty$. Let $P := \langle W \mid R, \rho \rangle$ be a presentation of the fundamental group of X. Then X is homotopy equivalent to a CW complex Z with the properties:* (i) $Z^2 = X_P \vee (bouquet\ of\ m_2 + |W|\ 2\text{-spheres})$; (ii) *$Z$ has $m_3 + m_1 - m_0 + 1 + |R|$ 3-cells; and* (iii) *Z has m_k k-cells for all $k \geq 4$.*

Proof. Let T be a maximal tree in X. Then T has m_0 vertices and $(m_0 - 1)$ 1-cells. So $X' := X/T$ has one vertex, $(m_1 - m_0 + 1)$ 1-cells, and m_k k-cells for $k \geq 2$. By the proof of 4.1.14, X' is homotopy equivalent to a complex Z of the form

$$Z = X_P \vee \left(\bigvee_{\beta \in \mathcal{B}} S_\beta^2 \right) \cup \bigcup \{ \tilde{e}_\alpha^3 \mid \alpha \in \mathcal{A} \} \cup \bigcup \{\text{cells of dimension} \geq 3\}$$

where $|\mathcal{B}| = |W| + m_2$, $|\mathcal{A}| = m_1 - m_0 + 1 + |R|$, and, for $k \geq 3$, Z has m_k k-cells. By 4.1.9, X is homotopy equivalent to Z. $\qquad\square$

Appendix: the equivariant case

Equivariant[2] analogs of 4.1.7 and 4.1.8 can be useful. We briefly describe how to get them.

A *G-set* is a set equipped with a left G-action by permutations. A *G-space* is a space equipped with a left G-action by homeomorphisms. A map $f : A \to B$ between G-spaces is a *G-map* (or an *equivariant map*) if $f(g.a) = g.f(a)$ for all $a \in A$ and $g \in G$.

Let Y be obtained from A by attaching $B^n(\mathcal{A})$ using $f : S^{n-1}(\mathcal{A}) \to \mathcal{A}$ as in Sect. 1.2. If $\mathcal{A}$ is a G-set, $B^n(\mathcal{A})$ becomes a G-space. If $\mathcal{A}$ is also a G-space and f is a G-map then Y becomes a G-space. Moreover, the stabilizer of each n-cell of (Y, A) acts trivially on that cell. It follows that if $(X, \{X^n \mid n \geq 0\})$

---

[2] The word *equivariant* refers to properties compatible with a given action of a group on a space.

is a rigid G-CW complex then for $n \geq 1$ the G-space X^n is obtained from the G-space X^{n-1} by attaching $B^n(\mathcal{A}_n)$ where the simultaneous attaching map is a G-map; here $\mathcal{A}_n$ is the G-set of n-cells. Conversely, any CW complex built in this way is a rigid G-CW complex.

The terms defined in Sect. 1.3 all have obvious G-analogs: *G-homotopy*, *G-homotopy equivalence*, *G-homotopy type*, *G-strong deformation retract*, etc. The theorems on extending maps and homotopies, 1.2.23 and 1.3.10, and the Homotopy Extension Theorem 1.3.15 have G-analogs for rigid G-CW complexes.

When an adjunction space $Y \cup_f X$ is formed from G-spaces using a G-map, it becomes a G-space; if X and Y are rigid G-CW complexes and f is a cellular G-map, the G-version of 4.1.1 says that $Y \cup_f X$ is a rigid G-CW complex. In particular this applies to mapping cylinders. There are routine G-analogs of 4.1.2 and 4.1.3 for rigid G-CW complexes. The rigid G-CW pair (X, A) is *G-n-connected* if for each $0 \leq k \leq n$ and each G-set $\mathcal{A}$, every G-map $(B^k(\mathcal{A}), S^{k-1}(\mathcal{A})) \to (X, A)$ is G-homotopic rel $S^{k-1}(\mathcal{A})$ to a G-map whose image lies in A. This replaces the hypothesis of n-connectedness in 4.1.4 (iii) to give a G-analog of that proposition (for the rigid case). The G-analogs of 4.1.5–4.1.8 for rigid G-CW complexes are then routine.

Exercises

1. Define a map $f : S^1 \to S^1$ by $e^{2\pi it} \mapsto e^{6\pi it}$ if $0 \leq t \leq \frac{2}{3}$, and $e^{2\pi it} \mapsto e^{-6\pi it}$ if $\frac{2}{3} \leq t \leq 1$; this wraps the circle three times around itself, twice positively and once negatively. Let D be the CW complex with one vertex, one 1-cell, and one 2-cell attached by the map f. Show that D is contractible. (This space D is called the *dunce hat* and is a classic example in topology because it is contractible but not "collapsible" – see [42].)

2. Why did we not describe the dunce hat as the presentation complex given by $\langle t \mid t^2 t^{-1} \rangle$?

3. Prove a locally finite version of Theorem 4.1.5: i.e., let $\mathcal{K}$ be a locally finite cover of X by subcomplexes, etc.

4. Complete the last step in the proof of 4.1.4.

5. Let $X = A \cup B$ be a CW complex where A and B are subcomplexes. Prove that if A, B and $A \cap B$ are contractible then X is contractible.[3]

4.2 Cell trading

We describe a technique which is often used in topology to simplify a CW complex within its homotopy type. In Chapter 7 we will use it to simplify $K(G, 1)$-complexes.

---

[3] This is a very special case of Proposition 9.3.20.

Proposition 4.2.1. (Cell Trading Lemma) *Let (X, A) be an n-connected CW pair, where*

$$X = A \cup \left(\bigcup_{\alpha \in \mathcal{A}} e_\alpha^n \right) \cup \left(\bigcup_{\beta \in \mathcal{B}} e_\beta^{n+1} \right) \cup \left(\bigcup_{\gamma \in \mathcal{C}} e_\gamma^{n+2} \right) \cup (\text{cells of dimension } > n+2).$$

There is another CW complex X'' having A as a subcomplex such that (X'', A) is homotopy equivalent to (X, A) and

$$X'' = A \cup \left(\bigcup_{\beta \in \mathcal{B}} \tilde{e}_\beta^{n+1} \right) \cup \left(\bigcup_{\delta \in \mathcal{A} \amalg \mathcal{C}} \tilde{e}_\delta^{n+2} \right) \cup (\text{ cells of dimension } > n+2).$$

Moreover, in dimensions $> n + 2$, X'' and X have the same number of cells.

Remark 4.2.2. Since (X, A) is n-connected the characteristic map of each e_α^n is homotopic to a map into A. This suggests that in some sense those cells are unnecessary. Proposition 4.2.1 makes this precise: they can be "traded" for $(n + 2)$-cells.

Proof (of 4.2.1). We embed X as a subcomplex of a CW complex X'; the required X'' will be a quotient complex of X' by a quotient map which restricts to an embedding of A in X''. We will have homotopy equivalences

$$(X, A) \hookrightarrow (X', A) \rightarrow (X'', A).$$

Consider $S^{n-1} \subset S^n \subset S^{n+1} \subset B^{n+2}$. Let Y be B^{n+2} with a CW complex structure such that: S^{n-1} is a subcomplex, S^n is obtained from S^{n-1} by attaching two n-cells $d^n(+)$ and $d^n(-)$, S^{n+1} is obtained from S^n by attaching two $(n + 1)$-cells $d^{n+1}(+)$ and $d^{n+1}(-)$, and B^{n+2} is obtained from S^{n+1} by attaching one $(n + 2)$-cell d^{n+2}. See Fig. 4.2.

We first deal with the case where $\mathcal{A}$ has only one element α. Define $f_\alpha :$ $d^{n+1}(-) \rightarrow X$ to be a map such that $f_\alpha \mid d^n(+)$ is a characteristic map for e_α^n (identifying $d^n(+)$ with B^n) and $f_\alpha(d^n(-)) \subset A$; f_α exists because (X, A) is n-connected. Let $X' = X \cup_{f_\alpha} Y$. Since $d^{n+1}(-)$ is a strong deformation retract of Y, Proposition 1.3.10 implies that X is a strong deformation retract of X'. So $(X, A) \hookrightarrow (X', A)$ is a homotopy equivalence.

Abusing notation, we write $X' = X \cup d^{n+1}(+) \cup d^{n+2}$. Let A' be the subcomplex $A \cup e_\alpha^n \cup d^{n+1}(+)$. Since $d^n(-)$ is a strong deformation retract of $d^{n+1}(+)$, A is a strong deformation retract of A'. Let $q : A' \rightarrow A$ be a retraction such that (inclusion) $\circ q : A' \rightarrow A'$ is homotopic to $\mathrm{id}_{A'}$. The required X'' is $A \cup_q X'$.

Applying 4.1.7 to the commutative diagram

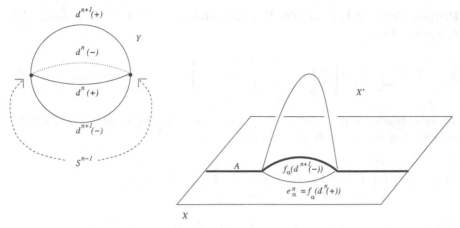

Fig. 4.2.

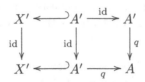

we get a homotopy equivalence $X' \to X''$ whose restriction to A is $q \mid A = \mathrm{id}_A$. Thus, by 4.1.5, we have a homotopy equivalence $(X', A) \to (X'', A)$. In the passage from X to X'' we have "lost" an n-cell e_α^n and "acquired" an $(n+2)$-cell $\tilde{e}_\alpha^{n+2}$, namely the image of d^{n+2} in X''.

Now, if $\mathcal{A}$ has more than one element, we treat each n-cell e_α^n this way, using $\coprod\limits_{\alpha \in \mathcal{A}} Y_\alpha$ instead of Y. $\qquad\qquad\qquad\qquad\qquad\qquad\qquad\qquad$ $\square$

Remark 4.2.3. The Relative Hurewicz Theorem (4.5.1 below) provides a homological procedure for verifying the n-connectedness hypothesis in 4.2.1 when X and A are simply connected.

4.3 Domination, mapping tori, and mapping telescopes

The *mapping torus* of a cellular map $h : Y \to Y$ is the quotient CW complex $T(h)$ obtained from the mapping cylinder, $M(h)$, by identifying the "top" and "bottom", i.e., (in the notation of Sect. 4.1) $T(h) = M(h)/\sim$, where $\sim$ is generated by: $i(y) \sim j(y)$ for all $y \in Y$. It is a CW complex by 4.1.1. See Fig. 4.3.

The space $T(h)$ has a covering space $\mathrm{Tel}(h)$, the *mapping telescope* of h. To form $\mathrm{Tel}(h)$, take $\coprod\limits_{m \in \mathbb{Z}} M(h)_m$, where each $M(h)_m = M(h)$, and identify

$j(y) \in M(h)_m$ with $i(y) \in M(h)_{m+1}$ for each m (see Fig. 4.3); Tel(h) is a free $\mathbb{Z}$-CW complex, where $n \in \mathbb{Z}$ takes each $M(h)_m$ by the "identity" map to $M(h)_{m+n}$. Clearly, the quotient $\mathbb{Z}\backslash$Tel(h) is homeomorphic to $T(h)$.

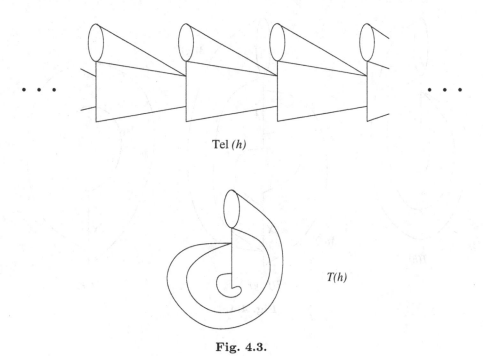

Tel *(h)*

T(h)

Fig. 4.3.

Now suppose h is the composition $Y \xrightarrow{g} X \xrightarrow{f} Y$ where f and g are cellular maps. Let $k = g \circ f : X \to X$.

Proposition 4.3.1. $T(h)$ *is homotopy equivalent to* $T(k)$.

Proof. We form an intermediate space $T(f, g)$, the quotient CW complex of $M(f) \coprod M(g)$ obtained by identifying $y \in Y \subset M(f)$ with $y \in Y \subset M(g)$ and $x \in X \subset M(f)$ with $x \in X \subset M(g)$, for each $y \in Y$ and each $x \in X$; see Fig. 4.4. The restriction of the quotient map $M(f) \coprod M(g) \to T(f, g)$ to $M(f)$ or to $M(g)$ is an embedding, so we write $M(f) \subset T(f, g)$ and $M(g) \subset T(f, g)$. Apply 4.1.7 to the commutative diagram

$$
\begin{array}{ccc}
T(f, g) \longleftarrow M(f) & \xrightarrow{\text{id}} & M(f) \\
\text{id} \downarrow \qquad \text{id} \downarrow & & \downarrow \text{collapse} \\
T(f, g) \longleftarrow M(f) & \xrightarrow[\text{collapse}]{} & Y
\end{array}
$$

to get $T(f,g)$ homotopy equivalent to $T(h)$. Similarly, $T(f,g)$ is homotopy equivalent to $T(k)$.   □

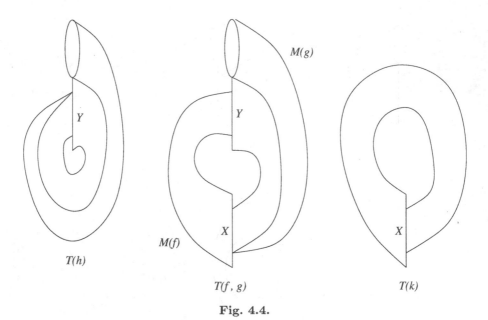

Fig. 4.4.

In the same way, there is a CW complex $\mathrm{Tel}(f,g)$ intermediate between $\mathrm{Tel}(h)$ and $\mathrm{Tel}(k)$. The space $\mathrm{Tel}(f,g)$ is the quotient of $\coprod_{m\in\mathbb{Z}} (M(f)_m \coprod M(g)_m)$, where each $M(f)_m = M(f)$ and each $M(g)_m = M(g)$, obtained by identifying $x \in M(g)_m$ with $x \in M(f)_m$, and $y \in M(f)_m$ with $y \in M(g)_{m+1}$, for all $m \in \mathbb{Z}$. See Fig. 4.5. As with $\mathrm{Tel}(h)$, $\mathrm{Tel}(f,g)$ is a free $\mathbb{Z}$-CW complex, where $n \in \mathbb{Z}$ takes $M(g)_m$ [resp. $M(f)_m$] by the "identity" map to $M(g)_{m+n}$ [resp. $M(f)_{m+n}$]. The quotient $\mathbb{Z}\backslash\mathrm{Tel}(f,g)$ is homeomorphic to $T(f,g)$.

By a proof similar to that of 4.3.1, we see that $\mathrm{Tel}(f,g)$ is homotopy equivalent to $\mathrm{Tel}(h)$ and to $\mathrm{Tel}(k)$. Indeed the proof gives homotopy equivalences of the nicest kind, namely:

Proposition 4.3.2. *There are homotopy equivalences ϕ, ψ, $\bar{\phi}$ and $\bar{\psi}$ making the following diagram commute, where the vertical arrows are the covering projections obtained by factoring out the free $\mathbb{Z}$-actions:*

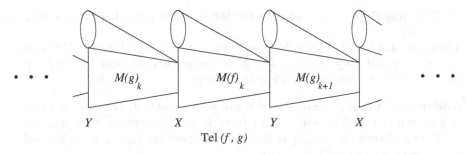

$$\cdots \qquad M(g)_k \qquad M(f)_k \qquad M(g)_{k+1} \qquad \cdots$$

$$Y \qquad\qquad X \qquad\qquad Y \qquad\qquad X$$

$$\mathrm{Tel}\,(f,\,g)$$

Fig. 4.5.

$$\mathrm{Tel}(h) \xleftarrow{\ \bar\phi\ } \mathrm{Tel}(f,g) \xrightarrow{\ \bar\psi\ } \mathrm{Tel}(k)$$
$$\downarrow \qquad\qquad\qquad \downarrow \qquad\qquad\qquad \downarrow$$
$$T(h) \xleftarrow{\ \phi\ } T(f,g) \xrightarrow{\ \psi\ } T(k).$$

$\square$

When $k = g \circ f : X \to X$ is homotopic to id_X, we call $g : Y \to X$ a *domination* and we say that Y *dominates* X; f is then a right homotopy inverse for g. Moreover, $h \simeq h \circ h$ (in words: h is a *homotopy idempotent*[4]) because $h = f \circ g \simeq f \circ k \circ g = f \circ (g \circ f) \circ g = h \circ h$. In this case we get

Proposition 4.3.3. $T(k)$ *is homotopy equivalent to* $X \times S^1$.

Proof. First, note that $T(\mathrm{id}_X) = X \times S^1$ and that $k \simeq \mathrm{id}_X$. We saw in Sect. 4.1 that there is a homotopy equivalence $M(k) \to M(\mathrm{id}_X)$ which restricts to the identity on the two copies of X, $i(X) \subset M(k)$ and $j(X) \subset M(k)$. By 4.1.7, the induced map $T(k) \to T(\mathrm{id}_X)$ is a homotopy equivalence. $\square$

As with 4.3.2, the proof of 4.3.3 gives more:

Proposition 4.3.4. *There are homotopy equivalences ξ and $\bar\xi$ making the following diagram commute, where the vertical arrows are the covering projections obtained by factoring out the $\mathbb{Z}$-actions:*

$$\mathrm{Tel}(k) \xrightarrow{\ \bar\xi\ } \mathrm{Tel}(\mathrm{id}_X) - X \times \mathbb{R}$$
$$\downarrow \qquad\qquad\qquad \downarrow$$
$$T(k) \xrightarrow{\ \xi\ } T(\mathrm{id}_X) = X \times S^1.$$

$\square$

---

[4] Homotopy idempotents are discussed in Sect. 9.2.

Collecting these results, we get a fundamental theorem about domination:

Theorem 4.3.5. *Let* $X \xrightarrow{f} Y \xrightarrow{g} X$ *be cellular maps between CW complexes, such that* $g \circ f \simeq \mathrm{id}_X$. *Then* X *is homotopy equivalent to* $\mathrm{Tel}(f \circ g)$, *and* $X \times S^1$ *is homotopy equivalent to* $T(f \circ g)$. $\square$

Addendum 4.3.6. *Assume* X *and* Y *are path connected. For suitable choice of base points* $z \in T(k)$ *and* $x \in X$, *there is an isomorphism* $\pi_1(T(k), z) \to \pi_1(X, x) \times \mathbb{Z}$ *such that* $\mathrm{Tel}(k)$ *is the covering space corresponding to the subgroup which goes to* $\pi_1(X, x) \times \{0\}$.

Proof. Combine the commutative diagrams in 4.3.2 and 4.3.4. Pick a base point for $\mathrm{Tel}(f, g)$, and pick all other base points so that all maps are base point preserving. By 3.3.2, $\phi_\#$, $\psi_\#$, $\xi_\#$, $\bar{\phi}_\#$, $\bar{\psi}_\#$ and $\bar{\xi}_\#$ are isomorphisms. The claim follows by looking at the corresponding commutative diagram of groups. $\square$

A CW-complex X is *finitely dominated* if it is dominated by a finite CW complex.

Corollary 4.3.7. *If* X *is finitely dominated then* $X \times S^1$ *has the homotopy type of a finite CW complex and* X *has the homotopy type of a finite-dimensional CW complex.* $\square$

Source Note: 4.3.1 was observed by Mather in [111].

Exercises

1. Write down a presentation of $\pi_1(T(h), y)$.
2. Describe $\pi_1(\mathrm{Tel}(h), \bar{y})$ as a subgroup of $\pi_1(T(h), y)$.
3. Let X consist of circles of radius 1 centered at $(\pm 2, 0) \in \mathbb{R}^2$ together with the arc $[-1, 1] \times \{0\}$, and let $h : X \to X$ be the "reflection in the y-axis" homeomorphism. Show that the mapping torus $T(h)$ is homeomorphic to the space obtained by gluing the boundary of a Möbius strip to a torus along a meridian circle of the torus. Compute the fundamental group of $T(h)$ from these points of view, and relate this to Exercise 5 of Sect. 3.1.

4.4 Review of homotopy groups

The nth *homotopy group* of a pointed space (X, x) is the set of homotopy classes of maps $(S^n, v) \to (X, x)$, where v is a fixed base point for S^n. If f is such a map, its (pointed) homotopy class is denoted by $[f]$, and the set of all such $[f]$ is denoted by $\pi_n(X, x)$. This is consistent with our previous usage when $n = 0$ or 1. Similarly, if (X, A, x) is a pointed pair, with $x \in A$ as base point, and if $n \geq 1$, we denote by $\pi_n(X, A, x)$ the set of homotopy classes,

$[f]$, of maps $f : (B^n, S^{n-1}, v) \to (X, A, x)$; $\pi_n(X, A, x)$ is the nth (relative) homotopy group of (X, A, x). The group structure is described in 4.4.5 below.[5]

4.4.1. The sets $\pi_n(X, x)$ and $\pi_n(X, A, x)$ are pointed sets with the homotopy class of the constant map at x as base point.

4.4.2. $\pi_k(X, x)$ [resp. $\pi_k(X, A, x)$] is trivial (i.e., consists of the base point alone) for all $0 \le k \le n$ iff X is n-connected [resp. (X, A) is n-connected].

4.4.3. $\pi_0(X, x)$ is canonically identified with the pointed set of path components of X, the base point being the path component containing x.

4.4.4. $\pi_n(X, \{x\}, x)$ is canonically identified with $\pi_n(X, x)$ via $[f] \to [\bar{f}]$ where $f : (B^n, S^{n-1}, v) \to (X, \{x\}, x)$, and $\bar{f} : (S^n, v) \to (X, x)$ is obtained from f using the canonical homeomorphism $B^n/S^{n-1} \to S^n$ of Convention 2.5.16. Thus elements of $\pi_n(X, x)$ can also be regarded as homotopy classes of maps $(B^n, S^{n-1}) \to (X, x)$.

4.4.5. There is a standard group structure on $\pi_n(X, x)$ when $n \ge 1$ [resp. $\pi_n(X, A, x)$ when $n \ge 2$] which is abelian when $n \ge 2$ [resp. $n \ge 3$].[6] The identity (or zero) element of this group is the base point chosen in 4.4.1. The details, especially the reason why the group is abelian, will be needed in the next section, so we recall them in a convenient form. As usual, we identify (B^n, S^{n-1}) with $(I^n, \overset{\bullet}{I}{}^n)$. An *island* on I^n is a product $\prod_{i=1}^{n} J_i$ where each J_i is a closed non-trivial interval lying in $\overset{\circ}{I}{}^1$. Similarly, an *island* on $\overset{\bullet}{I}{}^{n+1}$ is a product $\prod_{i=1}^{n+1} J_i$ where exactly n of the J_i's are non-trivial closed intervals in I^1, and the remaining J_i is a one-point set. A map $f : (\overset{\bullet}{I}{}^{n+1}, v) \to (X, x)$ is *concentrated* on the island W if $f(\overset{\bullet}{I}{}^{n+1} - W) = \{x\}$. Any map $f : (\overset{\bullet}{I}{}^{n+1}, v) \to (X, x)$ can be replaced, up to homotopy rel $\{v\}$, by a map concentrated on W. If $[f_1]$ and $[f_2]$ lie in $\pi_n(X, x)$, assume they are concentrated on disjoint islands W_1 and W_2; when $n \ge 2$ then $[f_1][f_2]$ is represented by any map $(\overset{\bullet}{I}{}^{n+1}, v) \to (X, x)$ agreeing with f_1 on W_1 and with f_2 on W_2, and sending the rest of $\overset{\bullet}{I}{}^{n+1}$ to $\{x\}$, while for $n = 1$, one requires that the f_1-island should lie to the left of the f_2-island. The group properties are easily checked. When $n \ge 2$, $\pi_n(X, x)$ is therefore abelian; then one writes $[f_1] + [f_2]$ rather than $[f_1][f_2]$. Of course there is no such homeomorphism when $n = 1$. For relative homotopy groups, represent $[f] \in \pi_n(X, A, x)$ by $f : B^n(= I^n) \to X$ where f maps the face

---

[5] Details can be found in many books on algebraic topology, for example [146, Chap. 7, Sect. 2] or [82, Chap. 2].

[6] In general, there is no useful group structure on $\pi_1(X, A, x)$ when A has more than one point.

$F_{n,-1}$ (see Sect. 2.5) into A and maps the rest of $\overset{\bullet}{I}{}^n$ to $\{x\}$. Now require each (n-dimensional) island W to meet $F_{n,-1}$ in an $(n-1)$-dimensional island disjoint from $\overset{\bullet}{F}_{n,-1}$. The group operation is defined as before. For an abelian multiplication the previous argument requires the dimension of $F_{n,-1}$ to be at least 2, i.e., $n \geq 3$.

4.4.6. A map $f : (X, x) \to (Y, y)$ [resp. $f : (X, A, x) \to (Y, B, y)$] induces a homomorphism $f_{\#} : \pi_n(X, x) \to \pi_n(Y, y)$ [resp. $f_{\#} : \pi_n(X, A, x) \to \pi_n(Y, B, y)$] of groups when its domain and range are groups, and a function of pointed sets when no group structure is present. This satisfies the usual properties of a covariant functor: $(g \circ f)_{\#} = g_{\#} \circ f_{\#}$ and $(\mathrm{id})_{\#} = \mathrm{id}$.

4.4.7. All this agrees with what we discussed previously for the fundamental group $\pi_1(X, x)$ in Sect. 3.3.

4.4.8. Define $\partial_{\#} : \pi_n(X, A, x) \to \pi_{n-1}(A, x)$ by $\partial_{\#}([f]) = [f \mid S^{n-1}]$. When $n \geq 2$, $\partial_{\#}$ is a homomorphism.

4.4.9. If $i : A \hookrightarrow X$ and $j : X(= (X, \{x\})) \to (X, A)$ are inclusions, the following *homotopy sequence of* (X, A, x) is exact:

$$\cdots \xrightarrow{\ \partial_{\#}\ } \pi_n(A, x) \xrightarrow{\ i_{\#}\ } \pi_n(X, x) \xrightarrow{\ j_{\#}\ } \pi_n(X, A, x) \xrightarrow{\ \partial_{\#}\ } \pi_{n-1}(A, x) \longrightarrow \cdots$$

and if $f : (X, A, x) \to (Y, B, y)$ is a map, f induces a commutative diagram of maps from the homotopy sequence of (X, A, x) to that of (Y, B, y). Note: when n is low, "kernel" means "pre-image of the base point," so "exact" still makes sense.

4.4.10. If $p : (\bar{X}, \bar{x}) \to (X, x)$ is a covering projection, then $p_{\#} : \pi_n(\bar{X}, \bar{x}) \to \pi_n(X, x)$ is an isomorphism for all $n \geq 2$.

4.4.11. A map $p : E \to B$ is a *fiber bundle* iff there is a space F, an open cover $\{U_\alpha\}$ of B, and, for each α, a homeomorphism h_α making the following diagram commute:

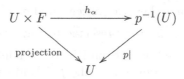

The spaces E, B and F are called, respectively, the *total space, base space* and *fiber* of the fiber bundle. Let $e \in E$ be a base point, write $b = p(e)$ and $F_e = p^{-1}(b)$ (F_e is a copy of F). If B is paracompact and Hausdorff there is a natural isomorphism $\pi_*(E, F_e, e) \simeq \pi_*(B, b)$ leading to the long exact sequence

$$\cdots \longrightarrow \pi_n(F_e, e) \xrightarrow{\text{(inclusion)}_{\#}} \pi_n(E, e) \xrightarrow{\ p_{\#}\ } \pi_n(B, b) \longrightarrow \pi_{n-1}(F_e, e) \longrightarrow \cdots$$

Exercises

1. Deduce the following Whitehead Theorem from 4.1.4: Let (X, x) and (Y, y) be path connected pointed CW complexes and let $f : (X, x) \to (Y, y)$ be a map such that $f_\# : \pi_n(X, x) \to \pi_n(Y, y)$ is an isomorphism for all n. Then f is a homotopy equivalence. More precisely, prove that a "pointed homotopy inverse" g exists for f such that $g \circ f$ and $f \circ g$ are homotopic to the appropriate identity maps rel base points. (Such a map f is a *pointed homotopy equivalence*.)

2. Write down and prove a version of the Whitehead Theorem of Exercise 1 in which X and Y are only required to have the homotopy types of CW complexes.

3. Give a counterexample to the "Whitehead theorem" in Exercise 1 when X does not have the homotopy type of a CW complex.

4. Prove that a map between path connected pointed CW complexes $f : (X, x) \to (Y, y)$ is an n-equivalence iff $f_\# : \pi_i(X, x) \to \pi_i(Y, y)$ is an isomorphism for all $i \leq n - 1$.

4.5 Geometric proof of the Hurewicz Theorem

Let (X, A) be a pair of CW complexes and let x be a vertex of A. When $\pi_n(X, A, x)$ is a group,[7] the *Hurewicz homomorphism* $h_n : \pi_n(X, A, x) \to H_n(X, A; \mathbb{Z})$ is defined as follows: if $f : (B^n, S^{n-1}, v) \to (X, A, x)$ is a map,

$$h_n([f]) = \left\{ \sum_{\alpha \in \mathcal{A}} [B^n : e^n_\alpha : f] e^n_\alpha \right\}; \text{ here, } B^n \text{ is given the usual CW structure}$$

(one vertex, one $(n-1)$-cell, and one n-cell), $\mathcal{A}$ indexes the n-cells of X which are not in A, and $\{\cdot\}$ marks the homology class of a relative n-cycle. When $A = \{x\}$ we write $h_n : \pi_n(X, x) \to H_n(X; \mathbb{Z})$, identifying $H_n(X; \mathbb{Z})$ with $H_n(X, \{x\}; \mathbb{Z})$ since $n \geq 1$. Note that h_1 has already appeared, as h, in 3.1.19.

We say X is *n-acyclic* if $\tilde{H}_k(X; \mathbb{Z}) = 0$ for all $k \leq n$; we say (X, A) is *n-acyclic* if $H_k(X, A; \mathbb{Z}) = 0$ for all $k \leq n$. It is often much easier to prove n-acyclicity than n-connectedness, so the following is a powerful tool:

Theorem 4.5.1. (Relative Hurewicz Theorem) *Let X and A be simply connected CW complexes and let $n \geq 2$. Then (X, A) is $(n-1)$-connected iff (X, A) is $(n-1)$-acyclic, and if this holds then $h_n : \pi_n(X, A, x) \to H_n(X, A; \mathbb{Z})$ is an isomorphism.*

The special case $A = \{x\}$ is older and is important:

Theorem 4.5.2. (Hurewicz Theorem) *Let X be a simply connected CW complex and let $n \geq 2$. Then X is $(n-1)$-connected iff X is $(n-1)$-acyclic, and if this holds then $h_n : \pi_n(X, x) \to H_n(X; \mathbb{Z})$ is an isomorphism.*

In preparation for the proof, we state two senses in which the Hurewicz homomorphism is natural. The proof is left as an exercise:

---

[7] i.e., $n \geq 2$ if $A \neq \{x\}$, $n \geq 1$ if $A = \{x\}$.

Proposition 4.5.3. (i) *The following diagram commutes*

$$
\begin{array}{ccccccccc}
\cdots \longrightarrow & \pi_n(A, x) & \xrightarrow{i_\#} & \pi_n(X, x) & \xrightarrow{j_\#} & \pi_n(X, A, x) & \xrightarrow{\partial_\#} & \pi_{n-1}(A, x) & \longrightarrow \cdots \\
& \downarrow{h_n} & & \downarrow{h_n} & & \downarrow{h_n} & & \downarrow{h_{n-1}} & \\
\cdots \longrightarrow & H_n(A; \mathbb{Z}) & \xrightarrow{i_*} & H_n(X; \mathbb{Z}) & \xrightarrow{j_*} & H_n(X, A; \mathbb{Z}) & \xrightarrow{\partial_*} & H_{n-1}(A; \mathbb{Z}) & \longrightarrow \cdots
\end{array}
$$

(ii) *If $f : (X, A, x) \to (Y, B, y)$ is a map, the following diagram commutes:*

$$
\begin{array}{ccc}
\pi_n(X, A, x) & \xrightarrow{f_\#} & \pi_n(Y, B, y) \\
\downarrow{h_n} & & \downarrow{h_n} \\
H_n(X, A; \mathbb{Z}) & \xrightarrow{f_*} & H_n(Y, B; \mathbb{Z}).
\end{array}
$$

$\square$

It is convenient to prove 4.5.2 first, as one case of that theorem is needed in the proof of 4.5.1.

Proof (of 4.5.2). We first deal with the special case in which (X, x) is a bouquet of n-spheres, $\left(\bigvee_\alpha S_\alpha^n, w \right)$. We are to prove $h_n : \pi_n \left(\bigvee_\alpha S_\alpha^n, w \right) \to H_n \left(\bigvee_\alpha S_\alpha^n \right) \cong \bigoplus_\alpha \mathbb{Z}$ has trivial kernel, since the rest of the theorem is obvious in this case. We pause for some preparations.

Recall from Sect. 4.4 the definition of an *island* on S^n; loosely, it is a nicely placed n-ball. An *archipelago* in S^n is a finite set of pairwise disjoint islands. The map $f : S^n \to \bigvee_{\alpha \in \mathcal{A}} S_\alpha^n$ is *concentrated* on the archipelago $\{W_\gamma \mid \gamma \in \mathcal{C}\}$ if (i) for each $\gamma \in \mathcal{C}$ there is some $\alpha \in \mathcal{A}$ such that $f(W_\gamma) \subset S_\alpha^n$, and (ii) $f \left(S^n - \left(\bigcup_\gamma W_\gamma \right) \right) = \{w\}$. $\square$

Proposition 4.5.4. *For any map $f : (S^n, v) \to \left(\bigvee_{\alpha \in \mathcal{A}} S_\alpha^n, w \right)$ there is an archipelago $\{W_\gamma \mid \gamma \in \mathcal{C}\}$ in S^n such that f is homotopic rel $\{v\}$ to a map g concentrated on $\{W_\gamma\}$.*

Proof. For each $\alpha \in \mathcal{A}$, pick $z_\alpha \neq v$ lying in S_α^n. By 1.3.7, there is a neighborhood N_α of $z_\alpha \in S_\alpha^n - \{w\}$ and a homotopy $D^\alpha : S_\alpha^n \times I \to S_\alpha^n$, rel $\{v\}$, such that $D_0^\alpha = \mathrm{id}$, $D_1^\alpha(V_\alpha) = \{w\}$ and $D_1^\alpha(N_\alpha) = S_\alpha^n$, where $V_\alpha := S_\alpha^n - N_\alpha$.

Let $U = \cup \{V_\alpha \mid \alpha \in \mathcal{A}\}$. Then $\mathcal{U} := \{U\} \cup \left\{ \bigvee_\alpha S_\alpha^n - \{v\} \right\}$ is an open cover of

$\bigvee_{\alpha \in \mathcal{A}} S_\alpha^n$. The homotopies D^α give a homotopy $D : \left(\bigvee_{\alpha \in \mathcal{A}} S_\alpha^n \right) \times I \to \bigvee_{\alpha \in \mathcal{A}} S_\alpha^n$,

rel $\{v\}$, such that $D_0 = \mathrm{id}$, $D_t(S_\alpha^n) = S_\alpha^n$ for all $\alpha \in \mathcal{A}$ and $0 \le t \le 1$, and $D_1(U) = \{w\}$.

We will use $\overset{\bullet}{I}{}^{n+1}$ as the domain of f. We give $\bigvee_{\alpha \in \mathcal{A}} S_\alpha^n$ the CW complex

structure consisting of one vertex, w, and an n-cell, S_α^n, for each α. By 2.4.13, there exists k such that each cell of $\overset{\bullet}{I}{}_k^{n+1}$ lies in some element of $f^{-1}\mathcal{U}$. f is homotopic to $D_1 \circ f$ which maps each n-cell of $\overset{\bullet}{I}{}_k^{n+1}$ into S_α^n for some α. By 1.4.4, $D_1 \circ f$ is homotopic to a cellular map f' such that f' takes the entire $(n-1)$-skeleton of $\overset{\bullet}{I}{}^{n+1}$ to w, and maps each n-cell of $\overset{\bullet}{I}{}_k^{n+1}$ into some S_α^n.

The n-cells of $\overset{\bullet}{I}{}_k^{n+1}$ are not pairwise disjoint, so they do not form an archipelago. We remedy this by passing to $\overset{\bullet}{I}{}_{k+2}^{n+1}$. The effect is to partition each n-cell e_γ^n of $\overset{\bullet}{I}{}_k^{n+1}$ into 4^n n-cells, 2^n of which are disjoint from $\overset{\bullet}{e}{}_\gamma^n$. The union of these 2^n cells is an n-cell $\tilde{e}_\gamma^n$, and there[8] is a homotopy $H : e_\gamma^n \times I \to e_\gamma^n$ such that $H_0 = \mathrm{id}$, $H_1(e_\gamma^n - \tilde{e}_\gamma^n) \subset \overset{\bullet}{e}{}_\gamma^n$ and $H_1(\tilde{e}_\gamma^n) = e_\gamma^n$. The required archipelago is $\{\tilde{e}_\gamma^n \mid \gamma \in \mathcal{C}\}$ where $\mathcal{C}$ indexes the n-cells of $\overset{\bullet}{I}{}_k^{n+1}$. The required map g is of $f' \circ H_1$. $\qquad\square$

Proposition 4.5.5. Let $n \ge 2$, let $\mathcal{A}$ be finite and let the map $f : S^n \to \bigvee_{\alpha \in \mathcal{A}} S_\alpha^n$ be concentrated on the archipelago $\{W_\gamma \mid \gamma \in \mathcal{C}\}$. Then f is homotopic to a map g concentrated on an archipelago $\{V_\alpha \mid \alpha \in \mathcal{A}\}$ such that $g(V_\alpha) \subset S_\alpha^n$ for each $\alpha \in \mathcal{A}$.

Proof. Pick an archipelago $\{V_\alpha \mid \alpha \in \mathcal{A}\}$ in S^n. By a simple application of 2.4.15, there is a homeomorphism $h : S^n \to S^n$ which is homotopic to the identity map, such that whenever f is non-constant on the island W_γ and maps W_γ into S_α^n then $h(W_\gamma) \subset V_\alpha$. The required map g is $f \circ h^{-1}$. $\qquad\square$

Proof (of 4.5.2 (concluded)). Let $f : (S^n, v) \to \left(\bigvee_{\alpha \in \mathcal{A}} S_\alpha^n, w \right)$ represent an

element of $\ker(h_n)$. Since S^n is compact we may proceed as if $\mathcal{A}$ were finite. By 4.5.4 and 4.5.5 there is an archipelago $\{V_\alpha\}$ so that $f(V_\alpha) \subset S_\alpha^n$ for each $\alpha \in \mathcal{A}$. The mapping degree $[S^n : S_\alpha^n : f]$ is the "degree" with which f maps V_α onto S_α^n, namely 0 since $[f] \in \ker(h_n)$. By 2.4.5, $f \mid V_\alpha$ is homotopic rel bd V_α to the constant map at v, for each α. Thus the special case is proved.

---

[8] The details here just involve elementary games with product cells in I_{k+2}^{n+1}.

We use this to prove the general case of 4.5.2. By induction on n, we may assume X is $(n-1)$-connected when $n \geq 2$; this is trivial when $n = 1$. Clearly $h_n : \pi_n(X, x) \to H_n(X; \mathbb{Z})$ is surjective (exercise). We are to prove $\ker(h_n)$ is trivial. Consider the commutative diagram

$$
\begin{array}{ccc}
\pi_n(X^n, x) & \xrightarrow{\ i_\# \ } & \pi_n(X, x) \\
\cong \Big\downarrow h'_n & & \Big\downarrow h_n \\
H_n(X^n; \mathbb{Z}) & \xrightarrow{\ i_* \ } & H_n(X; \mathbb{Z})
\end{array}
$$

where i denotes the inclusion map. Here, we have renamed one of the Hurewicz homomorphisms h'_n to distinguish it from h_n. Applying 4.2.1 inductively to the $(n-1)$-connected pair (X, x), we may assume $X^{n-1} = \{x\}$. The n-cells [resp. $(n+1)$-cells] of X are $\{e_\alpha^n \mid \alpha \in \mathcal{A}\}$ [resp. $\{e_\beta^{n+1} \mid \beta \in \mathcal{B}\}$]. Thus $X^n = \bigcup_\alpha e_\alpha^n = \bigvee_\alpha S_\alpha^n$. Let $g_\beta : S^n \to X^n$ be the attaching map for e_β^{n+1}. We may assume $g_\beta(v) = x$, so g_β defines $[g_\beta] \in \pi_n(X^n, x)$. We have $h'_n([g_\beta]) = \left\{ \sum_\alpha [e_\beta^{n+1} : e_\alpha^n] e_\alpha^n \right\}$. Let $f : (S^n, v) \to (X, x)$ represent an element $[f] \in \ker(h_n) \leq \pi_n(X, x)$. Write $f' : (S^n, v) \to (X^n, x)$ for the corestriction of f. In the above diagram, $h_n i_\#([f']) = h_n([f]) = 0$, so $h'_n([f']) \in \ker i_*$.

Thus there are integers n_β such that $h'_n([f']) = \left\{ \partial \left(\sum_\beta n_\beta e_\beta^{n+1} \right) \right\} =$

$\left\{ \sum_\beta n_\beta \sum_\alpha [e_\beta^{n+1} : e_\alpha^n] e_\alpha^n \right\} = \sum_\beta n_\beta h'_n([g_\beta])$. But h'_n is an isomorphism by the special case. So $[f'] = \sum_\beta n_\beta [g_\beta]$. And $i_\#([g_\beta]) = 0$ since g_β extends across e_β^{n+1}. So $i_\#([f']) = [f] = 0$. $\qquad\square$

Proof (of 4.5.1). The induction on n begins with $n = 2$. By 4.5.3 (i) we have a commutative diagram

$$
\begin{array}{ccccccc}
\pi_2(A, x) & \longrightarrow & \pi_2(X, x) & \longrightarrow & \pi_2(X, A, x) & \longrightarrow & 0 \\
\Big\downarrow h'_2 & & \Big\downarrow h''_2 & & \Big\downarrow h_2 & & \\
H_2(A; \mathbb{Z}) & \longrightarrow & H_2(X, \mathbb{Z}) & \longrightarrow & H_2(X, A; \mathbb{Z}) & \longrightarrow & 0
\end{array}
$$

with exact horizontal lines. By 4.5.2, the Hurewicz homomorphisms h'_2 and h''_2 are isomorphisms. So h_2 is an isomorphism.

Now we proceed by induction, assuming $n \geq 3$. As before, the only difficulty is in proving $\ker(h_n) = 0$. The proof of this is totally analogous to the proof of 4.5.2. We leave the details as an instructive exercise. $\qquad\square$

Example 4.5.6. Consider $X = S^1 \vee S^2$ with vertex v. Its universal cover $\tilde{X}$ is a line with a 2-sphere adjoined at each integer point. By 4.5.2 and 4.4.10,

$$H_2(\tilde{X}; \mathbb{Z}) \cong \bigoplus_{n=-\infty}^{\infty} \mathbb{Z} \cong \pi_2(\tilde{X}, \tilde{v}) \cong \pi_2(X, v).$$ Thus X is an example of a finite

CW complex whose second homotopy group is not finitely generated (as an abelian group). However, since the group of covering transformations of $\tilde{X}$ is infinite cyclic, $H_2(\tilde{X}; \mathbb{Z})$ can be regarded as a module over the group ring[9] $\mathbb{Z}C$ where C denotes the infinite cyclic group of covering transformations ($\cong \pi_1(X, v)$). This homology group is finitely generated as a $\mathbb{Z}C$-module, indeed cyclic, so the same is true of $\pi_2(\tilde{X}, \tilde{v}) \cong \pi_2(X, v)$. It is often desirable to view higher homotopy groups as modules over the group ring $\mathbb{Z}G$ where G is the fundamental group. In particular, when $\pi_n(X, v)$ is finitely generated as a $\mathbb{Z}G$-module one can kill it by attaching finitely many $(n + 1)$-cells to X.

Historical Note: The Relative Hurewicz Theorem appeared first in [88].

Exercises

1. Prove that h_n is a homomorphism.
2. Prove that under the hypotheses of 4.5.2 h_n is onto.
3. Fill in the missing details in the proof of 4.5.1.
4. Let X and Y be path connected CW complexes. Prove that if $f : (X, x) \to (Y, y)$ induces isomorphisms on fundamental group and on $\mathbb{Z}$-homology of universal covers, then f is a homotopy equivalence. (*Hint:* See Exercise 1 of Sect. 4.4.)
5. Let $X = A \cup B$ be a CW complex where A, B and $A \cap B$ are contractible subcomplexes. Prove that X is contractible using the Van Kampen theorem, the Mayer-Vietoris sequence, the Hurewicz Theorem and the Whitehead Theorem. (Compare Exercise 5 in Sect. 4.1.)
6. Find an example of a path connected CW complex X where h_1 is an isomorphism but h_2 is not surjective.

---

[9] See Sect. 8.1 for more on group rings.

5

Elementary Geometric Topology

The topics discussed here will arise in the course of what follows. This chapter can be used for reference.

5.1 Review of topological manifolds

A *manifold* or *topological manifold* is a metrizable space M such that for some n, called the *dimension* of M, every point has a neighborhood homeomorphic to $\mathbb{R}^n$ or to $\mathbb{R}^n_+$. Before sense can be made of this definition, one needs some fundamental theorems about the topology of $\mathbb{R}^n$ (5.1.1–5.1.6). We begin with *Invariance of Domain*:

Theorem 5.1.1. *Let $h : U \to \mathbb{R}^n$ be an embedding where U is an open subset of $\mathbb{R}^n$. Then $h(U)$ is open in $\mathbb{R}^n$.*

Remark 5.1.2. There is an elementary but long proof of this theorem in the literature – see [51, p. 358]. A more sophisticated proof, using singular homology, deduces 5.1.1 from the following three important lemmas. The proofs of the lemmas, and the deduction of 5.1.1 from them, are exercises for readers who know singular homology; alternatively, see [146, Sect. 4.7].

Lemma 5.1.3. *Let $A \subset S^n$ be homeomorphic to B^k, where $0 \leq k \leq n$. Then $\tilde{H}_i^\Delta(S^n - A; \mathbb{Z}) = 0$ for all i. [As usual $\tilde{H}_*^\Delta$ denotes reduced singular homology.]* □

Lemma 5.1.4. *Let $S \subset S^n$ be homeomorphic to S^k for $0 \leq k \leq n - 1$. Then $\tilde{H}_i^\Delta(S^n - S; \mathbb{Z})$ is trivial if $i \neq n - k - 1$ and is isomorphic to $\mathbb{Z}$ if $i = n - k - 1$.* □

Lemma 5.1.5. (Jordan-Brouwer Separation Theorem) *Let $C \subset S^n$ be homeomorphic to S^{n-1}. Then $S^n - C$ has two path components, each of whose frontier is C.* □

Much easier than these results is the following proposition which is essentially Exercise 3 in Sect. 2.2:

Proposition 5.1.6. (a) *Let* $m \neq n$. *No non-empty open subset of* $\mathbb{R}^n$ *is homeomorphic to an open subset of* $\mathbb{R}^m$.

(b) *Let* $x \in \mathbb{R}^{n-1} \subset \mathbb{R}_+^n$. *No neighborhood of* x *in* $\mathbb{R}_+^n$ *is homeomorphic to* $\mathbb{R}^n$. □

Let M be a manifold of dimension n. The *boundary* of M is $\{x \in M \mid$ some neighborhood of x is homeomorphic to $\mathbb{R}_+^n\}$; this subset is denoted by ∂M. The set $M - \partial M$, abbreviated to $\overset{o}{M}$, is the *interior* of M; it consists of all the points of M having neighborhoods homeomorphic to $\mathbb{R}^n$. The manifold M is *closed* if M is compact and ∂M is empty. M is *open* if each path component of M is non-compact and ∂M is empty.[1]

Proposition 5.1.6 has a number of consequences. If M is non-empty then $\overset{o}{M}$ is non-empty. Hence, by considering a neighborhood of a point of $\overset{o}{M}$, one sees that the dimension of M is well defined. Moreover, $\partial M \cap \overset{o}{M} = \emptyset$. A manifold of dimension n is called an *n-manifold*; a manifold of dimension 2 is also called a *surface*. Note that since ∂M is topologically characterized, a homeomorphism between manifolds carries boundary to boundary and interior to interior. If M carries the structure of a CW complex, the dimension (in the sense of Sect. 1.2) of that CW complex is n.

Here are some easily deduced properties of an n-manifold M: ∂M is an $(n-1)$-manifold; $\partial(\partial M) = \emptyset$; ∂M is a closed subset of M; ∂M is nowhere dense in M; $\overset{o}{M}$ is open in M; $\overset{o}{M}$ is dense in M. Manifolds are locally compact.

Remark 5.1.7. Consider the special cases of 5.1.3 and 5.1.4 in which $n = 3$. If A is an arc (i.e., is homeomorphic to B^1) and S is homeomorphic to S^2, we have $H_1^\Delta(S^3 - A) = 0 = H_1^\Delta(S^3 - S)$. However, there is an arc $A \subset S^3$ (the Fox-Artin Arc) whose complement is not simply connected; and there is a 2-sphere $S \subset S^3$ (the Alexander Horned Sphere) one of whose complementary components is not simply connected. For more on these, see, for example, [138].

Let M be a path connected n-manifold and let $B \subset \overset{o}{M}$ be an n-ball. By 5.1.1, int B is an open subset of M, but it does not follow that $M - $ int B is a manifold.[2] An n-ball B in M is *unknotted* if $M - $ int B is a manifold. If B

---

[1] The reuse of the terms "interior", "closed" and "open" is unfortunate but entrenched. More troublesome is the reuse of "boundary"; in addition to the two uses of that word in this book (here and in Sect. 2.1) many authors also use it for what we have called the "frontier". As we remarked in Sect. 1.2, the word "boundary" is also used for the subset $\overset{\bullet}{e}$ of a cell e in a CW complex.

[2] For example B might be the closure of the component of $(S^3 - $ Alexander Horned Sphere) which is simply connected – see 5.1.7.

and C are unknotted n-balls in M there is a homeomorphism[3] of M taking B onto C; this is a consequence of the Generalized Schoenflies Theorem (see [45]). Unknotted balls are needed for the "connected sum construction": let M_1 and M_2 be path connected n-manifolds, let B_1 and B_2 be unknotted n-balls[4] in $\overset{\circ}{M}_1$ and $\overset{\circ}{M}_2$ respectively with boundaries S_1 and S_2, and let $h : S_1 \to S_2$ be a homeomorphism between these $(n-1)$-spheres. The relation $x \sim h(x)$ for all $x \in S_1$ defines an equivalence relation on $\mathrm{cl}(M_1 - \mathrm{int}\ B_1) \coprod \mathrm{cl}(M_2 - \mathrm{int}\ B_2)$ and the resulting quotient space is called the *connected sum* of M_1 and M_2, denoted $M_1 \# M_2$. It is a manifold because the balls are unknotted. Up to homeomorphism this is independent of the choices of B_1, B_2 and h. See Fig. 5.1.

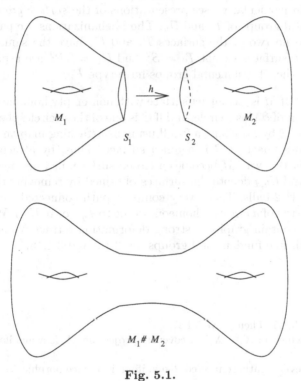

Fig. 5.1.

Example 5.1.8. The *closed orientable surface* T_g is defined for $g \in \mathbb{N}$ by: $T_0 = S^2$, $T_1 = T^2$ (the 2-torus), $T_{g+1} = T_g \# T_1$. This number g is the *genus* of the surface. This surface T_g is also called a *sphere with g handles*. The *closed*

---

[3] Indeed, B can be carried onto C by an ambient isotopy, i.e., a homotopy through homeomorphisms starting at id_M.

[4] It is easy to find unknotted n-balls in $\mathbb{R}^n$, hence in $\overset{\circ}{M}$ for any M.

non-orientable surface U_h is defined for $h \geq 1$ by: $U_1 = \mathbb{R}P^2$ (the projective plane), $U_{h+1} = U_h \# U_1$. This surface U_h is called[5] a *sphere with h crosscaps*. It is well-known, see for example [110] or [151], that every closed path connected surface is homeomorphic to some T_g or some U_h. These surfaces are CW complexes as follows. Let $K(g)$ be the presentation complex of

$$\langle a_1, b_1, \cdots, a_g, b_g \mid [a_1, b_1][a_2, b_2] \cdots [a_g, b_g] \rangle$$

where $[a, b]$ denotes the commutator $aba^{-1}b^{-1}$, and let $L(h)$ be the presentation complex of

$$\langle a_1, \cdots, a_h \mid a_1^2 a_2^2 \cdots a_h^2 \rangle.$$

The underlying spaces of the CW complexes $K(g)$ and $L(h)$ are T_g and U_h respectively. In particular we see presentations of the *surface groups*, namely the fundamental groups of T_g and U_h. The abelianizations are pairwise non-isomorphic, so no two of the surfaces T_g and U_h have the same homotopy type. All these surfaces except $T_0 = S^2$ and $U_1 = \mathbb{R}P^2$ are aspherical (see Exercise 7) so their fundamental groups have type F.

Example 5.1.9. If M is a compact surface with non-empty boundary then each path component of ∂M is a circle, and if C is one of those circles then the space obtained from M by attaching a 2-cell using as attaching map an embedding $S^1 \to M$ whose image is C is again a surface. Thus, by attaching finitely many 2-cells in this way M becomes a closed surface, i.e., becomes a T_g or a U_h. Let $T_{g,d}$ and $U_{h,d}$ denote the surfaces obtained by removing the interiors of d unknotted[6] 2-balls. Then every compact path connected surface whose boundary consists of d circles is homeomorphic to $T_{g,d}$ or to $U_{h,d}$. When $d > 0$, $T_{g,d}$ and $U_{h,d}$ contain graphs as strong deformation retracts; hence they are aspherical with free fundamental groups, by 3.1.9 and 3.1.16.

Exercises

1. Prove 5.1.3–5.1.5. Then prove 5.1.1.
2. Show that when $p : \bar{X} \to X$ is a covering projection, $\bar{X}$ is a manifold iff X is a manifold.
3. Show that every path connected 1-manifold is homeomorphic to S^1, I, $\mathbb{R}$ or $[0, \infty)$.
4. Show that if M is an n-manifold then $M \# S^n$ is homeomorphic to M.
5. Show that $T_{g+1,d}$ is homeomorphic to $T_{r,k} \# T_{g+1-r,d-k+2}$.
6. For n-manifolds M_1 and M_2, show that $\pi_1(M_1 \# M_2) \cong \pi_1(M_1) * \pi_1(M_2)$ when $n > 2$. Discuss the cases $n \leq 2$.
7. For $g \geq 3$ show that T_g is a $(g - 1)$ to 1 covering space of T_2.

---

[5] One expresses $T_{g+1} = T_g \# T_1$ [resp. $U_{h+1} = U_h \# U_1$] by saying T_{g+1} is obtained from T_g by *attaching a handle* [resp. U_{h+1} is obtained from U_h by *attaching a crosscap*].

[6] By the Schoenfliesz Theorem (see [138]) every 2-ball in a surface is unknotted.

8. Show that T_g is a 2 to 1 covering space of U_{g+1}.
9. By considering the universal cover of $K(2)$ (see Example 5.1.8) show that the universal cover of T_2 is homeomorphic to $\mathbb{R}^2$. Deduce that the universal covers of all the path connected closed surfaces except S^2 and $\mathbb{R}P^2$ are homeomorphic to $\mathbb{R}^2$. (In the terminology of Ch. 7, all such surfaces are aspherical.)
10. Let G act freely and cocompactly on a path connected orientable open surface S. Prove that $H_1(S; \mathbb{Z})$ is finitely generated as a $\mathbb{Z}G$-module if G is finitely presented, and it is not finitely generated as a $\mathbb{Z}G$-module if G is finitely generated but does not have type FP_2 over $\mathbb{Z}$.

5.2 Simplicial complexes and combinatorial manifolds

This is an exposition of simplicial complexes, their underlying polyhedra, joins, and combinatorial manifolds. It is intended to be both an exposition and a place to refer back to for definitions as needed.

An *abstract simplicial complex*, K, consists of a set V_K of *vertices* and a set S_K of finite non-empty subsets of V_K called *simplexes*; these satisfy: (i) every one-element subset of V_K is a simplex, and (ii) every non-empty subset of a simplex is a simplex. An *n-simplex* of K is a simplex containing $(n+1)$-vertices (in which case n is the *dimension* of the simplex). The *empty* abstract simplicial complex, denoted by $\emptyset$, has $V_\emptyset = S_\emptyset = \emptyset$. We say K is *finite* if V_K is finite (in which case S_K is finite), K is *countable* if V_K is countable (in which case S_K is countable), and K is *locally finite* if each vertex lies in only finitely many simplexes. The *dimension* of K is the supremum of the dimensions of its simplexes.

If K and L are abstract simplicial complexes, a *simplicial map* $\phi : K \to L$ is a function $V_K \to V_L$ taking simplexes of K onto simplexes of L. If ϕ has a two-sided inverse which is simplicial, then ϕ is a *simplicial isomorphism*.

One associates a CW complex with the abstract simplicial complex K as follows. Let W be the real vector space $\prod_{v \in V_K} \mathbb{R}$; i.e., the cartesian product of copies of $\mathbb{R}$ indexed by V_K, with the usual coordinatewise addition and scalar multiplication. Topologize every finite-dimensional linear subspace U of W by giving it the appropriate euclidean topology. Give W the weak topology with respect to this family of subspaces (which is suitable in the sense of Sect. 1.1 for defining a weak topology). This is called the *finite topology*[7] on W. Abusing notation, let v also denote the point of W having entry 1 in the v-coordinate, and all other entries 0. For each simplex $\sigma = \{v_0, \cdots, v_n\}$ of K, let $|\sigma|$ be the (closed) convex hull in W of $\{v_0, \cdots, v_n\}$. Let $|K|^n = \bigcup \{|\sigma| \mid \sigma$ is an n-simplex of $K\}$. Let $|K| = \bigcup_{n \geq 0} |K|^n$. Then $(|K|, \{|K|^n\})$, with the topology

---

[7] The finite topology does not make W a topological vector space, but it makes each finite-dimensional linear subspace a topological vector space.

inherited from W, is a CW complex. The details are an exercise; see also [51, pp. 171–172]. This $|K|$ is the *geometric realization* of K.

If $\phi : K \to L$ is a simplicial map, there is an associated map $|\phi| : |K| \to |L|$ which maps the vertex of $|K|$ corresponding to $v \in V_K$ to the vertex of $|L|$ corresponding to $\phi(v) \in V_L$, and is affine on each $|\sigma|$. Clearly $|\phi|$ is continuous, and if ϕ is a simplicial isomorphism, $|\phi|$ is a homeomorphism.

The notations $|\sigma|$ and $|K|$ are sometimes used in a slightly different way. Assume (i) V_K is a subset of $\mathbb{R}^N$ such that the vertices of each simplex of K are affinely independent. With σ as above, define $|\sigma|$ to be the (closed) convex hull of σ in $\mathbb{R}^N$. Write $|\overset{\circ}{\sigma}|$ for $\left\{ \sum_{i=0}^{n} t_i v_i \mid 0 < t_i < 1 \text{ and } \sum_{i=0}^{n} t_i = 1 \right\}$, the *open convex hull* of σ. Assume (ii) that whenever $\sigma \neq \tau \in S_K$, $|\overset{\circ}{\sigma}| \cap |\overset{\circ}{\tau}| = \emptyset$. Define $|K| = \bigcup \{ |\sigma| \mid \sigma \in S_K \}$ and $|K|^n = \bigcup \{ |\sigma| \mid \sigma \text{ is a } k\text{-simplex of } K \text{ and } k \leq n \}$. Then $|K|$ (with topology inherited from $\mathbb{R}^N$) and $\{ |\overset{\circ}{\sigma}| \mid \sigma \in S_K \}$ satisfy all but one of the requirements in Proposition 1.2.14 for $(|K|, \{ |K|^n \})$ to be a CW complex: the sole problem is that the topology which $|K|$ inherits from $\mathbb{R}^N$ might not agree with the weak topology with respect to $\{ |\sigma| \mid \sigma \in S_K \}$. Assume (iii) that $\{ |\sigma| \mid \sigma \in S_K \}$ is a locally finite family of subsets of the space $|K|$ (where $|K|$ has the inherited topology). If K satisfies these assumptions (i), (ii) and (iii), we call K a *simplicial complex in* $\mathbb{R}^N$, and we call $|K|$ its *underlying polyhedron*. The cell $|\sigma|$ of $|K|$ is often called a *simplex* of $|K|$; context prevents this double use of the word from causing problems. Note that $|K|$ might not be closed in $\mathbb{R}^N$; it is closed iff the set of simplexes $|\sigma|$ is a locally finite family of subsets of $\mathbb{R}^N$ (rather than of $|K|$).

A space Z is *triangulable* if there is an abstract simplicial complex K such that Z is homeomorphic to $|K|$, and K is called[8] a *triangulation* of Z.

Example 5.2.1. The half-open interval $(0, 1]$ in $\mathbb{R}$ is triangulable: take $V_K = \left\{ \dfrac{1}{n} \mid n \in \mathbb{N} \right\}$ and S_K to be the set of pairs $\left\{ \dfrac{1}{n}, \dfrac{1}{n+1} \right\}$ together with V_K. But the subspace $V_K \cup \{0\}$ of $\mathbb{R}$ is not triangulable.

Proposition 5.2.2. *If K is a simplicial complex in $\mathbb{R}^N$, the weak topology on $|K|$ with respect to $\{ |\sigma| \mid \sigma \in S_K \}$ agrees with the topology inherited from $\mathbb{R}^N$. Moreover, $(|K|, \{ |K|^n \})$ is a countable locally finite CW complex.*

Proof. $|K|_{\text{weak}}$ is a CW complex. Suppose it is not locally finite. Then for some simplex σ of K, there is an infinite collection $\{ \tau_\alpha \}$ of simplexes of K with $|\sigma| \cap |\tau_\alpha| \neq \emptyset$ for all α. Since $|\sigma|$ is compact, (iii), above, implies that there are open sets (in $\mathbb{R}^N$) $U_1, \cdots, U_k$ whose union contains $|\sigma|$ and meets only finitely many of $\{ |\tau_\alpha| \}$. This is a contradiction. So $|K|_{\text{weak}}$ is locally compact and Hausdorff. The "identity" map $|K|_{\text{weak}} \to |K|_{\text{inherited}}$ is continuous, and

---

[8] The word "triangulation" is also used for a homeomorphism $h : |K| \to Z$, or sometimes just for the CW complex $|K|$.

$|K|_{\text{inherited}}$ is clearly locally compact and Hausdorff. By 1.2.11, $|K|_{\text{weak}}$ is Hausdorff. By 10.1.8, below, $|K|_{\text{weak}}$ is locally compact, and by 10.1.6, below, this implies that the "identity" is a homeomorphism. Thus the two topologies agree. □

Remark 5.2.3. Conveniently, $|\overset{\circ}{\sigma}|$ as defined above coincides with $\overset{\circ}{e}$ as defined in Sect. 1.2 when $e = |\sigma|$ is a cell of the CW complex $|K|$.

Let K and L be abstract simplicial complexes: L is a *subcomplex* of K if $V_L \subset V_K$ and $S_L \subset S_K$. Then $|L|$ is a subcomplex of $|K|$ in the sense of Sect. 1.2. We say that L is a *full* subcomplex of K if, in addition, any set of vertices of L which is a simplex of K is a simplex of L. Then $|L|$ is a full subcomplex of $|K|$ in the sense of Sect. 1.5.

If L is a subcomplex of K, the *simplicial neighborhood of L in K* is the subcomplex $N(L)$, or $N_K(L)$, generated by all simplexes of K which have a vertex in L. Then $|N(L)| = N(|L|)$ in the sense of Sect. 11.4.

If K and L are abstract simplicial complexes, their *join*, $K * L$, is the abstract simplicial complex with[9] $V_{K*L} = V_K \amalg V_L$ and $S_{K*L} = S_K \amalg S_L \amalg \{\sigma \amalg \tau \mid \sigma \in S_K, \tau \in S_L\}$. If K [resp. L] $= \emptyset$, it is implied that $K * L = L$ [resp. K].

If X and Y are non-empty spaces, their *topological join* is the adjunction space $X*Y = (X \amalg Y) \cup_f (X \times I \times Y)$ where $f : X \times \{0, 1\} \times Y \to X \amalg Y$ is defined by $f(x, 0, y) = x$ and $f(x, 1, y) = y$. To see what this means geometrically, consider an equivalent definition: $X * Y$ is the quotient space $(X \times I \times Y)/\sim$ where the equivalence relation $\sim$ is generated by: $(x, 0, y_1) \sim (x, 0, y_2)$ for all $y_1, y_2 \in Y$; and $(x_1, 1, y) \sim (x_2, 1, y)$ for all $x_1, x_2 \in X$. Thus $X * Y$ contains a "line segment" joining each $x \in X$ to each $y \in Y$; two such "line segments" meet, if at all, at one end point; these "line segments" vary continuously in x and in y.

Example 5.2.4. Let $X = \{((1 - a), a, 0, 0) \in \mathbb{R}^4 \mid 0 \le a \le 1\}$ and let $Y = \{(0, 0, (1 - b), b) \in \mathbb{R}^4 \mid 0 \le b \le 1\}$. Then $X * Y$ can be identified with $\{((1 - t)(1 - a), (1 - t)a, t(1 - b), tb) \in \mathbb{R}^4 \mid 0 \le a, b, t \le 1\}$. Fixing a and b we get a line segment joining $((1 - a), a, 0, 0) \in X$ to $(0, 0, (1 - b), b) \in Y$, and the various line segments meet as described above. In this example, $X * Y$ is the standard 3-simplex in $\mathbb{R}^4$.

If $A \subset X$ and $B \subset Y$ are non-empty closed sets, we consider $A * B$ to be a subspace of $X * Y$, namely the image of $A \times I \times B$ in $X * Y$. The quotient and inherited topologies agree for reasons explained in Sect. 1.1.

If X is the empty space, $X * Y$ is defined to be Y. Similarly, if Y is empty, $X * Y$ is X. If, in the last paragraph, A [resp. B] is empty, this suggests identifying the resulting $A * B$ with the obvious copy of B [resp. A] in $X * Y$, and this will always be understood.

---

[9] If $V_K \cap V_L = \emptyset$ the $\amalg$ symbol could be replaced by $\cup$.

We seek a relationship between $|K| * |L|$ and $|K * L|$. For this, we consider $|K| \subset W_1$ and $|L| \subset W_2$, as above, where W_1 and W_2 are real vector spaces each having the finite topology. Let $W = W_1 \times W_2$, again with the finite topology. Then $|K * L| \subset W$. Identifying W_1 with $W_1 \times \{0\}$ and W_2 with $\{0\} \times W_2$, we have $|K| \subset |K * L| \supset |L|$. Define a function $\phi : |K| \times I \times |L| \to |K * L|$ by the formula $\phi(x, t, y) = (1 - t)x + ty$ where addition and scalar multiplication take place in W.

Proposition 5.2.5. *If K or L is locally finite then ϕ is continuous, and ϕ induces a homeomorphism $\Phi : |K| * |L| \to |K * L|$.*

Proof. Regard I as a CW complex in the usual way. By 1.2.19, $|K| \times I \times |L|$ is a CW complex, so in order to show that ϕ is continuous we need only show that for any $\sigma \in S_K$ and $\tau \in S_L$, $\phi| : |\sigma| \times I \times |\tau| \to W$ is continuous. The domain and the image of this restriction lie in finite-dimensional subspaces of W, so $\phi|$ is certainly continuous. One easily checks that ϕ respects the equivalence relation $\sim$ (see the alternative definition of $X * Y$, above) and that the induced map $\Phi : |K| * |L| \to |K * L|$ is injective. Moreover, $\phi(|\sigma| \times I \times |\tau|)$ clearly equals $|\sigma \amalg \tau|$ so Φ is surjective. Moreover, this analysis shows that Φ maps each cell of the CW complex $|K| * |L|$ (see the first definition of $X * Y$, and 4.1.1) homeomorphically onto a cell of $|K * L|$. Thus Φ^{-1} is continuous. $\square$

When K has just one vertex v, the join $K * L$ is called the *cone* on L with vertex v and *base* L; it is denoted $v * L$. Similarly, if X is the one-point space $\{v\}$, the topological join $X * Y$ is called the *cone* on Y with *vertex* v and *base* Y; one writes $v * Y$ for this. A special case of 5.2.5 is

Corollary 5.2.6. $\Phi : v * |L| \to |v * L|$ *is a homeomorphism.* $\square$

Here are some basic exercises about topological joins of balls and spheres:

Proposition 5.2.7. *There are homeomorphisms of pairs as follows:*

$$(B^m * B^n, (S^{m-1} * B^n) \cup (B^m * S^{n-1})) \cong (B^{m+n+1}, S^{m+n})$$
$$(B^m * S^n, S^{m-1} * S^n) \cong (B^{m+n+1}, S^{m+n}).$$

If $p \in \overset{\circ}{B}^m$, then $p \in \partial(B^m * B^n)$ and $p \in (B^m * S^n)^{\circ}$. $\square$

If $\sigma, \tau \in S_K$ and $\sigma \subset \tau$, we say that σ is a *face* of τ. Consider the set of all simplexes of K of which σ is a face: the subcomplex generated by these simplexes is called the *star* of σ in K and is denoted by $\mathrm{st}_K \sigma$. The *link* of σ in K is the subcomplex, $\mathrm{lk}_K \sigma$, of $\mathrm{st}_K \sigma$ generated by those simplexes of $\mathrm{st}_K \sigma$ which contain no vertex of σ. See Fig. 5.2.

This section, so far, has consisted of two parts, one on simplicial complexes and the other on joins. These are brought together in the next proposition. If σ is a simplex of K, $\bar{\sigma}$ denotes the abstract simplicial complex whose vertices are the vertices of σ and whose simplexes are the faces of σ.

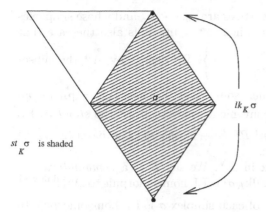

$st_K \sigma$ is shaded

Fig. 5.2.

Proposition 5.2.8. *Let σ be a simplex of the abstract simplicial complex K. Then $\bar{\sigma} * \mathrm{lk}_K \sigma = \mathrm{st}_K \sigma$. More precisely, the vertices of $\bar{\sigma}$ and of $\mathrm{lk}_K \sigma$ form two disjoint subsets of V_K, and the function sending each vertex to itself maps $\bar{\sigma} * \mathrm{lk}_K \sigma$ by a simplicial isomorphism onto the subcomplex $\mathrm{st}_K \sigma$ of K.* $\square$

The importance of 5.2.8 is that, with 5.2.5, it implies $|\sigma| * |\mathrm{lk}_K \sigma|$ is homeomorphic to $|\mathrm{st}_K \sigma|$. When K is a combinatorial n-manifold (see below) or a CW n-manifold (see Sect. 15.1), each $|\mathrm{lk}_K \sigma|$ will be homeomorphic to a sphere or a ball of dimension $n - \dim \sigma - 1$, so that, by 5.2.7 and 5.2.8, $|\mathrm{st}_K \sigma|$ will be an n-ball. To express this better we need some preliminaries.

Let K be a simplicial complex in $\mathbb{R}^N$. A *simplicial subdivision* of K is a simplicial complex K' in $\mathbb{R}^N$ such that $|K| = |K'|$ and $|K'|$ is a subdivision of $|K|$ (in the sense of Sect. 2.4). Let L be a simplicial complex in $\mathbb{R}^M$. A map $f : |K| \to |L|$ is *piecewise linear* (abbreviated to PL) if there is a simplicial subdivision K' of K such that for each simplex σ' of K', the restriction of f to $|\sigma'|$ is affine when regarded as a map into $\mathbb{R}^M$. This property of f depends only on the underlying polyhedra $|K|$ and $|L|$, in the sense that if $\bar{K}$ and $\bar{L}$ are simplicial complexes in $\mathbb{R}^N$ and $\mathbb{R}^M$ respectively such that $|K| = |\bar{K}|$ and $|L| = |\bar{L}|$, then the above map f is also PL when written $f : |\bar{K}| \to |\bar{L}|$. Clearly we have:

Proposition 5.2.9. *If $f : |K| \to |L|$ is a homeomorphism and if f is also a PL map, then f^{-1} is a PL map.* $\square$

Piecewise linearity is really a local property, like differentiability. By 5.2.9, the inverse of a PL homeomorphism is a PL homeomorphism.[10]

Recall that Δ^n is the (closed) convex hull of the points $p_0, \cdots, p_n \in \mathbb{R}^{n+1}$ where p_i has $(i+1)^{\text{th}}$ coordinate 1 and all other coordinates 0. Let $\mathbf{n}$ be the

---

[10] Contrast this with the differentiable map $f : \mathbb{R} \to \mathbb{R}$, $f(x) = x^3$, which is a homeomorphism whose inverse is not differentiable at 0.

simplicial complex in $\mathbb{R}^{n+1}$ whose vertices are $p_0, \cdots, p_n$ and whose simplexes are the non-empty sets of vertices. Then $\Delta^n = |\mathbf{n}|$. It is also the case that

$$\Delta^n = \left\{ \sum_{i=0}^{n} t_i p_i \mid 0 \leq t_i \leq 1 \ \text{and} \ \sum_i t_i = 1 \right\}. \ \text{We denote by} \ \overset{\bullet}{\Delta}{}^n \ \text{the subset}$$

of Δ^n consisting of points for which some $t_i = 0$. Then $\overset{\bullet}{\Delta}{}^n = |\overset{\bullet}{\mathbf{n}}|$ for an obvious subcomplex $\overset{\bullet}{\mathbf{n}}$ of $\mathbf{n}$. It is convenient to also call Δ^n the *standard* PL *n-ball* and to call $\overset{\bullet}{\Delta}{}^n$ the *standard PL* $(n-1)$-*sphere*. The pairs $(\Delta^n, \overset{\bullet}{\Delta}{}^n)$ and (B^n, S^{n-1}) are homeomorphic.

Let K be a simplicial complex in $\mathbb{R}^N$. We say K is a *combinatorial n-manifold* if for each simplex σ of K, $|\mathrm{lk}_K \sigma|$ is PL homeomorphic to $\Delta^{n-\dim \sigma - 1}$ or to $\overset{\bullet}{\Delta}{}^{n-\dim \sigma}$: in words, the link of each simplex σ is PL homeomorphic to the standard PL ball or PL sphere of dimension $n - \dim \sigma - 1$. Those simplexes whose links are PL homeomorphic to a standard ball define a subcomplex, ∂K, of K called the *combinatorial boundary* of K. We say K is a *closed* combinatorial n-manifold if $\partial K = \emptyset$ and K is finite.

We leave it to the reader to formulate and prove a PL version of 5.2.7 (e.g., $\overset{\bullet}{\Delta}{}^m * \Delta^n$ is PL homeomorphic to $\overset{\bullet}{\Delta}{}^{m+n+1}$).

Proposition 5.2.10. *If K is a combinatorial n-manifold, then for every simplex σ of K, $|\mathrm{st}_K \sigma|$ is PL homeomorphic to Δ^n. In particular, $|K|$ is a topological n-manifold, and $|\partial K| = \partial |K|$.*   □

Theorem 5.2.11. *Let K be a simplicial complex in $\mathbb{R}^N$ with the property that every point $x \in |K|$ has a (closed) neighborhood PL homeomorphic to Δ^n. Then K is a combinatorial n-manifold.*   □

If K is a combinatorial manifold, its geometric realization $|K|$ is a *piecewise linear manifold*, abbreviated to *PL manifold*.

Remark 5.2.12. There is an enormous literature behind the last few paragraphs, expounded in part in [90], [136] and [96]. It is difficult to construct a topological n-manifold which is not homeomorphic to a piecewise linear n-manifold, though such manifolds exist in all dimensions ≥ 4. Each topological manifold of dimension ≤ 3 is homeomorphic to a piecewise linear manifold which is unique up to PL homeomorphism; see for example [121]. Each closed topological manifold of dimension ≥ 6 admits the structure of a CW complex; see Essay III of [96].

Exercises

1. If K is a simplicial complex in $\mathbb{R}^N$, let $|K|_1$ be its geometric realization and let $|K|_2$ be its underlying polyhedron. Prove that there is a homeomorphism $h : |K|_1 \to |K|_2$ such that (in the notation of this section) $h(v) = v$ and for each $\sigma \in S_K$, h maps $|\sigma|_1$ onto $|\sigma|_2$ affinely.

2. Prove 5.2.7, and also a PL version of 5.2.7.
3. Prove that $|K|$ is a CW complex.
4. An abstract simplicial complex K is *connected* if for any vertices v, w there is a sequence of vertices $v = v_0, v_1, \cdots, v_n = w$ where, for every i, $\{v_i, v_{i+1}\}$ is a simplex of K. Prove that K is connected iff $|K|$ is path connected.
5. Prove that K is a locally finite [resp. finite] abstract simplicial complex iff $|K|$ is a locally finite [resp. finite] CW complex.
6. Prove that every n-dimensional locally finite abstract simplicial complex is isomorphic to a simplicial complex in $\mathbb{R}^{2n+1}$.
7. Define the *simplicial boundary* of $N_K(L)$ to be the subcomplex $\dot{N}_K(L)$ of $N_K(L)$ consisting of simplexes which do not have a vertex in L. Give an example where $|\dot{N}_K(L)| \neq \mathrm{fr}_{|K|}(|N_K(L)|)$.
8. Show that I^n is PL homeomorphic to Δ^n.
9. Show that $\mathbb{R}P^n$, S^n and T^n are homeomorphic to PL n-manifolds.
10. Show that the surfaces $T_{g,d}$ and $U_{h,d}$ of 5.1.9 are homeomorphic to PL 2-manifolds.
11. Construct a non-triangulable CW complex.
12. Prove 5.2.8.
13. Prove that any cone $v * Y$ is contractible.
14. The Simplicial Approximation Theorem says that if K and L are simplicial complexes in some $\mathbb{R}^N$, and if K is finite, then given a map $f : |K| \to |L|$ there is a simplicial subdivision K' of K and a simplicial map $\phi : K' \to L$ such that f is homotopic to $|\phi|$. This is proved in many books on algebraic topology, e.g., [146]. Using this, prove that every CW complex X has the homotopy type of some $|J|$ where J is an abstract simplicial complex. (*Hint*: First assume X is finite-dimensional and work by induction on dimension, using 4.1.8.)

5.3 Regular CW complexes

In this section we introduce a class of CW complexes in which cells are homeomorphic to balls.

Suppose that the space Y is obtained from the space A by attaching n-cells. We say that Y is obtained from A by *regularly* attaching n-cells if the characteristic map for each n-cell e^n_α can be chosen to be a homeomorphism $B^n_\alpha \to e^n_\alpha$. A CW complex X is *regular* if (referring to the definition in Sect. 1.2) each X^n is obtained from X^{n-1} by regularly attaching n-cells. In a regular CW complex, every n-cell is homeomorphic to B^n.

Example 5.3.1. The usual CW complex structure on S^1 consisting of one vertex and one 1-cell is not regular, but one easily finds a regular subdivision having two vertices and two 1-cells. Other examples are given in the Exercises.

Note the agreement of some of our notations in regular CW complexes: the cell e^n is an n-ball, so $\overset{\circ}{e}{}^n$ coincides with the manifold interior $(e^n)^\circ$, while $\overset{\bullet}{e}{}^n$ coincides with ∂e^n.

We derive some special properties of regular CW complexes (5.3.2 and 5.3.5).

Proposition 5.3.2. *Let e be a cell of the regular CW complex X, and let $C(e)$ be the carrier of e. Then, as spaces, $C(e) = e$. In other words, each cell of X is a subcomplex of X.*

For the proof of 5.3.2 we need two lemmas.

Lemma 5.3.3. *There is no embedding of S^n in $\mathbb{R}^n$.*

Proof. If there were such an embedding $h : S^n \to \mathbb{R}^n$ then, by 5.1.1 and 1.3.7, h would map every proper open subset of S^n onto an open subset of $\mathbb{R}^n$, hence $h(S^n)$ would be open in $\mathbb{R}^n$. But $h(S^n)$ is compact, hence closed in $\mathbb{R}^n$. And $h(S^n) \neq \mathbb{R}^n$ since the latter is not compact. Thus we would have a proper non-empty closed-and-open subset of $\mathbb{R}^n$ contradicting the fact that $\mathbb{R}^n$ is connected (being obviously path connected). $\square$

Lemma 5.3.4. *Let e_α^{n-1} and e_β^n be cells of the regular CW complex X. If $\overset{\circ}{e}_\alpha \cap \overset{\bullet}{e}_\beta \neq \emptyset$ then $e_\alpha \subset \overset{\bullet}{e}_\beta$.*

Proof. We have $\overset{\bullet}{e}_\beta^{\,n} \subset X^{n-1}$, and $\overset{\circ}{e}_\alpha^{\,n-1}$ is open in X^{n-1}, so $\overset{\circ}{e}_\alpha^{\,n-1} \cap \overset{\bullet}{e}_\beta^{\,n}$ is open in $\overset{\bullet}{e}_\beta^{\,n}$. Now $\overset{\bullet}{e}_\beta^{\,n}$ is homeomorphic to S^{n-1}, so, by 5.1.1, $\overset{\circ}{e}_\alpha^{\,n-1} \cap \overset{\bullet}{e}_\beta^{\,n}$ is open in $\overset{\bullet}{e}_\alpha^{\,n-1}$. But, since $\overset{\bullet}{e}_\beta^{\,n}$ is compact, $\overset{\circ}{e}_\alpha^{\,n-1} \cap \overset{\bullet}{e}_\beta^{\,n}$ is also closed in $\overset{\circ}{e}_\alpha^{\,n-1}$, and is non-empty. As in the proof of 5.3.3, this implies $\overset{\circ}{e}_\alpha^{\,n-1} \subset \overset{\bullet}{e}_\beta^{\,n}$, from which the Lemma follows. $\square$

Proof (of 5.3.2). We work by induction on the dimension, n, of e. If $n = 0$, the Proposition is trivial. Consider a cell e_α^{n-1} in $C(e)$. Then $\overset{\circ}{e}_\alpha^{\,n-1} \cap \overset{\bullet}{e} \neq \emptyset$, so, by 5.3.4, $e_\alpha^{n-1} \subset \overset{\bullet}{e}$. Since $C(e)$ consists of e together with the carriers of such e_α^{n-1}, the induction hypothesis completes the proof. $\square$

It follows that if $h_\alpha : (B^n, S^{n-1}) \to (e_\alpha^n, \overset{\bullet}{e}_\alpha^{\,n})$ is a characteristic map for an n-cell in a regular CW complex X, h_α^{-1} carries the CW complex structure $C(e_\alpha^n)$ on e_α^n back to a CW complex structure on B^n; this structure has exactly one n-cell, and $h_\alpha^{-1}(C(\overset{\bullet}{e}_\alpha^{\,n}))$ is a CW complex structure on S^{n-1}. This implies that there is an $(n-1)$-cell e_β^{n-1} of X with $e_\beta^{n-1} \subset \overset{\bullet}{e}_\alpha^{\,n}$. In fact, by induction one deduces that for each n-cell e_α^n of the regular CW complex X and for each $k < n$ there is a k-cell e^k with $e^k \subset e_\alpha^n$.

Recall that a cell e_1 is a *face* of a cell e_2 if $e_1 \subset e_2$. We have just seen that in a regular CW complex an n-cell has faces of all lower dimensions. Easy examples show that non-regular CW complexes need not have this property.

Proposition 5.3.5. *Let X be a regular CW complex, and let e_α^{k-2} be a face of e_β^k. Then e_α^{k-2} is a face of exactly two $(k-1)$-dimensional faces of e_β^k.*

Proof. Since $\overset{\bullet}{e}{}^k_\beta$ is homeomorphic to S^{k-1}, this follows from the following lemma. □

Lemma 5.3.6. *Let Y be a regular CW complex structure on an n-manifold. Every cell of Y is a face of an n-cell of Y. Every $(n-1)$-cell of Y is a face of at most two n-cells of Y. An $(n-1)$-cell, e, of Y is a face of exactly one n-cell of Y iff $e \subset \partial Y$. If e is a face of two n-cells of Y, then $\overset{\circ}{e} \subset \overset{\circ}{Y}$.*

Proof. We saw in Sect. 5.1 that every cell of Y has dimension $\leq n$. If some cell were not a face of an n-cell, there would be $k < n$ and a k-cell $\tilde{e}$ of Y which is not a face of any higher-dimensional cell of Y, implying $\overset{\circ}{\tilde{e}}$ open in Y, contradicting 5.1.6(a). If the $(n-1)$-cell e is a face of exactly one n-cell, then, since Y is regular, each $x \in \overset{\circ}{e}$ has a neighborhood in Y homeomorphic to $\mathbb{R}^n_+$. Thus $\overset{\circ}{e} \subset \partial Y$, and, since ∂Y is closed in Y, $e \subset \partial Y$. On the other hand, if e is a face of two n-cells, then every $x \in \overset{\circ}{e}$ clearly has a neighborhood homeomorphic to $\mathbb{R}^n$, so, by 5.1.6(b), $\overset{\circ}{e} \subset \overset{\circ}{Y}$. The proof that e is not a face of more than two n-cells is left as an exercise. □

The *abstract first derived* (or *barycentric subdivision*) of a regular CW complex X is the abstract simplicial complex sd X whose vertices are the cells of X and whose simplexes are those finite sets of cells $\{e_0, \cdots, e_n\}$ which can be ordered so that, for each $i < n$, e_i is a proper face of e_{i+1} (i.e., $e_i \subset \overset{\bullet}{e}_{i+1}$).

Convention 5.3.7. *We will always list the vertices of a simplex of sd X in order of increasing dimension (of the cells).*

Proposition 5.3.8. *When X is a regular CW complex, there is a homeomorphism $h : |sd\ X| \to X$ such that for every simplex $\{e_0, \cdots, e_k\}$ of sd X, $h(|\{e_0, \cdots, e_k\}|^\circ) \subset \overset{\circ}{e}_k$. Moreover, h^{-1} is cellular.*

Proof. Observe that for any cell e of X, sd $C(e)$ is the cone $e * [sd\ C(\overset{\bullet}{e})]$ where $C(\cdot)$ denotes "carrier"; note the different meanings of e! By induction on n we define homeomorphisms $h_n : |sd\ X^n| \to X^n$, each extending its predecessor: the required h agrees with h_n on $|sd\ X^n| \subset |sd\ X|$. In an obvious sense $|sd\ X^0| = X^0$; let $h_0 = $ id. Assume that h_0 has been extended to $h_{n-1} : |sd\ X^{n-1}| \to X^{n-1}$. Let e be an n-cell of X. Then h_{n-1} maps $|sd\ C(\overset{\bullet}{e})|$ homeomorphically onto $C(\overset{\bullet}{e})$. By 5.3.2, this is a homeomorphism between $(n-1)$-spheres. By the above remark about cones, it extends to a homeomorphism $|sd\ C(e)| \to C(e)$ between n-balls. Our h_n is defined to agree with this homeomorphism on $|sd\ C(e)|$. □

Corollary 5.3.9. *Every regular CW complex is triangulable.*[11] □

---

[11] In Exercise 11 of Sect. 5.2 the reader was asked for a CW complex (non-regular in view of 5.3.9) which is not triangulable.

The homeomorphism h of 5.3.8 defines a subdivision of X in the sense of Sect. 2.4 whose cells are $\{h(|\{e_0, \cdots, e_k\}|)\}$. Call such a subdivision a *first derived* (or *barycentric subdivision*) of X. Once h is chosen, it is convenient to denote by $\hat{e}_k$ the point of $\overset{\circ}{e}_k$ which is the image under h of the vertex e_k of $|\text{sd } X|$ and to call $\hat{e}_k$ a *barycenter* of e_k. In fact, one often identifies $|\text{sd } X|$ with X via h, thinking of the barycenters $\hat{e}_k$ as being the vertices of $|\text{sd } X|$; the cell (simplex) $|\{e_0, \cdots, e_k\}|$ of $|\text{sd } X|$ is then identified with a certain subset of e_k. See Fig. 5.3.

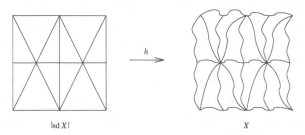

$$|\text{sd } X| \qquad\qquad X$$

Fig. 5.3.

We compute incidence numbers in regular CW complexes.

Proposition 5.3.10. *Let X be an oriented regular CW complex. For cells e_β^{n-1} and e_α^n of X, the incidence number $[e_\alpha^n : e_\beta^{n-1}]$ is ± 1 if e_β^{n-1} is a face of e_α^n, and is 0 otherwise.*

Proof. When $n = 1$, this is true by definition; see Sect. 2.5. Let $n > 1$. When e_β^{n-1} not a face of e_α^n, the incidence number is 0 by 2.5.8. Let Y be the subcomplex $C(\overset{\bullet}{e}{}_\alpha^n)$ of X, and let Z be the CW complex structure on S^{n-1} consisting of one 0-cell $\tilde{e}^0$ and one $(n-1)$-cell $\tilde{e}^{n-1}$. Then Y is homeomorphic to S^{n-1} by 5.3.2. There is a cellular map $f : Y \to Z$ taking $\overset{\circ}{e}{}_\beta^{n-1}$ homeomorphically onto $\overset{\circ}{\tilde{e}}{}^{n-1}$ and taking the rest of Y to $\tilde{e}^0$. Then $f_\#(e_\beta^{n-1}) = [e_\beta^{n-1} : \tilde{e}^{n-1} : f]\tilde{e}^{n-1}$, and $f_\#$ takes all other generators of $C_{n-1}(Y; \mathbb{Z})$ to $0 \in C_{n-1}(Z; \mathbb{Z})$. Inspection of the diagram in Sect. 2.5 defining mapping degrees reveals that $[e_\beta^{n-1} : \tilde{e}^{n-1} : f]$ is the degree of a homeomorphism $S^{n-1} \to S^{n-1}$ and is therefore ± 1, by 2.4.20. Since $H_{n-1}(Y; \mathbb{Z}) \cong \mathbb{Z}$, there is a cellular $(n-1)$-cycle in Y which generates $H_{n-1}(Y; \mathbb{Z})$ and whose image generates $H_{n-1}(Z; \mathbb{Z}) \cong \mathbb{Z}$. Thus, if we use homeomorphisms to identify Y and Z with S^{n-1}, f has degree ± 1. Now the diagram in Sect. 2.5 defining incidence numbers shows that $[e_\alpha^n : e_\beta^{n-1}]$ is this degree, namely ± 1. $\square$

Exercises

1. Find regular CW complex structures which are not triangulations (see 5.3.9) for: S^n and closed surfaces.
2. Show that the property of being a regular CW complex is preserved under the following constructions: disjoint union, finite product, subcomplex, and covering space.
3. Give an example where the universal cover of a non-regular CW complex is regular.
4. Prove that in an n-manifold no $(n-1)$-cell is a face of more than two n-cells.
5. Let X be a presentation complex for the presentation $\langle W \mid R, \rho \rangle$ of $G = \pi_1(X, v)$. Prove that the universal cover $\tilde{X}$ is a regular CW complex iff no element of W represents $1 \in G$, and no proper subword of a relation in $\rho(R)$ is conjugate to (i.e., cyclically equivalent to) a relation in $\rho(R)$. Prove that if $\tilde{X}$ is not regular, then there is a sequence of Tietze transformations leading to another presentation $\langle W' \mid R', \rho' \rangle$ of $\pi_1(X, v)$ so that if X' is the associated presentation complex $\tilde{X}'$ is regular. Prove that if W and R are finite, then this sequence is finite, and each presentation in the sequence is finite.
6. If Y is a regular G-CW complex show that $|\text{sd } Y|$ is a rigid G-CW complex.
7. Show that if Y is a subcomplex of X then sd Y is a full subcomplex of sd X.

5.4 Incidence numbers in simplicial complexes

Recall that Δ^n denotes the standard n-simplex (or PL n-ball).

The ith *face* Δ_i^{n-1} is the convex hull of the vertices $\{p_0, \cdots, \hat{p}_i, \cdots, p_n\}$ where $\hat{p}_i$ means "suppress p_i". We denote by $f_i : \Delta^{n-1} \to \Delta_i^{n-1}$ the affine homeomorphism which maps the vertices of $\Delta^{n-1} \subset \mathbb{R}^n$ to the vertices of Δ_i^{n-1} in order-preserving fashion. We also regard Δ^n as a CW complex via $\Delta^n = |\mathbf{n}|$ (see Sect. 5.2). Note that $\overset{\bullet}{\Delta}{}^n = \bigcup_{i=0}^{n} \Delta_i^{n-1}$.

We wish to identify Δ^n with B^n permanently (just as we identified I^n with B^n in Sect. 2.2), so for each n we choose, once and for all, a homeomorphism $d_n : B^n \to \Delta^n$. For $n = 0$, there is only one choice. For $n > 0$, the choice is made by induction to satisfy:

Convention 5.4.1. *If, in the CW complex Δ^n with $n > 0$, the n-cell Δ^n is oriented by $d_n : (B^n, S^{n-1}) \to (\Delta^n, \overset{\bullet}{\Delta}{}^n)$ and the $(n-1)$-cell Δ_0^{n-1} is oriented by $f_0 \circ d_{n-1} : (B^{n-1}, S^{n-2}) \to (\Delta_0^{n-1}, \overset{\bullet}{\Delta}{}_0^{n-1})$, then $[\Delta^n : \Delta_0^{n-1}] = 1$.*

As in 2.5.17 we have:

Proposition 5.4.2. *If, in the CW complex Δ^n, the n-cell Δ^n is oriented by d_n and the $(n-1)$-cell Δ_i^{n-1} is oriented by $f_i \circ d_{n-1}$, then $[\Delta^n : \Delta_i^{n-1}] = (-1)^i$.* $\square$

Next, consider the geometric realization $|K|$ of an abstract simplicial complex K. If σ is a simplex of K, an ordering $v_0, \cdots, v_n$ of the vertices of σ specifies an orientation of the cell $|\sigma|$, namely the orientation defined by the characteristic map $(\Delta^n, \overset{\bullet}{\Delta}{}^n) \to (|\sigma|, |\overset{\bullet}{\sigma}|)$ which is the affine homeomorphism taking the vertex p_i of Δ^n to the vertex v_i of σ. By 2.5.2 we have:

Proposition 5.4.3. *Under this rule, two orderings of the vertices of σ determine the same orientation of $|\sigma|$ iff they differ by an even permutation.* $\square$

Corollary 5.4.4. *Let $v_0, \cdots, v_n$ be vertices of a simplex σ of K, and, for $0 \leq i < j \leq n$, let $v_i, \cdots, v_j$ be vertices of a simplex τ. This ordering $v_0, \cdots, v_n$ specifies orientations of σ and τ, and an odd permutation of $v_i, \cdots, v_j$ changes both orientations.* $\square$

An *ordering* of an abstract simplicial complex K is a partial ordering of the set of vertices V_K whose restriction to every simplex of K gives a total ordering of the vertices of that simplex. A CHOICE OF ORDERING OF K WILL BE UNDERSTOOD TO SPECIFY AN ORIENTATION OF THE CW COMPLEX $|K|$ AS FOLLOWS: if σ is an n-simplex with vertices, in order, $v_0, \cdots, v_n$ the orientation of $|\sigma|$ is defined (as above) by choosing as characteristic map $(\Delta^n, \overset{\bullet}{\Delta}{}^n) \to (|\sigma|, |\overset{\bullet}{\sigma}|)$ the affine homeomorphism taking the vertex p_i of Δ^n to the vertex v_i of σ.

By Proposition 5.4.2, we get:

Proposition 5.4.5. *Let K be an ordered abstract simplicial complex and let σ be an n-simplex of K. If we write $|\sigma| = |\{v_0, \cdots, v_n\}|$, with vertices listed in order, this orientation convention leads to the following formula for the boundary homomorphism in the cellular chain complex $C_*(|K|; R) : \partial(\sigma) =$*
$$\sum_{i=0}^{n} (-1)^i |\{v_0, \cdots, \hat{v}_i, \cdots, v_n\}|.$$
$\square$

This famous formula shows the precise relationship between cellular homology and "simplicial homology." There are several versions of the latter, fully explained in [84, Sect. 2.3 and Sect. 3.2].

Example 5.4.6. If X is a regular CW complex, there is a *natural ordering* on sd X explained in Sect. 5.3 with the definition of sd X.

Exercise

1. Prove 5.4.2.

PART II: FINITENESS PROPERTIES OF GROUPS

In Part II we apply the algebraic topology of Part I to study finiteness properties and dimensions of groups. This is the aspect of our subject which does not involve cohomology, or issues "at infinity."

Topology models a group G by a $K(G,1)$-complex, i.e., a path connected CW complex with contractible universal cover and fundamental group isomorphic to G. Any two such have the same homotopy type. However, the properties of G discussed involve studying how "nice" a $K(G,1)$-complex can be shown to exist for a given G.

We compare these topological properties of G with the corresponding homological notions, where the role of a $K(G,1)$-complex is replaced by that of a free RG-resolution of the trivial RG-module R. (Here, R is the "ground ring," typically $\mathbb{Z}$ or a field.)

We prove the important Bestvina-Brady Theorem which gives insight into the subtle differences between the topological and homological approaches.

To illustrate the ideas in Part II, we apply them in several important situations: Coxeter groups, Thompson groups, and (briefly) outer automorphism groups of free groups.

6

The Borel Construction and Bass-Serre Theory

In this chapter we discuss an important topological method for dealing with the rigid action of a group G on a CW complex X. This "Borel construction" is presented in Sect. 6.1; once it is in place, the Rebuilding Lemma 6.1.4 shows us how to alter X, without altering its homotopy type, to have more desirable properties. The important special case where X is a tree is discussed in §6.2. The reader may find it helpful to read Sects. 6.1 and 6.2 in parallel.

6.1 The Borel construction, stacks, and rebuilding

If one understands an n-connected rigid G-CW complex Y, in particular if one understands the quotient $G\backslash Y$ and the stabilizers of the cells of Y, one may be able to deduce information about the group G. The particular case in which Y is a tree, called Bass-Serre Theory, is discussed in the next section: in that case one can sometimes deduce that G is decomposable as a free product with amalgamation or as an HNN extension. But the method is much more general, as we now explain.

Starting with a simply connected rigid G-CW complex Y which might not be free, the method provides a path connected CW complex Z with fundamental group G and a map $q : Z \to G\backslash Y$ called a "stack." A stack is like a fiber bundle[1] but the fibers over different cells may have different homotopy types. The method of improving spaces within their homotopy types introduced in Sect. 4.1 can be applied to the "fibers" over cells, by induction on the dimensions of the cells, to rebuild Z, thus producing a "better" $q' : Z' \to G\backslash Y$ which is equivalent to q. A first application of this rebuilding process to group theory is given in the proof of Theorem 6.1.5.

Recall that the left G-CW complex Y is rigid iff for each cell $\tilde{e}$ of Y, the stabilizer $G_{\tilde{e}}$ acts trivially on $\tilde{e}$. We saw in 3.2.2 that the quotient $G\backslash Y =: V$

---

[1] See 4.4.11.

of a rigid G-action receives a CW complex structure[2] from Y whose cells are the images of the cells of Y under the quotient map $p : Y \to V$.

Starting with a pointed CW complex (X, x_0) such that $\pi_1(X, x_0) \cong G$, we permanently identify G with $\pi_1(X, x_0)$ by some chosen isomorphism, so that the universal cover $\tilde{X}$, with some base vertex $\tilde{x}_0$ over x_0, is a free G-CW complex. The *diagonal action* of G on the product CW complex $\tilde{X} \times Y$ is defined by $g.(x, y) = (g.x, g.y)$. It is a free action so, by 3.2.1, the quotient map $r : \tilde{X} \times Y \to G\backslash(\tilde{X} \times Y) =: Z$ is a covering projection which imposes a natural CW complex structure on Z. The quotient maps p and r induce a commutative diagram:

$$\begin{array}{ccc} \tilde{X} \times Y & \xrightarrow{\text{projection}} & Y \\ \downarrow r & & \downarrow p \\ Z & \xrightarrow{\quad q \quad} & V \end{array}$$

Clearly, q is a quotient map. This procedure is (a special case of) the *Borel construction*. One easily checks:

Proposition 6.1.1. *The map q is cellular and takes each cell of Z onto a cell of V. For each subcomplex U of V, $q^{-1}(U)$ is a subcomplex of Z.* $\square$

Let V have base vertex v. Choose a vertex $\tilde{v}$ of Y as base point, such that $p(\tilde{v}) = v$. The base point of $\tilde{X} \times Y$ is $(\tilde{x}_0, \tilde{v})$, and its r-image is the base point z of Z. Since $\tilde{X} \times Y$ is a simply connected covering space, 3.2.3 implies $\pi_1(Z, z) \cong G$.

Let E_n be the set of n-cells of V. For each $e \in E_n$ make the following choices: (i) an n-cell $\tilde{e}$ of Y such that $p(\tilde{e}) = e$, (ii) a point $\tilde{u}_e \in \overset{\circ}{\tilde{e}}$, (iii) a characteristic map $h_{\tilde{e}} : (B^n, S^{n-1}) \to (\tilde{e}, \overset{\bullet}{\tilde{e}})$. (Of course, when $n = 0$, $\tilde{u}_w = \tilde{w}$ for each $w \in E_0$, and there is only one possible $h_{\tilde{w}}$.) Let $u_e = p(\tilde{u}_e)$, let h_e be the characteristic map $p \circ h_{\tilde{e}}$ for e, and let $X_e = q^{-1}(u_e)$. Then $q^{-1}(V^0) = \coprod_{w \in E_0} X_w$. We are going to show that $q^{-1}(V^{n+1})$ is homeomorphic to the adjunction complex

$$q^{-1}(V^n) \cup_f \left(\coprod_{e \in E_{n+1}} X_e \times B^{n+1} \right), \text{ where } f : \coprod_{e \in E_{n+1}} X_e \times S^n \to q^{-1}(V^n) \text{ is de-}$$

fined on each $X_e \times S^n$ to agree with the map f_e in the following commutative diagram (details to be explained):

$$
\begin{array}{ccccc}
\tilde{X} \times \{\tilde{u}_e\} \times S^n & \xrightarrow{r_e \times \mathrm{id}} & X_e \times S^n & \xrightarrow{\text{projection}} & S^n \\
\downarrow{\scriptstyle\text{projection}} & & \downarrow{\scriptstyle f_e} & & \downarrow{\scriptstyle h_e|} \\
\tilde{X} \times S^n & & & & \\
\downarrow{\scriptstyle \mathrm{id}\times(h_{\tilde{e}}|)} & & & & \\
\tilde{X} \times Y^n & \xrightarrow[r|]{} & q^{-1}(V^n) & \xrightarrow[q|]{} & V^n
\end{array}
$$

The map $r_e : \tilde{X} \times \{\tilde{u}_e\} \to X_e := q^{-1}(u_e)$ is the quotient map of the action of $G_{\tilde{u}_e}$. Since the given G-action on Y is rigid, $G_{\tilde{u}_e} = G_{\tilde{e}}$. Letting $x_e = r(\tilde{x}_0, \tilde{u}_e) \in X_e$, we have:

Proposition 6.1.2. *For each cell e of V, X_e is homeomorphic to $G_{\tilde{e}} \backslash \tilde{X}$, hence $\pi_1(X_e, x_e) \cong G_{\tilde{e}}$.* □

Since r_e is a quotient map, so is the map $r_e \times$ id in the diagram; to see this, apply 1.3.11 $(n+1)$ times to conclude that $r_e \times \mathrm{id} : \tilde{X} \times \{\tilde{u}_e\} \times B^{n+1} \to X_e \times B^{n+1}$ is a quotient map, and then restrict. The map $r_e \times \mathrm{id}$ is clearly surjective. Rigidity implies that there is a unique function, hence a map, f_e, making the left half of the diagram commute. It is then obvious that the whole diagram commutes and that f_e is cellular.

For the same reasons, there is a map $H_e : X_e \times B^{n+1} \to q^{-1}(V^{n+1})$ making the following diagram commute:

$$
\begin{array}{ccccc}
\tilde{X} \times \{\tilde{u}_e\} \times B^{n+1} & \xrightarrow{r_e \times \mathrm{id}} & X_e \times B^{n+1} & \xrightarrow{\text{projection}} & B^{n+1} \\
\downarrow{\scriptstyle\text{projection}} & & \downarrow{\scriptstyle H_e} & & \downarrow{\scriptstyle h_e} \\
\tilde{X} \times B^{n+1} & & & & \\
\downarrow{\scriptstyle \mathrm{id}\times h_{\tilde{e}}} & & & & \\
\tilde{X} \times Y^{n+1} & \xrightarrow[r|]{} & q^{-1}(V^{n+1}) & \xrightarrow[q|]{} & V^{n+1}
\end{array}
$$

Assembling the maps H_e, $e \in E_{n+1}$, we obtain the desired structure theorem:

Theorem 6.1.3. *The map* $q^{-1}(V^n) \coprod \left(\displaystyle\coprod_{e \in E_{n+1}} X_e \times B^{n+1} \right) \to q^{-1}(V^{n+1})$
which agrees with inclusion on $q^{-1}(V^n)$ and with H_e on $X_e \times B^{n+1}$ induces a homeomorphism

$$
s_{n+1} : q^{-1}(V^n) \cup_f \left(\coprod_{e \in E_{n+1}} X_e \times B^{n+1} \right) \to q^{-1}(V^{n+1})
$$

for each $n \geq 0$, such that s_{n+1} maps each cell of the adjunction complex homeomorphically onto a cell of $q^{-1}(V^{n+1})$. Moreover, the following diagram commutes:

$$q^{-1}(V^n) \amalg \left(\coprod_{e \in E_{n+1}} X_e \times B^{n+1} \right)$$

$$\downarrow \text{quotient} \qquad \searrow t$$

$$q^{-1}(V^n) \cup_f \left(\coprod_{e \in E_{n+1}} X_e \times B^{n+1} \right) \qquad V^{n+1}$$

$$\downarrow s_{n+1} \qquad \nearrow q|$$

$$q^{-1}(V^{n+1})$$

where t agrees with q on $q^{-1}(V^n)$ and with $h_e \circ$ projection on $X_e \times B^{n+1}$.

Proof. The function s_{n+1} is clearly a continuous bijection. Moreover, it maps cells bijectively onto cells. Thus $s_{n+1}^{-1}|$ is continuous on each cell, which, by 1.2.12, is enough to imply continuity of s_{n+1}^{-1} (exercise). The second part is clear. $\square$

At the risk of repetition, the point of this theorem is that it gives a useful decomposition of the space Z in the Borel construction.

The Borel Construction is important and will be used in several ways in this book, so we need vocabulary to describe the result. Let $\pi : A \to C$ be a cellular map between CW complexes. Let $h_e : (B^n, S^{n-1}) \to (e, \overset{\bullet}{e})$ be a characteristic map for the cell e of C, and for each such cell e let F_e be a CW complex. We call $\pi : A \to C$ a *stack* of CW complexes with *base space* C, *total space* A and *fiber* F_e *over* e, if for each $n \geq 1$ (denoting the set of n-cells of C by E_n) there is a cellular map $f_n : \coprod_{e \in E_n} F_e \times S^{n-1} \to \pi^{-1}(C^{n-1})$ and a homeomorphism $k_n : \pi^{-1}(C^{n-1}) \cup_{f_n} \left(\coprod_{e \in E_n} F_e \times B^n \right) \to \pi^{-1}(C^n)$ satisfying:

(i) k_n agrees with inclusion on $\pi^{-1}(C^{n-1})$, (ii) k_n maps each cell onto a cell, and (iii) the following diagram commutes:

$$\pi^{-1}(C^{n-1}) \coprod \left(\coprod_{e \in E_n} F_e \times B^n \right)$$

$$\Big\downarrow \text{quotient} \qquad \searrow u$$

$$\pi^{-1}(C^{n-1}) \cup_{f_n} \left(\coprod_{e \in E_n} F_e \times B^n \right) \qquad C^n$$

$$\Big\downarrow k_n \qquad \nearrow \pi|$$

$$\pi^{-1}(C^n)$$

where u agrees with π on $\pi^{-1}(C^{n-1})$ and with $h_e \circ$ projection on $F_e \times B^n$. Thus, $\pi : A \to C$ is "built" by induction on the skeleta of C so that over the interior, $\overset{\circ}{e}$, of an n-cell e, π is "like"[3] the projection: $F_e \times \overset{\circ}{e} \to \overset{\circ}{e}$. An immediate consequence of Theorem 4.1.7 is:

Proposition 6.1.4. (Rebuilding Lemma) *If for each cell e of C we are given a CW complex F'_e of the same homotopy type as F_e, then there is a stack of CW complexes $\pi' : A' \to C$ with fiber F'_e over e, and a homotopy equivalence h making the following diagram commute up to homotopy over each cell:*[4]

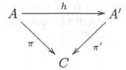

$$\square$$

The content of Theorem 6.1.3 is that $q : Z \to V$ is a stack of CW complexes with base space V, in which the fiber over the cell e is $G_{\tilde{e}} \backslash \tilde{X}$. We can use 6.1.4 to replace $G_{\tilde{e}} \backslash \tilde{X}$ by a more desirable CW complex of the same homotopy type and thereby produce a more desirable space Z' of the same homotopy type as Z. The first of a number of uses of this method appears in:

Theorem 6.1.5. *Let Y be a simply connected rigid G-CW complex. (i) If Y has finite 1-skeleton mod G and if the stabilizer of each vertex is finitely generated, then G is finitely generated. (ii) If Y has finite 2-skeleton mod G, if the stabilizer of each vertex is finitely presented, and if the stabilizer of each 1-cell is finitely generated, then G is finitely presented.*

---

[3] Readers may recognize the connection with fiber bundles and with block bundles in special cases.

[4] i.e. the homotopy can be chosen so that for each cell e of C its restriction to $\pi^{-1}(e) \times I$ has its image in e.

Proof. By 6.1.1, $q^{-1}(V^1)$ is a subcomplex of Z containing Z^1. By hypothesis in (i), for each $w \in E_0$, $\pi_1(X_w, x_w)$ is finitely generated. By 4.1.14, each X_w is homotopy equivalent to a CW complex X'_w having finite 1-skeleton. By 3.1.13 and 4.1.9, when $e \in E_1$, X_e is homotopy equivalent to a CW complex X'_e having one vertex. By applying 6.1.4 to the stack $q : Z \to V$, we obtain a stack $q' : Z' \to V$ in which Z' is homotopy equivalent to Z, and the fiber of q' over e [resp. over w] is X'_e [resp. X'_w].

The CW complexes Z' and $(q')^{-1}(V^1)$ have the same 1-skeleton. It is finite because: (a) Γ^0 is finite and $((q')^{-1}(V^0))^1$ is the finite graph $\coprod_{w \in E_0} (X'_w)^1$; (b) V^1 is finite and (since $X'_e \times B^1$ is a product and X'_e has one vertex) there is only one 1-cell of $(q')^{-1}(V^1)$ mapped onto each 1-cell of Γ. (Note that (1-cell of X'_e) $\times$ (vertex of B^1) does not give a new 1-cell of the adjunction complex.) Thus Z' has finite 1-skeleton. By 3.1.17, $G \cong \pi_1(Z, z) \cong \pi_1(Z', z')$ is finitely generated. Thus (i) holds.

The proof of (ii) is similar. We replace each X_w by X'_w, whose 2-skeleton is the wedge of a finite 2-complex and a bouquet of 2-spheres (see 4.1.14); for $e \in E_1$, we replace X_e by X'_e having finite 1-skeleton; for $e \in E_2$ we replace X_e by X'_e having a single vertex. The resulting CW complex Z' has finite 1-skeleton (for reasons already explained) and possibly infinite 2-skeleton. Indeed, its 2-cells are of three kinds: 2-cells of X'_w where $w \in E_0$, cells coming from (1-cells of X'_e) $\times B^1$ for $e \in E_1$, and cells coming from (vertex of X'_e) $\times B^2$ for $e \in E_2$. Thus all but finitely many of the 2-cells are 2-spheres (i.e., have trivial attaching maps) so they contribute nothing significant to the resulting presentation of $G \cong \pi_1(Z, z) \cong \pi_1(Z', z')$. □

Exercise

Prove that if $\pi : A \to C$ is a stack of CW complexes in which every fiber F_e is contractible, then π is a homotopy equivalence. In fact, π is a *hereditary homotopy equivalence*, meaning that for every subcomplex D of C, $\pi| : \pi^{-1}(D) \to D$ is a homotopy equivalence. (*Hint*: The proof of the Rebuilding Lemma allows us to take $A' = C$, $\pi' = \mathrm{id}_C$ and $h = \pi$.)

6.2 Decomposing groups which act on trees (Bass-Serre Theory)

This section is devoted to an important application of Theorem 6.1.3, the structure of groups which act rigidly on trees. The main theorem is Theorem 6.2.7. We begin by introducing the necessary background topics: graphs of groups, fundamental group based at a tree, and graphs of pointed CW complexes. In an appendix we discuss "generalized graphs of groups."

If e is an oriented 1-cell of a CW complex, we denote its initial and final points by $o(e)$ and $t(e)$ respectively (o for "origin" and t for "terminus"). A *graph of groups* is a system $(\mathcal{G}, \Gamma)$ consisting of: an oriented path connected graph Γ, a group $G(w)$ for each vertex w of Γ, a group $G(e)$ for each (oriented) 1-cell of Γ, and monomorphisms $G(o(e)) \xleftarrow{\phi_e^-} G(e) \xrightarrow{\phi_e^+} G(t(e))$. The group $G(w)$ [resp. $G(e)$] is called a *vertex group* [resp. *edge group*] of $(\mathcal{G}, \Gamma)$.

Let T be a maximal tree in Γ. As in Sect. 6.1, let E_0 [resp. E_1] be the set of vertices [resp. (oriented) 1-cells] of Γ. Associated with $(\mathcal{G}, \Gamma)$ and T is a group, denoted $\pi_1(\mathcal{G}, \Gamma; T)$, called the *fundamental group of* $(\mathcal{G}, \Gamma)$ *based at* T, namely: the quotient of the free product $\left(\underset{w \in E_0}{*} G(w) \right) * F(E_1)$ by the normal subgroup[5] generated by (i) elements of the form $e^{-1}.\phi_e^-(x).e.(\phi_e^+(x))^{-1}$ where $e \in E_1$ and $x \in G(e)$, and (ii) elements e, where e is a 1-cell of T.

This includes some well-known constructions as special cases:

Case 1: *Free products with amalgamation.* Here $\Gamma = T = B^1$, with two vertices, $w(-)$ and $w(+)$, and one 1-cell e. We are given monomorphisms $\phi_e^\pm : G(e) \to G(w(\pm))$. In this case

$$\pi_1(\mathcal{G}, \Gamma, T) = \langle G(w(-)), G(w(+)) \mid \phi_e^-(x)\phi_e^+(x)^{-1} \, \forall \, x \in G(e) \rangle.$$

This is often written[6] $G(w(-)) \underset{G(e)}{*} G(w(+))$.

Case 2: *HNN extensions.* Here $\Gamma = S^1$, with one vertex w and one 1-cell e, $T = \{w\}$, and we are given monomorphisms $\phi_e^\pm : G(e) \to G(w)$. In this case $\pi_1(\mathcal{G}, \Gamma; T) = \langle G(w), e \mid e^{-1}\phi_e^-(x)e(\phi_e^+(x))^{-1} \, \forall x \in G(e) \rangle$. This is often written[7] $G(w)\underset{\phi}{*}$ where $\phi : \phi_e^+(G(e)) \to \phi_e^-(G(e))$ is the isomorphism $\phi_e^- \circ (\phi_e^+)^{-1}$.

Indeed, the general case of $\pi_1(\mathcal{G}, \Gamma; T)$ is simply an iteration of these two special cases; for if $\Gamma = T$ then $\pi_1(\mathcal{G}, \Gamma; T)$ is just a free product with an amalgamation for each edge of T; and in the general case ($\Gamma \neq T$), $\pi_1(\mathcal{G}, \Gamma; T)$ is obtained from this "multiple" free-product-with-amalgamation by performing an HNN construction for each edge of Γ which is not an edge of T.

Proposition 6.2.1. (Britton's Lemma) *For each $w_0 \in E_0$, the homomorphism $\gamma_{w_0} : G(w_0) \to \pi_1(\mathcal{G}, \Gamma; T)$ induced by $G(w_0) \hookrightarrow \left(\underset{w \in E_0}{*} G(w) \right) * F(E_1)$ is a monomorphism.*

---

[5] As before, $F(E_1)$ denotes the free group generated by the set E_1.

[6] In general, if G_1, G_2 and A are groups and $\phi_i : A \rightarrowtail G_i$ are monomorphisms, one writes $G_1 *_A G_2$ for $\langle G_1, G_2 \mid \phi_1(a)\phi_2(a)^{-1} \, \forall \, a \in A \rangle$; in this abuse of notation one is thinking of $\phi_1(A)$, A and $\phi_2(A)$ as the same group and one is "amalgamating G_1 and G_2 along the common subgroup A."

[7] In general, if G is a group, A and B are subgroups, and $\phi : A \to B$ is an isomorphism, one writes $G*_\phi$ for $\langle G, t \mid t^{-1}at = \phi(a) \, \forall \, a \in A \rangle$. This group is the *HNN extension* of G by ϕ; the subgroup G (see 6.2.1) is the *base group* and t is called the *stable letter*. If $A = G$ this is an *ascending* HNN extension.

Remarks on the proof. The version for Cases 1 and 2, above, is found in [106, Chap. IV, Sect. 2]. The general case then follows by the above remarks; a direct proof is found in [142, p. 46]. All these involve reducing the length of a supposedly minimal non-trivial word in the kernel, thereby proving the kernel trivial. A topological proof is indicated in the next section; see Exercise 3 of Sect. 7.1.

Proposition 6.2.1 means that whenever we express a group G non-trivially as the fundamental group of a graph of groups, we are exhibiting a decomposition of G into "pieces" each of which is a subgroup of G. Our purpose here is to give such a decomposition (Theorem 6.2.7) whenever G acts rigidly on a tree.

We will need the "fundamental group of a CW complex based at a tree." Whenever $\bar{T}$ is a tree in a CW complex X, the quotient map $k : X \to X/\bar{T}$ is a homotopy equivalence by 4.1.9. Let $\bar{v}$ be the vertex of $X/\bar{T}$ such that $k(\bar{T}) = \bar{v}$. We define the *fundamental group of X based at $\bar{T}$*, denoted $\pi_1(X, \bar{T})$, to be the group $\pi_1(X/\bar{T}, \bar{v})$. For any vertex v of $\bar{T}$, k induces a canonical isomorphism $k_\# : \pi_1(X, v) \to \pi_1(X, \bar{T})$. The composition $\pi_1(X, v_1) \xrightarrow[k_{1\#}]{} \pi_1(X, \bar{T}) \xrightarrow[k_{2\#}^{-1}]{} \pi_1(X, v_2)$ is the isomorphism h_τ of 3.1.11, where τ is any edge path in $\bar{T}$ from v_1 to v_2. Note that if $\bar{T}^+$ is a maximal tree then $\pi_1(X, \bar{T})$ and $\pi_1(X, \bar{T}^+)$ are canonically isomorphic.

A *graph of pointed CW complexes* is a system $(\mathcal{X}, \Gamma)$ consisting of: an oriented path connected graph Γ, a pointed path connected CW complex $(X(w), x(w))$ for each vertex w of Γ, a pointed path connected CW complex $(X(e), x(e))$ for each 1-cell e of Γ, and cellular maps

$$(X(o(e)), x(o(e))) \xleftarrow{p_e^-} (X(e), x(e)) \xrightarrow{p_e^+} (X(t(e)), x(t(e))).$$

The *total complex*, $\mathrm{Tot}(\mathcal{X}, \Gamma)$, is the adjunction complex obtained from $\coprod\{X(w) \mid w \text{ is a vertex of } \Gamma\}$ by adjoining $\coprod\{X(e) \times B^1 \mid e \text{ is an edge of } \Gamma\}$ exactly as in 6.1.3, using the maps p_e defined by: $p_e(y, \pm 1) = p_e^\pm(y)$ for all $y \in X(e)$. The resulting map $q : \mathrm{Tot}(\mathcal{X}, \Gamma) \to \Gamma$ is a stack of CW complexes.

When each $p_{e\#}^\pm : \pi_1(X(e), x(e)) \to \pi_1(X(w), x(w))$ is a monomorphism, then $(\mathcal{X}, \Gamma)$ gives rise to a graph of groups $(\mathcal{G}, \Gamma)$, with vertex groups $\pi_1(X(w), x(w))$, edge groups $\pi_1(X(e), x(e))$, and monomorphisms $p_{e\#}^\pm$.

There is a map $s : \Gamma \to \mathrm{Tot}(\mathcal{X}, \Gamma)$ such that $s(w) = x(w)$ for each vertex w of Γ, and s maps each edge of Γ homeomorphically onto the image of $\{x(e)\} \times B^1$ in $\mathrm{Tot}(\mathcal{X}, \Gamma)$. In fact, $s(\Gamma)$ is a retract of $\mathrm{Tot}(\mathcal{X}, \Gamma)$. The maximal tree T is mapped by s isomorphically onto a tree $s(T)$ in $\mathrm{Tot}(\mathcal{X}, \Gamma)$.

Proposition 6.2.2. *Let each $p_{e\#}^\pm$ be a monomorphism. There is an isomorphism j making the following diagram commute for each vertex w of Γ:*

$$\pi_1(\mathrm{Tot}(\mathcal{X}, \Gamma), x(w)) \xleftarrow{\ \beta_w\ } \pi_1(X(w), x(w))$$

$$\cong \Big\downarrow k_\# \qquad\qquad\qquad\qquad \Big\downarrow \gamma_w$$

$$\pi_1(\mathrm{Tot}(\mathcal{X}, \Gamma), s(T)) \xrightarrow[\cong]{\ \ j\ \ } \pi_1(\mathcal{G}, \Gamma; T)$$

where β_w is induced by inclusion, and γ_w is as in 6.2.1.

Proof. The tree $s(T)$ can be extended, using 3.1.15, to a maximal tree T^+ in $\mathrm{Tot}(\mathcal{X}, \Gamma)$ such that, for each vertex w, $T^+ \cap X(w)$ is a maximal tree in $X(w)$ (considered as a subcomplex of $\mathrm{Tot}(\mathcal{X}, \Gamma)$). The presentation of $\pi_1(\mathrm{Tot}(\mathcal{X}, \Gamma), x(w))$ obtained by applying 3.1.16 to $\mathrm{Tot}(\mathcal{X}, \Gamma)$, using T^+, is seen by inspection to be a presentation of $\pi_1(\mathcal{G}, \Gamma; T)$. In fact, the proof of 3.1.15 shows that an isomorphism j exists as indicated. This is clear when the complexes each have one vertex. But that is enough. $\square$

Now we are ready for our application of 6.1.3. Let Y be a rigid G-tree.[8] This is a special case of what was considered in Sect. 6.1, so we carry over all the notation of that section except that we write Γ in place of V for $G\backslash Y$, since it is a graph. In particular, when $e \in E_1$, we have $f_e : X_e \times \{\pm 1\} \to q^{-1}(\Gamma^0)$ defined by the commutative diagram preceding Proposition 6.1.2. From that diagram one deduces several facts about the map f_e: (i) $f_e(X_e \times \{-1\}) \subset X_{o(e)}$. Define $f_e^- : X_e \to X_{o(e)}$ by $f_e^-(y) = f_e(y, -1)$. Then (ii) f_e^- is a covering projection; in fact, with obvious abuse of notation, it is the covering projection $G_{\tilde{e}}\backslash \tilde{X} \to G_{o(\tilde{e})}\backslash \tilde{X}$. Let $x'_{o(e)} = r(\tilde{x}_0, o(\tilde{e}))$. Then (iii) $f_e^-(x_e) = x'_{o(e)}$. Similarly, (iv) we have a pointed covering projection $f_e^+ : (X_e, x_e) \to (X_{t(e)}, x'_{t(e)})$ where $x'_{t(e)} = r(\tilde{x}_0, t(\tilde{e}))$.

Since $x'_{o(e)}$ and $x'_{t(e)}$ can be different from $x_{o(e)}$ and $x_{t(e)}$, we do not yet have a graph of pointed CW complexes. We alter things to make one. For each $e \in E_1$, pick a cellular path $\alpha(e)$ in $X_{o(e)}$ from $x'_{o(e)}$ to $x_{o(e)}$, and a cellular path $\beta(e)$ in $X_{t(e)}$ from $x'_{t(e)}$ to $x_{t(e)}$. Define $H : (X_{o(e)} \times \{0\}) \cup (\{x'_{o(e)}\} \times I) \to X_{o(e)}$ by $(x, 0) \mapsto x$ when $x \in X_{o(e)}$, and by $(x'_{o(e)}, \lambda) \mapsto \alpha(e)(\lambda)$ when $\lambda \in I$. By 1.3.15, H extends to $\bar{H} : X_{o(e)} \times I \to X_{o(e)}$. Let $g_e^- : (X_{o(e)}, x'_{o(e)}) \to (X_{o(e)}, x_{o(e)})$ be the map $g_e^-(x) = \bar{H}(x, 1)$. Similarly, define $g_e^+ : (X_{t(e)}, x'_{t(e)}) \to (X_{t(e)}, x_{t(e)})$. These $g_e^\pm$ are homotopic to the appropriate identity maps; by 1.3.16 we may assume they are cellular. On fundamental groups, we have $g_{e\#}^- = h_{\alpha(e)}$ and $g_{e\#}^+ = h_{\beta(e)}$ (in the notation of 3.1.11).

We now have a graph of pointed CW complexes $(\mathcal{X}, \Gamma)$: $(X(w), x(w))$ is (X_w, x_w), $(X(e), x(e))$ is (X_e, x_e), $p_e^- : (X_e, x_e) \to (X_{o(e)}, x_{o(e)})$ is the map $g_e^- \circ f_e^-$, and $p_e^+ = g_e^+ \circ f_e^+$. The induced homomorphisms $p_{e\#}^\pm$ are monomorphisms, by 3.4.6 and 3.1.11. Thus $(\mathcal{X}, \Gamma)$ induces a graph of groups $(\mathcal{G}, \Gamma)$ with $G(w) = \pi_1(X_w, x_w)$, $G(e) = \pi_1(X_e, x_e)$, $\phi_e^- = p_{e\#}^- : \pi_1(X_e, x_e) \to$

---

[8] In the literature one finds a "tree on which G acts by simplicial automorphisms without inversions" a special case of a rigid G-tree.

$\pi_1(X_{o(e)}, x_{o(e)})$, and $\phi_e^+ = p_{e\#}^+ : \pi_1(X_e, x_e) \to \pi_1(X_{t(e)}, x_{t(e)})$. By 6.2.2, the corresponding $\pi_1(\mathcal{G}, \Gamma; T)$ is isomorphic to $\pi_1(\text{Tot}(\mathcal{X}, \Gamma), s(T))$. Since $g_e^\pm$ are homotopic to the identity maps, 4.1.7 and 6.1.3 imply that $\pi_1(\text{Tot}(\mathcal{X}, \Gamma), s(T))$ is isomorphic[9] to $\pi_1(Z, z)$, where Z and $z = x_v$ are as in Sect. 6.1. And we saw in that section that $\pi_1(Z, z)$ is isomorphic to G.

In summary, we have expressed G as the fundamental group of a graph of groups $(\mathcal{G}, \Gamma)$. But we are not quite done. We would like to identify the vertex and edge groups with specific subgroups[10] of G.

Observe that $r^{-1}(X_{o(e)}) = \tilde{X} \times p^{-1}(o(e))$. Two of its path components are $\tilde{X} \times \{(o(e))^\sim\}$ and $\tilde{X} \times \{o(\tilde{e})\}$. We have a covering projection $r_1 := r \mid: \tilde{X} \times \{(o(e))^\sim\} \to X_{o(e)}$, and, using base points $(\tilde{v}, (o(e))^\sim)$ and $x_{o(e)} = r(\tilde{v}, (o(e))^\sim)$, we have an isomorphism $\chi_1 : G_{(o(e))^\sim} \to \pi_1(X_{o(e)}, x_{o(e)})$ as in 3.2.3. Similarly, $r_2 := r \mid: \tilde{X} \times \{o(\tilde{e})\} \to X_{o(e)}$ is a covering projection, and we have an isomorphism $\chi_2 : G_{o(\tilde{e})} \to \pi_1(X_{o(e)}, x'_{o(e)})$.

We picked a path $\alpha(e)$ in $X_{o(e)}$ from $x'_{o(e)}$ to $x_{o(e)}$. Let $\tilde{\alpha}(e)$ be the lift of this path to $\tilde{X} \times \{(o(e))^\sim\}$ whose final point is $(\tilde{v}, (o(e))^\sim)$. Since r maps the initial point of $\tilde{\alpha}(e)$ to $r(\tilde{v}, o(\tilde{e})) = x'_{o(e)}$, there must be a unique element $a(e) \in G$ such that the initial point of $\tilde{\alpha}(e)$ is $(a(e).\tilde{v}, (o(e))^\sim)$. Thus $a(e).o(\tilde{e}) = (o(e))^\sim$, and $G_{(o(e))^\sim} = a(e)G_{o(\tilde{e})}a(e)^{-1}$. Indeed, if $a'(e) \in G$ satisfies $a'(e).o(\tilde{e}) = (o(e))^\sim$, there is a path $\alpha'(e)$ from $x'_{o(e)}$ to $x_{o(e)}$ leading us to $a'(e)$ just as $\alpha(e)$ led us to $a(e)$.

Let c_g denote the conjugation isomorphism $h \mapsto g^{-1}hg$. One easily checks:

Proposition 6.2.3. *The following diagram commutes:*

$$\begin{array}{ccc} G_{o(\tilde{e})} & \xleftarrow{\;\;c_{a(e)}\;\;} & G_{(o(e))^\sim} \\[2mm] \downarrow{\scriptstyle \chi_2} & & \downarrow{\scriptstyle \chi_1} \\[2mm] \pi_1(X_{o(e)}, x'_{o(e)}) & \xrightarrow{\;\;h_{\alpha(e)}\;\;} & \pi_1(X_{o(e)}, x_{o(e)}) \end{array}$$

$\qquad\qquad\qquad\qquad\qquad\qquad\qquad\qquad\qquad\qquad\qquad\qquad\qquad$ $\square$

Similarly, corresponding to $\beta(e)$ is an element $b(e) \in G$ and an isomorphism $c_{b(e)} : G_{(t(e))^\sim} \to G_{t(\tilde{e})}$ for which the corresponding diagram commutes.

Define a new graph of groups $(\bar{\mathcal{G}}, \Gamma)$ by: $\bar{G}(w) = G_{\tilde{w}}$; $\bar{G}(e) = G_{\tilde{e}}$; $\bar{\phi}_e^-$ is the composition $G_{\tilde{e}} \hookrightarrow G_{o(\tilde{e})} \xrightarrow{c_{a(e)}^{-1}} G_{(o(e))^\sim}$; and $\bar{\phi}_e^+$ is the composition $G_{\tilde{e}} \hookrightarrow G_{t(\tilde{e})} \xrightarrow{c_{b(e)}^{-1}} G_{(t(e))^\sim}$. The meaning of the next proposition will become clear in the proof:

Proposition 6.2.4. *$(\bar{\mathcal{G}}, \Gamma)$ and $(\mathcal{G}, \Gamma)$ are isomorphic graphs of groups.*

---

[9] This is explained in more detail after 6.2.5, below.

[10] This is obviously desirable, and it will also enable us to avoid reliance on 6.2.1, which we have not yet proved.

Proof. The following diagram commutes, as well as a similar diagram in which $t(e)$ replaces $o(e)$, $b(e)$ replaces $a(e)$, and $+$ replaces $-$:

$$
\begin{array}{ccccc}
G_{\tilde{e}} & \hookrightarrow & G_{o(\tilde{e})} & \xrightarrow{c_{a(e)}^{-1}} & G_{(o(e))^{\sim}} \\
\downarrow{\scriptstyle \chi_3} & & \downarrow{\scriptstyle \chi_2} & & \downarrow{\scriptstyle \chi_1} \\
\pi_1(X_e, x_e) & \xrightarrow{f_{e\#}^{-}} & \pi_1(X_{o(e)}, x'_{o(e)}) & \xrightarrow{h_{\alpha(e)}} & \pi_1(X_{o(e)}, x_{o(e)})
\end{array}
$$

Here, the right hand square comes from 6.2.3. The left hand square obviously commutes.   □

It follows that $\pi_1(\bar{\mathcal{G}}, \Gamma; T)$ is isomorphic to G. But it does not follow that an isomorphism between those groups exists such that the canonical homomorphism $\gamma_w : G_{\tilde{w}} \to \pi_1(\bar{\mathcal{G}}, \Gamma; T)$ can be identified with $G_{\tilde{w}} \hookrightarrow G$ for all w. To achieve this, we place restrictions on our choice of the cells $\tilde{w}$ and $\tilde{e}$. (Until now, we have simply carried over the choices made in Sect. 6.1; each $\tilde{w}$ or $\tilde{e}$ was an arbitrary cell of Y "over" w or e.) We will need the following, which is proved using Zorn's Lemma:

Proposition 6.2.5. *There is a tree, $\tilde{T}$, in Y such that p maps $\tilde{T}$ isomorphically onto T.*   □

From now on, we choose each $\tilde{w}$ to be in $\tilde{T}$, and each $\tilde{e}$ to be in $\tilde{T}$ whenever e is in T. The effect is that whenever e is a cell of T, $o(\tilde{e}) = (o(e))^{\sim}$ and $t(\tilde{e}) = (t(e))^{\sim}$. For those cells e, $x_{o(e)} = x'_{o(e)}$ and $x_{t(e)} = x'_{t(e)}$; and we pick $\alpha(e)$ and $\beta(e)$ to be trivial, so that $a(e) = 1 = b(e)$. Thus $g_e^{\pm} = \mathrm{id}$ whenever the edge e lies in T.

The subcomplex $\{\tilde{x}_0\} \times \tilde{T} \subset \tilde{X} \times Y$ is a tree which contains all vertices of the form $(\tilde{x}_0, \tilde{w})$. Let $T_1 = r(\{\tilde{x}_0\} \times \tilde{T})$; T_1 is a tree in Z containing all vertices x_w; r maps $\{\tilde{x}_0\} \times \tilde{T}$ isomorphically onto T_1, and q maps T_1 isomorphically onto T.

By 4.1.8 and 6.1.3, there is a homotopy equivalence $h : Z \to \mathrm{Tot}(\mathcal{X}, \Gamma)$. To see this, apply 4.1.8 to the commutative diagram:

$$
\begin{array}{ccccc}
\coprod_{e \in E_1} X_e \times B^1 & \longleftarrow & \coprod_{e \in E_1} X_e \times S^0 & \xrightarrow{f} & \coprod_{v \in E_0} X_v \\
\downarrow{\scriptstyle \mathrm{id}} & & \downarrow{\scriptstyle \mathrm{id}} & & \downarrow{\scriptstyle \mathrm{id}} \\
\coprod_{e \in E_1} X_e \times B^1 & \longleftarrow & \coprod_{e \in E_1} X_e \times S^0 & \xrightarrow{g \circ f} & \coprod_{v \in E_0} X_v
\end{array}
$$

where f [resp. $g \circ f$] agrees with $f_e^{\pm}$ [resp. $g_e^{\pm} \circ f_e^{\pm}$] on $X_e \times \{\pm 1\}$. Indeed, h identifies T_1 with $s(T)$ in the natural way.

Now we can prove:

Proposition 6.2.6. *There is an isomorphism $\psi : \pi_1(\bar{\mathcal{G}}, \Gamma; T) \to G$ such that for every vertex w of Γ the following diagram commutes (where $\bar{\gamma}_w$ is analogous to γ_{w_0} in 6.2.1):*

$$
\begin{array}{ccc}
G_{\tilde{w}} & \hookrightarrow & G \\
\bar{\gamma}_w \downarrow & \nearrow \cong & \\
\pi_1(\bar{\mathcal{G}}, \Gamma; T) & \psi &
\end{array}
$$

Proof. The required ψ is read off from the following commutative diagram:

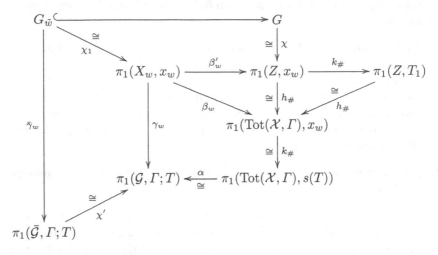

Here, χ' is the isomorphism arising from 6.2.4; β'_w is induced by inclusion. To see that ψ is independent of w, it is enough to check that the following diagram commutes, when w_1 and w_2 are vertices of Γ:

$$
\begin{array}{ccc}
G & \xrightarrow{\chi^{(1)}} & \pi_1(Z, x_{w_1}) \\
\chi^{(2)} \downarrow & \searrow^{h_{\alpha\#}} & \downarrow k_{\#}^{(1)} \\
\pi_1(Z, x_{w_2}) & \xrightarrow[k_{\#}^{(2)}]{} & \pi_1(Z, T_1)
\end{array}
$$

Here, α is any path in T_1 from x_{w_1} to x_{w_2}. We have seen that $h_{\alpha\#} = (k_{\#}^{(2)})^{-1} \circ k_{\#}^{(1)}$, so we need only prove that the other triangle commutes. If $g \in G$, and $\chi^{(1)}(g)$ is represented by the loop ω in Z at x_{w_1}, the lift, $\tilde{\omega}$, in $\tilde{X} \times Y$ with initial point $(\tilde{x}_0, \tilde{w}_1)$ has final point $(g.\tilde{x}_0, g.\tilde{w}_1)$. The path α lies in T_1; its lift $\tilde{\alpha}$ with initial point $(\tilde{x}_0, \tilde{w}_1)$ lies in $\{\tilde{x}_0\} \times \tilde{T}$ and so has final point $(\tilde{x}_0, \tilde{w}_2)$. Thus the lift of $\alpha^{-1}.\omega.\alpha$ with initial point $(\tilde{x}_0, \tilde{w}_2)$ has final point $(g\tilde{x}_0, g\tilde{w}_2)$. In other words, $\chi^{(2)}(g)$ is represented by $\alpha^{-1}.\omega.\alpha$. This means that $h_{\alpha\#} \circ \chi^{(1)}(g) = \chi^{(2)}(g)$. $\square$

Summarizing, we have shown how to decompose a group acting rigidly on a tree as the fundamental group of a graph of groups whose vertex groups and edge groups are stabilizers:

Theorem 6.2.7. *Let G act rigidly on the tree Y with quotient $p : Y \to G\backslash Y =: \Gamma$. Let T be a maximal tree in Γ and let $\tilde{T}$ be a tree in Y such that $p \mid: \tilde{T} \to T$ is an isomorphism. For each vertex w of Γ, let $\tilde{w} \in \tilde{T}$ be such that $p(\tilde{w}) = w$. For each 1-cell e of Γ pick a 1-cell $\tilde{e}$ of Y mapped by p onto e, subject to the rule that if $e \subset T$, then $\tilde{e} \subset \tilde{T}$. Orient each 1-cell e of Γ, and orient each $\tilde{e}$ to make $p \mid \tilde{e}$ orientation preserving. Let $a(e) \in G$ [resp. $b(e) \in G$] be such that $a(e).o(\tilde{e}) = (o(e))^\sim$ [resp. $b(e).t(\tilde{e}) = (t(e))^\sim$] subject to the rule that if $e \subset T$, $a(e) = b(e) = 1$. Let $(\bar{\mathcal{G}}, \Gamma)$ be the graph of groups with vertex groups $G_{\tilde{w}}$, edge groups $G_{\tilde{e}}$, and monomorphisms*

$$ G_{(o(e))^\sim} \xleftarrow{\;c_{a(e)}^{-1}\;} G_{o(\tilde{e})} \longleftarrow\hspace{-0.3em}\supset G_{\tilde{e}} \subset\hspace{-0.3em}\longrightarrow G_{t(\tilde{e})} \xrightarrow{\;c_{b(e)}^{-1}\;} G_{(t(e))^\sim}. $$

Then, letting $\bar{\gamma}_w : G_{\tilde{w}} \to \pi_1(\bar{\mathcal{G}}, \Gamma; T)$ denote the canonical homomorphism, there is an isomorphism $\psi : \pi_1(\bar{\mathcal{G}}, \Gamma; T) \to G$ such that for each vertex w of Γ, $\psi \circ \bar{\gamma}_w =$ inclusion: $G_{\tilde{w}} \hookrightarrow G$. In particular, $\bar{\gamma}_w$ is a monomorphism. $\square$

Remark 6.2.8. When the edge e is not in T, it is customary to choose $\tilde{e}$ subject to the rule: $o(\tilde{e}) = (o(e))^\sim$. Then $a(e) = 1$ for all e.

Example 6.2.9. The group $SL_2(\mathbb{Z})$ of 2×2 integer matrices of determinant 1 acts on the open upper half of the complex plane $\mathbb{C}$ by Möbius transformations. With respect to the hyperbolic metric $\frac{ds}{y}$ this action is by isometries and its kernel is the subgroup $\{\pm I\}$. The orbit of $\{z \mid \text{Im}(z) > 1\}$ consists of that set together with a collection of pairwise disjoint open (Euclidean) circular disks whose closures in $\mathbb{R}^2$ are tangent to the real axis $\mathbb{R}$ at rational points, one for each rational. Each such closure is also tangent to exactly two others (where we treat $\{z \mid \text{Im}(z) \geq 1\}$ as the closure of a circular disk of infinite radius "tangent at ∞"). See Fig. 6.1. The complementary region is thus invariant under the action of $SL_2(\mathbb{Z})$ and contains an invariant *Serre tree* as illustrated in Fig. 6.1. This tree has vertices of order 2 and order 3 with respective stabilizers $\mathbb{Z}_4$ and $\mathbb{Z}_6$. Each edge contains one vertex of each kind and has stabilizer of order 2. With a little thought about how edge-stabilizers inject into vertex-stabilizers, one deduces from Theorem 6.2.7 that $SL_2(\mathbb{Z})$ is isomorphic to the free product with amalgamation $\mathbb{Z}_4 *_{\mathbb{Z}_2} \mathbb{Z}_6$.

We now sketch the inverse construction: how to construct a rigid G-tree from a graph of groups.

Let $(\mathcal{G}, \Gamma)$ be a graph of groups, and let T be a maximal tree in Γ. Orient Γ. Form a graph of pointed CW complexes $(\mathcal{X}, \Gamma)$ by choosing pointed CW complexes $(X(w), x(w))$ and $(X(e), x(e))$, for every w and e, and pointed cellular maps $p_e^- : (X(e), x(e)) \to (X(o(e)), x(o(e)))$ and

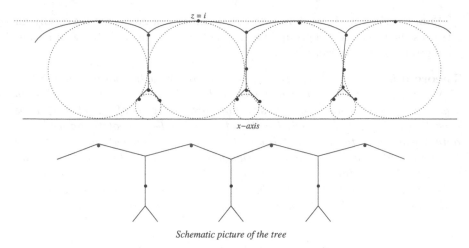

Schematic picture of the tree

Fig. 6.1.

$p_e^+ : (X(e), x(e)) \to (X(t(e)), x(t(e)))$ such that for suitable isomorphisms $\psi_w : \pi_1(X(w), x(w)) \to G(w)$ and $\psi_e : \pi_1(X(e), x(e)) \to G(e)$, $p_{e\#}^\pm$ can be identified with the monomorphisms $\phi_e^\pm$ in $\mathcal{G}$. Let $G = \pi_1(\mathcal{G}, \Gamma; T)$. By 6.2.2, G can be identified with $\pi_1(\mathrm{Tot}(\mathcal{X}, \Gamma), x(v))$ where v is the base vertex of Γ. Hence we have a free action of G on $U := (\mathrm{Tot}(\mathcal{X}, \Gamma))^\sim$. Applying Sect. 3.2, using 3.4.9 and 6.2.1, U is seen to be a quotient space obtained by gluing copies of $(X(e))^\sim \times B^1$ to copies of $(X(w))^\sim$ via the lifts of the p's, where e and w are variable. If we identify each copy of $(X(w))^\sim$ in U to a point, and each copy of $(X(e))^\sim \times B^1$ to a 1-cell, we obtain a graph Y and a commutative diagram

$$
\begin{array}{ccc}
U & \xrightarrow{\ s\ } & Y \\
{\scriptstyle r}\downarrow & & \downarrow{\scriptstyle p} \\
\mathrm{Tot}(\mathcal{X}, \Gamma) & \xrightarrow[\ q\ ]{} & \Gamma
\end{array}
$$

Here, r is the covering projection, s is the quotient map and p is a well defined map. Since s is easily seen to be a domination, and U is simply connected, Y is simply connected; i.e., Y is a tree, *the Bass-Serre tree* of $(\mathcal{G}, \Gamma)$. The free G-action on U induces a rigid G-action on Y; s is a G-map and p is the quotient map of the G-action. The tree T in Γ gives a tree $s(T)$ in $\mathrm{Tot}(\mathcal{X}, \Gamma)$ as before; this lifts to a tree $\tilde{s}(T)$ in U which is mapped isomorphically to a tree $\tilde{T}$ in Y. With choices as in Theorem 6.2.7, the resulting graph of groups $(\bar{\mathcal{G}}, \Gamma)$ is isomorphic to the original $(\mathcal{G}, \Gamma)$.

Example 6.2.10. A *Baumslag-Solitar group* is a group $BS(m, n)$ with presentation $\langle x, t \mid t^{-1} x^m t x^{-n} \rangle$ where $m, n \geq 1$. Clearly, $BS(m, n)$ is the HNN extension $\mathbb{Z} *_{\phi_{m,n}}$ where $\phi_{m,n} : m\mathbb{Z} \to n\mathbb{Z}$ is the isomorphism taking m to n.

Thus $BS(m,n)$ is the fundamental group of a circle of groups $(\mathcal{G}, \Gamma)$ where Γ has one vertex and one edge, and both the vertex group and the edge group are infinite cyclic. The homomorphisms $\phi_k : \mathbb{Z} \to \mathbb{Z}$, $1 \mapsto k$, are induced by covering projections $S^1 \to S^1$ when $k \geq 1$. Since all these covering projections lift to homeomorphisms of $\mathbb{R}$, there is an obvious homeomorphism $h : U \to Y \times \mathbb{R}$ where Y is the Bass-Serre tree of $(\mathcal{G}, \Gamma)$ and U is as in the above construction. Indeed, using h to make $Y \times \mathbb{R}$ into a G space, we have a commutative diagram of G-spaces:

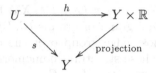

This is a topological description of U. But U is also a CW complex, the universal cover of $\mathrm{Tot}(\mathcal{X}, \Gamma)$, which in this case is the presentation complex of $BS(m,n)$. Exercise 8 of Sect. 3.2 asked for a description of the cells of the CW complex U in the case of $BS(1,2)$. It is instructive to "see" the topological product $Y \times \mathbb{R}$ in this CW complex.

The construction of Y from $(\mathcal{G}, \Gamma)$ is inverse to 6.2.7 in the following sense. In 6.2.7 we started with G acting on Y, together with a maximal tree $T \subset \Gamma$, and produced a decomposition of G as $\pi_1(\bar{\mathcal{G}}, \Gamma; T)$ whose vertex and edge groups are stabilizers of certain vertices and edges (determined by choices of $\tilde{T}$, of $\{\tilde{e} \mid e \in E_0 \cup E_1\}$ and of orientation of Γ). Conversely, we have sketched how one starts with $(\mathcal{G}, \Gamma)$ and T, and produces an action of $G := \pi_1(\mathcal{G}, \Gamma; T)$ on a tree Y, such that (with appropriate choices) the corresponding graph of groups $(\bar{\mathcal{G}}, \Gamma)$ is isomorphic to $(\mathcal{G}, \Gamma)$. Indeed, it can be shown that: starting with the G-tree Y, obtaining $(\bar{\mathcal{G}}, \Gamma)$, and then constructing a $\pi_1(\bar{\mathcal{G}}, \Gamma; T)$-tree $\bar{Y}$ as above, there is an equivariant isomorphism from $\bar{Y}$ to Y (i.e., an isomorphism commuting with the actions of the groups). For this reason, the theories of graphs of groups and of groups acting rigidly on trees are considered to be equivalent. For more details the reader is referred to [140]. A more algebraic version is described in [141] and [142].

Appendix: Generalized graphs of groups

A *generalized graph of groups* has the same definition as a graph of groups $(\mathcal{G}, \Gamma)$, except that the homomorphisms $\phi_e^{\pm}$ are not required to be monomorphisms. If T is a maximal tree in Γ, the *fundamental group* of $(\mathcal{G}, \Gamma)$ based at T is defined as before and is again denoted by $\pi_1(\mathcal{G}, \Gamma; T)$. A graph of pointed CW complexes $(\mathcal{X}, \Gamma)$ gives rise to a generalized graph of groups even when the homomorphisms $p_{e\#}^{\pm}$ (at the beginning of this section) are not monomorphisms, and the obvious analog of 6.2.2 holds. Generalized graphs of groups are useful in computing fundamental groups via a generalization of the Seifert-Van Kampen Theorem 3.1.18, as we now explain.

Let $\{X_\alpha\}_{\alpha \in \mathcal{A}}$ be a cover of the path connected CW complex X by path connected subcomplexes such that no point of X lies in more than two of the X_α's. Let Γ be the graph whose vertex set is $\mathcal{A}$, having an edge $e_{\alpha\beta\gamma}$ joining $\alpha \neq \beta$ for each path component $Y_{\alpha\beta\gamma}$ of $X_\alpha \cap X_\beta$. Then Γ is a path connected graph. Orient Γ.

By an obvious extension of 3.1.13, choose a maximal tree $\bar{T}$ in X such that each $\bar{T} \cap X_\alpha$ and each $\bar{T} \cap Y_{\alpha\beta\gamma}$ is a maximal tree in that subcomplex. Let $\phi_{\alpha\beta\gamma} : \pi_1(Y_{\alpha\beta\gamma}, \bar{T} \cap Y_{\alpha\beta\gamma}) \to \pi_1(X_\beta, \bar{T} \cap X_\beta)$ be induced by inclusion. Then we have a generalized graph of groups $(\mathcal{G}, \Gamma)$ whose vertex group at α is $\pi_1(X_\alpha, \bar{T} \cap X_\alpha)$, whose edge group over the edge labeled by $Y_{\alpha\beta\gamma}$ is $\pi_1(Y_{\alpha\beta\gamma}, \bar{T} \cap Y_{\alpha\beta\gamma})$, and whose structural homomorphisms for that edge are $\phi_{\alpha\beta\gamma}$ and $\phi_{\beta\alpha\gamma}$. Pick a maximal tree T in Γ.

Theorem 6.2.11. *[Generalized Van Kampen Theorem] Under these hypotheses the fundamental group of X is isomorphic to $\pi_1(\mathcal{G}, \Gamma; T)$.*

Proof. Let $(\mathcal{X}, \Gamma)$ be the generalized graph of path connected CW complexes having X_α over α, $Y_{\alpha\beta\gamma}$ over the edge joining α to β so indexed, and inclusions as structural maps. Let $\mathrm{Tot}(\mathcal{X}, \Gamma)$ be the total complex. The proof of 6.2.2 extends to show that $\pi_1(\mathcal{G}, \Gamma; T)$ is isomorphic to the fundamental group of $\mathrm{Tot}(\mathcal{X}, \Gamma)$. The space $\mathrm{Tot}(\mathcal{X}, \Gamma)$ presents itself naturally as a subcomplex of $X \times \Gamma$, consisting of X_α over the vertex α and $Y_{\alpha\beta\gamma} \times e_{\alpha\beta\gamma}$ over each edge $e_{\alpha\beta\gamma}$ joining α to β. Projection on the X factor gives us a map $p : \mathrm{Tot}(\mathcal{X}, \Gamma) \to X$. Consider the commutative diagram

$$
\begin{array}{ccccc}
\mathrm{Tot}(\mathcal{X}, \Gamma) & \longleftarrow \cup(Y_{\alpha\beta\gamma} \times e_{\alpha\beta\gamma}) & \xrightarrow{\ \mathrm{id}\ } & \cup(Y_{\alpha\beta\gamma} \times e_{\alpha\beta\gamma}) \\
\Big\downarrow{\scriptstyle \mathrm{id}} & \Big\downarrow{\scriptstyle \mathrm{id}} & & \Big\downarrow{\scriptstyle p|} \\
\mathrm{Tot}(\mathcal{X}, \Gamma) & \longleftarrow \cup(Y_{\alpha\beta\gamma} \times e_{\alpha\beta\gamma}) & \xrightarrow[\ p|\ ]{} & \cup Y_{\alpha\beta\gamma}
\end{array}
$$

The hypothesis that no point of X lies in more than two of the X_α's implies that the spaces $Y_{\alpha\beta\gamma}$ are pairwise disjoint and hence the indicated map $p|$ is a homotopy equivalence. The map of adjunction spaces coming from 4.1.7 is precisely p, which is therefore a homotopy equivalence. $\square$

Remark 6.2.12. As with ordinary graphs of groups there is an obvious epimorphism $\pi_1(\mathcal{G}, \Gamma; T) \twoheadrightarrow \pi_1(\Gamma, T)$. Since $\pi_1(\Gamma, T)$ is free this gives a lower bound for the number of generators of $\pi_1(X, v)$ in 6.2.11. In fact, since every epimorphism onto a free group has a right inverse we see that $\pi_1(\Gamma, T)$ is a retract[11] of $\pi_1(\mathcal{G}, \Gamma; T)$.

---

[11] The definition of *retract* in any category is analogous to that given in Sect. 1.3 for Spaces.

Remark 6.2.13. We end by discussing the relationship between graphs of groups and generalized graphs of groups. Let $(\mathcal{G}, \Gamma)$ be a generalized graph of groups, let T be a maximal tree in Γ and let $G := \pi_1(\mathcal{G}, \Gamma; T)$. Proceeding as in the "inverse construction" part of this section, one forms a graph of pointed CW complexes $(\mathcal{X}, \Gamma)$, identifying G with $\pi_1(\text{Tot}(\mathcal{X}, \Gamma), x(v))$; the universal cover of $\text{Tot}(\mathcal{X}, \Gamma)$ is denoted by U. The space U is built out of covering spaces, no longer universal covers, of vertex complexes and edge complexes, and, as before, one obtains a G-tree Y from this situation. The quotient $G \backslash Y$ is a copy of Γ. The vertex and edge stabilizers of this new graph of groups $(\bar{\mathcal{G}}, \Gamma)$ are the images in G of the vertex and edge groups of $\mathcal{G}$. This new decomposition of G may be quite uninteresting. For example, take Γ to be an edge, take both vertex groups to be trivial and take the edge group to be $\mathbb{Z}$ (as happens when the Seifert-Van Kampen Theorem is applied to S^2, using the two hemispheres as the "vertex spaces"). Then all the groups in $\bar{\mathcal{G}}$ are trivial.

Source Notes: The theory described in this section appeared in [141], translated into English as [142]. A more topological presentation, on which this section is based, appeared in [140].

Exercises

1. Prove 6.2.5.
2. Referring to Case 1 at the beginning of this section, describe the Bass-Serre tree in the case of a free product with amalgamation in which the monomorphism ϕ_e^+ is an isomorphism, and in the case where ϕ_e^+ and ϕ_e^- are both isomorphisms.
3. Describe the Bass-Serre tree of the circle of groups corresponding to the given presentation of the Baumslag-Solitar group $BS(m, n)$.
4. In an obvious way, $\mathbb{Z}$ can be decomposed as the fundamental group of a graph of groups, with one vertex and one edge, the vertex and edge groups being trivial. Thus if $\alpha : G \to \mathbb{Z}$ is an epimorphism, there is a corresponding decomposition of G with vertex and edge groups isomorphic to $\ker(\alpha)$. Give an example with G finitely presented such that $\ker(\alpha)$ is not finitely generated; give another graph of groups decomposition of the same group G with one vertex and one edge so that the vertex and edge groups are finitely generated.
5. Let $G = A \underset{C}{*} (B \underset{E}{*} D)$ be a decomposition of G using free products with amalgamation. Under what conditions does this decompose as the fundamental group of a graph of groups with graph $\bullet - \bullet - \bullet$ and with vertex groups isomorphic to A, B and D and edge groups isomorphic to C and E?
6. Prove 6.2.2 when $X(w)$ or $X(e)$ has more than one vertex.
7. Let $\bar{G} = G*_\phi$ be an HNN extension where ϕ is an automorphism of G. Prove that $\bar{G}$ is a semidirect product of G and $\mathbb{Z}$.

7

Topological Finiteness Properties and Dimension of Groups

This chapter is about topological models for a group G: these are the so-called $K(G,1)$-complexes with fundamental group G and contractible universal cover. We then proceed to use the method of Sect.6.1 to replace an arbitrary $K(G,1)$-complex by one with fewer cells or lower dimension.

7.1 $K(G,1)$ complexes

Let n be an integer ≥ -1. A space X is *n-connected* if[1] for every $-1 \leq k \leq n$, every map $S^k \to X$ extends to a map $B^{k+1} \to X$. Clearly (-1)-connected is the same as "non-empty" and "0-connected" is the same as "path connected." Note that when $n \geq 0$ a space X is n-connected iff for some (equivalently any) $x \in X$, the pair $(X, \{x\})$ is n-connected in the sense of Sect. 4.1. Note also that when X is non-empty and $n \geq 0$, X is n-connected iff for every $0 \leq k \leq n$, every map $S^k \to X$ is *homotopically trivial* (i.e., is homotopic to a constant map).

Proposition 7.1.1. *A CW complex is 1-connected iff it is simply connected.* □

Proposition 7.1.2. *Let the CW complex X be non-empty and let $n \geq 0$. X is n-connected iff the inclusion $X^n \hookrightarrow X$ is homotopic to a constant map. X is n-connected for all n iff X is contractible.*

Proof. For $n = 0$, the first sentence is clear. Let X be n-connected. By induction, $X^{n-1} \hookrightarrow X$ is homotopically trivial. Hence $X^n \hookrightarrow X$ is homotopically trivial iff a certain map $X^n/X^{n-1} \to X$ is homotopically trivial. But X^n/X^{n-1} is a wedge of n-spheres and X is n-connected. So $X^n \hookrightarrow X$ is homotopically trivial. Conversely, if $X^n \hookrightarrow X$ is homotopically trivial, then, by

---

[1] Here and elsewhere we depart from common usage by allowing $n = -1$. This convention will be seen to be helpful when we discuss homological and homotopical group theory. Our Sect. 2.7 on reduced homology was written with this in mind.

1.4.3, every map $S^k \to X$ is homotopically trivial whenever $k \leq n$, so X is n-connected. The second sentence follows from 4.1.4 and Remark 4.1.6. $\square$

It follows, in particular, that S^{n+1} is n-connected. By 1.4.3, a CW complex X is n-connected iff X^{n+1} is n-connected.

A space X is *n-aspherical* $(n \geq 0)$ if X is path connected and for every $2 \leq k \leq n$, every map $S^k \to X$ extends to a map $B^{k+1} \to X$. Thus "0-aspherical" and "1-aspherical" are the same as "path connected." We say X is *aspherical* if X is n-aspherical for all n.

Proposition 7.1.3. *A path connected CW complex X is n-aspherical iff its universal cover $\tilde{X}$ is n-connected. X is aspherical iff $\tilde{X}$ is contractible.*

Proof. Let $p : \tilde{X} \to X$ be the universal cover, and let $2 \leq k \leq n$. If X is n-aspherical and if $\tilde{f} : S^k \to \tilde{X}$ is a map then $p \circ \tilde{f}$ is homotopically trivial. By 2.4.6, the same is true of $\tilde{f}$. Conversely, if $\tilde{X}$ is n-aspherical, any map $f : S^k \to X$ lifts to a map $\tilde{f} : S^k \to \tilde{X}$ which extends to B^{k+1}, by 3.2.8. Hence f also extends to B^{k+1}. But $\tilde{X}$ is simply connected. So X is n-aspherical iff $\tilde{X}$ is n-aspherical iff $\tilde{X}$ is n-connected. The second sentence follows from the first together with 7.1.2. $\square$

Obviously, the product of aspherical [resp. n-aspherical, n-connected] CW complexes is aspherical [resp. n-aspherical, n-connected].

Let G be a group. A $K(G, 1)$-*complex* is a pointed aspherical CW complex[2] (X, x) such that $\pi_1(X, x)$ is isomorphic to G.

If (X, x) is a $K(G, 1)$-complex and y is another vertex of X, then (X, y) is a $K(G, 1)$-complex, by 3.1.11. Thus one may omit reference to the base point, saying "X is a $K(G, 1)$-complex".[3] With notation as in Sect. 3.2, we have:

Corollary 7.1.4. *A covering space of an n-aspherical [resp. aspherical] CW complex is n-aspherical [resp. aspherical]. If (X, x) is a $K(G, 1)$-complex and if $H \leq G$, then $(\bar{X}(H), \bar{x})$ is a $K(H, 1)$-complex.* $\square$

Proposition 7.1.5. *For any group G, there exists a $K(G, 1)$-complex X having only one vertex. Moreover, if (Y, y) is a path connected k-aspherical pointed CW complex such that $\pi_1(Y, y)$ is isomorphic to G, then there exists a $K(G, 1)$-complex whose $(k + 1)$-skeleton is Y^{k+1}.*

Proof. We will describe the $K(G, 1)$-complex X by induction on skeleta. The 2-skeleton (X^2, x) is built as in Example 1.2.17, reflecting some chosen presentation of G; X^2 is 1-aspherical. By induction, assume X^n has been constructed

---

[2] In the literature, $K(G, 1)$-complexes are sometimes called *Eilenberg-MacLane complexes of type $(G, 1)$*, or *classifying spaces for G*; see [149] for an explanation of the latter name. The notation BG is sometimes used instead of $K(G, 1)$.

[3] Of course, the G-action on $\tilde{X}$ by covering transformations depends on the base point $\tilde{x}$ and on the isomorphism chosen to identify G with $\pi_1(X, x)$. When we say (X, x) is a $K(G, 1)$-complex we tacitly assume these choices have been made.

and is $(n-1)$-aspherical, where $n \geq 2$. Let $\{f_\alpha : S^n \to X^n \mid \alpha \in \mathcal{A}_n\}$ be a set of maps such that every map $f : S^n \to X^n$ is homotopic to f_α for some $\alpha \in \mathcal{A}_n$. The $(n+1)$-skeleton X^{n+1} is obtained from X^n by attaching $B^{n+1}(\mathcal{A}_n)$ to X^n using $\{f_\alpha\}$. By 1.4.3, any map $S^n \to X^{n+1}$ is homotopic to some f_α and hence to a constant map; so X^{n+1} is n-aspherical. This completes the induction; X is aspherical. For the second part, the construction is similar, but one starts with (Y^n, y) where $n = \max\{k+1, 2\}$. $\square$

When (X, x) and (Y, y) are pointed CW complexes, let $[(Y, y), (X, x)]$ denote the set of homotopy classes of maps $(Y, y) \to (X, x)$. When H and K are groups, let $\hom(K, H)$ denote the set of homomorphisms $K \to H$. Then there is a natural function

$$h : [(Y, y), (X, x)] \to \hom(\pi_1(Y, y), \pi_1(X, x)) \quad \text{defined by} \quad [f] \mapsto f_\#.$$

Proposition 7.1.6. *If X is aspherical, h is a bijection.*

Proof. By 3.1.13, X and Y contain maximal trees. The quotient complexes of X and Y by these trees have the homotopy types of X and Y respectively, by 4.1.9. So we may assume X and Y have one vertex each.

We show h is surjective. Let $\phi : \pi_1(Y, y) \to \pi_1(X, x)$ be a homomorphism. We will construct a cellular map $f : (Y, y) \to (X, x)$ inducing ϕ. To define f on Y^1, we will first define[4] a map $\tilde{f}_1 : \tilde{Y}^1 \to \tilde{X}^1$ inducing a map $f_1 : Y^1 \to X^1$.

The isomorphism χ of 3.2.3 provides a canonical identification of $\pi_1(Y, y)$ [resp. $\pi_1(X, x)$] with $\tilde{Y}^0$ [resp. $\tilde{X}^0$]. Define $\tilde{f}_1| : \tilde{Y}^0 \to \tilde{X}^0$ to agree with ϕ, via this identification. For each 1-cell e^1_α of Y, pick a lift $\tilde{e}^1_\alpha$, a 1-cell of $\tilde{Y}^1$. Define $\tilde{f}_1$ to map $\tilde{e}^1_\alpha$ to a path in $\tilde{X}^1$ in such a way that $\tilde{f}_1 \mid \tilde{e}^1_\alpha$ extends the previously defined $\tilde{f}_1 \mid \overset{\bullet}{\tilde{e}}{}^1_\alpha$. For each 1-cell $\tilde{e}^1$ of $\tilde{Y}^1$, there is exactly one α and one $g \in \pi_1(Y, y)$ such that $g.\tilde{e}^1_\alpha = \tilde{e}^1$. Define $\tilde{f}_1$ on $\tilde{e}^1$ by[5] $\tilde{f}_1(z) = \phi(g) \circ \tilde{f}_1 \circ g^{-1}(z)$. Then, indeed, f_1 exists making the following diagram commute:

$$
\begin{array}{ccc}
\tilde{Y}^1 & \xrightarrow{\tilde{f}_1} & \tilde{X}^1 \\
{\scriptstyle p_Y}\downarrow & & \downarrow{\scriptstyle p_X} \\
Y^1 & \xrightarrow{f_1} & X^1
\end{array}
$$

Next, let e^2_β be a 2-cell of Y with attaching map $g_\beta : S^1 \to Y$. We have $g_\beta(S^1) \subset Y^1$; g_β is homotopic in Y^2 to a constant map, since g_β extends to a map on B^2. By 2.4.6, g_β lifts to $\tilde{g}_\beta : S^1 \to \tilde{Y}$. Since $\tilde{X}$ is simply connected, $\tilde{f}_1 \circ \tilde{g}_\beta$ extends to a map on B^2, hence so does $p_X \circ \tilde{f}_1 \circ \tilde{g}_\beta = f_1 \circ g_\beta$. As in 1.2.23, we can use this to extend f_1 to e^2_β. Doing this for each β, we get $f_2 : Y^2 \to X^2$ extending f_1. By induction, assume f_2 has been extended to

---

[4] Recall $\tilde{Y}^n$ means $(\tilde{Y})^n$.

[5] We are not distinguishing in notation between elements of the fundamental groups and the corresponding covering transformations.

a cellular map $f_n : Y^n \to X^n$ $(n \geq 2)$. Let e_γ^{n+1} be an $(n+1)$-cell of Y and let $g_\gamma : S^n \to Y$ be an attaching map. $f_n \circ g_\gamma$ extends to B^{n+1} since X is n-aspherical. Thus $f_{n+1} : Y^{n+1} \to X^{n+1}$ can be constructed extending f_n. Let $f : Y \to X$ be the resulting map. By 3.2.3 and our definition of $\tilde{f}$ on $\tilde{Y}^0$, $f_\# = \phi$.

The proof that h is injective is similar. Let $f, f' : (Y, y) \to (X, x)$ be maps such that $f_\# = f'_\# : \pi_1(Y, y) \to \pi_1(X, x)$. We are to build a homotopy $H : Y \times I \to X$ such that $H(\{y\} \times I) = \{x\}$, $H_0 = f$ and $H_1 = f'$. We start with H so defined on $(Y \times \{0, 1\}) \cup (\{y\} \times I)$, and we extend H inductively to $(Y \times \{0, 1\}) \cup (Y^n \times I)$ for each n, using 1.3.10. The product CW complex structure on $Y \times I$ has been described in Sect. 1.2. To deal with the case $n = 1$, we must extend H to $e_\alpha^1 \times I$ for each 1-cell e_α of Y. If $g_\alpha : \dot{I}^2 \to Y \times I$ is an attaching map for $e_\alpha^1 \times I$ then $g_\alpha(\dot{I}^2) \subset (Y \times \{0, 1\}) \cup (\{y\} \times I)$, so $H \circ g_\alpha$ is defined. Since $f_\# = f'_\#$, $H \circ g_\alpha$ extends to I^2. Hence, by 1.3.10, H extends to $(Y \times \{0, 1\}) \cup (Y^1 \times I)$. The extension to $(Y \times \{0, 1\}) \cup (Y^n \times I)$ when $n \geq 2$ proceeds as before, using the fact that X is aspherical. □

Thus, maps into a $K(G, 1)$-complex are classified up to pointed homotopy by homomorphisms of the fundamental group into G.

Corollary 7.1.7. *If (X, x) and (Y, y) are $K(G, 1)$ complexes, then there is a homotopy equivalence $f : (X, x) \to (Y, y)$ inducing any given isomorphism $\pi_1(X, x) \to \pi_1(Y, y)$. In particular, $K(G, 1)$-complexes are unique up to pointed homotopy equivalence.* □

The proof of 7.1.6 also proves:

Proposition 7.1.8. *Let X be n-aspherical. If Y has dimension $\leq n + 1$, then h is surjective. If Y has dimension $\leq n$, then h is bijective.* □

We now consider some useful methods for constructing $K(G, 1)$-complexes.

Theorem 7.1.9. *Let $(\mathcal{X}, \Gamma)$ be a graph of pointed CW complexes. Assume each $X(w)$ and each $X(e)$ is aspherical and that each $p_{e\#}^{\pm}$ is a monomorphism (on fundamental groups). Then $\mathrm{Tot}(\mathcal{X}, \Gamma)$ is aspherical.*

Proof. We saw in Sect. 6.2 (following 6.2.9) that $U := (\mathrm{Tot}(\mathcal{X}, \Gamma))^\sim$ is a quotient space obtained by gluing copies of $\tilde{X}(e) \times B^1$ to copies of $\tilde{X}(w)$ via the pointed lifts of the p's, where e and w are variable. By 4.1.7, U has the homotopy type of the subcomplex U', consisting of the $\tilde{X}(w)$'s with a copy of $\{\tilde{x}(e)\} \times B^1$ replacing each $\tilde{X}(e) \times B^1$; this is because each $\tilde{X}(e)$ is contractible. Then U' has the homotopy type of the quotient space U'' obtained by identifying each copy of each $\tilde{X}(w)$ to a point; this is because each $\tilde{X}(w)$ is contractible. But U'' is a graph and is simply connected since U is simply connected. Thus U is contractible. □

Next we turn to group extensions. Let $N \rightarrowtail G \overset{\pi}{\twoheadrightarrow} Q$ be a short exact sequence of groups. Let (Y, y) be a $K(N, 1)$ complex and let (Z, z) be a $K(Q, 1)$-complex. If Y and Z have particular features, the Borel Construction (Sect. 6.1) often gives a procedure for building a $K(G, 1)$-complex with similar features.

Theorem 7.1.10. *There is a $K(G, 1)$-complex W and a stack $W \to Z$ all of whose fibers are Y.*

Proof. As in that section, we start with an arbitrary $K(G, 1)$-complex (X, x) and we consider the diagonal left action of G on $\tilde{X} \times \tilde{Z}$ given by $g(x, z) = (gx, \pi(g)z)$. The quotient space $U = G \backslash \tilde{X} \times \tilde{Z}$ is a $K(G, 1)$-complex fitting into a commutative diagram

$$
\begin{array}{ccc}
\tilde{X} \times \tilde{Z} & \xrightarrow{\text{projection}} & \tilde{Z} \\
\downarrow & & \downarrow \\
U & \xrightarrow{\quad q \quad} & Z
\end{array}
$$

in which $q : U \to Z$ is a stack of CW complexes all of whose fibers are $N \backslash \tilde{X}$, a $K(N, 1)$-complex. The Rebuilding Lemma 6.1.4 yields a commutative diagram

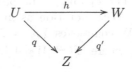

in which h is a homotopy equivalence (implying that W is $K(G, 1)$-complex) and q' is a stack of CW complexes all of whose fibers are Y. $\qquad\square$

Note that if Y and Z are finite [resp. finite-dimensional] then W is a finite [resp. finite-dimensional] $K(G, 1)$-complex.

Now we use cell trading to alter a $K(G, 1)$-complex with precise information on how the number of cells in each dimension is changed. Our goal is Theorem 7.1.13, the proof of which consists of a starting step (7.1.11) and an inductive step (7.1.12).

Proposition 7.1.11. *Let X and Z be homotopy equivalent path connected CW complexes. For each k, let X and Z have m_k and r_k k-cells, respectively. Then Z is homotopy equivalent to a CW complex Y such that $Y^1 = X^1$, Y has $(r_2 + m_1 + m_2 - m_0 + 1)$ 2-cells, $(r_3 + r_1 + m_2 - r_0 + 1)$ 3-cells, and r_k k-cells for all $k \geq 4$. Here, 0 and ∞ are permitted values of m_k and r_k.*

Proof. Let T be a maximal tree in X. We have $(X/T)^2 = X_P$ where $P := \langle W \mid R \rangle$ is a presentation of the fundamental group. By 4.1.16, Z is homotopy equivalent to a CW complex K whose 1-skeleton is $(X_P)^1$ such that K has

$|W| + |R| + r_2$ 2-cells, $r_3 + r_1 - r_0 + 1 + |R|$ 3-cells, and r_k k-cells when $k \geq 4$. Let $h : (X_P)^1 \to X^1$ be a homotopy equivalence (which exists by 4.1.9 and 3.1.13). Apply 4.1.7 to the following commutative diagram to get a CW complex Y homotopy equivalent to K with $Y^1 = X^1$.

$$
\begin{array}{ccc}
K \longleftarrow\!\!\!\supset (X_P)^1 & \xrightarrow{\;\mathrm{id}\;} & (X_P)^1 \\
\downarrow{\scriptstyle\mathrm{id}} \quad\quad \downarrow{\scriptstyle\mathrm{id}} & & \downarrow{\scriptstyle h} \\
K \longleftarrow\!\!\!\supset (X_P)^1 & \xrightarrow{\;h\;} & X^1
\end{array}
$$

The tree T has m_0 vertices and $m_0 - 1$ 1-cells. So $|W| = m_1 - m_0 + 1$ and $|R| = m_2$. Thus Y has $(r_2 + m_2 + m_1 - m_0 + 1)$ 2-cells and $(r_3 + r_1 - r_0 + 1 + m_2)$ 3-cells, $Y^1 = X^1$, and Y has r_k k-cells when $k \geq 4$. $\qquad\square$

Proposition 7.1.12. *Let $n \geq 2$. Let X be a $K(G,1)$-complex having m_k k-cells. Let Z be a $K(G,1)$-complex having r_k k-cells, such that $Z^{n-1} = X^{n-1}$. Let $s_2 = r_1 - r_0 + 1$ and, for $n \geq 3$, let $s_n = 0$. Then there exists a $K(G,1)$-complex Y with $Y^n = X^n$, such that Y has $r_{n+1} + m_n + s_n$ $(n+1)$-cells, $r_{n+2} + r_n + s_n$ $(n+2)$-cells, and r_k k-cells for each $k \geq n+3$. Here, $0 \leq m_k$, $r_k \leq \infty$.*

Proof. We first deal with the case $n = 2$. To begin, assume X has only one vertex (i.e., $m_0 = r_0 = 1$). Let P_1 and P_2 be presentations of G such that $X^2 = X_{P_1}$ and $Z^2 = X_{P_2}$. With notation as in the proof of 4.1.12, there are homotopy equivalent CW complexes $Y_{P_1} = X_{P_1} \cup \bigcup \{\tilde{e}_\beta^3 \mid \beta \in \mathcal{B}\}$ and $Y_{P_2} = X_{P_2} \cup \bigcup \{\tilde{e}_\alpha^3 \mid \alpha \in \mathcal{A}\}$ where $|\mathcal{B}| = |W_2| + |R_1|$ and $|\mathcal{A}| = |W_1| + |R_2|$. We have

$$
Z = X_{P_2} \cup \bigcup_{k=3}^{\infty} (r_k \text{ k-cells}).
$$

Let $Z^+ = X_{P_2} \cup \bigcup \{\tilde{e}_\alpha^3 \mid \alpha \in \mathcal{A}\} \cup \bigcup_{k=3}^{\infty} (r_k \text{ k-cells})$; that is, Z^+ is obtained from Z by attaching the 3-cells $\tilde{e}_\alpha^3$ to X_{P_2} as in Y_{P_2}. Since Z is 2-aspherical, Z^+ is homotopy equivalent to $Z \vee \bigvee \{S_\alpha^3 \mid \alpha \in \mathcal{A}\}$. By 4.1.7, $Z^+ = Y_{P_2} \cup \bigcup_{k=3}^{\infty} (r_k$ k-cells) is homotopy equivalent to $Y^+ = Y_{P_1} \cup \bigcup_{k=3}^{\infty} (r_k$ k-cells); this is because Y_{P_1} and Y_{P_2} have the same homotopy type.

We have $Y^+ = X_{P_1} \cup \bigcup \{\tilde{e}_\beta^3 \mid \beta \in \mathcal{B}\} \cup \bigcup_{k=3}^{\infty} (r_k$ k-cells). Since Y^+ and Z^+ have the same homotopy type, and Z is homotopy equivalent to $Z^+ \cup \{B_\alpha^4 \mid \alpha \in \mathcal{A}\}$, it follows that Z is homotopy equivalent to $Y = Y^+ \cup (|\mathcal{A}|$ 4-cells). This latter complex, Y, has $X_{P_1} = X^2$ for its 2-skeleton, has $r_3 + |W_2| + |R_1|$ 3-cells, has $r_4 + |W_1| + |R_2|$ 4-cells, and has r_k k-cells when $k \geq 5$.

If $m_0 > 1$, we "mod out" maximal trees in X and Z before applying the above argument, and, at the end, we alter Y so that $Y^2 = X^2$, just as in the proof of 7.1.11. Then $|W_1| = |W_2| = r_1 - r_0 + 1$, $|R_1| = m_2$, and $|R_2| = r_2$. This altered complex Y has $(r_3 + m_2 + r_1 - r_0 + 1)$ 3-cells, $(r_4 + r_2 + r_1 - r_0 + 1)$ 4-cells, and r_k k-cells when $k \geq 5$. This completes the case $n = 2$.

From now on, we assume $n \geq 3$. Since X and Z have the same 2-skeleton, $i : X^{n-1} \hookrightarrow Z$ defines an isomorphism $\phi : \pi_1(X, x) \to \pi_1(Z, x)$, where x is a vertex. By 7.1.6, there is a cellular map $f : (X, x) \to (Z, x)$ inducing ϕ. $f \mid X^{n-1}$ and i are homotopic, by 7.1.8. So, by the Homotopy Extension Theorem, we may assume that $f \mid X^{n-1} = i$.

Let $g = f\mid : X^n \to Z$. The key elements in our construction are to be found in the following commutative diagram:

$$
\begin{array}{ccc}
X^n & \xrightarrow{\,g\,} & Z \\
{\scriptstyle i'}\downarrow & & \uparrow{\scriptstyle p'} \\
M(g) & \xrightarrow[\,q'\,]{} & M
\end{array}
$$

Here, i' is the usual inclusion of the domain of g in $M(g)$; Z is identified with a subcomplex of $M(g)$ as usual. Since $g \mid X^{n-1}$ is an embedding, there is a natural copy of $Z \cup (X^{n-1} \times I)$ as a subcomplex of $M(g)$. Z lies at the "0-end" of $M(g)$ so that $Z \cap (X^{n-1} \times I) = X^{n-1} \times \{0\}$. The projection: $I \to \{0\}$ induces a homotopy equivalence $q : Z \cup (X^{n-1} \times I) \to Z$. See Fig. 7.1. The space M is $Z \cup_q M(g)$. The map q' is the restriction to $M(g)$ of the quotient map $Z \coprod M(g) \to M$. The map p' is the projection $M \to Z$ induced by the collapse $M(g) \to Z$.

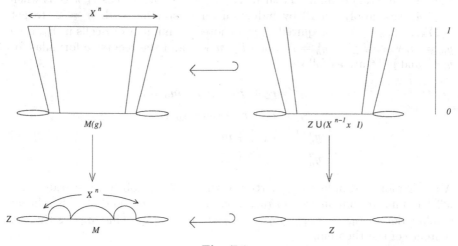

Fig. 7.1.

Applying 4.1.7 to the commutative diagram

$$M(g) \longleftarrow\!\!\!\!\!\!\!\supset Z \cup (X^{n-1} \times I) \xrightarrow{\;\text{id}\;} Z \cup (X^{n-1} \times I)$$

$$\text{id} \downarrow \qquad\qquad\qquad \downarrow \text{id} \qquad\qquad\qquad \downarrow q$$

$$M(g) \longleftarrow\!\!\!\!\!\!\!\supset Z \cup (X^{n-1} \times I) \xrightarrow[\;q\;]{} Z$$

we see that q' is a homotopy equivalence. Thus M is a $K(G, 1)$-complex.

Now, M is set up for cell-trading. We show that (M, X^n) is n-connected. We have $M^{n-1} = X^{n-1}$, so, for $i < n$, cellular maps $(B^i, S^{i-1}) \to (M, X^n)$ are maps into (X^n, X^n). Since $n \geq 3$, any cellular map $r : (B^n, S^{n-1}) \to (M, X^n)$ is pairwise homotopic to a map which is constant on S^{n-1} (since X^n is $(n-1)$-aspherical). Interpreting the latter as a map $S^n \to M$, it extends to B^{n+1}, since M is n-aspherical. Thus r is pairwise homotopic to a map into (X^n, X^n).

Consider the cells of M. There are the cells of X^n; there are the r_n n-cells of Z; there are $r_{n+1} + m_n$ $(n+1)$-cells (those of Z, and one for each n-cell of X^n); and there are the cells of Z of dimension $\geq n+2$. By 4.2.1, there is a CW complex Y homotopy equivalent to M having: the cells of X^n, $r_{n+1} + m_n$ $(n+1)$-cells, $r_{n+2} + r_n$ $(n+2)$-cells, and r_k k-cells for each $k \geq n+3$.   □

Theorem 7.1.13. *Let $n \geq 1$. Let X be a $K(G, 1)$-complex having m_k k-cells. Let Z be a $K(G, 1)$-complex having r_k k-cells. Then there exists a $K(G, 1)$-complex Y with $Y^n = X^n$ such that Y has r_k k-cells for each $k \geq n+3$. If $n \geq 2$, if X has finite n-skeleton and if Z has finite p-skeleton, then there is such a Y having finite p-skeleton.*

Proof. By 7.1.7, all $K(G, 1)$-complexes are homotopy equivalent. By 7.1.11, there is a $K(G, 1)$-complex Y_1 such that $Y_1^1 = X^1$ and Y_1 has r_k k-cells when $k \geq 4$. Now apply 7.1.12 by induction on n, starting with $n = 2$, to get $Y_2, Y_3, \ldots, Y_n =: Y$ as required. If y_n^i denotes the number of i-cells in Y_n, then $y_n^i = m_i$ when $i \leq n$, $y_n^i = r_i$ when $i \geq n+3$, and the recursive formulas for y_n^{n+1} and y_n^{n+2} are as follows:

$$y_1^2 = r_2 + m_1 + m_2 - m_0 + 1$$
$$y_1^3 = r_3 + r_1 + m_2 - r_0 + 1$$
$$y_n^{n+1} = y_{n-1}^{n+1} + m_n + s_n$$
$$y_n^{n+2} = y_{n-1}^{n+2} + y_{n-1}^n + s_n$$

A straightforward induction, starting with $n = 2$, establishes the statement: y_n^{n+1} is a linear combination of $r_0, \ldots, r_{n+1}, m_0, \ldots, m_n$; and y_n^{n+2} is a linear combination of $r_0, \ldots, r_{n+2}, m_0, \ldots, m_n$ (when $n \geq 2$). This implies the last sentence of the theorem.   □

Exercises

1. If (X, x) and (Y, y) are pointed CW complexes prove that every isomorphism $\pi_1(X, x) \to \pi_1(Y, y)$ is induced by a 2-equivalence $(X, x) \to (Y, y)$.
2. Describe in detail the covering projection $Y \to \mathrm{Tot}(\mathcal{X}, \Gamma)$ in the proof of 7.1.9.
3. Prove 6.2.1 (Britton's Lemma) using 7.1.9 and 3.4.9.
4. Write out explicitly what 7.1.9 says when Γ has two vertices and one edge; your answer should contain the phrases "free product with amalgamation" and "HNN extension."
5. Prove that if the CW complex X is n-connected then $\tilde{H}_k(X; R) = 0$ for all $k \leq n$. (*Hint*: Use Exercise 5 in Sect. 2.6.)
6. Prove that all the path connected closed surfaces except S^2 and $\mathbb{R}P^2$ are aspherical. *Hint*: compare 7.1.9.
7. Prove that if G is a countable group there exists a countable $K(G, 1)$-complex. *Hint*: The 2-skeleton can be countable; then use the Hurewicz Theorem to attach countably many cells in each dimension ≥ 3 to the universal cover.

7.2 Finiteness properties and dimensions of groups

Let G be a group. Even though $K(G, 1)$-complexes always exist and are unique up to homotopy equivalence, there may or may not exist $K(G, 1)$-complexes having special properties. For example, we say that G has *type* F_n if there exists a $K(G, 1)$-complex having finite n-skeleton.

Proposition 7.2.1. *Every group has type F_0; G has type F_1 iff G is finitely generated; G has type F_2 iff G is finitely presented; for $n \geq 2$, G has type F_n iff there exists a finite pointed n-dimensional $(n-1)$-aspherical CW complex (X, x) such that $\pi_1(X, x)$ is isomorphic to G.*

Proof. Every group has type F_0, by 7.1.5. Let G have type F_n and let (Z, z) be a $K(G, 1)$-complex with finite n-skeleton. For $n = 1$ [resp. $n = 2$], G is finitely generated [resp. finitely presented], by 3.1.17. For $n \geq 2$, $\pi_1(Z^n, z) \cong G$ by 3.1.7; Z^n is the required $(n-1)$-aspherical complex.

Conversely, if G is finitely generated [resp. finitely presented], we saw in 1.2.17 and 3.1.9 how to build a 2-dimensional CW complex X with one vertex x and $\pi_1(X, x) \cong G$, so that X^1 [resp. X] is finite. Attach cells of dimension ≥ 3 as in the proof of 7.1.5 to get the required $K(G, 1)$ having finite 1-skeleton [resp. 2-skeleton]. If $n \geq 2$ and (X, x) is a finite $(n-1)$-aspherical CW complex such that $\pi_1(X, x) \cong G$, attach cells of dimension $\geq n + 1$ as in 7.1.5 to build a $K(G, 1)$ having finite n-skeleton. $\square$

Here is a restatement of 7.2.1: G has type F_n iff there is an $(n-1)$-*connected free n-dimensional G-CW complex which is finite mod G.* This is clear when $n \geq 2$ and is an exercise when $n = 1$.

For every n, there are groups of type F_n which do not have type F_{n+1}. An efficient method of constructing such groups is given in Sect. 8.3. Such groups also occur "in nature," for example in certain families of S-arithmetic groups, see [1].

A CW complex each of whose skeleta is finite has *finite type*. We say that G has *type F_∞* if there exists a $K(G,1)$-complex of finite type.

Proposition 7.2.2. *G has type F_∞ iff G has type F_n for all n.*

Proof. "Only if" is trivial. To prove "if," apply 7.1.13 infinitely often, by induction on n, gradually building a $K(G,1)$-complex of finite type.   □

Next, we show that the properties F_n are invariant under passage to and from a subgroup of finite index.

Proposition 7.2.3. *Let $H \leq G$ and let $[G : H] < \infty$. Let G and H have type F_{n-1}. Then G has type F_n iff H has type F_n.*

Proof. "Only if" follows from 3.2.13, 7.1.4 and 7.2.1. When $n = 1$ [resp. $n = 2$], "if" is the well-known elementary fact that a group containing a finitely generated [resp. finitely presented] subgroup of finite index is itself finitely generated [resp. finitely presented]. It remains to prove "if" when $n \geq 3$.

Let X be a $K(G,1)$-complex having finite $(n-1)$-skeleton. Let $p : \bar{X} \to X$ be a finite-to-one covering projection, where $\bar{X}$ is a $K(H,1)$-complex: see 3.2.13 and 7.1.4. Note that $\bar{X}$ has finite $(n-1)$-skeleton. Since $n \geq 3$ and H has type F_n, Theorem 7.1.13 gives a $K(H,1)$-complex $\bar{Y}$ having finite n-skeleton such that $\bar{Y}^{n-1} = \bar{X}^{n-1}$. Thus $\bar{Y}^n$ is obtained from $\bar{X}^{n-1}$ by attaching finitely many n-cells, and $\bar{Y}^n$ is $(n-1)$-aspherical. Let $\{f_\alpha : S^{n-1} \to \bar{X}^{n-1} \mid \alpha \in \mathcal{A}\}$ be attaching maps for these n-cells, $\mathcal{A}$ being finite. Attach n-cells to X^{n-1} by the maps $p \circ f_\alpha : S^{n-1} \to X^{n-1}$. Let the resulting n-dimensional CW complex be Z^n. Let $\bar{Z}^n$ be the covering space of Z^n corresponding to the subgroup H (appropriate base points being understood everywhere). Then $\bar{Y}^n$ is a subcomplex of $\bar{Z}^n$, and $\bar{Y}^{n-1} = \bar{Z}^{n-1} = \bar{X}^{n-1}$. Since $\bar{Y}^n$ is $(n-1)$-aspherical, so is $\bar{Z}^n$, hence also Z^n, by 7.1.4. Thus G has type F_n, by 7.2.1.   □

Corollary 7.2.4. *Let $H \leq G$ and let $[G : H] < \infty$. For $0 \leq n \leq \infty$, G has type F_n iff H has type F_n.*

Proof. Apply 7.2.3 inductively. For the case $n = \infty$ then apply 7.2.2.   □

The one-point space is a $K(G,1)$-complex where G is the trivial group. Since the trivial group has finite index in every finite group, we have:

Corollary 7.2.5. *Every finite group has type F_∞.*   □

Among the notable groups of type F_∞ is Thompson's group F, discussed in later chapters. Many torsion free groups of type F_∞ turn out to have a stronger property, type F, which is discussed below. (The group F is an exception to this.[6])

We now turn to the question of minimizing the dimension of a $K(G,1)$-complex. The *geometric dimension* of G is ∞ if there does not exist a finite-dimensional $K(G,1)$-complex; otherwise it is the least integer d for which there exists a d-dimensional $K(G,1)$-complex.

Proposition 7.2.6. *G has geometric dimension 0 iff G is trivial. G has geometric dimension 1 iff G is free and non-trivial. If G has geometric dimension d, every subgroup of G has geometric dimension $\leq d$.*

Proof. The dimension 0 statement is clear. By 3.1.16, every 1-dimensional CW complex has free fundamental group, so if G has geometric dimension 1, G is free and non-trivial. Conversely, if G is free and non-trivial, we saw in Example 1.2.17 how to build a 1-dimensional CW complex whose fundamental group is G (by 3.1.8). Its universal cover is contractible, by 3.1.12, so it is a 1-dimensional $K(G,1)$.

Finally, let X be a d-dimensional $K(G,1)$-complex, and let $H \leq G$. By 3.2.11, X has a covering space $\bar{X}$ whose fundamental group is H. By 7.1.4, $\bar{X}$ is a d-dimensional $K(H,1)$-complex. $\qquad\square$

There is no question of replacing the inequality by equality in the last part of 7.2.6; just consider the trivial subgroup of the free group $\mathbb{Z}$. Even for subgroups of finite index there are limitations: we will see in 8.1.5 that a finite cyclic group has infinite geometric dimension, while its trivial subgroup (of finite index) has geometric dimension 0. Nevertheless we have:

Theorem 7.2.7. (Serre's Theorem) *Let G be torsion free, and let H be a subgroup of finite index having finite geometric dimension. Then G has finite geometric dimension.*

Proof. Let Y be a finite-dimensional $K(H,1)$-complex and let $H\bar{g}_1, \ldots, H\bar{g}_n$ be the cosets of H in G. Let $\tilde{Y}$ be the universal cover of Y. Let $\tilde{X} = \prod_{i=1}^{n}\tilde{Y}_i$ where each $\tilde{Y}_i = \tilde{Y}$. Then $\tilde{X}$ is a finite-dimensional contractible CW complex. We describe a (left) G-action on $\tilde{X}$. We have selected the coset representatives $\bar{g}_1, \ldots, \bar{g}_n$. A right action of G on the set $\{1, \cdots, n\}$ is defined by the formula $(i, g) \mapsto i.g$ where $\bar{g}_i g \in H\bar{g}_{i.g}$. Indeed, we can write $\bar{g}_i g = h(g,i)\bar{g}_{i.g}$, thus associating with $g \in G$ an n-tuple $(h(g,1), \cdots, h(g,n))$ of elements of H. The required left G-action on $\tilde{X}$ is

$$g.(y_1, \cdots, y_n) = (h(g,1)y_{1.g}, \cdots, h(g,n)y_{n.g}).$$

This action of G clearly makes $\tilde{X}$ into a rigid G-CW complex. It remains to prove that the action is free.

---

[6] This is a place where the two uses of the letter F might cause confusion: "type F" and "the group F".

Let $g.(y_1, \cdots, y_n) = (y_1, \cdots, y_n)$. Let m be such that $i.g^m = i$ for all i; since g permutes a finite set such an m exists. Then

$$g^m.(y_1, \cdots, y_n) = (h(g^m, 1)y_1, \cdots, h(g^m, n)y_n) = (y_1, \cdots, y_n).$$

So each $h(g^m, i) = 1$ because H acts freely on $\tilde{Y}$. So $\bar{g}_i g^m = \bar{g}_i$ for all i, implying $g^m = 1$. Thus $g = 1$ since G is torsion free. $\square$

Remark 7.2.8. The G-action on $\prod_{i=1}^{n} \tilde{Y}_i$ described in this proof does not restrict to the diagonal H-action.

One way of showing that a group G has geometric dimension $\leq d$ is to find some contractible d-dimensional free G-CW complex, since by 3.2.1 and 7.1.3 the quotient complex will be a d-dimensional $K(G, 1)$-complex. One way of showing that G has geometric dimension $\geq d$ is to show that $H_d(X; R) \neq 0$ for some $K(G, 1)$-complex X and some ring R, applying 7.1.7, 2.5.4 and 2.4.10. For example, recall that the d-torus T^d is the d-fold product of copies of S^1. As explained in Sect. 3.4, $\tilde{T}^d$ is homeomorphic to $\mathbb{R}^d$, so T^d is a $K(\mathbb{Z}^d, 1)$-complex.

Proposition 7.2.9. $H_d(T^d; \mathbb{Z}) \cong \mathbb{Z}$.

Proof. Give $\mathbb{R}$ the CW complex structure with vertex set $\mathbb{Z}$ and with 1-cells $[m, m+1]$ for each $m \in \mathbb{Z}$. Give $\mathbb{R}^n$ the product structure and regard T^n as the quotient complex of $\mathbb{R}^n$ by the obvious free action of $\mathbb{Z}^n$ on $\mathbb{R}^n$ (translation by integers in each coordinate); see 3.2.1. Then T^n has just one n-cell, e^n.

Orient the 0-cells of $\mathbb{R}$ by $+1$. Orient $[m, m+1]$ by the characteristic map $I^1 \to [m, m+1]$, $t \mapsto \frac{1}{2}(t + 2m + 1)$. Give $\mathbb{R}^n$ the product orientation. Then the $\mathbb{Z}^n$-action is orientation preserving and the quotient $q : \mathbb{R}^n \to T^n$ gives an orientation to T^n. It is enough to prove that $e^n \in C_n(T^n; \mathbb{Z})$ is a cycle.

Let $\tilde{e}^n = [0, 1]^n \in C_n(\mathbb{R}^n; \mathbb{Z})$. For $1 \leq i \leq n$ and $\epsilon = 0$ or 1, let $\tilde{e}_{i,\epsilon}^{n-1} = [0, 1]^{i-1} \times \{\epsilon\} \times [0, 1]^{n-i}$. By 2.5.17, $[\tilde{e}^n : \tilde{e}_{i,0}^{n-1}] = (-1)^i$ and $[\tilde{e}^n : \tilde{e}_{i,1}^{n-1}] = (-1)^{i+1}$. Clearly $e^n = q_\#(\tilde{e}^n)$. Let $h_i : \mathbb{R}^n \to \mathbb{R}^n$ be the translation $(x_1, \ldots, x_n) \mapsto (x_1, \ldots, x_{i-1}, x_i + 1, x_{i+1}, \ldots, x_n)$. Then $h_i(\tilde{e}_{i,0}^{n-1}) = \tilde{e}_{i,1}^{n-1}$. Moreover, as chains, $h_{i\#}(\tilde{e}_{i,0}^{n-1}) = \tilde{e}_{i,1}^{n-1}$. Since $q \circ h_i = q$, $q_\#(\tilde{e}_{i,1}^{n-1}) = q_\# \circ h_{i\#}(\tilde{e}_{i,0}^{n-1}) = q_\#(\tilde{e}_{i,0}^{n-1})$.

$$\begin{aligned}
\partial e^n &= \partial(q_\#(\tilde{e}^n)) \\
&= q_\#(\partial \tilde{e}^n), \quad \text{by 2.4.3} \\
&= q_\# \left(\sum_{i=1}^{n} (-1)^i (\tilde{e}_{i,0}^{n-1} - \tilde{e}_{i,1}^{n-1}) \right) \quad \text{by 2.5.17} \\
&= \sum_{i=1}^{n} (-1)^i (q_\#(\tilde{e}_{i,0}^{n-1}) - q_\#(\tilde{e}_{i,1}^{n-1})) \\
&= 0.
\end{aligned}$$

$\square$

Corollary 7.2.10. $\mathbb{Z}^d$ *has geometric dimension d.* $\square$

Combining this with 7.2.6 gives:

Corollary 7.2.11. *If G has a free abelian subgroup of rank d, then G has geometric dimension $\geq d$.* $\square$

In particular, 7.2.11 can be useful for showing that G has infinite geometric dimension.

We have mentioned (and we will prove in 8.1.5) that every non-trivial finite cyclic group has infinite geometric dimension. Hence, by 7.2.6:

Proposition 7.2.12. *Every group containing a non-trivial element of finite order has infinite geometric dimension.* $\square$

A group G has *type F* if there exists a finite $K(G,1)$-complex. Groups of type F discussed in this book include: finitely generated free groups, finitely generated free abelian groups, and torsion free subgroups of finite index in finitely generated Coxeter groups. Other important examples are: torsion free subgroups of finite index in arithmetic groups: see [29]; and torsion free subgroups of finite index in the outer automorphism group of a finitely generated free group: see [44].

If G has type F then G has type F_∞ and G has finite geometric dimension. It is natural to ask if the converse is true. We say that G has *type FD* if some (equivalently, any) $K(G,1)$-complex is finitely dominated.

Proposition 7.2.13. *G has type FD iff G has type F_∞ and G has finite geometric dimension.*

Proof. "If": Let X be a $K(G,1)$-complex of finite type and let Y be a d-dimensional $K(G,1)$-complex. By 7.1.7 there are maps $Y \xrightarrow{f} X \xrightarrow{g} Y$ such that $g \circ f \simeq \mathrm{id}_Y$. By 1.4.3, we may assume f and g are cellular, so that $f(Y) \subset X^d$. There are induced maps $Y \xrightarrow{f'} X^d \xrightarrow{g|} Y$ whose composition is homotopic to id_Y; and X^d is finite.

"Only if": Let X be a $K(G,1)$-complex and let $X \xrightarrow{f} Y \xrightarrow{g} X$ be homotopic to id_X, where Y is a finite CW complex. By 4.3.5, X is homotopy equivalent to $\mathrm{Tel}(f \circ g)$, which is finite-dimensional, so G has finite geometric dimension. To show that G is of type F_∞ we will show by induction that G is of type F_n for all n. By 7.2.2 this is enough. Certainly G is of type F_0. Assume G is of type F_{n-1}. Let X be a $K(G,1)$-complex such that X^{n-1} is finite and X is dominated by a finite complex. Then there is a finite subcomplex K of X and a homotopy $D : X \times I \to X$ such that $D_0 = \mathrm{id}_X$ and $D_1(X) \subset K$. Let L be a finite subcomplex[7] of X such that $D((X^{n-1} \cup K) \times I) \subset L$. We claim (X, L) is n-connected. To see this, let $\phi : (B^k, S^{k-1}) \to (X, L)$ be a cellular map

---

[7] We are using 1.2.13 and 1.4.3 repeatedly.

where $k \leq n$. Then $\phi(S^{k-1}) \subset X^{n-1}$, so $D_t \circ \phi(S^{k-1}) \subset L$ for all $t \in I$. Moreover, $D_1 \circ \phi(B^k) \subset K \subset L$. By 1.3.9, there is a strong deformation retraction $F : B^k \times I \times I \rightarrow B^k \times I$, of $B^k \times I$ onto $(B^k \times \{1\}) \cup (S^{k-1} \times I)$. The required homotopy $\Phi : B^k \times I \rightarrow X$, rel S^{k-1}, of ϕ into L is $\Phi(s,t) = D \circ (\phi \times \mathrm{id}) \circ F_t(s,0)$. By 4.2.1, X is homotopy equivalent to a CW complex Y such that $Y^n = L^n$ is finite. Thus G has type F_n. $\qquad\square$

Corollary 7.2.14. *Let G be a group and let H be a subgroup of finite index. If G has type FD, so has H. If H has type FD and if G is torsion free, then G has type FD.* $\qquad\square$

Note that the trivial group $\{1\}$ has type FD (indeed, type F) while nontrivial finite groups, in all of which $\{1\}$ has finite index, do not have type FD, by 7.2.12.

The question remains: does type FD imply type F? The general question of when a finitely dominated CW complex X is homotopy equivalent to a finite CW complex is understood: the only obstruction (Wall's finiteness obstruction) lies in the reduced projective class group $\tilde{K}_0(\mathbb{Z}[\pi_1(X)])$. See, for example, [29, Chap. VIII, Sect. 6]. Non-trivial obstructions occur, but it is unknown at time of writing whether the obstruction can be non-zero when X is aspherical.

The analog of 7.2.14 for type F is also unknown: obviously if $[G : H] < \infty$ and if G has type F then H has type F (by 7.1.4 and 3.2.13). But for torsion free G the converse is unknown.

Proposition 7.2.15. *If G has type FD, then $G \times \mathbb{Z}$ has type F.*

Proof. This follows from 4.3.7. In detail, let $X \xrightarrow{f} Y \xrightarrow{g} X$ be cellular maps, where X is a $K(G,1)$-complex, Y is a finite CW complex, and $g \circ f \simeq \mathrm{id}_X$. Then $X \times S^1$ is a $K(G \times \mathbb{Z}, 1)$-complex which, by 4.3.5, is homotopy equivalent to the finite mapping torus $T(f \circ g)$. $\qquad\square$

Corollary 7.2.16. *If G has type FD, then G is a retract of a group G' of type F; i.e., there are homomorphisms $G \rightarrow G' \rightarrow G$ whose composition is id_G, where the first arrow is an inclusion.* $\qquad\square$

Proposition 7.2.17. *If there is a $K(G,1)$-complex which is dominated by a d-dimensional CW complex then $G \times \mathbb{Z}$ has geometric dimension $\leq d+1$.*

Proof. Let Y dominate X, where X is a $K(G,1)$-complex and Y is d-dimensional. As in the proof of 7.2.15, $X \times S^1$ is homotopy equivalent to a $(d+1)$-dimensional CW complex, which is therefore a $K(G \times \mathbb{Z}, 1)$-complex. $\square$

Corollary 7.2.18. *If there is a $K(G,1)$-complex which is dominated by a d-dimensional CW complex then G has geometric dimension $\leq d+1$.* $\qquad\square$

Remark 7.2.19. The conclusion of 7.2.18 can be improved to "G has geometric dimension $\leq d$," except possibly when $d = 2$, where the situation is not yet understood. The proof of this can be found, for example, in [29, Chap. VIII, Sect. 7]. This proof is accessible to readers of the present chapter and is only omitted to save space. Note that it uses the Relative Hurewicz Theorem 4.5.1.

Here is a useful necessary and sufficient condition for type F_{n+1}:

Theorem 7.2.20. *Let $n \geq 1$, let the group G have type F_n, and let X be a $K(G,1)$-complex with finite n-skeleton. Then G has type F_{n+1} iff there is a $K(G,1)$-complex Y with finite $(n+1)$-skeleton and $Y^n = X^n$.*

Proof. "If" is clear. "Only if" follows from 7.1.13 when $n \geq 2$ and is obvious when $n = 1$.   □

In the next proof we suppress base points in homotopy groups to simplify notation:

Theorem 7.2.21. *Let $N \rightarrowtail G \twoheadrightarrow Q$ be an exact sequence of groups. If G has type F_n and if N has type F_{n-1} then Q has type F_n.*

Proof. This is obvious for $n \leq 2$ so we assume $n \geq 3$. Let Y be an $(n-2)$-aspherical finite $(n-1)$-dimensional CW complex whose fundamental group is isomorphic to Q, and let X be a $K(G,1)$-complex. As before, we consider the commutative diagram

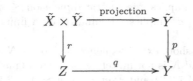

where r is the covering projection obtained from the diagonal action of G on $\tilde{X} \times \tilde{Y}$ (G acts on $\tilde{Y}$ via Q). Then q is a fiber bundle whose fiber is the $K(N,1)$-complex $N\backslash\tilde{X}$. It follows from the exact sequence in 4.4.11 that $q_\# : \pi_{n-1}(Z) \to \pi_{n-1}(Y)$ is an epimorphism. The map q is also a stack of CW complexes. Since N has type F_{n-1}, there is a $K(N,1)$-complex W having finite $(n-1)$-skeleton, which is of course homotopy equivalent to $N\backslash\tilde{X}$. By the Rebuilding Lemma 6.1.4 there is a diagram

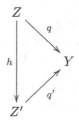

which commutes up to homotopy, where q' is a stack of CW complexes with fibers W, h is a homotopy equivalence, and Z' has finite $(n-1)$-skeleton. Since Z^{n-1} is $(n-2)$-aspherical with fundamental group isomorphic to G, the same is true of $(Z')^{n-1}$. Since G has type F_n, 7.2.20 implies that there is a $K(G,1)$-complex X' with finite n-skeleton whose $(n-1)$-skeleton is $(Z')^{n-1}$. The inclusion map $(Z')^{n-1} \hookrightarrow Z'$ induces an epimorphism on π_{n-1}, hence we can attach finitely many n-cells to Z' to kill $\pi_{n-1}(Z')$. Since q' induces an epimorphism on π_{n-1}, the same is true of Y. Thus, using 7.2.20 again, we see that Q has type F_n. $\square$

This theorem should be compared with Exercise 1 where it is asserted that if N and Q have type F_n then G has type F_n.

Exercises

1. Let $N \rightarrowtail G \twoheadrightarrow Q$ be a short exact sequence of groups. Prove that if N has type F_n and Q has type F_n then G has type F_n. *Hint*: see Theorem 7.1.10.
2. Devise similar exercises involving finite geometric dimension, type FD and type F.
3. Prove that if G is the fundamental group of a finite graph of groups whose vertex groups have type F_n and whose edge groups have type F_{n-1} then G has type F_n. *Hint*: Use 7.1.9 and 6.1.4.
4. How many other proofs of 7.2.9 can you find?
5. Sharpen 7.2.13 by specifying the dimensions: (i) "dominated by a finite d-dimensional complex" implies "geometric dimension $\leq d + 1$"; (ii) "F_d and geometric dimension $\leq d$" implies "dominated by a finite d-dimensional complex."
6. Give an example of a short exact sequence of groups $N \rightarrowtail G \twoheadrightarrow Q$ where G and Q have type F, and N is not finitely generated. (Thus one cannot expect a theorem in the spirit of 7.2.21 and Exercise 1 of Sect. 7.2 for this case.)

7.3 Recognizing the finiteness properties and dimension of a group

We can use the Borel Construction to build $K(G,1)$-complexes with good finiteness properties, under various hypotheses. We begin with a generalization of Theorem 6.1.5:

Theorem 7.3.1. *For $n \geq 1$, let Y be an $(n-1)$-connected rigid G-CW complex having finite n-skeleton mod G. If the stabilizer of each i-cell has type F_{n-i} for all $i \leq n-1$, then G is of type F_n.*

Proof. We leave the case $n = 1$ as an exercise. Starting with a $K(G,1)$-complex (X, v), we construct a commutative diagram as in Sect. 6.1

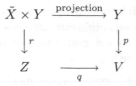

$$\begin{array}{ccc} \tilde{X} \times Y & \xrightarrow{\text{projection}} & Y \\ \Big\downarrow r & & \Big\downarrow p \\ Z & \xrightarrow{\quad q \quad} & V \end{array}$$

in which $q : Z \to \Gamma$ is a stack of CW complexes with fiber $G_{\tilde{e}} \backslash \tilde{X}$ over e. Since $\tilde{X}$ is contractible and Y is $(n-1)$-connected, Z is an $(n-1)$-aspherical complex whose fundamental group is G. Since $G_{\tilde{e}} \backslash \tilde{X}$ is a $K(G_{\tilde{e}}, 1)$-complex and $G_{\tilde{e}}$ has type F_{n-i} when e is an i-cell of Γ, the Rebuilding Lemma gives a CW complex Z' homotopy equivalent to Z and having finite n-skeleton (the details are similar to those in the proof of 6.1.5). By 7.2.1, the existence of such a space Z' is equivalent to G having type F_n.   □

Remark 7.3.2. We will see applications of this theorem later in the book. One important case, not covered here, is the case where G is hyperbolic. Then there is a naturally occurring finite-dimensional simplicial complex R, called the *Rips complex*, on which G acts so that $G \backslash R$ is a finite CW complex and the stabilizer of each cell is finite. It follows from 7.3.1 that hyperbolic groups have type F_∞ and (by 7.3.4 below) that torsion-free hyperbolic groups have type F. The details of this construction are found in [24].

A similar proof gives the following two theorems:

Theorem 7.3.3. *Let Y be a contractible rigid G-CW complex of dimension $\leq m$. For each i, let the stabilizers of the i-cells of Y have geometric dimension $\leq d_i$. Then G has geometric dimension $\leq \max\{d_i + i \mid 0 \leq i \leq m\}$.*   □

Theorem 7.3.4. *Let Y be a contractible rigid G-CW complex whose quotient $G \backslash Y$ is finite. Let the stabilizer of each cell of Y have type F [resp. FD]. Then G has type F [resp. FD].*   □

Exercises

1. Prove the $n = 1$ case of 7.3.1.
2. Prove 7.3.3.
3. Prove 7.3.4.
4. Sharpen 7.2.13: if G has geometric dimension d and has type F_d then G has type FD.

7.4 Brown's Criterion for finiteness

We saw in Sect. 7.2 that a group G has type F_n iff there is an $(n-1)$-connected free n-dimensional G-CW complex which is finite mod G. And in 7.3.1 we extended this by weakening "free" to "rigid" provided the stabilizers of cells have appropriate finiteness properties. In practice, it often happens that the desired hypotheses on stabilizers are satisfied but the G-CW complex is not finite mod G. Here we show how a weaker hypothesis than "finite mod G" can yield the same conclusion.

A sequence of groups and homomorphisms $G_1 \xrightarrow{f_2^1} G_2 \xrightarrow{f_3^2} \cdots$ is *ind-trivial*[8] if for each i there exists $j > i$ such that the composition $f_j^i : G_i \to G_j$ is trivial. This also makes sense when each G_i is a pointed set and each f_{i+1}^i is a pointed function. A G-*filtration* of a path connected G-CW complex Y is a countable collection of G-subcomplexes $K_0 \subset K_1 \subset \cdots$ such that $Y = \bigcup_{i=0}^{\infty} K_i$.
The G-filtration $\{K_i\}$ is *essentially $(n-1)$-connected* if for $0 \le k \le n-1$ the sequence $\{\pi_k(K_0, y) \to \pi_k(K_1, y) \to \cdots\}$ is ind-trivial, where $y \in K_0$ is a chosen base point and the unmarked morphisms are induced by inclusion. This definition is independent of $y \in K_0$. A necessary condition is that Y be $(n-1)$-connected.

Theorem 7.4.1. (Brown's Criterion) *Let the $(n-1)$-connected free G-CW complex Y admit a G-filtration $\{K_i\}$ where each $G \backslash K_i$ has finite n-skeleton. Then G has type F_n iff $\{K_i\}$ is essentially $(n-1)$-connected.*

Proof. "If": We begin with $n = 1$. Consider the commutative diagram

$$
\begin{array}{ccc}
K_0 & \hookrightarrow & K_{i_1} \\
\downarrow & & \downarrow \\
G \backslash K_0 & \longrightarrow & G \backslash K_{i_1}
\end{array}
$$

By essential 0-connectedness we can choose i_1 so that K_0 lies in the path component K'_{i_1} of K_{i_1} containing the base point y. Then $(K'_{i_1})^1$ is a path connected free G-graph which is finite mod G. Hence G has type F_1.

Next, let $n = 2$. We can choose i_2 so as to get a commutative diagram of groups

$$
\begin{array}{ccc}
\pi_1(K'_{i_1}, y) & \xrightarrow{\text{trivial}} & \pi_1(K'_{i_2}, y) \\
\downarrow & & \downarrow \\
\pi_1(G \backslash K'_{i_1}, z_1) & \longrightarrow & \pi_1(G \backslash K'_{i_2}, z_2) \\
\downarrow & \overset{\lambda_1}{\nearrow} & \downarrow \\
G & \xrightarrow{\text{id}} & G
\end{array}
$$

where the vertical exact sequences come from covering space theory, the top horizontal homomorphism is trivial, and the squares commute. It follows that

---

[8] Also known in the literature as *essentially trivial* – our term is explained in Chap. 11.

λ_1 exists making the triangles commute. Since $G\backslash K'_{i_2}$ is finite, it follows that G is a retract of a finitely presented group, hence (exercise) G is finitely presented, i.e., of type F_2.

Let (X, x) be a $K(G, 1)$-complex. For $n = 3$, we construct a commutative diagram

$$
\begin{array}{ccc}
(G\backslash K'_{i_2}, z_2) & \hookrightarrow & (G\backslash K'_{i_3}, z_3) \\
\uparrow {\scriptstyle f_1} & & \uparrow {\scriptstyle f_2} \\
(X^2, x) & \hookrightarrow & (X^3, x)
\end{array}
$$

where f_1 induces λ_1 on fundamental group, and (using 4.4.10) i_3 is such that $\pi_2(G\backslash K'_{i_2}, z_2) \to \pi_2(G\backslash K'_{i_3}, z_3)$ is trivial; compare the proof of 7.1.6.

Proceeding by induction, we get $f_n : (X^n, x) \to (G\backslash K'_{i_n}, z_n)$ which can be extended to $f : (X, x) \to (Z, z)$, where Z is obtained from $G\backslash K'_{i_n}$ by attaching suitable cells of dimension $> n$. This construction and 7.1.6 give a map $g : (Z, z) \to (X, x)$ such that $g \circ f \simeq \mathrm{id}_X$ rel$\{x\}$. Thus X is dominated by a CW complex with finite n-skeleton, so by (the proof of) 7.2.13, G has type F_n.

"Only if": Assume G has type F_n. Let X be a $K(G, 1)$-complex with finite n-skeleton. By 7.1.7, there is an n-equivalence $f : G\backslash Y^n \to X^n$ with n-inverse $g : X^n \to G\backslash Y^n$. These lift to G-maps $\tilde{f}$ and $\tilde{g}$ and there is a G-homotopy $H : Y^{n-1} \times I \to Y^n$ from the inclusion to the map $\tilde{g} \circ \tilde{f}|$. Since each K_i^n is cocompact, given i there exists $j \geq i$ such that H restricts and corestricts to a homotopy $\bar{H} : K_i^{n-1} \times I \to K_j^n$ from the inclusion to a map which factors through the $(n - 1)$-connected CW complex $\tilde{X}^n$. It follows that $\{K_i\}$ is essentially $(n - 1)$-connected, since $\pi_{n-1}(K_i^{n-1}, y) \to \pi_{n-1}(K_i^n, y)$ is onto. $\square$

An important generalization of Brown's Criterion is given in the Exercise below.

Source Notes: Brown's Finiteness Criterion first appeared in [30].

Exercise

Extend Brown's Criterion to the case where the G-action is not free but is such that the stabilizer of each i-cell has type F_{n-i}; see [30].

8

Homological Finiteness Properties of Groups

Here we introduce homology of groups and homological finiteness properties. The first two sections provide homological analogs of some of the topics in Chapter 7. A free (or projective) resolution of the trivial RG-module R plays the role of the universal cover of a $K(G, 1)$-complex. The properties FP_n and cohomological dimension are analogous to F_n and geometric dimension. This leads us to the Bestvina-Brady Theorem, which gives a method of constructing groups G for which the homological and topological properties are subtly different.

8.1 Homology of groups

Let G be a group and, as usual, let R be a commutative ring. The *group ring*[1] RG is $\bigoplus_{g \in G} R(g)$, where $R(g)$ is the free R-module generated by the one-element set $\{g\}$, with the multiplication $\left(\sum_{g \in G} r_g g \right) \left(\sum_{g' \in G} s_{g'} g' \right) = \sum_{g, g' \in G} r_g s_{g'} (gg')$. Here, our convention is to write $\sum_{g \in G} r_g g$ rather than $(r_g)_{g \in G}$ for an element of RG. It is straightforward to check that this multiplication is well defined and makes RG into a ring. This ring is commutative when G is abelian. The identity element in RG is $1 := \sum_{g \in G} r_g g$ where $r_1 = 1 \in R$ and $r_g = 0$ for all $g \neq 1 \in G$; this multiple use of the symbol 1 (the multiplicative identity elements of R, of G, and of RG) will not cause trouble. We write 0 for $\sum_{g \in G} 0g$.

---

[1] Strictly, the term "group ring" should only be used when the ring R is $\mathbb{Z}$; for other ground rings, the more correct term is "group algebra".

For a not-necessarily-commutative ring Λ with $1 \neq 0$ the *tensor product* $B \otimes_\Lambda A$ of a right Λ-module B and a left Λ-module A has the structure of an abelian group; it is generated by elements of the form $b \otimes a$ where $b \in B$ and $a \in A$ subject to bilinearity relations of the form $b\lambda \otimes a = b \otimes \lambda a$ where $\lambda \in \Lambda$. If Λ is commutative, the left action of Λ on $B \otimes_\Lambda A$ defined by $\lambda(b \otimes a) = b\lambda \otimes a$ makes $B \otimes_\Lambda A$ into a Λ-module.

When dealing with the case $\Lambda = RG$ we must elaborate on this. In general, RG is not a commutative ring; but R is commutative, so $B \otimes_{RG} A$ has a natural R-module structure, and this will always be understood in what follows.[2] It is often convenient to abbreviate $\otimes_{RG}$ to $\otimes_G$.

The left action of G on R defined by $g.1 = 1$ for all $g \in G$ makes R into an RG-module. Unless we say otherwise, it is this *trivial* action which is to be understood when we regard R as an RG-module.

A *free RG-resolution* of R is an exact sequence

$$\cdots \to F_2 \xrightarrow{\partial_2} F_1 \xrightarrow{\partial_1} F_0 \xrightarrow{\epsilon} R \to 0$$

of left RG-modules in which each F_i is free. Associated with this free resolution is the chain complex $\cdots \xrightarrow{\partial_2} F_1 \xrightarrow{\partial_1} F_0 \to 0$, whose homology modules in positive dimensions are trivial. Denoting this by (F, ∂), we see that ϵ induces an isomorphism $H_0(F) \to R$. In the spirit of Sect. 2.9, one thinks of the free resolution as the augmentation of the chain complex (F, ∂).

The basic fact about free resolutions, sometimes called the "Fundamental Lemma of Homological Algebra," is that they are unique up to chain homotopy equivalence. Versions of this purely algebraic theorem can be found in standard textbooks (e.g., [29, Chap. 1, Sect. 7] or [83, Chap. 4, Sect. 4]). We will state one such version without proof:

Theorem 8.1.1. *Let*

$$\cdots \to F_1 \to F_0 \xrightarrow{\epsilon} R \to 0$$

and

$$\cdots \to F_1' \to F_0' \xrightarrow{\epsilon'} R \to 0$$

be free RG-resolutions of R. For every isomorphism $\phi : H_0(F) \to H_0(F')$ there is a chain homotopy equivalence $\{\phi_i : F_i \to F_i'\}$ inducing ϕ, and $\{\phi_i\}$ is unique up to chain homotopy. □

By convention there is a "canonical" choice for ϕ in applications of 8.1.1, namely $H_0(F) \xrightarrow{\epsilon_*} R \xrightarrow{\epsilon_*'^{-1}} H_0(F')$.

Let M be a right RG-module. The *homology R-modules of G with coefficients in M* are computed from the R-chain complex

---

[2] The point is that when $R = \mathbb{Q}$, for example, we want to think of $B \otimes_{\mathbb{Q}G} A$ as a $\mathbb{Q}$-vector space, not just as an abelian group.

$$\cdots \to M \otimes_G F_2 \overset{\mathrm{id}\,\otimes\partial_2}{\longrightarrow} M \otimes_G F_1 \overset{\mathrm{id}\,\otimes\partial_1}{\longrightarrow} M \otimes_G F_0 \to 0.$$

They are denoted $H_*(G, M)$. By 8.1.1 (and the convention which follows it) there is no ambiguity in this definition.

These definitions and Theorem 8.1.1 are motivated by topological ideas. Let (X, v) be an oriented pointed CW complex whose universal cover is $(\tilde{X}, \tilde{v})$. Let $G = \pi_1(X, v)$. By 3.2.9 (see also 3.3.3) we may regard $\tilde{X}$ as a free left G-CW complex. Let $p : (\tilde{X}, \tilde{v}) \to (X, v)$ be the covering projection. For each n-cell e_α^n of X, pick an n-cell $\tilde{e}_\alpha^n$ of $\tilde{X}$ such that $p(\tilde{e}_\alpha^n) = e_\alpha^n$; in particular, pick $\tilde{v}$ for v. Orient $\tilde{e}_\alpha^n$ so that $p \mid: \tilde{e}_\alpha^n \to e_\alpha^n$ is orientation preserving. Each n-cell of $\tilde{X}$ over e_α^n has the form $g\tilde{e}_\alpha^n$, where $g \in G$ is unique. The CW complex $\tilde{X}$ is to be oriented so that the covering transformations $x \mapsto g.x$ are orientation preserving on cells.

For each $g \in G$ and each n, the covering transformation $x \mapsto g.x$ induces an isomorphism $g_\# : C_n(\tilde{X}; R) \to C_n(\tilde{X}; R)$. Thus we have a left RG-module structure on $C_n(\tilde{X}; R)$ given by $g.c = g_\#(c)$.

Proposition 8.1.2. *The oriented n-cells $\tilde{e}_\alpha^n$ of $\tilde{X}$ freely generate $C_n(\tilde{X}; R)$ as an RG-module. The boundary $\partial : C_n(\tilde{X}; R) \to C_{n-1}(\tilde{X}; R)$ is a homomorphism of left RG-modules. If X is aspherical, this gives a free RG-resolution of R:*

$$\cdots \overset{\partial_3}{\longrightarrow} C_2(\tilde{X}; R) \overset{\partial_2}{\longrightarrow} C_1(\tilde{X}; R) \overset{\partial_1}{\longrightarrow} C_0(\tilde{X}; R) \overset{\epsilon}{\longrightarrow} R \longrightarrow 0$$

where ϵ is defined by $\epsilon \left(\sum_{\alpha,g} m_{\alpha,g} g \tilde{e}_\alpha^0 \right) = \sum_{\alpha,g} m_{\alpha,g}.$ $\qquad\square$

Note that the underlying R-chain complex of this free resolution is the augmented cellular chain complex $C_*(\tilde{X}; R)$ and ϵ is the augmentation; so free RG-resolutions of R arise in topology as the augmented chain complexes of the universal covers of $K(G, 1)$-complexes. The uniqueness up to RG-chain homotopy of such resolutions follows from the algebraic Proposition 8.1.1, but this can also be seen topologically:

Proposition 8.1.3. *Let (X, x) and (Y, y) be $K(G, 1)$-complexes, with pointed universal covers $(\tilde{X}, \tilde{x})$ and $(\tilde{Y}, \tilde{y})$. Let the groups $\pi_1(X, x)$ and $\pi_1(Y, y)$ be identified with G via given isomorphisms so that $C_*(\tilde{X}; R)$ and $C_*(\tilde{Y}; R)$ are RG-chain complexes. Then these RG-chain complexes are canonically chain homotopy equivalent.*

Proof. By hypothesis, there is a given isomorphism $\phi : \pi_1(X, x) \to \pi_1(Y, y)$ inducing $\mathrm{id} : G \to G$. By 7.1.7, there is a cellular homotopy equivalence $f : (X, x) \to (Y, y)$ inducing ϕ, and f is unique up to pointed homotopy. Let $k : (Y, y) \to (X, x)$ be a cellular homotopy inverse for f. Let $F : k \circ f \simeq \mathrm{id}_X$

and $K : f \circ k \simeq \mathrm{id}_Y$ be cellular homotopies relative to the base points x and y. By repeated use of 3.3.4, one finds cellular homotopies $\tilde{F} : \tilde{k} \circ \tilde{f} \simeq \mathrm{id}_{\tilde{x}}$ and $\tilde{K} : \tilde{f} \circ \tilde{k} \simeq \mathrm{id}_{\tilde{y}}$ relative to the base points $\tilde{x}$ and $\tilde{y}$; here, $\tilde{F} : \tilde{X} \times I \to \tilde{X}$ and $\tilde{K} : \tilde{Y} \times I \to \tilde{Y}$ are lifts of F and K. Hence the chain maps $\tilde{f}_\#$ and $\tilde{k}_\#$ induced by $\tilde{f}$ and $\tilde{k}$, and the chain homotopies $\tilde{D}_F$ and $\tilde{D}_K$ induced by $\tilde{F}$ and $\tilde{K}$ (as described in 2.7.14) are homomorphisms of RG-modules. Thus $\tilde{f}_\#$ is a chain homotopy equivalence of RG-chain complexes.[3]   □

The chain homotopy equivalence constructed in the last proof induces $\tilde{f}_* : H_0(\tilde{X}; R) \to H_0(\tilde{Y}; R)$. By 8.1.1, it is unique up to chain homotopy.

Proposition 8.1.4. *Let (X, v) be a $K(G, 1)$-complex. Then $H_*(G, R) \cong H_*(X; R)$.*

Proof. The chain complex $(R \otimes_G C_*(\tilde{X}; R), \mathrm{id} \otimes \partial)$ is isomorphic[4] to the chain complex $(C_*(X; R), \partial)$.   □

Even when one knows that a free resolution "comes from" a $K(G, 1)$-complex in the sense of 8.1.2, it is sometimes easier to describe the free resolution than the complex. For example, let $G = \mathbb{Z}_n = \langle t \mid t^n \rangle$, the cyclic group of order n. Let $N = 1 + t + \cdots + t^{n-1} \in RG$. Consider:

$$\cdots \xrightarrow{\; t-1 \;} RG \xrightarrow{\; N \;} RG \xrightarrow{\; t-1 \;} RG \xrightarrow{\; \epsilon \;} R \longrightarrow 0.$$

Here, $t-1$ and N denote multiplication by those elements, and $\epsilon \left(\sum_{i=0}^{n-1} m_i t^i \right) = \sum_{i=0}^{n-1} m_i$. Obviously this sequence is exact, hence it is a free RG-resolution of R. Applying the functor $R \otimes_G \cdot$ we find that $H_*(G, R)$ is calculated from

$$\cdots \xrightarrow{\; 0 \;} R \xrightarrow{\; \times n \;} R \xrightarrow{\; 0 \;} R \longrightarrow 0.$$

Thus $H_k(G, R) \cong R/(n)$ when k is odd, and $H_k(G, R) \cong \{r \in R \mid nr = 0\}$ when k is even.

By taking $R = \mathbb{Z}$, and applying 8.1.4, we conclude:

Proposition 8.1.5. *For $n \geq 2$, $\mathbb{Z}_n$ has infinite geometric dimension.*   □

This plugs the gap in the proof of 7.2.12.

Indeed, there is a $K(\mathbb{Z}_n, 1)$-complex (X, v) such that $C_*(\tilde{X}; R) \xrightarrow{\epsilon} R$ is a free $R\mathbb{Z}_n$-resolution of R, but some careful work is needed to describe the attaching maps. The skeleta of such $K(\mathbb{Z}_n, 1)$-complexes are called *generalized lens spaces*. For the details, see [42].

---

[3] The word "canonical" indicates that the chain homotopy equivalence is determined up to chain homotopy by the hypotheses.

[4] Compare with 13.2.1.

Exercises

1. Prove 8.1.1.
2. Let $f : \pi_1(X_1, v_1) \to \pi_1(X_2, v_2)$ induce $\phi : G_1 \to G_2$ on fundamental group. Show that $f_{\#} : C_*(\tilde{X}_1; R) \to C_*(\tilde{X}_2; R)$ satisfies[5] $f_{\#}(g.c) = \phi(g).f_{\#}(c)$.

8.2 Homological finiteness properties

A module is *projective* if it is a direct summand of a free module. A *projective RG-resolution* of R has the same definition as "free RG-resolution" except that the modules F_i are only required to be projective. Proposition 8.1.1 also holds for projective resolutions; see [29, Chap. 1, Sect. 7].

A group G has (i) *type FP_n over R*; (ii) *type FP_∞ over R*; (iii) *type FP over R*; (iv) *type FL over R*; (v) *cohomological[6] dimension $\leq d$ over R* if there is a projective RG-resolution of R which is (i) finitely generated in dimensions $\leq n$; (ii) finitely generated in each dimension; (iii) finitely generated in each dimension and trivial in all but finitely many dimensions; (iv) same as (iii) but with all the modules free; (v) trivial in dimensions greater than d. When R is not mentioned $\mathbb{Z}$ is understood: "G has type FP_n," etc. Collectively (i)–(iv) are called *homological finiteness properties* of G (with respect to R) as distinct from the parallel properties F_n, F_∞, FD and F, which are called *topological finiteness properties of G*. Clearly each topological finiteness property implies the corresponding homological finiteness property with respect to $\mathbb{Z}$, and if G has geometric dimension $\leq d$ then G has cohomological dimension $\leq d$ with respect to $\mathbb{Z}$. The only one of these statements that is not immediate is "FD implies FP;" this follows from 7.2.13 and the corresponding statement for resolutions [29, Chap. 8, Sect. 6].

If G has type FP_1 over some R then G is finitely generated (see [29] or [14]). Except possibly for the case $d = 2$, cohomological dimension $\leq d$ implies geometric dimension $\leq d$ (see Theorem VIII.7.1 of [29]). When G is finitely presented, "FP_n implies F_n," "FP_∞ implies F_∞," "FP implies FD," and "FL implies F" are all proved in [29, Chap. 8, Sect. 7]. However, there are groups of type FP_2 which are not finitely presented; examples will be constructed in Sect. 8.3.

By standard homological algebra, if $\cdots \to F_1 \to F_0 \twoheadrightarrow \mathbb{Z}$ is a projective $\mathbb{Z}G$-resolution of the trivial $\mathbb{Z}G$-module $\mathbb{Z}$ then, for any commutative ring R, $\cdots \to R \otimes_{\mathbb{Z}} F_1 \to R \otimes_{\mathbb{Z}} F_0 \twoheadrightarrow R$ is a projective RG-resolution of the trivial RG-module R; the R-module structures come from "extension of scalars" (see 12.4.7 for a fuller discussion). Thus any homological finiteness property which holds over $\mathbb{Z}$ holds over all rings R, and the cohomological dimension over $\mathbb{Z}$ is an upper bound for the cohomological dimensions over all rings R. In Sect.

---

[5] The terms *semilinear* or *ϕ-linear* are sometimes used for this property.
[6] The connection with cohomology is explained in Sect. 13.10.

13.8 we will give an example having cohomological dimension 2 over $\mathbb{Q}$ and cohomological dimension 3 over $\mathbb{Z}$.

The *cohomological dimension* (*over* R) of the group G is the least integer d such that G has cohomological dimension $\leq d$, if there is such an integer. Otherwise the cohomological dimension is said to be infinite. Write $\mathrm{cd}_R G \leq \infty$ for this.

We end with a comparison of the properties F_n and FP_n. Paraphrasing 7.2.20 we have:

Theorem 8.2.1. *Let Z be an $(n-1)$-connected n-dimensional free G-CW complex which is finite mod G. Then G has type F_n; and G has type F_{n+1} iff it is possible to attach finitely many G-orbits of $(n+1)$-cells to Z to make an n-connected free G-CW complex.*

The homological analog is:

Theorem 8.2.2. *Let Z be an n-dimensional free G-CW complex which is finite mod G, and which is $(n-1)$-acyclic[7] with respect to R. Then G has type FP_n; and G has type FP_{n+1} iff $H_n(Z; R)$ is a finitely generated RG-module.*

Proof. There is an obvious exact sequence of RG-modules

$$0 \to K_n \to C_n(Z; R) \to \cdots \to C_0(Z; R) \xrightarrow{\epsilon} R$$

and K_n is the image of a free module F_{n+1}. Let $K_{n+1} = \ker(F_{n+1} \twoheadrightarrow K_n)$. Proceeding thus by induction we get a free resolution $\cdots \to F_{n+1} \to C_n(Z; R) \to \cdots \to C_0(Z; R) \twoheadrightarrow R$ of R which is finitely generated in dimensions $\leq n$. Thus G has type FP_n. If $H_n(Z; R) = Z_n(Z; R)$ is finitely generated over RG, the same procedure gives a free resolution of R of the form $\cdots \to F_{n+1} \to C_n(Z; R) \to \cdots \to C_0(Z; R) \twoheadrightarrow R$ which is finitely generated over RG in dimensions $\leq n+1$, implying that G has type FP_{n+1}. Conversely, if G has type FP_{n+1}, a technique in homological algebra ("Schanuel's Lemma") implies that $Z_n(Z; R) = H_n(Z; R)$ is a finitely generated RG-module; for details of this algebra see [29, Chap. 8, 4.3].   □

Remark 8.2.3. The omitted material in this proof is essentially the analog in homological algebra of cell trading as described in Sect. 4.2. Note that cell trading was used in the proof of 8.2.1 (since 7.1.13 was used in the proof of 7.2.20).

Remark 8.2.4. There is a homological analog of Brown's Criterion 7.4.1. See [30].

---

[7] "n-acyclic" was defined in Sect. 4.5; n-*acyclic with respect to R* is defined similarly using R-coefficients rather than $\mathbb{Z}$-coefficients.

Exercises

1. Prove that G is countable if and only if there is a $K(G, 1)$ of locally finite type.
2. Let the finitely presented group act freely and cocompactly on the CW complex X. Prove that $H_1(X; R)$ is finitely generated as an RG-module. (*Hint*: prove that $\pi_1(X, x)$ is finitely generated as a G-group.)

8.3 Synthetic Morse theory and the Bestvina-Brady Theorem

The main theorem (8.3.12) of this section gives an elegant topological recipe for constructing examples of groups with prescribed finiteness properties: F_n but not F_{n+1}, FP_n but not FP_{n+1}, and FP_n but not F_n. The examples in question are torsion free subgroups of right-angled Artin groups, the latter being of type F. The method uses a crude form of Morse Theory in which the homotopy type of a CW complex is analyzed by means of a well chosen real valued function. For this the CW complex should be "affine" as we now explain.

A. Affine CW complexes:

A (polyhedral) *convex cell* in $\mathbb{R}^N$ is a non-empty compact set $C \subset \mathbb{R}^N$ which is the solution set of a finite number of linear equations $f_i(x) = 0$ and linear inequalities $g_j(x) \geq 0$. A *face* of C is obtained by changing some of the inequalities to equalities. Thus a face of C is also a convex cell. A *vertex* of C is a face consisting of one point. If the affine subspace spanned by C is n-dimensional then C is a *convex n-cell*. Note that the intersection or the product of convex cells is a convex cell.

We list some well-known properties of a convex cell C which are proved in many books (e.g., [90, Chap. 1]): (i) the faces of C are determined by C and not by the choice of the f_i's and g_j's used to define C; (ii) C is the convex hull of its vertices; (iii) the image of C under an affine map $\mathbb{R}^N \to \mathbb{R}^M$ is a convex cell; (iv) there is a simplicial complex K in $\mathbb{R}^N$ such that $C = |K|$; (v) writing $\overset{\bullet}{C}$ for the union of all proper faces of the compact convex n-cell C, there is a PL homeomorphism $(C, \overset{\bullet}{C}) \to (\Delta^n, \overset{\bullet}{\Delta}{}^n)$ – i.e., C is a PL n-ball.

An *affine homeomorphism* $C \to D$ between convex n-cells $C \subset \mathbb{R}^N$ and $D \subset \mathbb{R}^M$ is a homeomorphism which extends to an affine homeomorphism between the affine subspaces spanned by C and D.

Up to now we have required that the domain of a characteristic map for an n-cell in a CW complex be B^n or I^n. It is convenient to change that here: we allow any compact convex n-cell to be the domain. An *affine CW complex* is a regular CW complex X with two further properties: (i) the intersection of any two cells is either empty or is a face of both; and (ii) there is a maximal set $\mathcal{C}$ of characteristic maps for the cells of X whose domains are compact convex cells; if e_β is a face of e_α, and if h_β and h_α are characteristic maps in

$\mathcal{C}$ with domains C_β and C_α, we require that $h_\alpha^{-1} \circ h_\beta$ map C_β by an affine homeomorphism onto a face of C_α. Once we choose a characteristic map h_α for each cell e_α of X so that this relation among them holds, we can then use the Axiom of Choice to find a maximal such set $\mathcal{C}$.

Example 8.3.1. If K is a simplicial complex, then in an obvious way $|K|$ is an affine CW complex.

Subcomplexes of affine CW complexes are affine CW complexes. The product of affine CW complexes (with topology understood in the sense of k-spaces) is an affine CW complex; see 1.2.19.

We may treat the cells of an affine CW complex as if they were compact convex cells, and the "convex structure" on a cell passes unambiguously to its faces.

B. Morse Theory:

Let X be an affine CW complex. A *Morse function* on X is a map $f : X \to \mathbb{R}$ which is affine on each cell, is non-constant on each cell of positive dimension, and takes X^0 to a closed discrete subset of $\mathbb{R}$. When J is a closed subset of $\mathbb{R}$, we write $X_J := f^{-1}(J)$.

Proposition 8.3.2. *Let $J \subset J'$ be closed connected subsets of $\mathbb{R}$ such that $X_{J'} - X_J$ contains no vertices of X. Then X_J is a strong deformation retract of $X_{J'}$.*

Proof. The general case is easily adapted from the case we consider: $J = (-\infty, 1]$ and $J' = (-\infty, 2]$. If e is an n-cell of X not lying in X_J then the set $e \cap f^{-1}([1,2])$ inherits a convex structure from e. The subset $e \cap f^{-1}(2)$ is a face, and the hypothesis about vertices ensures that $e \cap f^{-1}(2)$ is an $(n-1)$-dimensional face. The convexity of e makes it possible to carry over the "radial projection" proof[8] of 1.3.9 to get a strong deformation retraction of $e \cap f^{-1}([1,2])$ onto the part of its boundary obtained by removing the interior of the face $e \cap f^{-1}(2)$. When X is n-dimensional, the desired strong deformation retract is defined by applying the previous sentence to each such n-cell e, then to the $(n-1)$-cells, etc. Indeed, the details are quite similar to those in the proof of 1.3.12. When X is infinite-dimensional, a modification along the lines of the proof of 1.3.14 is needed.   □

It follows from this proposition that one may analyze the homotopy type of X by examining how $f^{-1}((-\infty, t])$ changes as t passes[9] through a point of

---

[8] If $h : C \to e$ is the characteristic map, where $C \subseteq \mathbb{R}^N$, the "light source" of 1.3.9 should be chosen appropriately in $\mathbb{R}^N$; the deformation should be performed on C and then carried over, via h, to e.

[9] We call this "synthetic Morse Theory." In real Morse Theory X is a complete Riemannian manifold, f is a smooth map with isolated critical points, the role of the vertices here is played by the critical points of f, the connectivity of a descending link here corresponds to the index of a critical point, and the strong

$f(X^0)$. This is done by looking at "descending links" of vertices, as we now explain.

Extending the notion of "link" from simplicial complexes (Sect. 5.2), we define the *link* of a vertex v in a convex cell e to be the union of all the faces of e which do not contain v. We denote this subcomplex of e by $\mathrm{lk}_e v$. The *link of v in X* is $\mathrm{lk}_X v = \bigcup\{\mathrm{lk}_e v \mid e$ is a cell of X containing $v\}$. The *ascending link of v in X* (with respect to the Morse function f) is:

$$\mathrm{lk}_X^\uparrow v = \bigcup\{\mathrm{lk}_e v \mid v \in e \text{ and } f \mid e \text{ has a minimum at } v\}$$

The descending link $\mathrm{lk}_X^\downarrow v$ is defined similarly, replacing "minimum" by "maximum." See Fig. 8.1.

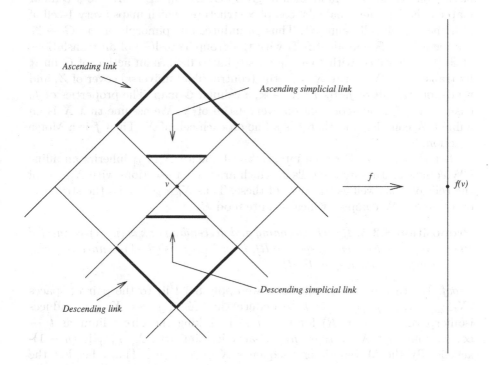

Fig. 8.1.

Proposition 8.3.3. *Let $J \subset J'$ be closed connected subsets of $\mathbb{R}$ such that $\inf J = \inf J'$ and $J' - J$ contains only one point of $f(X^0)$, namely t. Then $X_J \cup \bigcup\{$cone on $\mathrm{lk}_X^\downarrow v \mid f(v) = t\}$ is a strong deformation retract of $X_{J'}$.*

deformation retraction in 8.3.2 would be achieved in real Morse Theory by deforming along the gradient flow lines defined by f. However, the basic topological idea of analyzing change in homotopy type is the same. For real Morse Theory, see [120].

Proof. The proof is similar to that of 8.3.2, but the deformations described there are only applied to cells which do not contain a vertex v such that $f(v) = t$. □

Corollary 8.3.4. (i) *If each descending link in 8.3.3 is $(n-1)$-connected then $(X_{J'}, X_J)$ is n-connected.*
(ii) *If each descending link in 8.3.3 is $(n-1)$-acyclic with respect to the ring R then $(X_{J'}, X_J)$ is n-acyclic with respect to R.* □

C. Finiteness properties and Morse Theory:

We apply this to group theory, first in a rather general setting, later in a situation of special interest. Let (Z, z) be a finite aspherical CW complex having one vertex and fundamental group G. Let $f_0 : Z \to S^1$ be a cellular map (S^1 having the usual CW complex structure) which maps every 1-cell of Z homeomorphically onto S^1. Thus f_0 induces an epimorphism $\phi : G \to \mathbb{Z}$. We regard $\mathbb{Z} \subset \mathbb{R}$ and identify $\mathbb{R}$ with the group $\mathrm{Transl}(\mathbb{R})$ of all translations of $\mathbb{R}$ (r is identified with $t \mapsto t + r$), so that ϕ defines an action of G on $\mathbb{R}$ by translations. We write $X = \tilde{Z}$, the (contractible) universal cover of Z, and we denote the lift of f_0 by $f : X \to \mathbb{R}$, a cellular G-map. The properties of f_0 ensure that f is non-constant on every 1-cell of X. We assume that X is an affine CW complex and that f is affine on each cell of X. Then f is a Morse function.

Let $H = \ker(\phi)$. For each integer $k \geq 1$ the set $X_{[-k,k]}$ inherits an affine CW complex structure the cells of which are the intersections with $X_{[-k,k]}$ of the cells of X as well as the faces of these. Thus $X_{[-k,k]}$ admits the structure of a free H-CW complex which is finite mod H.

Proposition 8.3.5. *If each ascending and descending link is $(n-1)$-connected [resp. $(n-1)$-acyclic with respect to R], then $X_{[-k,k]}$ is $(n-1)$-connected [resp. $(n-1)$-acyclic with respect to R].*

Proof. For the $(n-1)$-acyclic case, we apply 8.3.4(ii) to the pair of spaces $(X_{(-\infty,t]}, X_{(-\infty,k]})$ for $t \geq k$ to deduce that $X_{(-\infty,k]} \hookrightarrow X_{(-\infty,t]}$ induces isomorphisms on $H_k(\cdot; R)$ for $k \leq n - 1$. Taking the direct limit as $t \to \infty$, we find that $X_{(-\infty,k]}$ is $(n-1)$-acyclic. Similarly, $X_{[-k,\infty)}$ is $(n-1)$-acyclic. By the Mayer-Vietoris sequence $X_{[-k,k]}$ is $(n-1)$-acyclic. For the π_1-case (omitting base points) and $k \leq t < \infty$, the inclusion map induces $\pi_1(X_{[-k,k]}) \to \pi_1(X_{[-k,t]})$. By 8.3.4 (using the fact that $f(X^0)$ is discrete) and the homotopy exact sequence 4.4.9, this is an isomorphism. Taking the direct limit with respect to t, it follows that $\pi_1(X_{[-k,k]}) \to \pi_1(X_{[-k,\infty)})$ is an isomorphism. Similarly $\pi_1(X_{[-k,k]}) \to \pi_1(X_{(-\infty,k]})$ is an isomorphism. Since $\pi_1(X)$ is trivial, so is $\pi_1(X_{[-k,k]})$ by the Seifert-Van Kampen Theorem 3.1.18. Then, under the appropriate hypothesis, the Hurewicz Theorem 4.5.2 gives $(n-1)$-connectedness of $X_{[-k,k]}$. □

Corollary 8.3.6. *Under the same hypotheses on ascending and descending links the group H has type F_n [resp. type FP_n with respect to R].* □

D. Review of CAT(0) geometry:

In order to check that a group G is of type F one (usually) constructs a finite $K(G, 1)$-complex X. It is often the case that one has a good idea for X, but one has difficulty proving that the universal cover $\tilde{X}$ is contractible. The classical Cartan-Hadamard Theorem in differential geometry says that a simply connected open manifold equipped with a complete Riemannian metric of non-positive sectional curvature is contractible. Since this curvature hypothesis is local, it follows that any closed Riemannian manifold with non-positive sectional curvature has a contractible universal cover. CAT(0) geometry extends this idea in useful ways. Here, we recall only what we need. A general reference for CAT(0) geometry is [24].

A *proper* CAT(0) *space* is a metric space (M, d) having the following properties: (i) it is a non-empty geodesic metric space: this means that an isometric copy of the closed interval $[0, d(a, b)]$ called a *geodesic segment* joins any two points $a, b \in M$; (ii) for any geodesic triangle Δ in M with vertices a, b, c let Δ' denote a triangle in the Euclidean plane with vertices a', b', c' and corresponding side lengths of Δ' and Δ equal; let ω and ω' isometrically parametrize geodesic segments from b to c and from b' to c' respectively; then for any $0 \leq t \leq d(b, c)$, $d(a, \omega(t)) \leq |a' - \omega'(t)|$; and (iii) d is *proper*, i.e., the closed ball around any $a \in M$ of any radius r is compact. A metric satisfying (i) and (ii) is a *CAT(0) metric*.

In a CAT(0) space the geodesic segment from a to b is unique and varies continuously with a and b. This implies (see I.3.13 in [24]):

Proposition 8.3.7. *Fix a base point $a \in M$. For any $b \in M$ let the path $\omega_{[b,a]} : [0, d(b, a)] \to M$ isometrically parametrize the geodesic segment from b to a. Then the function $H : M \times I \to M$, $(b, t) \mapsto \omega_{[b,a]}(td(b, a))$ is a contraction of M to the point a in which the track of every point b is the geodesic segment from b to a.* $\qquad \square$

It follows that if a finite path connected CW complex Z has fundamental group isomorphic to G, and if $\tilde{Z}$ admits a CAT(0) metric, then Z is a $K(G, 1)$-complex and G has type F. We now describe a criterion for this.

In the definition of an affine CW complex X, whenever e_β is a face of e_α with characteristic maps $h_\alpha : C_\alpha \to e_\alpha$ and $h_\beta : C_\beta \to e_\beta$, it was required that $h_\alpha^{-1} \circ h_\beta$ map C_β by an affine homeomorphism onto a face of C_α. If we strengthen this by consistently requiring $h_\alpha^{-1} \circ h_\beta$ to map C_β isometrically onto a face of C_α, then X is a *metric CW complex*. If, in addition, every convex domain-cell C_α is a cube I^n for some n, then X is a *cubical complex* and each n-cell is called an *n-cube* of X. The cubes of X thus have well-defined euclidean metrics.[10] In cubical complexes there is another kind of link. Extending the

---

[10] See I.7.33 of [24] for more on cubical complexes. A path connected cubical complex X supports a metric as follows: any two points a, b can be joined by a path consisting of straight segments each lying in a cube and hence having a length; the length of the path is then the sum of the lengths of the segments, and the

definition of Δ^{N-1} (Sect. 2.2), let $\frac{1}{2}\Delta^{N-1}$ denote the convex hull of $\{\frac{1}{2}u_i\}$ where $u_1, \ldots, u_N$ are the coordinate unit vectors in $\mathbb{R}^N$. We define $\frac{1}{2}\Delta^{N-1}$ to be the *simplicial link* of the vertex 0 in the cube $[0,1]^N$; simplicial links of the other vertices of the cube are defined similarly. If v is a vertex of the cubical complex X, the *simplicial link of v in X* is the union of the simplicial links of v in all the cubes of X which have v as a vertex. Clearly, this is the geometric realization of an abstract simplicial complex, namely, $|\text{simplk}_X v|$. Indeed, with respect to a Morse function there are subcomplexes $|\text{simplk}_X^\uparrow v|$ and $|\text{simplk}_X^\downarrow v|$ corresponding to $\text{lk}_X^\uparrow v$ and $\text{lk}_X^\downarrow v$ in the obvious way (see Fig. 8.1) and $\text{lk}_X v$ and $|\text{simplk}_X v|$ are homeomorphic by a homeomorphism which matches ascending [resp. descending] links (exercise).

An *abstract graph* is an abstract simplicial complex of dimension ≤ 1. A *flag complex* is an abstract simplicial complex L with the property that a finite set of vertices of L is a simplex iff each pair of those vertices is a 1-simplex. There is an obvious bijection between flag complexes and abstract graphs given by $L \mapsto L^1$. Note that an abstract first derived of any regular CW complex is a flag complex.

The key theorem which connects these concepts is due to Gromov (see II.5.20 of [24]):

Theorem 8.3.8. *Let X be a simply connected cubical rigid G-complex which is finite mod G. If the simplicial link of every vertex of X is a flag complex, then X admits a G-invariant CAT(0) metric which agrees with the (given) Euclidean metric on each cube.*

E. Prescribing finiteness properties:

We can now discuss an important special case in which there is a converse to 8.3.6, namely, that if the ascending and descending links are not $(n-1)$-connected [resp. not $(n-1)$-acyclic with respect to R] then the group H does not have type F_n [resp. does not have type FP_n with respect to R]. That occupies the rest of the section.

Starting again, let L be a non-empty finite flag complex and let G be the corresponding *right-angled Artin group*, i.e., the group presented by $\langle S \mid R \rangle$ where S is the set of vertices of L and $R = \{s_i s_j s_i^{-1} s_j^{-1} \mid s_i \text{ and } s_j \text{ span a 1-simplex of } L\}$. A finite $K(G,1)$-complex Z is found as follows. Let $T = \prod_{w \in L^0} S_w^1$

where each S_w^1 is a circle with the usual CW complex structure of one vertex and one 1-cell; T has the product CW complex structure, and is a torus. Each simplex σ of L defines a subtorus of T consisting of all points whose coordinates[11] not in σ are the base point of S^1. The union of all such subtori is Z. By 3.1.8, the fundamental group of Z is isomorphic to G.

---

distance between a and b is the inf of the lengths of these paths. This is the CAT(0) metric referred to in 8.3.8.

[11] Each coordinate of a point in T is labeled by a vertex of L. The (abstract) simplex σ is a finite set of vertices.

Let X be the universal cover of Z. The CW complex X is a cubical complex. Indeed, if the simplex σ of L has dimension k, the associated subtorus in Z has dimension $k+1$, and by 3.4.9 its pre-image in X consists of a collection of pairwise disjoint copies of $\mathbb{R}^{k+1}$ with the usual cubical structure; these are called *sheets*[12] in X.

Proposition 8.3.9. *The simplicial link of each vertex of X is a flag complex.*

Before proving this we observe that since $S^{k+1} \cong S^k * S^0$ there is a *canonical triangulation* K^{k+1} of S^{k+1} defined inductively by $K^{k+1} = K^k * K^0$ where K^0 consists of two points, and $*$ denotes join.

Proof (of 8.3.9). Let v be a vertex of X which lies in the $(k+1)$-dimensional sheet Σ. Then $|\mathrm{simplk}_\Sigma(v)|$ is a canonical triangulation of S^k. This is the part of $|\mathrm{simplk}_X(v)|$ in Σ. Each vertex of $\mathrm{simplk}_X(v)$ is associated with a vertex $\psi(v)$ of L, and ψ defines a simplicial map $\mathrm{simplk}_X(v) \to L$ such that the pre-image of every n-simplex in $\mathrm{st}_L \psi(v)$ consists of n-simplexes and their faces. Since L is flag it follows that $\mathrm{simplk}_X(v)$ is flag. $\square$

Together with 8.3.8 this proves:

Corollary 8.3.10. Z *is a* $K(G,1)$*-complex.* $\square$

The homomorphism $\phi : G \to \mathbb{Z}$ which takes each generator to $1 \in \mathbb{Z}$ is induced by a map $f_0 : Z \to S^1$ which takes every 1-cell of Z homeomorphically to S^1, and one sees easily that f_0 can be chosen so that its lift $f : X \to \mathbb{R}$ is affine on each cube. Thus f is a Morse function.

Proposition 8.3.11. *Each ascending and descending link of a vertex of X is homeomorphic to $|L|$.*

Proof. We will show that for every vertex v of X both $\mathrm{simplk}^{\uparrow}_X v$ and $\mathrm{simplk}^{\downarrow}_X v$ are isomorphic to L as abstract simplicial complexes. Orient the circles S^1_w of Z so that ϕ takes the corresponding generator of the fundamental group of Z to $1 \in \mathbb{Z}$. Thus T is oriented, hence also Z; Z has just one vertex z. Let v be a vertex of X, and let $p : X \to Z$ be the universal cover. For each vertex w of L, v lies in an oriented line $\lambda_w \subset X$ which p maps (as a covering projection) to the circle S^1_w in T, all other coordinates being the base point. The Morse function f maps λ_w homeomorphically in an orientation-preserving fashion to $\mathbb{R}$. The link of v in λ_w is a 0-sphere consisting of points u^+_w and u^-_w mapped by f above and below $f(v)$ respectively. If the product of $S^1_{w_1}$ and $S^2_{w_2}$ in T lies in Z then λ_{w_1} and λ_{w_2} span an oriented plane containing v which is partitioned into cubes, and the simplicial link of v in this plane is the join of the corresponding 0-spheres (the canonical triangulation of S^1). Clearly, the ascending link is the join of $u^+_{w_1}$ and $u^+_{w_2}$ while the descending link is the join

---

[12] In terms of CAT(0) geometry these sheets are euclidean convex subspaces which meet orthogonally in subsheets.

of $u_{w_1}^-$ and $u_{w_2}^-$. Thus we get a 1-simplex in $\mathrm{simplk}_X^\uparrow v$ and in $\mathrm{simplk}_X^\downarrow v$ for each 1-simplex of L. Proceeding thus, we find isomorphisms from L to both $\mathrm{simplk}_X^\uparrow v$ and $\mathrm{simplk}_X^\downarrow v$.   □

Theorem 8.3.12. (Bestvina-Brady Theorem *Let L be a finite non-empty flag complex, let G be the corresponding right-angled Artin group, let $\phi : G \to \mathbb{Z}$ be the epimorphism taking all generators to $1 \in \mathbb{Z}$, and let $H = \ker(\phi)$. Then*

(i) *H has type F_n iff $|L|$ is $(n-1)$-connected;*
(ii) *H has type FP_n with respect to R iff $|L|$ is $(n-1)$-acyclic with respect to R.*

For the proof we need:

Lemma 8.3.13. *Let v be a vertex of X and let U be the union of the sheets containing v.*

(a) *U is an open cone; in fact there is a homeomorphism from U to the space $\mathrm{lk}_X v \times [0, \infty)/\mathrm{lk}_X v \times \{0\}$ taking v to the quotient point so that f is increasing [resp. decreasing] on cone rays passing through points of $\mathrm{lk}_X^\uparrow v$ [resp. $\mathrm{lk}_X^\downarrow v$].*
(b) *For any vertex v of X, $|\mathrm{simpl}_X v|$ is a strong deformation retract of the space $X - \{v\}$.*

Proof. (a) is clear. For (b) we use 8.3.7. The deformation of all points of X along geodesics ending at v gives the required strong deformation retract of $X - \{v\}$ onto $|\mathrm{simpl}_X v|$.   □

Proof (of Theorem 8.3.12). The "if" statements follow from 8.3.6 and 8.3.11. We begin with the "only if" part of (i). We have a filtration $\{X_{[-k,k]}\}_{k \in \mathbb{N}}$ of X by H-subspaces which are compact mod H. We will show that if the space $|L|$ is not $(n-1)$-connected then this filtration is not essentially $(n-1)$-connected, hence (by 7.4.1) that H does not have type F_n. Fix an integer $m > 0$. Let v be a vertex of X with $f(v) > m$. Let $h : S^{n-1} \to |\mathrm{simplk}_X^\downarrow v|$ be homotopically non-trivial; by 8.3.11 there is such a map. Write $J := (-\infty, f(v))$. Then $|\mathrm{simplk}_X^\downarrow v| \subset X_J$. By 8.3.13(a), there is a homotopy $H : S^{n-1} \times I \to X_J$ with $H_0 = h$ and $H_1(S^{n-1}) \subset X_{\{0\}} = f^{-1}(0)$; this is because pushing outwards from v in each relevant sheet lowers the f-value of points in $|\mathrm{simpl}_X^\downarrow v|$. We claim that H_1 is not homotopically trivial in $X - \{v\}$, so certainly not homotopically trivial in $X_{[-m,m]}$. Since m is arbitrary this will be enough.

Suppose H_1 could be extended to $\bar{H} : B^n \to X - \{v\}$. Then we would have $H(S^{n-1} \times I) \cup \bar{H}(B^n)$ lying in $X - \{v\}$, so 8.3.13(b) would imply that h is homotopically trivial in $|\mathrm{simplk}_X v|$. The analysis of the links in the proof of 8.3.11 shows that $|\mathrm{simplk}_X^\downarrow v|$ is a (simplicial) retract of $|\mathrm{simplk}_X v|$. Thus h would be homotopically trivial in $|\mathrm{simplk}_X^\downarrow v|$, a contradiction.

The "only if" part of (ii) is proved similarly: the map h is replaced by an R-cycle which does not bound in $|\text{simplk}_X^{\downarrow} v|$. The argument shows that the filtration $\{X_{[-k,k]}\}$ is not essentially $(n-1)$-acyclic. We deduce that H does not have type FP_n by the following lemma.   □

Lemma 8.3.14. *Let K be an $(n-1)$-acyclic (with respect to R) free G-CW complex and let $\{K_i\}$ be a G-filtration where each $G\backslash K_i$ has finite n-skeleton. If G has type FP_n then $\{K_i\}$ is essentially $(n-1)$-acyclic with respect to R.*

Proof. Let $\{F_*, \partial\}$ be a free RG-resolution of R which is finitely generated in dimensions $\leq n$. By 8.1.1 there are mutually inverse chain homotopy equivalences $C_*(K; R) \underset{g}{\overset{f}{\rightleftarrows}} F_*$. Let D be a chain homotopy between $g \circ f$ and id. For any j there exists k such that $D(C_{n-1}(K_j; R)) \leq C_n(K_k; R)$. Now $f_\#(Z_{n-1}(K; R)) \leq \partial(F_n)$, so $g_\# f_\#(Z_{n-1}(K; R)) \leq g_\# \partial(F_n) = \partial(g_\#(F_n))$. Since F_n is finitely generated, k can be chosen so large that $g_\#(F_n) \leq C_n(K_k; R)$. We write $i : K_j \hookrightarrow K_k$. Then for any $z \in Z_{n-1}(K_j; R)$, $i_\#(z) = g_\# f_\#(z) + \partial D z \in \partial C_n(K_k; R)$.   □

Remark 8.3.15. Lemma 8.3.14 generalizes 8.2.2, and is a partial homological version of Brown's Criterion 7.4.1. For a full version see [30].

Remark 8.3.16. There is yet another kind of finiteness property of groups. We have seen the properties F_n and FP_n over the ring R. We say that G has type FP'_n over R if there is an n-dimensional free G-CW complex X which is $(n-1)$-acyclic over R and is finite mod G. Clearly FP'_n implies FP_n, and FP'_1 is equivalent to "finitely generated" (as are FP_1 and F_1). FP'_2 and FP_2 are equivalent (exercise). It is unknown if FP_n implies FP'_n when $n \geq 3$.

The proof of 8.3.12(ii) actually gives: $|L|$ is $(n-1)$-acyclic $\Rightarrow H$ has type $FP'_n \Rightarrow H$ has type $FP_n \Rightarrow |L|$ is $(n-1)$-acyclic. Thus for any $n \geq 2$, Theorem 8.3.12 gives a method of constructing groups of type FP'_n which are not of type F_n.

Remark 8.3.17. In [10] similar methods are used to construct a group of type FP (indeed, of type FP') which is not finitely presented.

Source Notes: This section is based on [10] as simplified in [36].

Exercise

1. Prove that if v is a vertex of a cubical complex X then $\text{lk}_X v$ and $|\text{simpl}_X v|$ are homeomorphic by a homeomorphism which matches ascending [resp. descending] links.
2. Give an example of a regular CW complex which does not admit the structure of an affine CW complex. *Hint:* (i) in that definition might not hold.
3. Prove that FP_n and FP'_n are equivalent for $n \leq 2$.

9

Finiteness Properties of Some Important Groups

In this chapter we introduce Coxeter groups, Thompson's groups, and (in outline) outer automorphism groups of free groups, and we apply the theory developed so far to these. The recent literature in group theory contains similar treatments of many other important classes of groups, so our examples should be seen merely as case studies.

9.1 Finiteness properties of Coxeter groups

In this section we define Coxeter groups. We show that finitely generated Coxeter groups are of type F_∞, while their torsion free subgroups of finite index are of type F. In Sects. 13.9 and 16.6 we will refine the methods used here to produce interesting examples.

A *Coxeter system* (G, S) consists of a group G and a set S of generators of G, each of order 2, such that if $m : S \times S \to \mathbb{N} \cup \{\infty\}$ is the function $m(s_1, s_2) =$ the order of $s_1 s_2$, then G has a presentation $\langle S \mid (s_1 s_2)^{m(s_1, s_2)}$ $\forall s_1, s_2 \in S \rangle$, with the convention that when $m(s_1, s_2) = \infty$ the corresponding relation is absent. A group G is called a *Coxeter group* if there exists a Coxeter system (G, S). WE WILL ALWAYS ASSUME THAT S IS FINITE.

A subgroup of G is a *standard (parabolic) subgroup* if it is generated by a subset of S. If $T \subset S$ we denote by $\langle T \rangle$ the corresponding standard subgroup. A *standard coset* is a coset $g\langle T \rangle$ with $g \in G$ and $T \subset S$. The *length* of $g \in G$ is the length of the shortest word in the free group $F(S)$ which represents g in G; it is denoted by $l(g)$. A shortest word representing an element of G is said to be *reduced*.

Proposition 9.1.1. *If $g = s_1 \cdots s_d$ with $d > l(g)$, there are indices $i < j$ such that $g = s_1 \cdots \hat{s}_i \cdots \hat{s}_j \cdots s_d$ (i.e., suppress s_i and s_j).*

Proof. This is a basic algebraic fact about Coxeter systems, in fact a characterizing property. See, for example, p. 53 of [31]. ☐

Corollary 9.1.2. *The function $T \mapsto \langle T \rangle$ from subsets of S to standard subgroups of G is a bijection. Moreover, $T_1 \subset T_2$ iff $\langle T_1 \rangle \subset \langle T_2 \rangle$.*

Proof. The required inverse is $H \mapsto H \cap S$. If H is a standard subgroup then $H = \langle H \cap S \rangle$. If $T \subset S$, then $T \subset \langle T \rangle \cap S$. It remains to show that $\langle T \rangle \cap S \subset T$. If $g \in \langle T \rangle \cap S$ then $g = t_1 \ldots t_r$ where each $t_i \in T$, and $l(g) = 1$, so 9.1.1 implies that $g = t_i \in T$ for some i. $\square$

Recall from Sect. 5.4 the notion of "ordered abstract simplicial complex". Examples come from partially ordered sets (posets): if $(A, \le)$ is a poset the *associated ordered abstract simplicial complex*, also denoted by A, has A as its set of vertices and the strictly ordered $(k+1)$-tuples $a_{i_0} < a_{i_1} < \cdots < a_{i_k}$ in A as its k-simplexes. This passage from posets to complexes is an important way of passing from discrete mathematics to topology. We will apply it to the poset D of finite standard cosets formed from the Coxeter system (G, S); the partial ordering is by inclusion. The corresponding ordered abstract simplicial complex D is the *Davis complex* of (G, S). There is an obvious left action of G on D by simplicial automorphisms which are order-preserving on simplexes.

Theorem 9.1.3. (Davis' Theorem) *The G-CW complex $|D|$ is rigid, contractible, and finite mod G. The stabilizer of each cell is finite.*

Proof. Since G acts by order-preserving simplicial automorphisms, the action on $|D|$ is rigid. If $\{g_0 \langle T_0 \rangle, \cdots, g_k \langle T_k \rangle\}$ are the vertices, in order, of a simplex of D then it is easy to see that it can be written as $g_0\{\langle T_0 \rangle, \cdots, \langle T_k \rangle\}$; here each $T_i \subset S$ generates a finite standard subgroup, and $\langle T_i \rangle \subset \langle T_{i+1} \rangle$. By 9.1.2 it follows that $T_i \subset T_{i+1}$ for all i. Finiteness mod G follows. The stabilizer of this simplex is $g_0 \langle T_0 \rangle g_0^{-1}$ which is finite. It only remains to prove that $|D|$ is contractible. This requires some preliminary discussion. We conclude this proof after Lemma 9.1.6.

Let F be the full subcomplex of D generated by the finite standard subgroups of G. Every simplex of F is a face of one whose initial vertex is $\langle \emptyset \rangle = \{1\}$, the trivial standard subgroup. In other words, F is the cone $\{1\} * K$ whose vertex is $\{1\}$ and whose base K is the full subcomplex of D generated by the non-trivial finite standard subgroups.[1] As in 5.2.6, $|F|$ is the topological cone $\{1\} * |K|$ and is a finite *fundamental domain* in $|D|$, i.e., a finite subcomplex whose G-translates cover $|D|$. The subcomplex $g|F|$ is a cone with vertex $\{g\}$ and base $|gK|$.

When $\langle T \rangle$ is a non-trivial finite standard subgroup of G, let F_T denote the subcomplex of F generated by all (ordered) simplexes whose initial vertex is $\langle T \rangle$; this is a full subcomplex of F. Clearly, $|F_T| = \bigcap_{s \in T} |F_{\{s\}}|$, and since F_T is a cone with vertex $\langle T \rangle$, $|F_T|$ is contractible. Let $F_{\sigma(T)} = \bigcup_{s \in T} F_{\{s\}}$, i.e., the

---

[1] K is sometimes called the *nerve* of (G, S).

largest subcomplex of F each of whose simplexes lies in $F_{\{s\}}$ for some $s \in T$. Then $|F_{\sigma(T)}| = \bigcup_{s \in T} |F_{\{s\}}|$. Note that $F_{\sigma(T)}$ is not defined if $T = \emptyset$.

Lemma 9.1.4. *When $\langle T \rangle$ is non-trivial and finite, $|F_{\sigma(T)}|$ is contractible.*

Proof. Whenever $U = \{s_0, \cdots, s_r\}$ is a subset of T, $\langle U \rangle$ is a finite standard subgroup of $\langle T \rangle$, so F_U is a cone. Thus $\{|F_{\{s\}}| \mid s \in T\}$ is a cover of $|F_{\sigma(T)}|$ by cones such that each intersection is also a cone. It is a general fact (Exercise 5 in Sect. 4.1) that if a CW complex Z admits a cover by subcomplexes such that every intersection of members of the cover is contractible then Z is contractible. Hence $|F_{\sigma(T)}|$ is contractible. $\qquad\square$

For $g \in G$, let $B(g) = \{s \in S \mid l(gs) < l(g)\}$. When $g \neq 1$, $B(g) \neq \emptyset$.

Proposition 9.1.5. *For every $g \in G$, the standard subgroup $\langle B(g) \rangle$ is finite.*

Proof. Write $d = l(g)$. It is enough to show that every reduced word in $\langle B(g) \rangle$ has length $\leq d$. Let $h = t_1 \cdots t_k$ be a word of minimal length in the elements of $B(g)$ and let $g = s_1 \cdots s_d$ be reduced. By 9.1.1, $l(h) = k$. For $k \leq d$, we claim that $g = s_1 \cdots \hat{s}_{i_1} \cdots \hat{s}_{i_k} \cdots s_d t_1 \cdots t_k$; i.e., $t_1 \cdots t_k$ can appear at the right hand end of a reduced representation of g; this is proved by induction on k using the facts $l(g) = d$ and $l(h) = k$. The details are left to the reader. We now show that $k \leq d$. Suppose $k > d$. By the Claim we have $g = t_1 \cdots t_d$, so $g t_{d+1} = t_1 \cdots t_{d+1}$. Since $t_{d+1} \in B(g)$, the right side is not reduced, a contradiction. $\qquad\square$

At this point, some notation is useful. Order the members of G, $1 = g_0, g_1, g_2, \ldots$ so that $l(g_i) \leq l(g_j)$ whenever $i < j$. Let A_n be the subcomplex $\bigcup_{i=0}^{n} g_i F$ of D, i.e., the largest subcomplex each of whose simplexes lies in some $g_i F$.

Lemma 9.1.6. $A_n \cap g_{n+1} F = g_{n+1} F_{\sigma(B(g_{n+1}))}$ *and therefore* $|A_n| \cap |g_{n+1} F|$ *is contractible.*

Proof. First, $B(g_{n+1}) \neq \emptyset$, so the right side is defined. The inclusion $\supset$ holds because if $s \in B(g_{n+1})$ then $g_{n+1} s = g_i$ for some $i \leq n$, and so any simplex of $g_{n+1} F_{\{s\}}$ with initial vertex $g_{n+1} s$ must lie in $A_i \subset A_n$, as well as in $g_{n+1} F$. For the inclusion $\subset$, let $\langle T \rangle$ be a vertex of F and let $g_{n+1} \langle T \rangle \in A_n$. Then $g_{n+1} \langle T \rangle = g_i \langle T' \rangle$ for some finite standard subgroup $\langle T' \rangle$ and some $i \leq n$, hence g_{n+1} is not of minimal length in the coset $g_{n+1} \langle T \rangle$, by Exercise 1. Write $g_{n+1} = kh$ where k is an element of minimal length in $g_{n+1} \langle T \rangle$ and $h \in \langle T \rangle$. Then $h \neq 1$ and, by an application of 9.1.1, $l(g_{n+1}) = l(k) + l(h)$. Write $h = t_1 \ldots t_m$ where $m = l(h)$ and each $t_i \in T$. Then $l(h t_m) = l(h) - 1$. So $l(g_{n+1} t_m) = l(k h t_m) = l(k) + l(h t_m) = l(g_{n+1}) - 1$. Thus $t_m \in B(g_{n+1})$, so $g_{n+1} \langle T \rangle \in g_{n+1} F_{\sigma(B(g_{n+1}))}$. This proves $\subset$ for vertices, which is enough since both sides are full subcomplexes of D. $\qquad\square$

Proof (of Theorem 9.1.3 (concluded)). For each n, $|A_n| = \bigcup_{i=0}^{n} g_i|F|$ is contractible; this is clear for $n = 0$, and follows for $n > 0$ by induction using 9.1.6. Since $|D| = \bigcup_{i=0}^{\infty} g_i|F|$, $|D|$ is m-connected for all m and hence contractible, by 7.1.2. This completes the proof of Theorem 9.1.3.   □

By 7.3.1, 7.2.2, 7.2.5 and 9.1.3 we get:

Theorem 9.1.7. *Every finitely generated Coxeter group is of type F_∞. Every torsion free subgroup of finite index is of type F.*

This leads to the question: do finitely generated Coxeter groups have torsion free subgroups of finite index? The positive answer rests on two propositions for which we give references. As usual, $GL_n(\mathbb{R})$ denotes the group of all invertible real $n \times n$ matrices.

Proposition 9.1.8. *Let (G, S) be a Coxeter system and let S have n elements. Then G is isomorphic to a subgroup of $GL_n(\mathbb{R})$.*

Proof. See [31, Chap. 2, Sect. 5]. In fact there is a "canonical representation" of G in $GL_n(\mathbb{R})$.   □

Proposition 9.1.9. (Selberg's Lemma) *Every finitely generated subgroup of $GL_n(\mathbb{C})$ has a torsion free subgroup of finite index.*

Proof. See, for example, [132, p. 326].   □

By 9.1.8 and 9.1.9 we can round out 9.1.7:

Proposition 9.1.10. *A finitely generated Coxeter group has a torsion free subgroup of finite index.*   □

Theorem 9.1.11. *Let (G, S) be a Coxeter system and let $d \ (\geq 1)$ be the largest number such that there is a d-element subset T of S with $\langle T \rangle$ finite. Then every torsion free subgroup of finite index in G has geometric dimension $\leq d$ and has type F.*

Proof. The dimension of $|K|$ is $d-1$, so the dimension of $|D|$ is d. The torsion free subgroup H acts freely on D, and $G\backslash|D|$ is finite.   □

Source Notes: The treatment here is based on [46]. I am indebted to John Meier for help with this section.

Exercise

1. Show that a standard coset in a finitely generated Coxeter group contains a unique element of minimal length.
2. Show that every standard subgroup of a Coxeter group is a Coxeter group with respect to the Coxeter system defined by it.

9.2 Thompson's group F and homotopy idempotents

There are several groups named for Richard Thompson; the one discussed here is[2]

$$F := \langle x_0, x_1, \ldots \mid x_n^{x_i} = x_{n+1} \ \forall \ 0 \le i < n \rangle.$$

Although this elegant presentation has infinitely many generators and relations, F is clearly generated by x_0 and x_1. Using Tietze transformations it can be shown that F admits a finite presentation as follows: for $n = 2$, 3 or 4 define $x_n = x_{n-1}^{x_{n-2}}$; then F is presented by

$$\langle x_0, x_1 \mid x_2^{x_0} = x_3, x_3^{x_1} = x_4 \rangle.$$

From the infinite presentation we see that F admits a *shift homomorphism* $\phi : F \to F$, $x_i \mapsto x_{i+1}$ for all i. Then $\phi^2(x) = \phi(x)^{x_0}$ for all $x \in F$ so ϕ is an *idempotent up to conjugacy*. Indeed, the triple (F, ϕ, x_0) is the universal example of an idempotent up to conjugacy in the sense of the following (whose proof is immediate):

Proposition 9.2.1. *Let G be a group. When an endomorphism $\psi : G \to G$ and an element $g_0 \in G$ are such that $\psi^2(g) = \psi(g)^{g_0}$ for all $g \in G$, then there is a unique homomorphism $\rho_\psi : F \to G$ such that $\rho_\psi(x_0) = g_0$ and $\rho_\psi \circ \phi = \psi \circ \rho_\psi$.* $\square$

We now prepare to represent F as a group of homeomorphisms of $\mathbb{R}$ (Corollary 9.2.3). Here, it is convenient to consider the homeomorphisms of $\mathbb{R}$ as acting on the right; thus we write $(t)h$ rather than $h(t)$, so $(t)(hk)$ means $((t)h)k$. Let $\bar{x}_n$ be the piecewise linear homeomorphism of $\mathbb{R}$ which is the identity on $(-\infty, n]$, has slope 2 on $[n, n+1]$, and has slope 1 on $[n+1, \infty)$. Let T be the unit translation homeomorphism of $\mathbb{R}$, $t \mapsto t + 1$. Then $\bar{x}_n^T = \bar{x}_{n+1}$. The group $\bar{F}$ of homeomorphisms of $\mathbb{R}$ generated by $\{\bar{x}_n \mid n \ge 0\}$ admits an endomorphism $\bar{\phi} : \bar{F} \to \bar{F}$, $\bar{x} \mapsto \bar{x}^T$. One checks that $\bar{x}_n^{\bar{x}_i} = \bar{x}_{n+1}$ when $i < n$, so $\bar{\phi}^2(\bar{x}) = \bar{\phi}(\bar{x})^{\bar{x}_0}$. Thus, by 9.2.1, there is an epimorphism $\rho : F \to \bar{F}$, $x_n \mapsto \bar{x}_n$, and $\rho \circ \phi = \bar{\phi} \circ \rho$. We will now show that ρ is an isomorphism, and at the same time we will establish a "normal form" for the elements of F.

Consider elements $x \in F$ with the following properties:

(1) $x = x_{i_1} \ldots x_{i_k} x_{j_m}^{-1} \ldots x_{j_1}^{-1}$ with $i_1 \le \ldots \le i_k$, $j_1 \le \ldots \le j_m$, where $k \ge 0$ and $m \ge 0$; and

(2) when x_i and x_i^{-1} both occur in this product then x_{i+1} or x_{i+1}^{-1} also occurs.

Note that when (1) holds but not (2) then there is a subproduct of the form $x_i \phi^{i+2}(y) x_i^{-1}$ which is equal to $\phi^{i+1}(y)$; thus any x satisfying (1) can be re-expressed to satisfy both (1) and (2). We say that x is in *normal form* if it is so expressed. Existence and uniqueness of normal forms come from:

---

[2] Notation: a^b means $b^{-1}ab$ and $[a, b]$ means $aba^{-1}b^{-1}$.

Proposition 9.2.2. *Every $x \in F$ can be expressed as a product of the x_i's and their inverses to satisfy (1) and (2) in exactly one way.*

Proof. By 9.2.1 it is enough to show that any $\bar{x} \in \bar{F}$ has a unique normal form as above. Let $\bar{x} = \bar{x}_{i_1} \ldots \bar{x}_{i_k} \bar{x}_{j_m}^{-1} \ldots \bar{x}_{j_1}^{-1}$ be the image under ρ of a normal form in F (this is what we mean by a normal form in $\bar{F}$). Assume all subscripts $\geq i$. Then the right derivative of $\bar{x}$ at the point $t = i$ is 2^n where n is the $\bar{x}_i$ exponent sum in the normal form. This will hold for any other normal form of $\bar{x}$.

Suppose there were two different normal forms for $\bar{x}$, and among all such choose $\bar{x}$ and the two different normal forms for $\bar{x}$ so that the sum L of their lengths is as small as possible. Certainly $L > 0$. Let i be the smallest subscript occurring in either normal form. They cannot both start with $\bar{x}_i$ (by minimality) or end with $\bar{x}_i^{-1}$. Since they have the same $\bar{x}_i$ exponent sum, one must have the form $\bar{x}_i \bar{y} \bar{x}_i^{-1}$ while the other, $\bar{z}$, only has subscripts $> i$. So $\bar{y} = \bar{z}^{\bar{x}_i} = \bar{\phi}(\bar{z})$ as homeomorphisms. Hence the formal expressions $\bar{y}$ and $\bar{\phi}(\bar{z})$ are equal by our minimality supposition. But then $\bar{y}$ only involves subscripts $\geq i + 2$, and this contradicts the fact that $\bar{x}_i \bar{y} \bar{x}_i^{-1}$ is a normal form.   □

Corollary 9.2.3. *The epimorphism $\rho : F \to \bar{F}$ is an isomorphism.*   □

Since $\bar{F}$ is obviously torsion free and $\bar{\phi}$ is obviously injective (it comes from a conjugation) we have:

Corollary 9.2.4. *F is torsion free and ϕ is injective.*   □

It follows that $F_1 = \phi(F)$ is a copy of F with presentation

$$\langle x_1, x_2, \ldots \mid x_n^{x_i} = x_{n+1} \ \forall \ 1 \leq i < n \rangle.$$

Hence $\phi(F_1) \leq F_1$ and we have:

Proposition 9.2.5. *F is the ascending HNN extension of F_1 by $\phi \mid F_1 : F_1 \rightarrowtail F_1$ with stable letter x_0.*   □

Repeating with respect to $F_2 = \phi(F_1)$, etc., we see that F is an infinitely iterated HNN extension where the intersection of all the base groups $F_1, F_2, \ldots$ is trivial.

Proposition 9.2.6. *F contains a free abelian subgroup of infinite rank. Hence F has infinite geometric dimension.*

Proof. We work in $\bar{F}$. The homeomorphisms (of $\mathbb{R}$) $\bar{x}_{2i} \bar{x}_{2i+1}^{-1}$ where $i = 0, 1, 2, \ldots$ have disjoint supports and hence generate an abelian subgroup. It is easy to see that they freely generate. The last sentence follows from 7.2.11.   □

In Sect. 9.3 we will show that the group F has type F_∞.

Theorem 9.2.7. *Every quotient of F by a non-trivial normal subgroup is abelian, hence is a quotient of $\mathbb{Z}^2$.*

Proof. Let $N \lhd F$. Consider a non-trivial element of N with normal form $x_{i_1} \ldots x_{i_k} x_{j_m}^{-1} \ldots x_{j_1}^{-1}$. By conjugating and inverting as needed we may assume either $i_1 < j_1$ or $m = 0$ (i.e., no "negative part"). Rewrite this as $x_{i_1}^r x_{i_{r+1}} \ldots x_{i_k} x_{j_m}^{-1} \ldots x_{j_1}^{-1}$ where either $k < r + 1$ (meaning the whole "positive part" is $x_{i_1}^r$) or $k \geq r + 1$ and $i_{r+1} > i_1$. Of course, $r \geq 1$. We define the length[3] of this element to be $k + m$. The length is not changed by inversion and is not increased by the conjugations referred to above. Thus we may make the additional assumption that the non-trivial element of N under consideration has minimal length among all non-trivial members of N. We call it $x = x_{i_1}^r x_{i_{r+1}} \ldots x_{i_k} x_{j_m}^{-1} \ldots x_{j_1}^{-1}$.

Claim. When $x_n^p = v$ mod N, and all the subscripts in the normal form of v are larger than n, then for all $s \geq 0$ we have $\phi^s(v) = x_n^p$ mod N.

Proof. (of Claim). $x_n^p = x_n^{-s} v x_n^s$ mod N, and $x_n^{-s} v x_n^s = \phi^s(v)$.

We will use the Claim twice. First, writing $w = x_{j_1} \ldots x_{j_m} x_{i_k}^{-1} \ldots x_{i_{r+1}}^{-1}$ we have $x_{i_1}^r = w$ mod N. By the Claim, $\phi^s(w) = x_{i_1}^r$ mod N for all $s \geq 0$. Given $n \geq 0$ there exists $s \geq 0$ such that $x_n^{-1} \phi^s(w) x_n = \phi^{s+1}(w)$. Thus $x_n^{-1} x_{i_1}^r x_n = x_{i_1}^r$ mod N, so $x_{i_1}^r$ commutes with F mod N.

If $n > i_1$ then $x_{n+r} = x_{i_1}^{-r} x_n x_{i_1}^r = x_n$ mod N. By the Claim $x_n = x_{n+r+s}$ mod N for all $s \geq 0$. In particular $x_m = x_{m+1}$ mod N when m is large. Finally, let $0 \leq p < q$. Then for sufficiently large n we have $x_p^{-n} x_q x_p^n = x_{q+n} = x_{q+n+1}$ mod N, and $x_{q+n+1} = x_p^{-n}(x_p^{-1} x_q x_p) x_p^n$. So $x_q = x_p^{-1} x_q x_p$ mod N; i.e., F/N is abelian. $\square$

Corollary 9.2.8. *If G is a group and $\rho : F \to G$ is a homomorphism then either ρ is a monomorphism or $\rho(x_n) = \rho(x_1)$ for all $n \geq 1$.*

Proof. The abelianization $F \to F/[F : F]$ has this property, so the statement follows from 9.2.7. $\square$

The group F and the shift homomorphism ϕ play an important role in understanding homotopy idempotents. Recall that when $g : Y \to X$ is a domination and $g \circ f \simeq \mathrm{id}_X$ then $f \circ g : Y \to Y$ is a homotopy idempotent; i.e., writing $h = f \circ g$, $h \simeq h^2$. A homotopy idempotent $h : Y \to Y$ *splits* (or *is splittable*) if there exist a space X and maps f and g as above so that $h \simeq f \circ g$ and $g \circ f \simeq \mathrm{id}_X$. Otherwise, we say that h *does not split* (or *is unsplittable*).

Assume Y is a path connected CW complex with base vertex y and that $h : Y \to Y$ is a homotopy idempotent. By 3.3.1 we may alter h by a homotopy to make it a pointed map $h : (Y, y) \to (Y, y)$. If $H : h \simeq h^2$ is a homotopy and

---

[3] More generally the *length* of a normal form in F is the number of "letters" $x_i^{\pm}$ occurring in it, well-defined because of the uniqueness of normal forms.

$\omega : I \to Y$ is the loop $\omega(t) = H(\omega, t)$, then on fundamental group we have $h_{\#}^2([\sigma]) = h_{\#}([\sigma])^{[\omega]}$ for all loops σ in Y at y. So $h_{\#}$ is an idempotent up to conjugacy and, by 9.2.1, we have a canonical commutative diagram

$$
\begin{array}{ccc}
F & \xrightarrow{\ \rho\ } & \pi_1(Y, y) \\
\phi \downarrow & & \downarrow h_{\#} \\
F & \xrightarrow[\ \rho\ ]{} & \pi_1(Y, y)
\end{array}
$$

Theorem 9.2.9. *Let $h : (Y, y) \to (Y, y)$ be such that[4] $h \underset{\omega}{\simeq} h^2$ where ω is a loop at y. The following are equivalent (where $h_{\#} : \pi_1(Y, y) \to \pi_1(Y, y)$ is the induced homomorphism):*

(i) *h splits;*
(ii) *h is homotopic to $h' : (Y, y) \to (Y, y)$ such that[5] $h' \underset{y}{\simeq} (h')^2$;*

(iii) $\text{image}(h_{\#}) = \text{image}(h_{\#}^2)$;
(iv) $h_{\#}([\omega]) = h_{\#}^2([\omega])$.

Proof. (i) $\Rightarrow$ (ii): Since h splits there is a homotopy commutative diagram of base point preserving maps

$$
\begin{array}{ccc}
Y & \xleftarrow{\ h\ } & Y \\
g \downarrow & \ \searrow^{f}\ & \downarrow g \\
X & \xleftarrow[\ \text{id}\ ]{} & X
\end{array}
$$

Thus $k := g \circ f : (X, x) \to (X, x)$ is a pointed homotopy equivalence (by Remark 4.1.6) and $k \simeq \text{id}_X$. Let k' be a pointed homotopy inverse for k. Then $g \circ (f \circ k') \underset{x}{\simeq} \text{id}_X$. Define $h' := f \circ k' \circ g$. Then the following diagram commutes in Pointed Homotopy

$$
\begin{array}{ccc}
(Y, y) & \xleftarrow{\ h'\ } & (Y, y) \\
g \downarrow & \ \searrow^{f \circ k'}\ & \downarrow g \\
(X, x) & \xleftarrow[\ \text{id}_X\ ]{} & (X, x)
\end{array}
$$

---

[4] In general, given pointed maps $f_0, f_1 : (Z', z') \to (Z'', z'')$ and a loop ω at z'' we say $f_0 \underset{\omega}{\simeq} f_1$ iff there is a homotopy $K : f_0 \simeq f_1$ such that $K(z', t) = \omega(t)$. By the Homotopy Extension Property this only depends on $[\omega] \in \pi_1(Z'', z'')$ (when we are dealing with pointed CW complexes). If ω is the constant loop at z'' we write $f_0 \underset{z''}{\simeq} f_1$.

[5] Such a map h' is called a *pointed homotopy idempotent*.

so h' is a pointed homotopy idempotent. Since $k' \simeq \mathrm{id}$, $h' \simeq f \circ g \simeq h$.

(ii) $\Rightarrow$ (iii): Let $h' \underset{\alpha}{\simeq} h$ where h' is a pointed homotopy idempotent and α is a loop at y. Then $h'_{\#} = (h'_{\#})^2 : \pi_1(Y, y) \to \pi_1(Y, y)$. Hence (exercise) $h^2_{\#}(\cdot) = [h(\alpha)]^{-1} h_{\#}(\cdot)[h(\alpha)]$, from which it follows that $h_{\#}$ and $h^2_{\#}$ have the same image.

(iii) $\Rightarrow$ (iv): For some $[\sigma] \in \pi_1(Y, y)$ $h_{\#}([\omega]) = h^2_{\#}([\sigma])$ so $h^2_{\#}([\omega]) = h^3_{\#}([\sigma]) = h_{\#}(h^2_{\#}([\sigma])) = h_{\#}([\omega]^{-1} h_{\#}([\sigma])[\omega]) = h([\omega])$.

(iv) $\Rightarrow$ (i): We sketch this proof leaving the details as an exercise. Define $X := \mathrm{Tel}(h)$, define $g : Y \to X$ by $g(y) = i(y) \in M(h)_0 \subset \mathrm{Tel}(h)$, and define $f : X \to Y$ to be the map[6] induced by a homotopy $H : h \underset{\omega}{\simeq} h^2$ which agrees with h on every "integer" copy of Y in X. Then $f \circ g = h$. It is to be shown that if $h_{\#}([\omega]) = h^2_{\#}([\omega])$ then $g \circ f \simeq \mathrm{id}_X$.

Step 1. The hypothesis $h_{\#}([\omega]) = h^2_{\#}([\omega])$ implies $(g \circ f)_{\#}$ is an epimorphism on π_1 where the base point of X is $g(y)$.

Step 2. For any finite CW complex K we have $g_{\#} = \mathrm{id} : [K, X] \to [K, X]$ where $[K, X]$ denotes the set of homotopy classes of maps $K \to X$; this is because X is the mapping telescope of $h \simeq h^2$ (see Exercise 1).

Step 3. The previous steps imply that for all n $(g \circ f)_{\#} : \pi_n(X, x) \to \pi_n(X, x)$ is an isomorphism, hence by the Whitehead Theorem (Exercise 1, Sect. 4.4) $g \circ f$ is a homotopy equivalence. It follows that X is finitely dominated.

Step 4. Step 2 holds for finitely dominated CW complexes, hence $g \circ f$ is homotopic to id_X. $\qquad\square$

Remark 9.2.10. Note that ρ depends on $[\omega]$ coming from a homotopy $H : h \simeq h^2$, whereas the issue of whether h splits is not related to a particular H and ω.

Theorem 9.2.11. (Freyd-Heller Theorem) *h splits iff ρ is not a monomorphism. In particular, if Z is a $K(F, 1)$-complex and $h : Z \to Z$ is induced by the shift $\phi : F \to F$ then h is a homotopy idempotent which does not split.*

Proof. This follows from 9.2.8 and 9.2.9. $\qquad\square$

Source Notes: Thompson's Group F was first studied by Richard Thompson; his work appeared in [112]. The group was later rediscovered, independently, by Freyd and Heller [65] and by Dydak and Minc [57]. (The paper [65] circulated in preliminary preprint form for many years before its publication.) The material on homotopy idempotents in this section is based on [65]. For more on this group see the source notes for Sect. 9.3. [37] is a useful expository reference.

---

[6] More precisely, if $\pi : X \to \mathbb{R}$ is the obvious map and $y \in Y$, write $[y, t]$ for the point of X whose Y-coordinate is y and which projects under π to t. Then the formula for f is $f([y, t]) = H_{t-[t]}(y)$.

Exercises

1. Let $X = \text{Tel}(h)$. With notation as in the proof of 9.2.9, prove the following *Lemma*: Let $j : X \to X$ be a map such that $j \mid Y$ is homotopic to inclusion $Y \hookrightarrow X$. Then, for every finite CW complex K, $j_\# : [K, X] \to [K, X]$ is the identity function.

2. Show that the following are presentations[7] of F (where x_n is defined to be $x_{n-1}^{x_{n-2}}$):

    (i) $\langle x_0, x_1 \mid x_2^{x_0} = x_3, x_3^{x_1} = x_4 \rangle$ and

    (ii) $\langle x_0, x_1 \mid x_2^{x_0} = x_3, x_3^{x_0} = x_3^{x_1} \rangle$.

9.3 Finiteness properties of Thompson's Group F

In this section we prove that Thompson's Group F has type F_∞ (9.3.19).

A. Binary trees and subdivisions of I:

By a *complete rooted binary tree* we mean a pointed oriented tree (Y, y_0) where y_0 meets two edges and all other vertices meet three edges. The base point y_0 is the *root*. Each edge is oriented away from the root; in other words, reduced edge paths starting at y_0 are in agreement with the chosen orientations on edges. We impose an additional labeling on the edges of Y: each vertex is the initial point of exactly two (oriented) edges; one of those edges is labeled 0 and the other is labeled 1. One should think of Y as embedded in $\mathbb{R}^2$ with all oriented edges pointing downward, the 0-edge to the left and the 1-edge to the right; see Fig. 9.1.

By a *finite tree* we mean a finite rooted subtree (T, y_0) of (Y, y_0) with inherited orientation and labeling such that y_0 has degree 2 or 0 (the latter only in the *trivial tree* $\{y_0\}$), while all other vertices have degree 3 or 1 (the latter vertices being called *leaves* of T).

By a *binary subdivision* of the closed unit interval I we mean a partition (in the sense of calculus) of I into subintervals whose end points have the form $\frac{m}{2^n}$ where m and n are (non-negative) integers. The set of subintervals in such a subdivision is to be ordered by the $\mathbb{R}$-ordering of their left endpoints.

Each finite tree T determines a binary subdivision of I and a bijection between the leaves of T and the subintervals in the subdivision as follows: if T is the trivial tree, the corresponding subdivision has just the one interval I, with bijection $\{y_0\} \to \{I\}$. Assume the subdivision and bijection defined for all trees having k leaves. Let T', with $(k + 1)$ leaves, be obtained from T by including in T' the two edges of Y which touch the ith leaf of T (as determined by the T-bijection) but which are not in T; then the subdivision corresponding to T' is obtained from the one corresponding to T by bisecting

---

[7] Because of the word length of relations when written in terms of x_0 and x_1, these are known as "the (10,18) presentation" and "the (10,14) presentation" of F respectively.

the ith subinterval of the latter; this enlargement replaces the ith leaf of T by two leaves ordered by "$0, 1$," and the ith subinterval of the T-subdivision by two subintervals of half the size ordered by "left,right." This together with the T-bijection defines the T'-bijection.

If T is a finite tree, there is a unique reduced edge path from y_0 to each leaf, and, using the 0- and 1-labels on edges, this gives an unambiguous description of a leaf by a finite string of 0's and 1's. Using the lexicographic ordering on these strings and the ordering inherited from $\mathbb{R}$ on the subintervals of the T-subdivision, the T-bijection is clearly order-preserving.

We will call a binary subdivision which arises from a finite tree, as above, a *tree subdivision* of I. Not every binary subdivision is a tree subdivision.

B. The group F as dyadic PL homeomorphisms of I:

Let $PL_2(I)$ denote the group of all piecewise linear increasing homeomorphisms of I whose points of non-differentiability are dyadic rational numbers (i.e., of the form $\frac{m}{2^n}$) and all of whose slopes are integer powers of 2. An ordered pair of finite binary trees (S, T) is *balanced* if S and T have the same number of leaves. Such an ordered pair defines an element h of $PL_2(I)$: the domain and codomain copies of I are subdivided by S and T as in Part A, and h is defined to take the ith subinterval of the domain by an affine increasing homeomorphism to the ith subinterval of the codomain. Of course another balanced pair (S', T') can define the same $h \in PL_2(I)$; for example, if (S, T) defines h, and $S \subset S'$, then there is a finite binary tree $T' \supset T$ such that (S', T') also defines h.

Proposition 9.3.1. *If $h \in PL_2(I)$ then h is defined by some balanced pair of finite binary trees.*

Proof. Let K be a dyadic subdivision of I such that h is affine on each interval in K. Let n be such that the subdivision K' with vertices $\{\frac{m}{2^n} \mid 0 \leq m \leq 2^n\}$ subdivides K. Choose k so that the subdivision K'' having vertices $\{\frac{j}{2^k} \mid 0 \leq j \leq 2^k\}$ subdivides $h(K')$. Then K' and K'' are tree subdivisions, as is $h^{-1}(K'')$, and $h : h^{-1}(K'') \to K''$ is a simplicial isomorphism. $\qquad\square$

We now define an isomorphism between F and $PL_2(I)$. As in Sect. 9.2 we consider homeomorphisms of I as acting on the right in defining the group structure. Define $\tilde{x}_0$ and $\tilde{x}_1$ to be the elements of $PL_2(I)$ defined by the balanced pairs indicated in Fig. 9.1.

One easily checks that $x_0 \mapsto \tilde{x}_0$ and $x_1 \mapsto \tilde{x}_1$ extends to a homomorphism $\tilde{\rho} : F \to PL_2(I)$.

Theorem 9.3.2. $\tilde{\rho} : F \to PL_2(I)$ *is an isomorphism.*

Proof. The proof that $\tilde{\rho}$ is a monomorphism is similar to the proof of the corresponding statement in 9.2.3. That $\tilde{\rho}$ is an epimorphism follows from 9.3.1. $\square$

$$\tilde{x}_0 \qquad\qquad\qquad \tilde{x}_1$$

Fig. 9.1.

In view of 9.3.2 we will identify F with $PL_2(I)$ via $\tilde{\rho}$ from now on.

C. An F-poset:

Let $\mathcal{T}$ denote the set of all finite trees; recall that this means: finite rooted subtrees of the complete rooted binary tree (Y, y_0) as described above; in particular, $\mathcal{T}$ is closed under finite union and finite intersection. If T has n leaves and if $1 \le i \le n$ the ith *simple expansion* of $T \in \mathcal{T}$ is the smallest finite tree $e_i(T)$ containing T such that the ith leaf of T is not a leaf of $e_i(T)$; i.e., $e_i(T)$ is the union of T and a "caret" at the ith leaf of T. If $i > n$ it is convenient to define $e_i(T) = T$. Thus $e_i : \mathcal{T} \to \mathcal{T}$. An *expansion* is a finite composition of simple expansions. The *length* of an expansion e is $\le k$ if e is the composition of at most k simple expansions. Note that the union of two finite trees is an expansion of both.

Lemma 9.3.3. *If $i < j$, $e_i e_j = e_{j+1} e_i$.* □

Lemma 9.3.4. *Given expansions e and e' there exist expansions $\bar{e}$ and $\bar{e}'$ such that $\bar{e} \circ e = \bar{e}' \circ e'$.*

Proof. When e and e' have length ≤ 1 this follows from 9.3.3. The general case is done by induction on the sum of the lengths of e and e'. □

By 9.3.1, any $g \in F$ is defined by some balanced pair (U, V): g maps the intervals of the U-subdivision of I affinely onto those of the V-subdivision of I. If $e(U)$ is an expansion of U it follows that g is also defined by the balanced pair $(e(U), e(V))$. Thus if $(g, S) \in F \times \mathcal{T}$ there exist e and T such that the balanced pair $(e(S), T)$ defines g. This leads us to define an equivalence relation on $F \times \mathcal{T}$ by $(g_1, S_1) \sim (g_2, S_2)$ if there exist an expansion e and $T \in \mathcal{T}$ such that $(e(S_i), T)$ is a balanced pair defining g_i for $i = 1, 2$ (same e and T for S_1 and S_2). Using 9.3.4 one easily proves:

Lemma 9.3.5. *This is indeed an equivalence relation on $F \times \mathcal{T}$.* □

We write $B := F \times \mathcal{T}/\sim$.

Lemma 9.3.6. *The left action of F on $F \times \mathcal{T}$ $g(h, T) = (gh, T)$ induces a left action of F on B.* □

We write $[h, T]$ for the member of the F-set B defined by (h, T). The F-action is $g[h, T] = [gh, T]$. That the simple expansion functions $e_i : \mathcal{T} \to \mathcal{T}$ define functions $E_i : B \to B$ where $E_i([h, T]) = [h, e_i(T)]$ follows from:

Lemma 9.3.7. *If* $(g_1, S_1) \sim (g_2, S_2)$ *then for all* i $(g_1, e_i(S_1)) \sim (g_2, e_i(S_2))$.

Proof. There exist e and T such that $(e(S_i), T)$ is a balanced pair representing g_i for $i = 1, 2$. We write $(g_1, S_1) \underset{k}{\sim} (g_2, S_2)$ if the length of this e is $\leq k$. The lemma is proved by induction on k using 9.3.3. $\square$

Lemma 9.3.8. *Each* E_i *is an* F-*function; i.e.,* $E_i(g[h, T]) = gE_i([h, T])$. $\square$

Define $f : B \to \mathbb{N}$ by $f([h, T]) =$ the number of leaves of T. This is well defined.

Lemma 9.3.9. *Let* $n \geq i$. *The function* E_i *maps* $f^{-1}(n)$ *bijectively onto* $f^{-1}(n + 1)$.

Proof. That E_i is injective is clear. The proof that E_i is surjective should be clear from Example 9.3.10, below. $\square$

Let $B_n = f^{-1}([1, n])$. Then 9.3.9 says that when $n \geq i$, E_i maps $B - B_{n-1}$ bijectively onto $B - B_n$. We call E_i a *simple expansion operator* on $B - B_{n-1}$ and we call its inverse $C_i : B - B_n \to B - B_{n-1}$ a *simple contraction operator*.

Example 9.3.10. Let T have more than i leaves. If the ith and $(i + 1)^{\text{th}}$ leaves form a caret (i.e., $T = e_i(T')$ for some $T' \in \mathcal{T}$) then $C_i([h, T]) = [h, T']$. If these leaves do not form a caret, let S be any tree with the same number of leaves as T such that $S = e_i(S')$ for some $S' \in \mathcal{T}$. Let $g \in F$ be represented by (S, T). Then $[g, S] = [1, T]$, so $[hg, S] = [h, T]$ and $C_i([h, T]) = [hg, S']$.

We make B into an F-poset by defining $[g, S] \leq [h, T]$ if for some $i_1, i_2, \ldots, i_r$, with $r \geq 0$, $E_{i_r} \circ \cdots \circ E_{i_1}([g, S]) = [h, T]$. Each B_n is an F-sub-poset of B.

Lemma 9.3.11. *The* F-*action on the set* B *is a free action.* $\square$

D. Finiteness Properties of F**:**
As usual we reuse the symbol B for the associated ordered abstract simplicial complex defined by the poset B.

Proposition 9.3.12. *The induced* F-*action on* $|B|$ *is a free action.*

Proof. The stabilizers of vertices are trivial by 9.3.11. Since the action of F on $[h, S]$ preserves the number of leaves of S, the stabilizer of each simplex of S is also trivial. $\square$

Proposition 9.3.13. *The poset* B *is a directed set.*

Proof. Let $\{b_1, \ldots, b_k\} \subset B$. Write $b_i = [h_i, S_i]$. Then S_i expands to S_i' such that $(h_i, S_i') \sim (1_F, T_i')$ for some T_i'. Write $b_i' = [h_i, S_i'] = [1_F, T_i']$. Then $b_i \leq b_i'$ for all i. Let $T = \bigcup_{i=1}^{k} T_i'$ and write $b = [1_F, T]$. Then $b_i' \leq b$ for all i.   □

Proposition 9.3.14. *If a poset P is a directed set then $|P|$ is contractible.*

Proof. Whenever K is a finite subcomplex of P there exists $v \in P$ such that the cone $v * |K|$ is a subcomplex of $|P|$. Thus the homotopy groups of $|P|$ are trivial, so, by the Whitehead Theorem, $|P|$ is contractible.   □

Corollary 9.3.15. $|B|$ *is contractible.*   □

The function $f : B \to \mathbb{N}$ extends affinely to a Morse function[8] (also denoted by) $f : |B| \to \mathbb{R}$. Then $|B_n| = f^{-1}((-\infty, n])$.

Proposition 9.3.16. *The CW complex $F \backslash |B_n|$ is finite.*

Proof. Let $[h, S] \in B_n$. Its F-orbit contains $[1, S]$ and there are only finitely many finite trees having at most n leaves. This shows that the 0-skeleton of $F \backslash |B_n|$ is finite. The rest is clear.   □

Proposition 9.3.17. *For each integer k there is an integer $m(k)$ such that if b is a vertex of B and $f(b) \geq m(k)$, then the downward link $\mathrm{lk}^{\downarrow}_{|B|} b$ is k-connected.*

We postpone the proof of 9.3.17 until the next subsection.

Proposition 9.3.18. *For $n \geq m(k)$, $|B_n|$ is k-connected. Hence $\{|B_n|\}$ is essentially k-connected for all k.*

Proof. By 9.3.17 and 8.3.4 we conclude that $(|B_n|, |B_{n-1}|)$ is $(k+1)$-connected if $n \geq m(k)$. When combined with the Whitehead Theorem and 9.3.15 this proves what is claimed.   □

By 7.4.1 and 7.2.2 we conclude:

Theorem 9.3.19. (Brown-Geoghegan Theorem) *Thompson's Group F has type F_∞.*   □

E. Analysis of the downward links:
    It remains to prove 9.3.17. We begin with two topics of general interest (9.3.20 and 9.3.21).
    If $\mathcal{U} = \{X_\alpha\}$ is a cover of the CW complex X by subcomplexes, the *nerve* of $\mathcal{U}$ is the abstract simplicial complex $N(\mathcal{U})$ having a vertex v_α for each X_α, and a simplex $\{v_{\alpha_0}, \ldots, v_{\alpha_k}\}$ whenever $\bigcap_{i=0}^{k} X_{\alpha_i} \neq \emptyset$. The following property of nerves is widely used in topology.

---
[8] See Section 8.3.

Proposition 9.3.20. *If the cover $\mathcal{U}$ is finite and if $\bigcap\limits_{i=0}^{k} X_{\alpha_i}$ is contractible whenever it is non-empty, then $|N(\mathcal{U})|$ and X are homotopy equivalent.*

Proof. There is a vertex $v_{\alpha_0,\dots,\alpha_k}$ of the first derived $sd|N(\mathcal{U})|$ for each simplex $\{v_{\alpha_0},\dots,v_{\alpha_k}\}$ of $N(\mathcal{U})$. Pick a point $x_{\alpha_0,\dots,\alpha_k} \in \bigcap\limits_{i=0}^{k} X_{\alpha_i}$. Define a map $\alpha : sd\,|N(\mathcal{U})| \to X$ taking each vertex $v_{\alpha_0,\dots,\alpha_k}$ to the point $x_{\alpha_0,\dots,\alpha_k}$; the contractibility hypothesis makes it possible to extend α so that the simplex whose first[9] vertex is $v_{\alpha_0,\dots,\alpha_k}$ is mapped into $\bigcap\limits_{i=0}^{k} X_{\alpha_i}$. A straightforward generalization of the proof of 4.1.5 shows that α is a homotopy equivalence. $\square$

The particular nerve to which this will be applied is the abstract simplicial complex[10] L_n whose simplexes are the sets of pairwise disjoint adjacent pairs in the ordered set $(1, 2, \dots, n)$.

Proposition 9.3.21. *For any integer $k \geq 0$ there is an integer $m(k)$ such that $|L_n|$ is k-connected when $n \geq m(k)$.*

Proof. By induction on k we prove a sharper statement:
Claim: given $k \geq 0$ there are integers $m(k)$ and $q(k)$ such that when $n \geq m(k)$ the k-skeleton $|L_n|^k$ is homotopically trivial by means of a homotopy $H^{(k)}$ in which, for every simplex σ of $|L_n|^k$, $H^{(k)}(|\sigma| \times I)$ is supported by a subcomplex having $\leq q(k)$ vertices. For $k = 0$ we can take $m(0) = 5$ and $q(0) = 3$. Assume the Claim holds when k is replaced by $k - 1 \geq 0$. For any k-simplex σ, $H^{(k-1)}(|\overset{\bullet}{\sigma}| \times I)$ is supported by a subcomplex J having at most $r = (k+1)(q(k-1))$ vertices. If $n \geq 2r + 2$ there is a vertex v of L_n such that J lies in the link of v. Extend $H^{(k-1)}$ to $H^{(k)}$ in two steps: first, $H^{(k)}$ is to be the identity map on $|\sigma| \times \{0\}$ and constant on $|\sigma| \times \{1\}$; second, (since no new vertices were involved in this first extension) we can extend $H^{(k)}$ to $|\sigma| \times I$ by coning at v. Letting $m(k) = \max\{m(k-1), 2r+2\}$ and $q(k) = r+1$, the induction is complete. $\square$

We now consider $\mathrm{lk}^{\downarrow}_{|B|} b$ where k is given and $f(b) = n$.

Proposition 9.3.22. *There is a finite cover $\mathcal{U}$ of $\mathrm{lk}^{\downarrow}_{|B|} b$ such that all non-empty intersections of members of $\mathcal{U}$ are contractible, and the nerve of $\mathcal{U}$ is isomorphic to L_n.*

For this we need a lemma:

---

[9] We are using the ordering Convention preceding 5.3.7.

[10] If Γ_n is the abstract graph whose vertices are $\{1, 2, \dots, n\}$ and whose 1-simplexes are the adjacent pairs $\{j, j+1\}$ then L_n is often called the *matching complex* of Γ_n.

Lemma 9.3.23. *A set $\{C_{i_1}(b), \ldots, C_{i_r}(b)\}$ of simple contractions of b, written so that $i_1 < \ldots < i_r$, has a lower bound with respect to $\leq$ iff the pairs $\{i_j, i_{j+1}\}$ are pairwise disjoint. If there is a lower bound there is a greatest lower bound.*

Proof. If there is a lower bound $[h, T]$ then by equivariance we need only consider the case $h = 1$. Then b is an expansion of $[1, T]$ and so has the form $[1, S]$. Since $[1, T] \leq C_{i_j}(b) \leq [1, S]$ for every j, each $C_{i_j}(b)$ must have the form $[1, S_{i_j}]$ where $e_{i_j}(S_{i_j}) = S$. The required conclusion follows. Moreover,
$$\left[1, \bigcap_j S_{i_j}\right]$$
is the greatest lower bound. The converse is clear.   □

Proof (of 9.3.22). Consider all the simple contractions $b_1, \ldots, b_{n-1}$ of b. The downward link $\mathrm{lk}^{\downarrow}_{|B|} b = |B_{<b}|$ and is covered by the subcomplexes $|B_{\leq b_i}|$. By 9.3.23 the nerve of this cover is L_n. Moreover, if $\bigcap_j |B_{\leq b_{i_j}}| \neq \emptyset$ then this intersection is $|B_{\leq \bar{b}}|$ where $\bar{b}$ is the greatest lower bound of the b_{i_j}'s. This is contractible by 9.3.14.   □

Proof (of 9.3.17). Combine 9.3.20, 9.3.21 and 9.3.22.   □

Other interesting properties of the group F can be found in Sect. 13.11 and in 16.9.7.

Source Notes: The poset B has a richer interpretation as the poset of bases of the free Cantor algebra (or Jónnson-Tarski algebra) on one generator. For this, see [30] and [81]. The proof of Theorem 9.3.19 given here is adapted from [30]. For the first (quite different) proof see [32]. Other interesting proofs are in [150] and [62].

Exercises

1. Give an example of a binary subdivision of I which is not a tree subdivision.
2. In the light of 9.3.1, characterize the group $\bar{F}$ in Sect. 9.2 intrinsically as a group of homeomorphisms of $\mathbb{R}$.
3. Prove the lemmas in Sect. C.

9.4 Thompson's simple group T

All members of $PL_2(I)$ $(= F)$ fix $\{0, 1\} \subseteq I$. Thus they induce homeomorphisms of the circle S^1 (via the quotient map $q : I \to S^1$, $t \mapsto e^{2\pi i t}$) which fix the point $1 \in S^1$. In this section we will identify the group of all such homeomorphisms of the circle with the group F.

A *dyadic rotation* is a homeomorphism $S^1 \to S^1$ of the form $\rho_x : e^{it} \mapsto e^{i(t+x)}$ where $x \in I$ is a dyadic rational number. We write T or $PL_2(S^1)$ for the group of homeomorphisms of S^1 generated by F (identified as above) and the dyadic rotations. Thus F is the subgroup of T which fixes $1 \in S^1$. This group T is another Thompson group of great interest. It was the first known infinite finitely presented simple group.

A *dyadic point* of S^1 is a point of the form $e^{2\pi i x}$ where x is a dyadic rational number. A *dyadic subdivision* of S^1 consists of a finite set V of dyadic points of S^1 with $|V| \geq 2$. This set V defines a set of $|V|$ intervals[11] in S^1: their end-points lie in V and their interiors are disjoint from V. Clearly we have:

Proposition 9.4.1. *If $f \in T$, there are dyadic subdivisions V_1 and V_2 of S^1 such that f maps V_1 bijectively onto V_2 preserving cyclic order in S^1, and f has constant derivative on each (open) interval defined by V_1 as above.* $\square$.

In words: the members of T are the "orientation-preserving, dyadic, piecewise linear" homeomorphisms of S^1.

The group T is clearly countable. But we can prove much more:

Theorem 9.4.2. *The group T is of type F_∞.*

Proof. Since we have a convention $\mathbb{R}^n \subseteq \mathbb{R}^{n+1}$, we can write $\Delta^n \subseteq \Delta^{n+1}$ and define $\Delta^\infty := \varinjlim_n \Delta^n$ to be the "infinite simplex." This is the geometric realization of the abstract simplicial complex having the obvious orthogonal basis of unit vectors in $\mathbb{R}^\infty := \varinjlim_n \mathbb{R}^n$ as vertices, and all finite subsets of vertices as simplexes. We identify the (countable) set of vertices of Δ^∞ with the dyadic points of S^1 (by some bijection). Since T permutes the dyadic points, this defines a rigid action of T on Δ^∞ by simplicial isomorphisms. The stabilizer of each n-simplex is clearly isomorphic to F^{n+1}, the $(n+1)$-fold product of copies of F. And Δ^∞ has one cell mod T in each dimension. Thus the theorem follows from 7.3.1 and 9.3.19. $\square$

This proof used our previous knowledge that F has type F_∞. There is also a direct proof of 9.4.2 which runs parallel to the proof of 9.3.19 given here; see [30].

Theorem 9.4.3. *The group T is simple.*

Proof. Let $g \neq 1 \in T$. We show that T itself is the only normal subgroup of T containing g. Since $g \neq 1$ there is some $a \in S^1$ such that $g(a) \neq a$. Let J be an open interval in S^1 containing a such that $g(J) \cap J = \emptyset$. Choose $h \in T$ so that h is supported on J and $h \neq 1$. Then ghg^{-1} is supported on gJ. The interval J can be chosen so small that there is an interval K in S^1 disjoint from $J \cup g(J)$. Then $k := h^{-1}(ghg^{-1}) = (h^{-1}gh)g^{-1}$ fixes K pointwise and is a non-trivial member of $NC(g)$, the normal closure of g in T. If $b \in K$ then k

---

[11] $|V|$ denotes the number of members of V.

fixes a neighborhood of b and hence k is conjugate to an element of F'. (Recall that $F' = \{f \in F \mid f = \text{id near } 0\}$.) Thus $NC(g) \cap F'$ is a non-trivial normal subgroup of F. Hence, by 9.2.7, $F' \le NC(g)$. Indeed, every conjugate of F' lies in $NC(g)$.

Claim. Any $q \in T$ has the form $q = q_1 q_2$ where q_1 and q_2 lie in conjugates of F'.

Proof (of Claim). Assume $q \ne 1$ and let $c \in S^1$ be such that $q(c) \ne c$. Choose open intervals L and M in S^1 such that $c \in M \subseteq L$, $q(c) \in q(M) \subseteq L$ and $M \cap q(M) = \emptyset$. Choose $p \in T$ supported in L such that $p = q$ on M. Then qp^{-1} fixes M pointwise and $q = (qp^{-1})p$. The interval L can be chosen so that its complement contains an open interval (which p fixes pointwise). Thus the Claim is proved.

The Theorem follows from the Claim since, as we have seen, qp^{-1} and p lie in the normal closure of g. ☐

Remark 9.4.4. The simple group T of type F_∞ has torsion; indeed T contains a copy of every finite cyclic group. But there are simple groups G of type F which have finite 2-dimensional $K(G,1)$-complexes. Such a group can occur as the automorphism group of a product of two trees, and can have the form $F_1 *_A F_2$ where F_1 and F_2 are finitely generated free groups and A has finite index ≥ 2 in both. These were discovered by Burger and Mozes; see [35].

Source Notes: Thompson's original work, written up in [112], emphasized T and a related finitely presented infinite simple group V rather than F. The (larger) group V can be described roughly as the group of all piecewise linear, piecewise continuous dyadic bijections of I. Thus $F \le T \le V$. See [30] for more on T and V. [37] is also a useful reference. The proof of 9.4.2 given here is from [32]; see also [30]. The proof of 9.4.3 given here was suggested by M. Brin.

Exercise

Let $\tilde{T}$ be the group of all PL homeomorphisms of $\mathbb{R}$ satisfying $h(x + 1) = h(x) + 1$, whose points of non-differentiability are dyadic rationals and whose slopes (derivatives) are powers of 2. Show that there is a short exact sequence $\mathbb{Z} \rightarrowtail \tilde{T} \twoheadrightarrow T$. Prove that $\tilde{T}$ is a torsion free group of type F_∞.

9.5 The outer automorphism group of a free group

In the previous sections we have discussed finiteness properties and dimension in detail for finitely generated Coxeter groups and for two Thompson groups, as well as for certain torsion free subgroups of these. In this section we give a brief indication of how to handle the same issues for outer automorphism groups of free groups.

An automorphism $\phi : G \to G$ of a group G is *inner* if it is the conjugation by an element of G; i.e., for some $h \in G$ and all $g \in G$, $\phi(g) = hgh^{-1}$. The group of all automorphisms of G is denoted by $\text{Aut}(G)$ and the subgroup of all inner automorphisms of G by $\text{Inn}(G)$. The latter is a normal subgroup and the quotient group $\text{Aut}(G)/\text{Inn}(G)$ is denoted by $\text{Out}(G)$, the *outer automorphism group* of G. In this section we discuss the topological finiteness properties and geometric dimension of $\text{Out}(F_n)$ where, as usual, F_n denotes a free group of rank n.

Let $\Gamma_n = \bigvee_{i=1}^{n} S_i^1$, the wedge of n circles, with wedge point x. We orient the edges of Γ_n and we identify F_n with $\pi_1(\Gamma_n, x)$, where the edge loops defined by the (oriented) edges represent the free generators. From now on, $n \geq 2$ is fixed.

A *marked graph (of rank n)* is a pair (Γ, h) where Γ is a path connected finite graph each of whose vertices has valence ≥ 3 (i.e., each vertex belongs to at least three edges) and $h : \Gamma_n \to \Gamma$ is a cellular homotopy equivalence. Two such, (Γ, h) and (Γ', h'), are *equivalent* if there is a homeomorphism $k : \Gamma \to \Gamma'$ such that $k \circ h$ is homotopic to h'. We write $[\Gamma, h]$ for the equivalence class of (Γ, h).

The marked graphs of rank n are to form the set of vertices, K_n^0, of a flag complex K_n: we need only describe the 1-simplexes.

A subcomplex Φ of a graph Γ is a *forest* if each path component of Φ is a tree. The quotient graph Γ_Φ is then obtained from Γ by "collapsing" each path component of Φ to a point. The quotient map $q_\Phi : \Gamma \to \Gamma_\Phi$ is called a *forest collapse*. By an obvious extension of 4.1.9, q_Φ is a homotopy equivalence. Two vertices $[\Gamma, h]$ and $[\Gamma', h']$ form a 1-simplex of K_n iff for some non-trivial forest collapse Φ of, say, Γ we have $[\Gamma', h'] = [\Gamma_\Phi, q_\Phi \circ h]$. Note that a non-trivial forest collapse changes a graph's homeomorphism type, so these two vertices are different. We have thus defined K_n^1 and hence the flag complex K_n.

Next, we define a right action of $\text{Out}(F_n)$ on K_n by simplicial automorphisms. Each $g \in \text{Out}(F_n)$ can be represented by a homotopy equivalence $f_g : \Gamma_n \to \Gamma_n$, well-defined up to (unpointed) homotopy equivalence. The action on K_n^0 is given by $[\Gamma, h]g = [\Gamma, h \circ f_g]$. Clearly, this takes 1-simplexes to 1-simplexes and so defines a right action of $\text{Out}(F_n)$ on K_n.

The following proposition is an exercise:

Proposition 9.5.1. a) *The simplicial complex K_n has dimension $2n - 3$; in fact every simplex is a face of a $(2n - 3)$-simplex.*
b) *The stabilizer of every simplex is finite.*
c) *K_n is finite mod $\text{Out}(F_n)$.* □

A more difficult theorem is:

Theorem 9.5.2. *The space $|K_n|$ is contractible.* □

We will not prove 9.5.2. Proofs can be found in [44] and [78]. The proof in [44] runs parallel, in broad outline, to the proof given in Sect.9.1 for the corresponding Coxeter group statement 9.1.3. One builds $|K_n|$ as a sequence of finite subcomplexes $|K_{n,k}|$, where $|K_{n,k+1}|$ is obtained from $|K_{n,k}|$ by gluing a contractible complex to $|K_{n,k}|$ along a contractible subcomplex.

Proposition 9.5.3. *The group* $\mathrm{Out}(F_n)$ *has a torsion free subgroup of finite index.*

Proof. Every homotopy equivalence $f : (\Gamma_n, x) \to (\Gamma_n, x)$ induces an automorphism of F_n on fundamental group and an automorphism of $\mathbb{Z}^n$ on first homology (with $\mathbb{Z}$-coefficients). This correspondence defines a canonical epimorphism $\mathrm{Out}(F_n) \twoheadrightarrow GL_n(\mathbb{Z})$. The Proposition therefore follows from 9.1.9.
□

Proposition 9.5.4. *The group* $\mathrm{Out}(F_n)$ *contains a free abelian subgroup of rank* $2n - 3$.

Proof. Let $y_1, \ldots, y_n$ denote free generators of F_n. The automorphisms in $\mathrm{Aut}(F_n)$ $y_i \mapsto y_1 y_i$ and $y_i \mapsto y_i y_1$ for $2 \leq i \leq n$ determine the required free abelian subgroup of $\mathrm{Out}(F_n)$. The details are an exercise.   □

The previous statements can be summarized as follows:

Theorem 9.5.5. (Culler-Vogtmann Theorem) *The group* $\mathrm{Out}(F_n)$ *is of type* F_∞. *Every torsion free subgroup has geometric dimension* $\leq 2n - 3$. *If* H *is a torsion free subgroup of finite index then there is a finite* $(2n - 3)$-*dimensional* $K(H, 1)$-*complex, and the geometric dimension of* H *is precisely* $2n - 3$.   □

Source Notes: Theorem 9.5.5 is due to Culler and Vogtmann [44].

Exercises

1. Prove 9.5.1. *Hint*: Part a) is a consequence of the following which should be proved:

   Lemma. *Let* Γ *be a finite graph in which every vertex has valence* ≥ 3 *and* $\pi_1(\Gamma, v) \cong F_n$. *Then any maximal tree in* Γ *has at most* $2n - 3$ *edges. Moreover, there exists such a graph for which maximal trees have precisely* $2n - 3$ *edges.*

2. Prove that $\mathrm{Out}(F_2)$ has a free subgroup of finite index.
3. Fill in the details of the proof of 9.5.4.

PART III: LOCALLY FINITE ALGEBRAIC TOPOLOGY FOR GROUP THEORY

Part III is a continuation of Part I. It deals with CW complexes whose skeleta are locally finite, and with cellular homology theory based on infinite (or "locally finite") chains, as distinct from the "finite chain" theory of Chap. 2. We introduce two cellular cohomology theories to match the two homology theories. We can then define homology and cohomology of ends. Just as the homology in Chap. 2 is a homotopy invariant, the new homology is a proper homotopy invariant; the basics of proper homotopy are presented in Chap. 10.

Locally Finite CW Complexes and Proper Homotopy

10.1 Proper maps and proper homotopy theory

In this section we set up the foundations of proper homotopy theory. Our exposition runs parallel to Chap. 1: first the general topology of spaces and proper maps, then proper homotopy theory of locally finite CW complexes.

A map $f : X \to Y$ between topological spaces is *closed* if for each closed subset A of X, $f(A)$ is closed in Y.

Lemma 10.1.1. *If $f : X \to Y$ is a closed map, if $A \subset Y$, and if U is an open subset of X such that $f^{-1}(A) \subset U$, then there is an open set V in Y such that $A \subset V$ and $f^{-1}(V) \subset U$.*

Proof. The required V is $Y - f(X - U)$. □

A map $f : X \to Y$ is *perfect* if f is a closed map and $f^{-1}(y)$ is compact for all $y \in Y$.

Proposition 10.1.2. *Let $f : X \to Y$ be a perfect surjection, and let the space X be locally compact. Then Y is locally compact.*

Proof. Let $y \in Y$. Since X is locally compact and $f^{-1}(y)$ is compact, there is a compact set $N \subset X$ such that $f^{-1}(y) \subset \text{int } N$. By 10.1.1, there is an open neighborhood V of y such that $f^{-1}(y) \subset f^{-1}(V) \subset N$. $f(N)$ is compact, and is a neighborhood of y. □

A map $f : X \to Y$ is *proper* if for each compact subset C of Y, $f^{-1}(C)$ is compact. Proper maps are mainly of interest when X and Y are locally compact and Hausdorff, because in such spaces there is a rich supply of compact sets with non-empty interior. For example, if X and Y are path connected locally compact metric spaces, if a sequence x_n in X converges to ∞ (i.e., is not supported in any ball of finite radius) and if f is proper, then the sequence $f(x_n)$ converges to ∞ in Y.

Proposition 10.1.3. *Let $f : X \to Y$ be a proper map where X and Y are Hausdorff. If Y is either first countable or locally compact then f is perfect.*

Proof. We need only prove that f is closed. First, let Y be locally compact. Let A be a closed non-empty subset of X. Let $y \in \mathrm{cl}_Y f(A)$, and let N be a compact neighborhood of y in Y. Then $f^{-1}(N) \cap A$ is compact, so $N \cap f(A)$ is compact, hence $N \cap f(A)$ is closed in Y, so $y \in f(A)$.

If Y is first countable, the argument is similar. With A and $y \in \mathrm{cl}_Y f(A)$ as before, there is a sequence $\langle y_n \rangle$ in $f(A)$ converging to y. Let C be the compact set $\{y_n \mid n \geq 1\} \cup \{y\}$. Let $x_n \in A$ be such that $f(x_n) = y_n$. Then $\langle x_n \rangle$ is a sequence in the compact Hausdorff space $f^{-1}(C)$. Consider $B = \{x_n \mid n \geq 1\}$. If B were infinite and had no limit point in $f^{-1}(C)$ then B would be an infinite compact discrete space which is impossible (compare the proof of 1.2.6). So either B is finite or B has a limit point in $f^{-1}(C)$. Either way, $y \in f(A)$. $\square$

From 10.1.2 and 10.1.3 we get:

Corollary 10.1.4. *If $f : X \to Y$ is a proper surjection, where X is locally compact Hausdorff and Y is first countable Hausdorff, then Y is locally compact.* $\square$

Corollary 10.1.5. *Let $f : X \to Y$ be a map, where X and Y are locally compact Hausdorff. f is proper iff f is perfect.*

Proof. If f is proper then, by 10.1.3, f is perfect . Let f be perfect, and let A be a compact subset of Y. Each $y \in A$ has a compact neighborhood N_y in Y. Since $f^{-1}(y)$ is compact, there is a compact subset M_y of X such that $f^{-1}(y) \subset \mathrm{int}\, M_y$. By 10.1.1, there is an open set V_y in Y such that $y \in V_y \subset N_y$ and $f^{-1}(y) \subset f^{-1}(V_y) \subset M_y$. Finitely many members of $\{V_y \mid y \in Y\}$ cover A, so finitely many M_y's cover the closed set $f^{-1}(A)$, so $f^{-1}(A)$ is a closed subset of a compact space, so $f^{-1}(A)$ is compact. $\square$

Corollary 10.1.6. *A proper bijective [resp. surjective] map between locally compact Hausdorff spaces is a homeomorphism [resp. a quotient map].* $\square$

We now relate this general topology to CW complexes. Let Y be obtained from A by attaching n-cells. Let $f : S^{n-1}(\mathcal{A}) \to A$ be a simultaneous attaching map. We say Y *is obtained from A by properly attaching n-cells* if f is proper.

Proposition 10.1.7. *This definition is independent of the choice of simultaneous attaching map. When A is locally compact Hausdorff, Y is locally compact Hausdorff iff f is proper.*

Proof. For the first part, f is proper iff each compact subset of A meets only finitely many n-cells of Y. This latter condition is independent of f.

For the second part, Y is Hausdorff, by 1.2.2, and $S^{n-1}(\mathcal{A})$ is clearly Hausdorff. Assume first that f is proper. By 10.1.3, the map f is perfect. The space $A \coprod B^n(\mathcal{A})$ is locally compact. So, by 10.1.2, Y is locally compact.

Conversely, let Y be locally compact. Suppose f is not proper. Then there is a compact subset K of A such that $f^{-1}(K)$ meets infinitely many S_α^{n-1} (for the restriction of f to each S_α^{n-1} is certainly proper). Since Y is locally compact, there is a compact subset $N \subset Y$ such that $K \subset \operatorname{int} N$. Letting q be as in 1.2.3, $q^{-1}(\operatorname{int} N)$ meets infinitely many $B_\alpha^n - S_\alpha^{n-1}$, so int N meets infinitely many $\overset{\circ}{e}\,_\alpha^n$. Thus N contains an infinite closed subset inheriting the discrete topology from Y. This contradicts the compactness of N. $\qquad\square$

A CW complex is *locally finite* if each cell is disjoint from all but finitely many cells of X.

Proposition 10.1.8. *A CW complex is locally compact iff it is locally finite.*

Proof. Let the CW complex X be locally compact. Suppose X is not locally finite. Let e_α be a cell which meets infinitely many cells. Since e_α is compact, there is a compact subset N of X such that $e_\alpha \subset \operatorname{int} N$. The weak topology has the property that for each cell e, an open set either meets $\overset{\circ}{e}$ or is disjoint from e. So int N meets $\overset{\circ}{e}_\beta$ for infinitely many indices β. One finishes as in the proof of 1.2.13 by using this fact to produce an infinite closed discrete subspace of N, contradicting the compactness of N.

Conversely, let X be locally finite, and let $x \in X$. There is a unique cell e^m such that $x \in \overset{\circ}{e}\,^m$, an open subset of X^m. By local finiteness there is an open neighborhood V_m of x in $\overset{\circ}{e}\,^m$ such that every cell of X which meets V_m contains x. As in the proof of 1.2.11, extend V_m skeleton by skeleton to form sets V_n open in X^n, with $V_n \subset V_{n+1} \subset \ldots$, such that each V_n only meets cells of X containing x. Since X is locally finite, it follows by induction that each cl V_n is compact. Moreover, local finiteness implies that for some k, $V_k = \bigcup_{n \geq m} V_n$ so that cl V_k is a compact neighborhood of x in X. $\qquad\square$

Note that a CW complex can be locally finite and infinite dimensional. For a path connected example, consider $[0, \infty)$ with the CW complex structure consisting of a vertex at each $n \in \mathbb{N}$, and a 1-cell for each closed interval $[n, n+1]$; to this, attach an n-cell, for each n, by the constant attaching map taking S^{n-1} to the vertex n.

If $\{X_\alpha \mid \alpha \in \mathcal{A}\}$ is a set of locally finite CW complexes, so is $\coprod_\alpha X_\alpha$ (compare 1.2.18). If X and Y are locally finite CW complexes, so too is $X \times Y$ (compare 1.2.19). If A is a subcomplex of the CW complex X then the inclusion map $A \hookrightarrow X$ is proper. A *locally finite CW pair* is a CW pair (X, A) in which X is locally finite. The analog of 1.2.22 is:

Proposition 10.1.9. *If $\{A_\alpha\}$ is a family of pairwise disjoint finite subcomplexes of the locally finite CW complex X, and if there exist pairwise disjoint open sets $U_\alpha \subset X$ such that for each α, $A_\alpha \subset U_\alpha$, then the quotient complex is locally finite, and the quotient map $X \to X/\sim$ is proper and cellular.* $\qquad\square$

The building of proper maps with CW complex domains is done by analogy with 1.2.23:

Proposition 10.1.10. *Let (X, A) be a locally finite CW pair, and let the n-cells of X which are not in A be indexed by $\mathcal{A}$. Let $\{h_\alpha : B_\alpha^n \to X \mid \alpha \in \mathcal{A}\}$ be a set of characteristic maps for those n-cells. Let $f_{n-1} : (X^{n-1} \cup A) \to Z$ and $g : B^n(\mathcal{A}) \to Z$ be proper maps, such that $f_{n-1} \circ (h_\alpha \mid S_\alpha^{n-1}) = g \mid S_\alpha^{n-1}$. Then the unique map (given by 1.2.23) $f_n : (X^n \cup A) \to Z$ having restrictions $f_n \mid X^{n-1} \cup A = f_{n-1}$ and $f_n \circ h_\alpha = g \mid B_\alpha^n$ is proper.* □

We now turn to proper homotopy. Consider maps $f_0, f_1 : (X, A) \to (Y, B)$ and let $X' \subset X$. f_0 is *properly homotopic to* f_1 *relative to* X', denoted $f_0 \underset{p}{\simeq} f_1$ rel X', if there exists a proper map $F : (X \times I, A \times I) \to (Y, B)$ which is a homotopy relative to X' from f_0 to f_1. Of course, this can only happen if f_0 and f_1 are themselves proper (see Exercise 2). The main propositions in Sect. 1.3 all have proper analogs, which, for the most part, we leave to the reader to state. A proper map $f : (X, A) \to (Y, B)$ is a *proper homotopy equivalence* if there is a homotopy inverse $g : (Y, B) \to (X, A)$ which is proper such that $g \circ f \underset{p}{\simeq} \mathrm{id}_{(X,A)}$ and $f \circ g \underset{p}{\simeq} \mathrm{id}_{(Y,B)}$. We call such a map g a *proper homotopy inverse* for f, and we say that (X, A) and (Y, B) have the *same proper homotopy type*. A subspace $A \subset X$ is a *proper strong deformation retract* of X if there is a proper strong deformation retraction of X to A.

If $f : X \to Y$ is a map where X is compact and Y is Hausdorff, then f is proper; whereas if Y is compact and X is not compact then f cannot be proper. Thus ordinary homotopy theory and proper homotopy theory only coincide on compact spaces. In particular, a Hausdorff space has the proper homotopy type of a point iff it is both compact and contractible. We will see that the proper analog of contractibility is "having the proper homotopy type of $[0, \infty)$."

A pair of spaces (X, A) has the *proper homotopy extension property with respect to a space* Z if every proper map $(X \times \{0\}) \cup (A \times I) \to Z$ extends to a proper map $X \times I \to Z$. The proofs of 1.3.15 and 1.3.16 can be adapted in an obvious way to give:

Theorem 10.1.11. *If (X, A) is a locally finite CW pair, then (X, A) has the proper homotopy extension property with respect to any space. Indeed, a proper map $F : (X \times \{0\}) \cup (A \times I) \to Z$ extends to a proper map $\tilde{F} : X \times I \to Z$ such that for every cell e_α of X which is not in A, $\tilde{F}_1(e_\alpha) = F_0(e_\alpha) \cup F(\overset{\bullet}{e}_\alpha \times I)$.* □

Next, we discuss cellular approximation. A CW complex X is *strongly locally finite* if $\{C(e) \mid e$ is a cell of $X\}$ is a locally finite cover[1] of X. Clearly "strongly locally finite" implies "locally finite." By induction on dimension one proves:

---

[1] Recall that $C(A)$ is the carrier of $A \subset X$.

Proposition 10.1.12. *If X is finite-dimensional and locally finite, then X is strongly locally finite.* $\square$

Example 10.1.13. Here is an example of a CW complex X which is locally finite but not strongly locally finite. X^0 is a single vertex. Assuming X^n defined, with one cell e^i in each dimension $i \leq n$, attach an $(n+1)$-cell trivially at a point of $\overset{\circ}{e}{}^n$ to form X^{n+1}. Note that X has finite type. In fact (exercise) X has a subdivision which is strongly locally finite. Thus, while local finiteness is a topologically invariant property (by 10.1.8), strong local finiteness is not.

Theorem 10.1.14. *[Proper Cellular Approximation Theorem] Let $f : X \to Y$ be a proper map between CW complexes, with X locally finite and Y strongly locally finite, and let A be a subcomplex of X such that $f \mid A$ is cellular. Then f is properly homotopic, rel A, to a proper cellular map.*

Proof. To see that the proof of 1.4.3 gives this, use 1.4.4 and the following useful criterion 10.1.15. $\square$

Proposition 10.1.15. *Let X and Y be locally compact Hausdorff spaces, let $F : X \times I \to Y$ be a homotopy, and let F_0 be proper. Let $\mathcal{K}$ be a locally finite cover of Y by compact sets such that for each $x \in X$ there exists $K_x \in \mathcal{K}$ for which $F(\{x\} \times I) \subset K_x$. Then F is a proper homotopy.*

Proof. Suppose $F^{-1}(L)$ is not compact, where L is a compact subset of Y. Let $p : X \times I \to X$ be projection. The closed set $A \subset X \times I$ is compact iff $p(A)$ is compact. Thus $J := pF^{-1}(L)$ is not compact. Since I is compact, p is closed, so J is closed in X. The family $\{K_x \mid x \in J\}$ is infinite, for otherwise $F_0(J)$ would lie in a compact set $K_{x_1} \cup \ldots \cup K_{x_n}$, implying J compact (being a closed subset of the compact set $\bigcup_{i=1}^{n} F_0^{-1}(K_{x_i})$). Let $\{K_{z_n} \mid n \geq 1\}$ be an infinite subset of $\mathcal{K}$ such that each $z_n \in J$. For each n, $(\{z_n\} \times I) \cap F^{-1}(L) \neq \emptyset$, so $F(\{z_n\} \times I) \cap L \neq \emptyset$, so $K_{z_n} \cap L \neq \emptyset$. But, since L is compact and $\mathcal{K}$ is locally finite, only finitely many members of $\mathcal{K}$ meet L. This is a contradiction. $\square$

Here is an example to show that 10.1.14 does not hold when Y is merely locally finite:

Example 10.1.16. Let $X = \mathbb{N}$, a 0-dimensional CW complex, and let Y be the CW complex of Example 10.1.13. Let $f : X \to Y$ send the vertex n to a point of $\overset{\circ}{e}{}^n$. By 1.2.13, f is proper. However, there is no proper cellular map $X \to Y$. $\square$

We now discuss proper maps and covering spaces. Let $f : (X_1, v_1) \to (X_2, v_2)$ be a map of pointed path connected CW complexes. For $i = 1$

or 2, write $G_i = \pi_1(X_i, v_i)$, and let H_i be a subgroup of G_i. As usual we write $(\bar{X}_i(H_i), \bar{v}_i)$ for the pointed path connected covering space of X_i whose fundamental group is H_i, and $q_{H_i} : \bar{X}_i(H_i) \to X_i$ for the covering projection. By 3.3.4, there is a map $\bar{f}$ making the following diagram commute iff $f_\#(H_1) \leq H_2$.

$$
\begin{array}{ccc}
(\bar{X}_1(H_1), \bar{v}_1) & \xrightarrow{\bar{f}} & (\bar{X}_2(H_2), \bar{v}_2) \\
\downarrow{q_{H_1}} & & \downarrow{q_{H_2}} \\
(X_1, v_1) & \xrightarrow{f} & (X_2, v_2)
\end{array}
$$

Proposition 10.1.17. *Let f be proper, and let $f_\#(H_1) \leq H_2$. The lift $\bar{f}$ is proper iff H_1 has finite index in $f_\#^{-1}(H_2)$.*

Before proving 10.1.17, we recall a standard topological construction. If $g : Y \to B$ and $p : E \to B$ are maps of Hausdorff spaces, the *pull-back* of p by g consists of the space $g^*E := \{(y, e) \in Y \times E \mid g(y) = p(e)\}$ topologized as a (closed) subspace of the product space $Y \times E$ and maps p' and $\tilde{g}$ making the following diagram commute:

$$
\begin{array}{ccc}
g^*E & \xrightarrow{\tilde{g}} & E \\
\downarrow{p'} & & \downarrow{p} \\
Y & \xrightarrow{g} & B
\end{array}
$$

where the maps p' and $\tilde{g}$ are defined by: $p'(y, e) = y$ and $\tilde{g}(y, e) = e$.

Lemma 10.1.18. *If g is proper, so is $\tilde{g}$.*

Proof. Let C be a compact subset of E. Then $\tilde{g}^{-1}(C)$ is closed in g^*E, and $\tilde{g}^{-1}(C) \subset g^{-1}(p(C)) \times C$. Since g is proper, $g^{-1}(p(C)) \times C$ is compact. So $\tilde{g}^{-1}(C)$ is compact. □

Lemma 10.1.19. *If p is a covering projection, so is p'.*

Proof. Let $y \in Y$ and let U be a neighborhood of $p(y)$ in B which is evenly covered by p. Then $p^{-1}(U) = \bigcup\{U_\alpha \mid \alpha \in \mathcal{A}\}$ where $\mathcal{A}$ is an indexing set and $\{U_\alpha \mid \alpha \in \mathcal{A}\}$ consists of pairwise disjoint open subsets of E each mapped homeomorphically onto U by p. We claim that $g^{-1}(U)$ is evenly covered by p'. Indeed, $(p')^{-1}g^{-1}(U) = \bigcup_\alpha \{\tilde{g}^{-1}(U_\alpha) \mid \alpha \in \mathcal{A}\}$. These sets $\tilde{g}^{-1}(U_\alpha)$ are pairwise disjoint open subsets of g^*E. It is an exercise in the definitions to show that p' maps $\tilde{g}^{-1}(U_\alpha)$ bijectively to $g^{-1}(U)$. Projections in a cartesian product map open sets to open sets, so p' maps any open subset of $\tilde{g}^{-1}(U_\alpha)$ onto an open subset of $g^{-1}(U)$, hence onto an open subset of Y, since $g^{-1}(U)$ is open in Y. Thus p' maps $\tilde{g}^{-1}(U_\alpha)$ homeomorphically onto $g^{-1}(U)$ as claimed. □

Let base points $y \in Y$ and $e \in E$ be chosen so that $g(y) = p(e) = b$, and let $z := (y, e)$ be the base point of g^*E. We leave as an exercise:

Lemma 10.1.20. *If p is a covering projection then we have $p'_\#(\pi_1(g^*E, z)) = g_\#^{-1}p_\#(\pi_1(E, e))$.* □

Note that the covering space g^*E of Y need not be path connected. For example, let $Y = B = E = S^1$, let $p = g$ be the covering projection $e^{2\pi it} \mapsto e^{4\pi it}$ (a map of degree 2). Then it is not hard to see that $g^*E = S^1 \coprod S^1$ and p' is the "identity" on each path component.

Now we return to the map $f : (X_1, v_1) \to (X_2, v_2)$.

Proof (of 10.1.17). Let P be the path component of $f^*\bar{X}_2(H_2)$ containing the base point. Consider the following diagram (in which base points are suppressed):

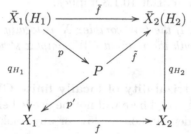

By 10.1.19, 10.1.20 and 3.4.8, $p' : P \to X_1$ is equivalent to $q_{f_\#^{-1}(H_2)}$: $\bar{X}_1(f_\#^{-1}(H_2)) \to X_1$ (in the sense of Sect. 3.4). Since $f_\#(H_1) \leq H_2$, the lift p of q_{H_1} exists making $q_{H_1} = p' \circ p$. The top triangle commutes by 3.3.4. Thus, the whole diagram commutes. The map $\tilde{f}$ is proper by 10.1.18; hence if p is proper $\bar{f}$ is proper. By 3.2.13, we conclude that if $[f_\#^{-1}(H_2) : H_1] < \infty$ then $\bar{f}$ is proper. On the other hand, if $[f_\#^{-1}(H_2) : H_1] = \infty$, then $\bar{f}^{-1}(\bar{v}_2) = p^{-1}\tilde{f}^{-1}(\bar{v}_2)$ contains a point-inverse of p, hence an infinite discrete closed subset, implying that $\bar{f}^{-1}(\bar{v}_2)$ is not compact. □

Corollary 10.1.21. *With hypotheses as in 10.1.17, let $F : X_1 \times I \to X_2$ be a proper homotopy such that $F_0 = f$. Let $\bar{F} : \bar{X}_1(H_1) \times I \to \bar{X}_2(H_2)$ be the lift of F such that $\bar{F}_0 = \bar{f}$. Then $\bar{F}$ is a proper homotopy iff $\bar{f}$ is a proper map. If $F(\{v_1\} \times I) = v_2$ then $\bar{F}(\{\bar{v}_1\} \times I) = \bar{v}_2$.*

Proof. The first part follows from 10.1.17. The second part follows from 2.4.6 □

Proposition 10.1.22. *Let $f : (X_1, v_1) \to (X_2, v_2)$ be a map of pointed path connected CW complexes, and let $f : X_1 \to X_2$ be a homotopy equivalence [resp. proper homotopy equivalence.] Then $f : (X_1, v_1) \to (X_2, v_2)$ is a homotopy equivalence [resp. proper homotopy equivalence] of pairs.*

Proof. See the remark following the statement of 4.1.5. The proof in the proper case is similar and is an exercise.   □

These results, together with 3.3.4, immediately imply:

Theorem 10.1.23. *Let* $f : (X_1, v_1) \to (X_2, v_2)$ *be a map of pointed path connected CW complexes. Let* $H_1 \leq \pi_1(X_1, v_1)$ *and* $H_2 \leq \pi_1(X_2, v_2)$ *be subgroups. If* $f : X_1 \to X_2$ *is a proper homotopy equivalence, and if* $f_\# : \pi_1(X_1, v_1) \to \pi_1(X_2, v_2)$ *takes* H_1 *isomorphically onto* H_2, *then the lift* $\bar{f} : (\bar{X}(H_1), \bar{v}_1) \to (\bar{X}(H_2), \bar{v}_2)$ *is a proper homotopy equivalence.*   □

One cannot expect a converse to this. For example, the map $S^1 \to S^1$, $e^{2\pi it} \mapsto e^{4\pi it}$ is not a homotopy equivalence, but its lift to the universal cover is the homeomorphism $\mathbb{R} \to \mathbb{R}$, $t \mapsto 2t$.

In this discussion of covering spaces we have not required that the CW complexes be locally finite. But 10.1.8 implies:

Proposition 10.1.24. *If the CW complex* X *is locally finite, and* $\bar{X}(H)$ *is a covering space of* X *(with the induced CW complex structure), then* $\bar{X}(H)$ *is locally finite.*   □

Appendix: Metrizability of locally finite CW complexes

The general topology used here will not be needed elsewhere in this book. For definitions the reader can consult [51] or any book on general topology. We sketch a proof of

Proposition 10.1.25. *Every locally finite CW complex is metrizable.*

Proof. It is enough to consider the case where X is path connected. Then X is countable by 11.4.3 below. By the Urysohn Metrization Theorem it is enough to show that X is regular and second countable. Since X is locally finite each X^n is obtained from X^{n-1} by properly attaching countably many n-cells. Let $\mathcal{A}_n$ index the n-cells of X. The nth simultaneous attaching map $S^{n-1}(\mathcal{A}_n) \to X^{n-1}$ provides a perfect surjection $B^n(\mathcal{A}_n) \coprod X^{n-1} \twoheadrightarrow X^n$. Letting $\mathcal{A}$ index all the cells we thus get a perfect surjection $h : \coprod_{\alpha \in \mathcal{A}} B^{n(\alpha)} \twoheadrightarrow X$. The domain of h is clearly regular and second countable and both these properties are transmitted by perfect surjections; see p. 235 of [51].   □

Source Notes: The first paper on proper homotopy theory I am aware of is [28]. For more on this see the source notes for Sect. 17.6. Example 10.1.13 is found in [145].

Exercises

1. If $f : X \to Y$ is a proper map and Y is locally compact show that X is locally compact.

2. Show that if $f : X \to Y$ is a proper map and A is a closed subset of X then $f| : A \to Y$ is proper.

3. Show that a non-locally finite CW complex is not first countable (and hence non-metrizable).

4. Prove Proposition 10.1.9.

5. Show that X in 10.1.13 has a strongly locally finite subdivision.

6. Show that a locally finite CW complex is locally finite-dimensional.

7. Prove Lemma 10.1.20.

8. Using 10.1.19, prove that if subgroups G_1 and G_2 have finite index in the group G then $G_1 \cap G_2$ has finite index in G. What can be deduced about the relationship between the indices $[G : G_1]$, $[G : G_2]$ and $[G : G_1 \cap G_2]$?

9. Construct a bijective map: $I \times [0, \infty) \to I \times [0, \infty)$ which is not a homeomorphism (compare 10.1.6).

10. If a CW complex X is locally finite show that each X^n is obtained from X^{n-1} by properly attaching n-cells.

11. Show that if K is a locally finite simplicial complex in the sense of Sect. 8.2 then $|K|$ is a locally finite CW complex.

12. Prove that a regular locally finite CW complex is strongly locally finite.

13. Give another proof of 8.1.3 by deducing it from 10.1.23.

10.2 CW-proper maps

A CW complex X has *locally finite type* if X^n is locally finite for each n. For example, if Z has finite type but is not finite, its universal cover $\tilde{Z}$ has locally finite type but is not locally finite. Between such spaces it is natural to consider maps which satisfy a weaker condition than "properness," namely "proper on each skeleton" (with an additional condition for non-cellular maps).

A map $f : X \to Y$ between CW complexes is *CW-proper* if for each n there exists k such that $f(X^n) \subset Y^k$ and $f | : X^n \to Y$ is proper. We define *CW-proper homotopy*, *CW-proper homotopy equivalence*, etc., by analogy with the proper case. Of course, a cellular map $f : X \to Y$ is CW-proper if $f | : X^n \to Y$ is proper for all n.

It is important to remember that IN THE CONTEXT OF FINITE-DIMENSIONAL CW COMPLEXES THERE IS NO DIFFERENCE BETWEEN "CW-PROPER" AND "PROPER."

If X is any infinite CW complex having finite type, a constant map $X \to Y$ is always CW-proper but never proper. The proper map in Example 10.1.16 is not CW-proper. We note:

Proposition 10.2.1. Let $f : X \to Y$ be a map between CW complexes, where X has finite type. Then f is CW-proper. $\square$

"CW-proper" is not a topologically invariant notion as the following example shows.

Example 10.2.2. Define the "infinite sphere," denoted S^∞ or $\varinjlim_n S^n$, to be the CW complex whose 0-skeleton is S^0, whose $(n-1)$-skeleton is homeomorphic to S^{n-1}, and whose n-skeleton is obtained from $(S^\infty)^{n-1}$ by attaching two n-cells using, for both, a homeomorphism $S^{n-1} \to (S^\infty)^{n-1}$ as attaching map. Clearly, the resulting $(S^\infty)^n$ is homeomorphic to S^n, so the inductive definition is complete. By 7.1.2, S^∞ is n-connected for all n, hence contractible. The same is true of $S^\infty \times \mathbb{R}$. The n-skeleton of S^∞ is finite, while the n-skeleton of $S^\infty \times \mathbb{R}$ is infinite. Hence the constant map $S^\infty \to \{\text{point}\}$ is CW-proper, while $S^\infty \times \mathbb{R} \to \{\text{point}\}$ is not. However, S^∞ and $S^\infty \times \mathbb{R}$ are known to be homeomorphic [79], though this fact from infinite-dimensional topology is far from obvious.

Theorem 10.2.3. (CW-proper Cellular Approximation Theorem) *Let $f : X \to Y$ be a CW-proper map between CW complexes of locally finite type, and let A be a subcomplex of X such that $f \mid A$ is cellular. Then f is CW-proper homotopic, rel A, to a CW-proper cellular map.*

Proof. Similar to the proof of 10.1.14. In other words, the theorem follows from the proof of 1.4.4, using 10.1.15 on each skeleton of $X \times I$.   □

Exercises

1. State and prove analogs of 10.1.10 and 10.1.11 in the context of this section.
2. If X_1 and X_2 have locally finite type, if $f : X_1 \to X_2$ is a CW-proper map and if $f_\#(H_1) \leq H_2$ prove that the lift $\bar{f} : \bar{X}_1(H_1) \to \bar{X}_2(H_2)$ is CW-proper iff H_1 has finite index in $f_\#^{-1}(H_2)$.
3. State and prove a CW-proper analog of 10.1.23, and an analog for CW-proper n-equivalences.

11

Locally Finite Homology

In this chapter we introduce a homology theory which agrees with the cellular homology theory of Chap. 2 on finite CW complexes, but which is a more sensitive detector of topological properties of infinite CW complexes having locally finite type. This new homology is not a homotopy invariant but it is a proper or CW-proper homotopy invariant in appropriate contexts. The "new homology modulo the old homology" defines homology at the end of a CW complex.

11.1 Infinite cellular homology

Let X be an oriented CW complex and let R be a ring. Let $R(e_\alpha^n)$ denote the free left R-module generated by the (oriented) n-cell e_α^n. Let $C_n^\infty(X; R) = \prod_{\alpha \in \mathcal{A}} R(e_\alpha^n)$, where $\mathcal{A}$ indexes the n-cells of X. It is convenient to denote elements of $C_n^\infty(X; R)$ by $\sum_{\alpha \in \mathcal{A}} m_\alpha e_\alpha^n$ where $m_\alpha \in R$ is the α-entry in the cartesian product, even though it may be the case that $m_\alpha \neq 0$ for infinitely many values of α. Elements of $C_n^\infty(X; R)$ are called *infinite cellular n-chains*[1] in X with coefficients in R. Recall from Sect. 2.3 and Sect. 2.6 that elements of $C_n(X; R)$ are written $\sum_{\alpha \in \mathcal{A}} m_\alpha e_\alpha^n$ where all but finitely many $m_\alpha = 0$. Thus we consider $C_n(X; R) \subset C_n^\infty(X; R)$. We sometimes call the elements of $C_n(X; R)$ *finite cellular n-chains*. In general, the term "n-chain" means "finite n-chain" unless the context clearly indicates otherwise. When $n < 0$, $C_n^\infty(X; R)$ is defined to be the trivial R-module 0.

We need a boundary homomorphism for infinite chains. The obvious candidate is $\partial : C_n^\infty(X; R) \to C_{n-1}^\infty(X; R)$ defined by

---

[1] Also known as *locally finite n-chains* when X is locally finite.

$$\partial\left(\sum_\alpha m_\alpha e_\alpha^n\right) = \sum_\beta \left(\sum_\alpha m_\alpha [e_\alpha^n : e_\beta^{n-1}]\right) e_\beta^{n-1}.$$

This might not be well defined, but, by 2.5.8, it is a well defined homomorphism when X^n is locally finite. FOR THE REST OF THIS SECTION WE WORK WITH CW COMPLEXES OF LOCALLY FINITE TYPE.

Proposition 11.1.1. *The composition*

$$C_n^\infty(X;R) \xrightarrow{\partial} C_{n-1}^\infty(X;R) \xrightarrow{\partial} C_{n-2}^\infty(X;R)$$

is zero, for all n. ☐

Thus $(C_*^\infty(X;R), \partial)$ is a chain complex. Its homology modules, denoted[2] by $H_*^\infty(X;R)$, are the *cellular homology modules based on infinite chains*.

Let $f : X \to Y$ be a CW-proper cellular map between oriented CW complexes of locally finite type. Define $f_\# : C_n^\infty(X;R) \to C_n^\infty(Y;R)$ by

$$f_\#\left(\sum_\alpha m_\alpha e_\alpha^n\right) = \sum_\beta \left(\sum_\alpha m_\alpha [e_\alpha^n : \tilde{e}_\beta^n : f]\right) \tilde{e}_\beta^n.$$

Since $f \mid: X^n \to Y^n$ is proper, 2.5.10 implies that $f_\#$ is well defined, and it is obviously a homomorphism.

We have commutative diagrams

$$
\begin{array}{ccc}
C_n^\infty(X;R) & \xrightarrow{\partial} & C_{n-1}^\infty(X;R) \\
\uparrow & & \uparrow \\
C_n(X;R) & \xrightarrow{\partial} & C_{n-1}(X;R)
\end{array}
\qquad
\begin{array}{ccc}
C_n^\infty(X;R) & \xrightarrow{f_\#} & C_n^\infty(Y;R) \\
\uparrow & & \uparrow \\
C_n(X;R) & \xrightarrow{f_\#} & C_n(Y;R)
\end{array}
$$

where the vertical arrows are inclusions. Clearly $(g \circ f)_\# = g_\# \circ f_\# : C_n^\infty(X;R) \to C_n^\infty(Y;R) \to C_n^\infty(Z;R)$ whenever f and g are CW-proper cellular maps. And $(\mathrm{id}_X)_\# = \mathrm{id} : C_n^\infty(X;R) \to C_n^\infty(X;R)$. Moreover:

Proposition 11.1.2. *With $f : X \to Y$ as above, the following diagram commutes:*

$$
\begin{array}{ccc}
C_n^\infty(X;R) & \xrightarrow{\partial} & C_{n-1}^\infty(X;R) \\
\downarrow{\scriptstyle f_\#} & & \downarrow{\scriptstyle f_\#} \\
C_n^\infty(Y;R) & \xrightarrow{\partial} & C_{n-1}^\infty(Y;R).
\end{array}
$$

☐

---

[2] Also known as *homology based on locally finite chains* and written $H_*^{lf}(X;R)$.

Thus f induces a chain map $C_*^\infty(X; R) \to C_*^\infty(Y; R)$ and hence a homomorphism of graded R-modules $f_* : H_*^\infty(X; R) \to H_*^\infty(Y; R)$. Note that when X has only finitely many cells in dimensions n and $n+1$, then $C_n^\infty(X; R) = C_n(X; R)$, $C_{n+1}^\infty(X; R) = C_{n+1}(X; R)$, hence $H_n^\infty(X; R) = H_n(X; R)$.

The analog of 2.7.1 is:

Proposition 11.1.3. *When X^1 is infinite, locally finite and path connected, then $H_0^\infty(X; R) = 0$.*

For the proof of 11.1.3, we need some preliminaries. A *proper edge ray* in an unoriented, path connected, locally finite graph Y is an infinite sequence $\tau := (\tau_1, \tau_2, \ldots)$ of edges in Y such that each $(\tau_1, \ldots, \tau_k)$, called the kth *initial segment*, is an edge path, and no edge appears infinitely often in τ. The *initial point* of τ is the initial point of τ_1; τ has no final point. The *distance* between vertices $u \neq v$ of Y is the least k for which there is an edge path $(\tau_1, \ldots, \tau_k)$ having initial point u and final point v; this only makes sense when Y is path connected. (If we define the distance from u to u to be 0, we have a metric on Y^0.) Let y be a base vertex for Y. The *ball of radius n about y* is the subgraph $B_n(y)$ whose vertices are those distant $\leq n$ from y, and whose 1-cells are those 1-cells of Y which can be oriented so as to have initial point distant $< n$ from y. Because Y is locally finite $B_n(y)$ is finite.

For a subgraph A of Y, a *complementary component* is a path component of $Y - A$; when A is a finite subgraph there are only finitely many of these. A complementary component of A is *bounded* if its closure in Y is compact; those which are not bounded are *unbounded* and there is at least one such when A is finite and Y is infinite. The closure of a complementary component is a path connected subgraph of Y.

Let Y_n be the union of $B_n(y)$ and all the bounded complementary components of $Y - B_n(y)$. Then Y_n is a finite subgraph of Y and all the complementary components of Y_n are unbounded. Pick $n_{k+1} > n_k$ so that $Y_{n_k} \subset B_{n_{k+1}}(y)$. This completes an inductive definition and proves:

Proposition 11.1.4. *If Y is an infinite, locally finite, path connected graph, there are finite subgraphs $Y_{n_1} \subset Y_{n_2} \subset \ldots$ such that every complementary component of each Y_{n_k} is unbounded, $\text{fr } Y_{n_k} \cap \text{fr } Y_{n_{k+1}} = \emptyset$ and $Y = \bigcup_k Y_{n_k}$.*

If $v \in \text{fr } Y_{n_k}$, then v is a vertex of Y, and, for suitable orientations, v is the initial point of a non-degenerate edge of Y_{n_k} and also of a non-degenerate edge which is not in Y_{n_k}. Every vertex of $Y - Y_{n_{k+1}}$ can be joined by an edge path in $Y - Y_{n_{k-1}}$ to a vertex of $\text{fr } Y_{n_k}$. Every vertex of $Y - Y_{n_k}$ can be joined by an edge path in $Y - Y_{n_k}$ to a vertex of $Y - Y_{n_{k+1}}$. $\square$

Proof (of 11.1.3). Pick subcomplexes Y_{n_k} as in 11.1.4. For each k and each vertex v of $Y_{n_{k+1}} - Y_{n_k}$, use 11.1.4 to pick (by induction) a proper edge ray $\tau_v := (\tau_1^{i_1}, \tau_2^{i_2}, \ldots)$ with initial point v where τ_j is an edge of X with the preferred orientation, $i_j = \pm 1$, and for each m all but finitely many τ_j's lie in $X^1 - Y_{n_m}$. Then $c := \sum_{j=1}^{\infty} i_j \tau_j$ is[3] an infinite cellular 1-chain in X such that $\partial c = v$. The key point is that an edge of X occurs in only finitely many of the chains c. Thus, given any infinite cellular 0-chain $d := \sum_{\alpha} m_{\alpha} v_{\alpha}$,

$$d = \sum_{\alpha} m_{\alpha}(\partial c_{\alpha}) = \partial \left(\sum_{\alpha} m_{\alpha} c_{\alpha} \right) \text{ for suitable[4] infinite 1-chains } c_{\alpha}. \qquad \square$$

The analog of 2.7.2 is:

Proposition 11.1.5. *If $X = \coprod_{\alpha} X_{\alpha}$, with each X_{α} of locally finite type and oriented, then $H_n^{\infty}(X; R)$ is isomorphic to $\prod_{\alpha} H_n^{\infty}(X_{\alpha}; R)$.* $\qquad \square$

In particular, we know $H_n^{\infty}(X; R)$ when we know $H_n^{\infty}(X_{\alpha}; R)$ for each path component X_{α} of X.

Proposition 11.1.6. *Let $[0, \infty)$ have the usual CW complex structure (vertices n, 1-cells $[n, n+1]$ for each $n \in \mathbb{N}$). For any orientation, $H_n^{\infty}([0, \infty); R) = 0$ for all n.*

Proof. For $n = 0$ this follows from 11.1.4. For $n > 1$ it is trivial. For $n = 1$, it is obvious that $\ker(\partial : C_1^{\infty}([0, \infty); R) \to C_0^{\infty}([0, \infty); R)) = 0$. $\qquad \square$

This should be regarded as the analog of 2.7.3: in proper homotopy theory a proper ray often has the role played by a base point in ordinary homotopy theory. We will see in Chap. 16 that the analog of a base point is a (proper) base ray.

The analogs of 2.7.5 and 2.7.6 are obvious:

Proposition 11.1.7. *If X has dimension d, then $H_n^{\infty}(X) = 0$ for all $n > d$.* $\square$

Proposition 11.1.8. *The inclusion $X^{n+1} \hookrightarrow X$ induces an isomorphism on H_i^{∞} for $i \leq n$.* $\qquad \square$

Define $Z_n^{\infty}(X; R) := \ker(\partial : C_n^{\infty}(X; R) \to C_{n-1}^{\infty}(X; R))$ and define $B_n^{\infty}(X; R) := \text{image}(\partial : C_{n+1}^{\infty}(X; R) \to C_n^{\infty}(X; R))$: these are the R-modules of *infinite n-cycles* and *infinite n-boundaries* respectively. Of course,

---

[3] To conform strictly to our previous notation, we should have written this summation over all the (oriented) 1-cells of X^1, not just the 1-cells τ_j, assigning coefficient 0 to the others. We will sometimes abuse notation in this way.

[4] We changed notation from X^1 to Y in the text leading up to the proof of 11.1.3 to distinguish the oriented graph X^1 from the unoriented graph Y. Recall from Sect. 3.1 that when Y is unoriented an edge is appropriately denoted by τ_j, but when the underlying 1-cell already has a preferred orientation it is better to use τ_j for that orientation and τ_j^{-1} for the opposite orientation.

$H_n^\infty(X; R) = Z_n^\infty(X; R)/B_n^\infty(X; R)$. Infinite cycles $z, z' \in Z_n^\infty(X; R)$ represent the same element of $H_n^\infty(X; R)$ iff $z - z' \in B_n^\infty(X; R)$, in which case z and z' are *properly homologous*.

The analog of 2.7.10 is:

Theorem 11.1.9. *Let $f, g : X \to Y$ be CW-proper cellular maps between oriented CW complexes of locally finite type. Assume either* (a) *f and g are CW-proper homotopic or* (b) *f and g are properly homotopic where X is locally finite and Y is strongly locally finite. Then $f_* = g_* : H_*^\infty(X; R) \to H_*^\infty(Y; R)$.*

Proof. Similar to that of 2.7.10: in place of 1.4.3, use 10.2.3 in Case (a) and 10.1.14 in Case (b). □

Using obvious analogs of the remarks which follow Theorem 2.7.10, we can regard $H_*^\infty(\cdot\,; R)$ as a covariant functor from the category $\mathcal{C}$ to the category of graded R-modules and homomorphisms, where $\mathcal{C}$ is either the category of oriented CW complexes of locally finite type and CW-proper homotopy classes of CW-proper maps, or the category of oriented strongly locally finite CW complexes and proper homotopy classes of proper maps. In particular we have this analog of 2.7.11:

Corollary 11.1.10. *Let $f : X \to Y$ be a map between oriented CW complexes. If either X and Y are of locally finite type and the map f is a CW-proper homotopy equivalence, or X and Y are strongly locally finite and the map f is a proper homotopy equivalence, then $f_* : H_*^\infty(X; R) \to H_*^\infty(Y; R)$ is an isomorphism.* □

Homeomorphisms are proper homotopy equivalences, so the second part of 11.1.10 gives a sense in which H_*^∞ is a topological invariant.[5]

The properties of "finite" cellular homology given in Sect. 2.8 hold for infinite cellular homology with almost no change. Let (X, A) be an oriented CW pair where X has locally finite type. From the short exact sequence of chain complexes

$$0 \longrightarrow C_*^\infty(A; R) \longrightarrow C_*^\infty(X; R) \longrightarrow C_*^\infty(X; R)/C_*^\infty(A; R) \longrightarrow 0$$

one obtains the relative homology modules $H_n^\infty(X, A; R)$ and an exact sequence

$$\cdots \longrightarrow H_n^\infty(A; R) \longrightarrow H_n^\infty(X; R) \longrightarrow H_n^\infty(X, A; R) \xrightarrow{\partial_*} H_{n-1}^\infty(A; R) \longrightarrow \cdots$$

---

[5] To make this precise, the reader should state analogs of 2.7.12 and 2.7.13.

Define
$$Z_n^\infty(X, A; R) := \{c \in C_n^\infty(X; R) \mid \partial c \in C_{n-1}^\infty(A; R)\}$$
and

$$B_n^\infty(X, A; R) := \{c \in C_n^\infty(X; R) \mid c \text{ is properly homologous to an element of } C_n^\infty(A; R)\}.$$

Then[6]
$$H_n^\infty(X, A; R) \cong Z_n^\infty(X, A; R)/B_n^\infty(X, A; R).$$

There are obvious versions of the Proper and CW-Proper Cellular Approximation Theorems for pairs, leading to versions of 11.1.9 and 11.1.10 for pairs. The analogs of naturality of ∂_*, excision and the Mayer-Vietoris sequence hold for cellular homology based on infinite chains.

If X has infinitely many vertices, the augmentation of X defined in Sect. 2.9 is not locally finite. Thus there is no useful notion of augmentation for $C_*^\infty(X; R)$, hence no useful analog of reduced homology.

A proper cellular map $f : (X, A) \to (Y, B)$ between CW pairs of locally finite type is a CW-*proper n-equivalence* if there is a CW-proper cellular map $g : (Y, B) \to (X, A)$ such that the compositions $g \circ f| : (X^{n-1}, A^{n-1}) \to (X, A)$ and $f \circ g| : (Y^{n-1}, B^{n-1}) \to (Y, B)$ are CW-proper homotopic to the respective inclusion maps. The map g is a CW-*proper n-inverse* for f. When X and Y are finite-dimensional we use the simpler terms *proper n-equivalence*, etc.

Proposition 11.1.11. *A CW-proper n-equivalence induces isomorphisms on* $H_k^\infty(\cdot; R)$ *for all* $k \le n - 1$.   □

Source Notes: An interesting treatment of homology and cohomology, which influenced the presentation here, is [110].

Exercises

1. In 10.1.23 assume instead that f is a CW-proper n-equivalence where $n \ge 2$. Prove that $\bar{f}$ is a CW-proper n-equivalence.
2. Given a finite subgraph A of an infinite locally finite path connected graph Y prove that there is a finite subgraph $B \supset A$ such that for every vertex v of Y which is not in B there is a proper edge ray in Y with initial point v which involves no edges of A.
3. Discover a Mayer-Vietoris sequence for H_*^∞.

---

[6] The alternative interpretation of $H_n(X, A; R)$ as $\tilde{H}_n(X/A; R)$ in Sect. 2.7 and Sect. 2.8 does not carry over, since, in general, X^n/A^n is not locally finite.

11.2 Review of inverse and direct systems

In this section we state the definitions and some properties of inverse and direct systems and limits. This material is part of category theory. Some readers may prefer to skip or skim this section, referring back to it later as required.[7]

A *pre-ordered set* is an ordered pair $(\mathcal{A}, \leq)$ where $\mathcal{A}$ is a set and $\leq$ is a reflexive transitive relation on $\mathcal{A}$; i.e., $\alpha \leq \alpha$; and $\alpha \leq \beta$ and $\beta \leq \gamma$ imply $\alpha \leq \gamma$. If $\leq$ also satisfies the law: $\alpha \leq \beta$ and $\beta \leq \alpha$ imply $\alpha = \beta$, then $(\mathcal{A}, \leq)$ is a *partially ordered set* (abbreviation: *poset*[8]). A *directed set* is a pre-ordered set $(\mathcal{A}, \leq)$ which satisfies the law: whenever α and β are in $\mathcal{A}$, there exists $\gamma \in \mathcal{A}$ such that $\alpha \leq \gamma$ and $\beta \leq \gamma$. Usually, we suppress $\leq$, saying "$\mathcal{A}$ is a directed set" etc.

Let $\mathcal{C}$ be a category. An *inverse system* $\{X_\alpha, f_\alpha^\beta; \mathcal{A}\}$ in $\mathcal{C}$ consists of: a directed set $\mathcal{A}$, an object X_α of $\mathcal{C}$ for each $\alpha \in \mathcal{A}$, and a morphism[9] of $\mathcal{C}$, $f_\alpha^\beta : X_\beta \to X_\alpha$, for each $\alpha \leq \beta \in \mathcal{A}$; these satisfy: (i) $f_\alpha^\alpha = \mathrm{id}_{X_\alpha}$, and (ii) whenever $\alpha \leq \beta \leq \gamma$, $f_\alpha^\gamma = f_\alpha^\beta \circ f_\beta^\gamma : X_\gamma \to X_\alpha$. The morphism f_α^β is called a *bond* of the inverse system.[10] An *inverse limit* of this inverse system consists of: an object X of $\mathcal{C}$, and a morphism $p_\alpha : X \to X_\alpha$ for each $\alpha \in \mathcal{A}$; these satisfy: (i) $p_\alpha = f_\alpha^\beta \circ p_\beta$ for all $\alpha \leq \beta$, and (ii) given an object Z and morphisms $g_\alpha : Z \to X_\alpha$ for all $\alpha \in \mathcal{A}$ such that $g_\alpha = f_\alpha^\beta \circ g_\beta$ whenever $\alpha \leq \beta$, there is a unique morphism $g : Z \to X$ such that $g_\alpha = p_\alpha \circ g$ for all $\alpha \in \mathcal{A}$.

Our main interest is in the categories Sets, R-modules (where R is a ring), and Groups.[11] We are about to construct inverse limits in those categories, and give a usable recognition theorem. However, two abstract remarks may be helpful here: first, because the definition of inverse limit involves a universal property, two inverse limits of the same inverse system are canonically isomorphic; secondly, while inverse limits do not always exist,[12] they do exist whenever $\mathcal{C}$ has arbitrary products, and any two morphisms of $\mathcal{C}$ which have the same domain and codomain have an equalizer.

If $\mathcal{C}$ is the category Sets, there is a particular inverse limit, X, known as "the" inverse limit, namely: $X = \{(x_\alpha) \in \prod_{\alpha \in \mathcal{A}} X_\alpha \mid f_\alpha^\beta(x_\beta) = x_\alpha$ for all $\alpha \leq \beta \in \mathcal{A}\}$, and $p_\alpha : X \to X_\alpha$ is the restriction to X of the projection

---

[7] Readers interested in a fuller treatment of the abstract theory can find it in [109, Chap. 1] or [3]. A more elementary treatment of the R-module case is in [110, Appendix].

[8] Some authors use the term "partially ordered" for "pre-ordered."

[9] It may help to consider the case $\mathcal{A} = \mathbb{N}$. The morphisms point from higher-indexed objects to lower-indexed objects.

[10] We will often abbreviate $\{X_\alpha, f_\beta^\alpha; \mathcal{A}\}$ to $\{X_\alpha, f_\beta^\alpha\}$ or even to $\{X_\alpha\}$.

[11] In later chapters we will also deal with the categories Spaces and Homotopy in this context.

[12] For example, inverse limits do not exist (in general) in the category of CW complexes and homotopy classes of maps.

function. This set X is often denoted by $\varprojlim\{X_\alpha, f_\alpha^\beta\}$. If $\mathcal{C}$ is R-modules or Groups, and $\{X_\alpha, f_\alpha^\beta\}$ is an inverse system in $\mathcal{C}$, "the" inverse limit in $\mathcal{C}$ is constructed by recognizing that $\varprojlim\{X_\alpha, f_\alpha^\beta\}$ is a submodule or subgroup of the product R-module or group $\prod_{\alpha \in \mathcal{A}} X_\alpha$, and each p_α is a homomorphism; i.e., "the" inverse limit in $\mathcal{C}$ is $\varprojlim\{X_\alpha, f_\alpha^\beta\}$ with the appropriate extra structure.

Here is a recognition criterion:

Proposition 11.2.1. *Let $\mathcal{C}$ be R-modules or Groups. Let Z be an object of $\mathcal{C}$ and let $g_\alpha : Z \to X_\alpha$ be a homomorphism for each $\alpha \in \mathcal{A}$, such that $g_\alpha = f_\alpha^\beta \circ g_\beta$ whenever $\alpha \le \beta$. Let $g : Z \to \varprojlim\{X_\alpha, f_\alpha^\beta\} =: X$ be the unique homomorphism such that $p_\alpha \circ g = g_\alpha$ for all $\alpha \in \mathcal{A}$. g is an isomorphism iff $\bigcap_\alpha \ker g_\alpha$ is trivial, and given $(x_\alpha) \in X \subset \prod_\alpha X_\alpha$ there exists $z \in Z$ such that $g_\alpha(z) = x_\alpha$ for all $\alpha \in \mathcal{A}$.* $\square$

Note that the conditions in 11.2.1 are equivalent to: $(Z, \{g_\alpha\})$ is an inverse limit of $\{X_\alpha, f_\alpha^\beta\}$.

We need a category whose objects are inverse systems of objects in (the category) $\mathcal{C}$. We get to the "right" definition, pro-$\mathcal{C}$, by first considering two less suitable competitors.

Let $\mathcal{C}$ be an arbitrary category and let $\mathcal{A}$ be a directed set. We form $\mathcal{C}_{\text{inv}}^{\mathcal{A}}$, *the category of inverse systems over $\mathcal{A}$*, as follows. The objects of $\mathcal{C}_{\text{inv}}^{\mathcal{A}}$ are the inverse systems $\{X_\alpha, f_\alpha^\beta; \mathcal{A}\}$ in $\mathcal{C}$. A morphism from $\{X_\alpha, f_\alpha^\beta; \mathcal{A}\}$ to $\{Y_\alpha, g_\alpha^\beta; \mathcal{A}\}$ is a set of morphisms of $\mathcal{C}$, $h_\alpha : X_\alpha \to Y_\alpha$ such that $h_\alpha \circ f_\alpha^\beta = g_\alpha^\beta \circ h_\beta$ for all $\alpha \le \beta \in \mathcal{A}$. In topological applications, however, one wants to deal with inverse systems indexed by different directed sets at the same time; thus the categories $\mathcal{C}_{\text{inv}}^{\mathcal{A}}$ are inadequate. So we next define a category inv-$\mathcal{C}$. The objects of inv-$\mathcal{C}$ are inverse systems $\{X_\alpha, f_{\alpha'}^\alpha; \mathcal{A}\}$ in $\mathcal{C}$, where $\mathcal{A}$ is no longer fixed. A morphism of inv-$\mathcal{C}$ from $\mathcal{X} := \{X_\alpha, f_{\alpha'}^\alpha; \mathcal{A}\}$ to $\mathcal{Y} := \{Y_\beta, g_{\beta'}^\beta; \mathcal{B}\}$ consists of: a function $\phi : \mathcal{B} \to \mathcal{A}$ and, for each $\beta \in \mathcal{B}$, a morphism of $\mathcal{C}$, $q_\beta : X_{\phi(\beta)} \to Y_\beta$, such that whenever $\beta \le \beta' \in \mathcal{B}$ there exists $\alpha \in \mathcal{A}$ with $\phi(\beta) \le \alpha$ and $\phi(\beta') \le \alpha$, making the following diagram[13] commute in $\mathcal{C}$:

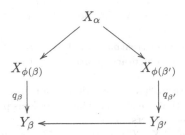

There is an obvious definition of composition in inv-$\mathcal{C}$, and the identity morphism of $\{X_\alpha, f_{\alpha'}^\alpha; \mathcal{A}\}$ consists of $\mathrm{id}_{\mathcal{A}}$, with each $q_\alpha = \mathrm{id}_{X_\alpha}$.

---

[13] Here and throughout, the unmarked arrows are bonds.

The desirable category, pro-$\mathcal{C}$, is a quotient category of inv-$\mathcal{C}$ so it has the same objects. If $\mathcal{X}$ and $\mathcal{Y}$ are objects of inv-$\mathcal{C}$, define an equivalence relation on the set of morphisms of inv-$\mathcal{C}$ from $\mathcal{X}$ to $\mathcal{Y}$ by: $(\phi, \{q_\beta\}) \sim (\phi', \{q'_\beta\})$ iff for each $\beta \in \mathcal{B}$ there exists $\alpha \in \mathcal{A}$ with $\alpha \geq \phi(\beta)$ and $\alpha \geq \phi'(\beta)$ such that the following diagram commutes in $\mathcal{C}$:

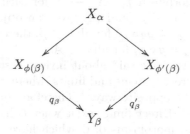

The equivalence classes so defined are the morphisms of pro-$\mathcal{C}$ from $\mathcal{X}$ to $\mathcal{Y}$. The composition of two morphisms of pro-$\mathcal{C}$ is defined to be the equivalence class of the composition of any of their inv-$\mathcal{C}$ representatives. The identity morphism of $\mathcal{X}$ in pro-$\mathcal{C}$ is the equivalence class of $\mathrm{id}_\mathcal{X}$ in inv-$\mathcal{C}$. An isomorphism in pro-$\mathcal{C}$ is called a *pro-isomorphism*. Two objects are *pro-isomorphic* if there is a pro-isomorphism from one to the other.

If $\mathcal{A}' \subset \mathcal{A}$ where $\mathcal{A}$ is pre-ordered, $\mathcal{A}'$ inherits a pre-ordering from $\mathcal{A}$; $\mathcal{A}'$ is *cofinal* in $\mathcal{A}$ if for each $\alpha \in \mathcal{A}$, there is $\alpha' \in \mathcal{A}'$ such that $\alpha \leq \alpha'$. If $\mathcal{A}$ is directed and $\mathcal{A}'$ is cofinal in $\mathcal{A}$ then $\mathcal{A}'$ is directed.

Whenever $\mathcal{A}'$ is cofinal in $\mathcal{A}$, an inverse system $\mathcal{X}\{X_\alpha, f_\beta^\alpha; \mathcal{A}\}$ gives rise to a *cofinal subsystem* $\mathcal{X}' := \{X_{\alpha'}, f_{\beta'}^{\alpha'}; \mathcal{A}'\}$ in which one retains only the sets and bonds indexed by elements of $\mathcal{A}'$. There is a *restriction morphism* $\mathcal{X} \to \mathcal{X}'$ in inv-$\mathcal{C}$ defined by $\mathcal{A}' \hookrightarrow \mathcal{A}$ and $\{\mathrm{id}_{X_{\alpha'}} \mid \alpha' \in \mathcal{A}'\}$.

Proposition 11.2.2. *The restriction morphism induces an isomorphism in* pro-$\mathcal{C}$. ☐

It is primarily because of 11.2.2 that pro-$\mathcal{C}$, rather than inv-$\mathcal{C}$, is the useful category.

If $\mathcal{C}$ is Sets or R-modules or Groups, a morphism $h : \mathcal{X} \to \mathcal{Y}$ of inv-$\mathcal{C}$ clearly induces a morphism $\varprojlim h : \varprojlim \mathcal{X} \to \varprojlim \mathcal{Y}$, obtained by the universal property of inverse limits. Moreover, if $h \sim h'$ then $\varprojlim h = \varprojlim h'$. In fact:

Proposition 11.2.3. *Let $\mathcal{C}$ be* Sets *or R-modules or* Groups *(or, indeed, any category with inverse limits). Then we have covariant functors as follows:* $\varprojlim : \mathcal{C}_{inv}^{\mathcal{A}} \to \mathcal{C}$ *for each directed set $\mathcal{A}$;* $\varprojlim :$ inv-$\mathcal{C} \to \mathcal{C}$; $\varprojlim :$ pro-$\mathcal{C} \to \mathcal{C}$. ☐

The category $\mathcal{C}$ can always be considered to be a full subcategory of pro-$\mathcal{C}$: each object of $\mathcal{C}$ is considered as an inverse system indexed by a one-element directed set. An inverse system in $\mathcal{C}$ is *stable* if it is pro-isomorphic to an object of $\mathcal{C}$.

We now turn to direct systems. A *direct system* $\{X_\alpha, f^\alpha_\beta; \mathcal{A}\}$ in $\mathcal{C}$ consists of: a directed set $\mathcal{A}$, an object X_α of $\mathcal{C}$ for each $\alpha \in \mathcal{A}$, and a morphism of $\mathcal{C}$, $f^\alpha_\beta : X_\alpha \to X_\beta$, for each $\alpha \leq \beta \in \mathcal{A}$; these satisfy: (i) $f^\alpha_\alpha = \mathrm{id}_{X_\alpha}$, and (ii) whenever $\alpha \leq \beta \leq \gamma$, $f^\alpha_\gamma = f^\beta_\gamma \circ f^\alpha_\beta : X_\alpha \to X_\gamma$. Again, f^α_β is called a *bond* of the direct system. A *direct limit* of this direct system consists of: an object X of $\mathcal{C}$ and a morphism $j_\alpha : X_\alpha \to X$ for each $\alpha \in \mathcal{A}$; these satisfy: (i) $j_\alpha = j_\beta \circ f^\alpha_\beta$ for all $\alpha \leq \beta$, and (ii) given an object Z and morphisms $g_\alpha : X_\alpha \to Z$ such that $g_\alpha = g_\beta \circ f^\alpha_\beta$ for all $\alpha \leq \beta$, there is a unique morphism $g : X \to Z$ such that $g_\alpha = g \circ j_\alpha$ for all $\alpha \in \mathcal{A}$.

Almost everything we have said about inverse systems and limits has a "dual" statement for direct systems and limits, where, roughly, "dual" means that all arrows point the opposite way.[14] Therefore, our discussion of direct limits will be shortened. Direct limits exist when $\mathcal{C}$ has arbitrary sums (coproducts), and any two morphisms of $\mathcal{C}$ which have the same domain and codomain have a coequalizer.

In the category Sets, "the" direct limit of $\{X_\alpha, f^\beta_\alpha\}$ is the quotient set, X, of $\coprod_{\alpha \in \mathcal{A}} X_\alpha$ under the equivalence relation obtained by identifying $x_\alpha \in X_\alpha$ with $x_\beta \in X_\beta$ whenever there exists $\gamma \geq \alpha, \beta$ such that $f^\alpha_\gamma(x_\alpha) = f^\beta_\gamma(x_\beta)$; here $j_\beta : X_\beta \to X$ is induced by the canonical inclusion $i_\beta : X_\beta \to \coprod_\alpha X_\alpha$. This set X is denoted $\varinjlim\{X_\alpha, f^\beta_\alpha\}$. If $\mathcal{C}$ is R-modules or Groups, and $\{X_\alpha, f^\beta_\alpha\}$ is a direct system in $\mathcal{C}$, there is an obvious R-module or group structure on $\varinjlim\{X_\alpha, f^\beta_\alpha\}$ with respect to which each j_α is a homomorphism; this[15] is a direct limit in $\mathcal{C}$.

The recognition criterion is:

Proposition 11.2.4. *Let $\mathcal{C}$ be R-modules or Groups. Let Z be an object of $\mathcal{C}$ and let $g_\alpha : X_\alpha \to Z$ be a homomorphism for each $\alpha \in \mathcal{A}$, such that $g_\alpha = g_\beta \circ f^\beta_\alpha$ whenever $\alpha \leq \beta$. Let $g : \varinjlim\{X_\alpha, f^\beta_\alpha\} =: X \to Z$ be the unique homomorphism such that $g \circ j_\alpha = g_\alpha$ for all $\alpha \in \mathcal{A}$. Then g is an isomorphism iff $Z = \cup\{\text{image } g_\alpha \mid \alpha \in \mathcal{A}\}$, and for each α $\ker g_\alpha \subset \cup\{\ker f^\alpha_\beta \mid \beta \geq \alpha\}$.* $\square$

We form the categories of direct systems: $\mathcal{C}^\mathcal{A}_{\mathrm{dir}}$, dir-$\mathcal{C}$, and most importantly ind-$\mathcal{C}$ by analogy with the inverse case. In particular, a morphism of dir-$\mathcal{C}$ from $\mathcal{X} := \{X_\alpha, f^{\alpha'}_\alpha; \mathcal{A}\}$ to $\mathcal{Y} := \{Y_\beta, f^{\beta'}_\beta; \mathcal{B}\}$ consists of: a function $\phi : \mathcal{A} \to \mathcal{B}$ and, for each $\alpha \in \mathcal{A}$, a morphism of $\mathcal{C}$, $q_\alpha : X_\alpha \to Y_{\phi(\alpha)}$, such that whenever $\alpha \leq \alpha' \in \mathcal{A}$ there exists $\beta \in \mathcal{B}$ with $\phi(\alpha) \leq \beta$ and $\phi(\alpha') \leq \beta$ making the following diagram commute in $\mathcal{C}$:

---

[14] Direct limits in $\mathcal{C}$ are precisely inverse limits in the opposite category $\mathcal{C}^{\mathrm{opp}}$, and vice versa.

[15] The category theoretic sum of the R-modules or groups $\{X_\alpha \mid \alpha \in \mathcal{A}\}$ does not have $\coprod_\alpha X_\alpha$ as its underlying set. Nonetheless, our construction does yield a direct limit in those categories.

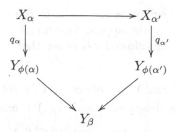

The morphisms of ind-$\mathcal{C}$ from $\mathcal{X}$ to $\mathcal{Y}$ are the equivalence classes generated by $(\phi, \{q_\alpha\}) \sim (\phi', \{q'_\alpha\})$ iff for each $\alpha \in \mathcal{A}$ there exists $\beta \in \mathcal{A}$ with $\beta \geq \phi(\alpha)$ and $\beta \geq \phi'(\alpha)$ such that the following diagram commutes in $\mathcal{C}$:

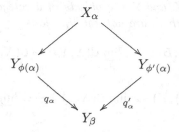

The terms *ind-isomorphism* and *ind-isomorphic* are analogous to their "pro" counterparts.

If $\mathcal{X}'$ is a cofinal (direct) subsystem of $\mathcal{X}$, there is a *corestriction morphism* $\mathcal{X}' \to \mathcal{X}$ in dir-$\mathcal{C}$ defined by $\mathcal{A}' \hookrightarrow \mathcal{A}$ and $\{\mathrm{id}_{X_{\alpha'}} \mid \alpha' \in \mathcal{A}'\}$.

Proposition 11.2.5. *The corestriction morphism induces an isomorphism in* ind-$\mathcal{C}$. $\qquad\Box$

Proposition 11.2.6. *Let $\mathcal{C}$ be Sets or R-modules or Groups. Then we have covariant functors as follows:* $\varinjlim : \mathcal{C}^{\mathcal{A}}_{dir} \to \mathcal{C}$ *for each directed set $\mathcal{A}$;* $\varinjlim :$ dir-$\mathcal{C} \to \mathcal{C}$; $\varinjlim :$ ind-$\mathcal{C} \to \mathcal{C}$. $\qquad\Box$

$\mathcal{C}$ is considered to be a full subcategory of ind-$\mathcal{C}$.

Sometimes, one can avoid reference to the complicated definition of a morphism in pro-$\mathcal{C}$ or ind-$\mathcal{C}$. Using the notation $\mathcal{K}(A, B)$ for the set of morphisms in the category $\mathcal{K}$ from the object A to the object B, we have:

Proposition 11.2.7. (a) *there is a natural bijection*

$$\mathrm{pro}\text{-}\mathcal{C}(\{X_\alpha, f^\alpha_{\alpha'}; \mathcal{A}\}, \{Y_\beta, g^\beta_{\beta'}; \mathcal{B}\}) \to \varprojlim_\beta \varinjlim_\alpha \mathcal{C}(X_\alpha, Y_\beta);$$

(b) *there is a natural bijection*

$$\mathrm{ind}\text{-}\mathcal{C}(\{X_\alpha, f^{\alpha'}_\alpha; \mathcal{A}\}, \{Y_\beta, g^{\beta'}_\beta; \mathcal{B}\}) \to \varprojlim_\alpha \varinjlim_\beta \mathcal{C}(X_\alpha, Y_\beta).$$

$\qquad\Box$

Here, inverse and direct limits are taken in the category Sets. The α or β under an arrow indicates the appropriate directed set $\mathcal{A}$ or $\mathcal{B}$. From now on this convention will be followed whenever there is ambiguity about the appropriate directed set.

Corollary 11.2.8. *If X and Y are objects of $\mathcal{C}$, pro-$\mathcal{C}(\{X_\alpha, f_{\alpha'}^\alpha; \mathcal{A}\}, Y)$ is in natural bijective correspondence with $\varinjlim_{\alpha} \mathcal{C}(X_\alpha, Y)$; and ind-$\mathcal{C}(X, \{Y_\beta, g_\beta^{\beta'}; \mathcal{B}\})$ is in natural bijective correspondence with $\varinjlim_{\beta} \mathcal{C}(X, Y_\beta)$.* $\square$

From the universal properties for inverse and direct limits, we get:

Corollary 11.2.9. *If X and Y are objects of a category $\mathcal{C}$ in which inverse and direct limits exist, there are natural bijections*

$$\text{pro-}\mathcal{C}(X, \{Y_\beta, g_{\beta'}^\beta; \mathcal{B}\}) \;\to\; \varprojlim_{\beta} \mathcal{C}(X, Y_\beta) \to \mathcal{C}(X, \varprojlim_{\beta}\{Y_\beta, g_{\beta'}^\beta; \mathcal{B}\})$$

$$\text{ind-}\mathcal{C}(\{X_\alpha, f_\alpha^{\alpha'}; \mathcal{A}\}, Y) \;\to\; \varprojlim_{\alpha} \mathcal{C}(X_\alpha, Y) \to \mathcal{C}(\varinjlim_{\alpha}\{X_\alpha, f_\alpha^{\alpha'}; \mathcal{A}\}, Y).$$

$\square$

In the literature, the inverse limit is also called the *limit* or *projective limit*, and the direct limit is also called the *colimit* or *inductive limit*. This explains the notations pro-$\mathcal{C}$ and ind-$\mathcal{C}$.

An *inverse sequence* in a category $\mathcal{C}$ is an inverse system whose directed set is $\mathbb{N}$ with the usual ordering. In such an inverse system, every f_m^n is fully determined (by composition) when the bonds $f_n^{n+1} : X_{n+1} \to X_n$ are specified. So we simplify notation further, writing f_n for f_n^{n+1}, and even writing "$\{X_n\}$ is an inverse sequence in $\mathcal{C}$" when the context makes clear what the bonds are. The definition of *direct sequence* is dual, and the dual remarks on notation apply.

Nearly all the inverse and direct systems occurring in this book are sequences. When dealing with cofinal subsequences of inverse or direct sequences we will use a less formal notation. If $\{G_n\}$ is an inverse or direct sequence and $\{n_k \mid k \in \mathbb{N}\}$ is a subsequence, we write $\{G_{n_k}\}$ for the corresponding cofinal subsequence of $\{G_n\}$.

Here is a convenient way of recognizing an isomorphism in pro-$\mathcal{C}$ between inverse sequences in $\mathcal{C}$. Let $\mathcal{X}\{X_m\}$ and $\mathcal{Y}\{Y_n\}$ be inverse sequences, $\{X_{m_k}\}$ and $\{Y_{n_k}\}$ cofinal subsequences, and $f_{n_k} : X_{m_k} \to Y_{n_k}$ morphisms of $\mathcal{C}$ which commute with the appropriate bonds. We may assume $k \leq m_k$ and $k \leq n_k$. Defining $\phi(k) = m_k$, the morphisms (bond $\circ f_{n_k}$) $: X_{m_k} \to Y_k$ define a morphism $f : \mathcal{X} \to \mathcal{Y}$ of inv-$\mathcal{C}$. Conversely, any morphism of inv-$\mathcal{C}$, $f : \mathcal{X} \to \mathcal{Y}$, yields subsequences and morphisms $f_{n_k} : X_{m_k} \to Y_{n_k}$ as above.

Proposition 11.2.10. *Let* $\mathcal{X}$ *and* $\mathcal{Y}$ *be inverse sequences. A morphism* $f :$ $\mathcal{X} \to \mathcal{Y}$ *of* inv-$\mathcal{C}$ *induces an isomorphism of* pro-$\mathcal{C}$ *iff for suitable subsequences as above there are morphisms* g_{m_k} *of* $\mathcal{C}$ *making the following diagram commute for all* k:

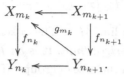

$\square$

Source Notes: The formalism of pro-categories was introduced into algebraic geometry by Grothendieck as part of his program to prove the Weil Conjectures. An early source applying these ideas to topology is [3]. This language was co-opted into shape theory (see Sect. 17.6) at roughly the same time in [128], [108], and [59]. The transfer to group theory set out in this book was made explicit in [70], [71] and [68].

Exercises

1. Give an example of a preordered set which is not a poset. *Hint*: Consider the set of open covers of a space, where $\mathcal{V} \leq \mathcal{U}$ means that $\mathcal{V}$ refines $\mathcal{U}$.
2. Express the definition of "inverse limit" by means of a universal property.
3. Write out the definition of composition in inv-$\mathcal{C}$ in detail.
4. Show that composition of morphisms of pro-$\mathcal{C}$ (as indicated in the text) is well defined.
5. Show by examples that $\varprojlim$ and $\varinjlim$ cannot be interchanged in 11.2.7.
6. State and prove the dual of 11.2.10.

11.3 The derived limit

We define $\varprojlim^1$ for inverse sequences and discuss some of the associated algebra. For topological motivation, proceed to Sect. 11.4, referring back to this section when necessary.

The categories $\mathcal{C}$ of most interest here are: Sets (= sets and functions), Groups (= groups and homomorphisms), Left R-modules (= left R-modules and homomorphisms) and Right R-modules. We will often combine the last two under the name "R-modules"; this should be interpreted consistently as Left R-modules or Right R-modules.

When $\mathcal{C}$ is R-modules or Groups, there is another sort of limit, called $\varprojlim^1$, which measures the amount of information lost on passing from an inverse sequence $\{X_n\}$ to its inverse limit $\varprojlim\{X_n\}$. For example, the inverse sequences of $\mathbb{Z}$-modules $\{\mathbb{Z} \xleftarrow{\times 2} \mathbb{Z} \xleftarrow{\times 2} \ldots\}$ and $\{0 \longleftarrow 0 \longleftarrow \ldots\}$ have trivial inverse limits, but are not isomorphic in pro-$(\mathbb{Z}$-modules), as can easily be checked.

It will turn out that $\varprojlim^1$ of the first sequence is very large, while $\varprojlim^1$ of the second is trivial.

Let $\{M_n\}$ be an inverse sequence of R-modules. Consider the *shift* homomorphism $s : \prod_{n=1}^{\infty} M_n \to \prod_{n=1}^{\infty} M_n$, $(x_1, x_2, \ldots) \mapsto (x_1 - f_1(x_2), x_2 - f_2(x_3), \ldots)$. Obviously, the kernel of s is $\varprojlim\{M_n\}$. The cokernel of s, $\prod M_n/\text{image } s$, is called $\varprojlim^1\{M_n\}$, the *first derived limit*. It is an R-module.

We can also define $\varprojlim^1$ in the case of groups. If $\{G_n\}$ is an inverse sequence of groups, $\varprojlim^1\{G_n\}$ is the quotient set of the set $\prod_{n=1}^{\infty} G_n$ under the equivalence relation generated by: $(x_n) \sim (y_n)$ iff there exists $(z_n) \in \prod G_n$ such that $y_n = z_n x_n f_n(z_{n+1}^{-1})$ for all n. In this case we regard $\varprojlim^1\{G_n\}$ as a pointed set whose base point, denoted by $[1]$, is the equivalence class of $(1, 1, \ldots)$. We say that $\varprojlim^1\{G_n\}$ is *trivial* if it is the pointed set $\{[1]\}$. When every G_n is abelian, our definition of $\varprojlim^1$ reduces to the previous one for $\mathbb{Z}$-modules, and $\varprojlim^1\{G_n\}$ then acquires the additional structure of a $\mathbb{Z}$-module.

If $\mathcal{C}$ is R-modules or Abelian Groups, $\varprojlim^1$ defines a covariant functor $\mathcal{C}_{\text{inv}}^{\mathbb{N}} \to \mathcal{C}$ in the obvious way.[16] Similarly, Groups$_{\text{inv}}^{\mathbb{N}} \to$ Pointed Sets.

We now discuss the vanishing and exactness properties of $\varprojlim^1$ both for R-modules and for groups. Since these properties for R-modules only depend on the underlying abelian group structure it is enough to consider the case of groups.

An inverse sequence $\{G_n\}$ of groups is *semistable* or *Mittag-Leffler* or *pro-epimorphic* if for each m there exists $\phi(m) \geq m$ such that for all $k \geq \phi(m)$, image $f_m^{\phi(m)} =$ image f_m^k. The point of this notion is:[17]

Proposition 11.3.1. $\{G_n\}$ *is semistable iff* $\{G_n\}$ *is isomorphic in the category* pro-Groups *to an inverse sequence whose bonds are epimorphisms.* $\square$

Theorem 11.3.2. *If* $\{G_n\}$ *is semistable then* $\varprojlim^1\{G_n\}$ *is trivial. If* $\varprojlim^1\{G_n\}$ *is trivial and each group* G_n *is countable then* $\{G_n\}$ *is semistable.*

Proof. A convenient notation for this proof is: if $G_r \leftarrow G_s$ is a bond and $g_s \in G_s$, then the image of g_s in G_r is denoted by g_r^s, and the image of G_s in G_r is denoted by G_r^s.

---

[16] There is also an induced covariant functor $\varprojlim^1 :$ towers-$\mathcal{C} \to \mathcal{C}$ (or towers-Groups $\to$ Pointed Sets) where towers-$\mathcal{C}$ is the full subcategory of pro-$\mathcal{C}$ generated by inverse sequences in $\mathcal{C}$. For the most part, we can avoid using this.

[17] This proposition justifies the term "pro-epimorphic" which, unfortunately, is not in standard use. Later we will need the companion notion "pro-monomorphic." It is good to keep this pairing of terms in mind. We will discuss "stable" inverse sequences in R-modules and in Groups; "semistable and pro-monomorphic" is equivalent to "stable." Note the general definition of "stable" in Sect. 11.2.

Assume $\{G_n\}$ is semistable and let $(x_n) \in \prod_n G_n$. With ϕ as in the definition of "semistable," define $\psi : \mathbb{N} \to \mathbb{N}$ by $\psi(1) = 1$ and $\psi(k+1) = \phi(\psi(k)+1)$. We may assume ϕ is strictly increasing, hence also ψ. By induction we choose $y_{\psi(k)} \in G_{\psi(k)} : y_{\psi(2)}$ is arbitrary and, for $k \geq 1$, $y_{\psi(k)}$ satisfies

$$y_{\psi(k)}^{\psi(k+1)} \left(y_{\psi(k)}^{\psi(k+2)} \right)^{-1} = x_{\psi(k)}^{\psi(k+1)} x_{\psi(k)}^{\psi(k+1)+1} \cdots x_{\psi(k)}^{\psi(k+2)-1}.$$

For $\psi(k-1) \leq m < \psi(k)$ define

$$z_m = x_m x_m^{m+1} \cdots x_m^{\psi(k+1)-1} y_m^{\psi(k+1)}.$$

One checks that $x_m = z_m (z_m^{m+1})^{-1}$; this is clear when $\psi(k-1) \leq m < m+1 < \psi(k)$ and it also holds when $m + 1 = \psi(k)$. Thus (x_n) is equivalent to [1], so $\varprojlim^1 \{G_n\}$ is trivial.

Conversely, assume $\varprojlim^1 \{G_n\}$ is trivial with each G_n countable, and suppose $\{G_n\}$ is not semistable. Then there is a sequence (n_k) such that, for all $k \geq 0$, $G_{n_0}^{n_{k+1}} \subsetneq G_{n_0}^{n_k}$. Write $H_k := G_{n_0}^{n_k}$. By Exercises 2 and 3, $\varprojlim_k^1 \{H_k\}$ is trivial and hence $\varprojlim_n^1 \{H_{k_n}\}$ is trivial for any subsequence (k_n). Choose $x_k \in H_k - H_{k+1}$. For every strictly increasing function $\alpha : \mathbb{N} \to \mathbb{N}$ there is $z_k^{(\alpha)} \in H_{\alpha(k)}$ such that $x_{\alpha(k)} = z_k^{(\alpha)}(z_{k+1}^{(\alpha)})^{-1}$. Since H_0 is countable and there are uncountably many such functions α there must be two, say α and β, such that $\alpha \neq \beta$ but $\alpha(1) = \beta(1)$. Let n be the greatest integer such that $\alpha(i) = \beta(i)$ for all $i \leq n$. Then $z_n^{(\alpha)} = z_n^{(\beta)}$ and $z_{n+1}^{(\alpha)} \neq z_{n+1}^{(\beta)}$, so $x_{\alpha(n)} \neq x_{\beta(n)}$. We may assume $\alpha(n+1) > \beta(n+1)$. Then $x_{\beta(n)} = z_n^{(\alpha)}(z_{n+1}^{(\beta)})^{-1}$ where $z_{n+1}^{(\beta)} \in H_{\beta(n+1)} \subset H_{\beta(n)+1}$, $z_n^{(\alpha)} \in H_{\alpha(n)} \subset H_{\beta(n)+1}$ and $x_{\beta(n)} \notin H_{\beta(n)+1}$. This is a contradiction. $\qquad \square$

Remark 11.3.3. A counterexample to the second part of 11.3.2 when the groups G_n are uncountable can be found in [58].

A diagram of pointed sets and functions $(A, a) \xrightarrow{f} (B, b) \xrightarrow{g} (C, c)$ is *exact* at (B, b) if image f = kernel g where kernel g is defined[18] to be $g^{-1}(c)$. Homomorphisms of groups $G' \to G \to G''$ are *exact* at G if they are exact as functions of pointed sets, where each group has its identity element as base point. A *short exact sequence* of inverse sequences of groups consists of: inverse sequences $\{G_n'\}$, $\{G_n\}$ and $\{G_n''\}$ of groups, and, for each n, an exact sequence

$$\{1\} \to G_n' \xrightarrow{i_n} G_n \xrightarrow{p_n} G_n'' \to \{1\}$$

such that, for all n, $f_n \circ i_{n+1} = i_n \circ f_n'$ and $f_n'' \circ p_{n+1} = p_n \circ f_n$. In the language of Sect. 11.2, $\{i_n\}$ and $\{p_n\}$ are morphisms of $(\text{Groups})_{\text{inv}}^{\mathbb{N}}$.

---

[18] Warning: In Pointed Sets, if kernel $g = \{b\}$, it does not follow that g is injective.

Proposition 11.3.4. *In this situation, there is an exact sequence of pointed sets*

$$\{1\} \longrightarrow \varprojlim\{G'_n\} \xrightarrow{\varprojlim\{i_n\}} \varprojlim\{G_n\} \xrightarrow{\varprojlim\{p_n\}} \varprojlim\{G''_n\}$$

$$\xrightarrow{\ \delta\ } \varprojlim{}^1\{G'_n\} \xrightarrow{\varprojlim^1\{i_n\}} \varprojlim{}^1\{G_n\} \xrightarrow{\varprojlim^1\{p_n\}} \varprojlim{}^1\{G''_n\} \longrightarrow \{1\}.$$

Here, δ maps (x''_n) to the equivalence class of $(x'_n) \in \prod_n G'_n$, where $x'_n = i_n^{-1}(x_n \cdot f_n(x_{n+1}^{-1}))$, x_n being any member of $p_n^{-1}(x''_n)$. This exact sequence is natural with respect to morphisms of $(\text{Groups})_{inv}^{\mathbb{N}}$.

Proof. This is a long but straightforward check that δ is well defined and kernel = image in each position. See page 168 of [109] for details. □

Notice the similarity between Proposition 11.3.4 and the derivation of the long exact sequence in homology from a short exact sequence of chain complexes (Sect. 2.1). In fact, in the category Abelian Groups or the category R-modules there is a simple proof of 11.3.4. One regards

$$\cdots \longrightarrow 0 \longrightarrow \prod_n G_n \xrightarrow{\ s\ } \prod_n G_n \longrightarrow 0$$

as a chain complex whose H_1 is $\varprojlim\{G_n\}$ and whose H_0 is $\varprojlim^1\{G_n\}$. The hypothesis of 11.3.4 is interpreted as a short exact sequence of such chain complexes, and the conclusion of 11.3.4 is just the corresponding homology exact sequence. In particular in the R-module case, this proof establishes that δ is a homomorphism of R-modules. In fact, we have:

Corollary 11.3.5. *If we start with a short exact sequence of inverse sequences of R-modules in Proposition 11.3.4, the resulting "six-term" exact sequence consists of R-modules and homomorphisms.* □

The next proposition is left as an exercise.[19]

Proposition 11.3.6. *The restriction morphism $\{G_n\} \to \{G_{n_k}\}$ in inv-$\mathcal{C}$ induces a bijection of pointed sets $\varprojlim_n^1\{G_n\} \to \varprojlim_k^1\{G_{n_k}\}$. In the case of R-modules this bijection is an isomorphism.* □

We now consider how $\varprojlim$ and $\varprojlim^1$ behave with respect to homology. Consider an inverse sequence of chain complexes of R-modules $(C^{(i)}, \partial^{(i)})$ and chain maps $C^{(i+1)} \xrightarrow{\phi_i} C^{(i)}$. Then $(C, \partial) := (\varprojlim_i\{C^{(i)}\}, \varprojlim_i \partial^{(i)})$ and $(C', \partial') := (\varprojlim_i^1\{C^{(i)}\}, \varprojlim_i^1 \partial^{(i)})$ are chain complexes.

---

[19] In the case of finitely presented groups a proof of 11.3.6 using a topological interpretation of $\varprojlim^1$ is sketched in Remark 16.1.3.

Proposition 11.3.7. *Let $C'_n = 0$ for all n. Then for each n there is a short exact sequence*

$$0 \to \varprojlim_i{}^1 H_{n+1}(C^{(i)}) \xrightarrow{\bar{a}} H_n(C) \xrightarrow{\bar{b}} \varprojlim_i H_n(C^{(i)}) \to 0$$

This sequence is natural with respect to morphisms of $(R\text{-chain complexes})^{\mathbb{N}}_{inv}$.

Proof. Since $C'_n = 0$ for all n we have a short exact sequence of chain complexes

$$0 \longrightarrow C \longrightarrow \prod_{i=1}^{\infty} C^{(i)} \xrightarrow{s} \prod_{i=1}^{\infty} C^{(i)} \longrightarrow 0$$

giving, as in Sect. 2.1, a long exact sequence

$$\cdots \longrightarrow \prod_{i=1}^{\infty} H_{n+1}(C^{(i)}) \xrightarrow{s_{n+1}} \prod_{i=1}^{\infty} H_{n+1}(C^{(i)}) \longrightarrow H_n(C)$$

$$\longrightarrow \prod_{i=1}^{\infty} H_n(C^{(i)}) \xrightarrow{s_n} \cdots$$

Interpreting $\ker(s_n)$ and $\mathrm{coker}(s_{n+1})$ as a $\varprojlim$ and a $\varprojlim^1$ we get the required sequence. $\qquad\square$

A similar proof using the short exact sequence

$$0 \longrightarrow \prod_{i=1}^{\infty} C^{(i)} \xrightarrow{s} \prod_{i=1}^{\infty} C^{(i)} \longrightarrow C' \longrightarrow 0 \text{ gives:}$$

Proposition 11.3.8. *Let $C_n = 0$ for all n. Then for each n there is a short exact sequence*

$$0 \to \varprojlim_i{}^1 H_n(C^{(i)}) \xrightarrow{a} H_n(C') \xrightarrow{b} \varprojlim_i H_{n-1}(C^{(i)}) \to 0.$$

This sequence is natural with respect to morphisms of $(R\text{-chain complexes})^{\mathbb{N}}_{inv}$. $\square$

Remark 11.3.9. Sometimes, one wants explicit descriptions of a and b in 11.3.8. The homomorphism a is induced by inclusion. To describe b, we start with $(x_i) \in \ker \partial' = Z_n(C')$ representing an element of $H_n(C')$. There exists $(c_i) \in \prod_i C^{(i)}_{n-1}$ such that $\partial x_i = c_i - \phi_i(c_{i+1})$. Since $0 = \partial c_i - \phi_i(\partial c_{i+1})$, $(\partial c_i) \in \varprojlim_i B_{n-2}(C^{(i)}) = 0$, hence $\partial c_i = 0$ for all i. So $(c_i) \in \prod_i Z_{n-1}(C^{(i)})$, and $\phi_i(c_{i+1}) = c_i - \partial x_i$. Thus (c_i) defines an element of $\varprojlim_i Z_{n-1}(C^{(i)})/B_{n-1}(C^{(i)})$.

The homomorphism b maps $\{(x_i)\}$ to $(\{c_i\})$, where $\{\cdot\}$ denotes a homology class. The analogous description of $\bar{a}$ and $\bar{b}$ is left to the reader.

For matters discussed in this book it will not be important to compute $\varprojlim^1\{G_n\}$. The important question is: is it trivial or non-trivial. For information on the structure of $\varprojlim^1\{G_n\}$ in the abelian case see [94]. In the non-abelian case, the pointed set $\varprojlim^1\{G_n\}$ is either trivial or uncountable; some related issues of topological interest are discussed in [99] and [69].

Although the vanishing of $\varprojlim^1$ of an inverse sequence of R-modules depends only on the underlying abelian group structure, the fact that the bonds are homomorphisms of R-modules can play a useful role. For example:

Proposition 11.3.10. *Let $\{M_n\}$ be an inverse sequence in the category R-modules, where each M_n is finitely generated. If R is a field, then $\{M_n\}$ is semistable, hence $\varprojlim^1\{M_n\} = 0$.*

Proof. For each m, the R-dimensions of the vector spaces image $(M_n \to M_m)$ are non-increasing as n increases. Since M_m has finite R-dimension, this function of n becomes constant. Hence $\{M_n\}$ is semistable. Apply 11.3.2. $\square$

Next, we discuss $\varprojlim^1$ of sequences of finitely generated torsion modules. If M is an R-module, the *torsion submodule* is tor $M\{m \in M \mid rm = 0$ for some non-zero $r \in R\}$. M is *torsion* if tor $M = M$; M is *torsion free* if tor $M = 0$; M is *cyclic* if M is isomorphic to $R/(r)$ where $0 \neq r \in R$ and (r) is the ideal generated by r; equivalently, M is cyclic if it is generated by a single non-zero element.

Proposition 11.3.11. *Let R be a PID and let M be a finitely generated R-module. Then M is isomorphic to $F \oplus$ tor M, where F is the direct sum of ρ copies of R, tor M is the direct sum of τ cyclic modules $R/(r_i)$, and r_i divides r_{i+1} when $i < \tau$. The integers ρ and τ are unique, and the elements $r_1, \cdots, r_\tau$ are unique up to multiplication by invertible elements of R. Moreover, tor M has the following minimal property: any decreasing chain of submodules contains only finitely many members.*

Proof. The first part is well known. For the last part, there exists $r \in R$ which annihilates tor M. So tor M is a module over the ring $R/(r)$. This latter ring satisfies the descending chain condition (see page 243 of [156]), hence so does tor M; see p. 158 of [156]. $\square$

The integer ρ in 11.3.11 is called the *rank* of the free R-module F. An infinitely generated free R-module is said to have *infinite rank* or rank ∞.

Proposition 11.3.12. *If $\{M_n\}$ is an inverse sequence of finitely generated torsion R-modules, where R is a PID, then $\{M_n\}$ is semistable. Hence $\varprojlim_n^1\{M_n\} = 0$.*

Proof. This is immediate from the last sentence of 11.3.11. $\square$

Of course when $R = \mathbb{Z}$ (an important case) the modules M_n in 11.3.12 are finite. We note a consequence of 11.3.2:

Proposition 11.3.13. *If $\{G_n\}$ is an inverse sequence of finite groups, then $\varprojlim\limits_n {}^1\{G_n\}$ is trivial.* $\square$

Turning to direct sequences, the analog of 11.3.4 is much simpler:

Proposition 11.3.14. *Let $\{G'_n\}$, $\{G_n\}$ and $\{G''_n\}$ be direct sequences of groups. For each n, let*

$$\{1\} \longrightarrow G'_n \xrightarrow{\;i_n\;} G_n \xrightarrow{\;p_n\;} G''_n \longrightarrow \{1\}$$

be a short exact sequence such that i_n and p_n commute with the bonds, i.e., a short exact sequence of direct sequences. Then there is a short exact sequence of groups

$$\{1\} \longrightarrow \varinjlim\{G'_n\} \xrightarrow{\;\lim\{i_n\}\;} \varinjlim\{G_n\} \xrightarrow{\;\lim\{p_n\}\;} \varinjlim\{G''_n\} \longrightarrow \{1\}.$$

This sequence is natural with respect to morphisms of $(\text{Groups})_{\mathrm{dir}}^{\mathbb{N}}$. $\square$

An R-module M is *finitely presented* if there is a short exact sequence $0 \to K \to F \to M \to 0$ in which F is free while both F and K are finitely generated. When R is a PID, M is finitely presented iff M is finitely generated.

Proposition 11.3.15. *Let $\{M_n\}$ and $\{N_n\}$ be direct sequences of finitely presented R-modules having isomorphic direct limits. Then $\{M_n\}$ and $\{N_n\}$ are ind-isomorphic.* $\square$

The following is well-known (compare 4.1.7 in [146]):

Proposition 11.3.16. *Let $\{(C^{(i)}, \partial^{(i)}), \phi_i\}$ be a direct sequence of R-chain complexes and chain maps, and let (C, ∂) be the direct limit chain complex. Then for each n, there is an isomorphism $c : \varinjlim\limits_i H_n(C^{(i)}) \to H_n(C)$, which is natural with respect to morphisms of $(R\text{-chain complexes})_{\mathrm{dir}}^{\mathbb{N}}$. Explicitly, if an element of $\varinjlim\limits_i H_n(C^{(i)})$ is represented by $\{z\} \in H_n(C^{(j)})$ where $z \in Z_n(C^{(j)})$, then its image under c is represented by the image of z in C under the canonical homomorphism $C^{(j)} \to C$.* $\square$

Source Notes: $\varprojlim{}^1$ in the context of non-abelian groups first appeared in [21]. Theorem 11.3.2 appeared in [73] in the abelian case; the general case appeared in [67].

Exercises

1. Prove 11.3.1.

2. Let $\{G_n\}$ and $\{H_n\}$ be inverse sequences of groups and assume there are epi-morphisms $G_n \twoheadrightarrow H_n$ which commute with the bonds. Prove that if $\varprojlim^1\{G_n\}$ is trivial then $\varprojlim^1\{H_n\}$ is trivial.

3. Let the bonds in $\{G_n\}$ be monomorphisms and let $\varprojlim^1\{G_n\}$ be trivial. Then every cofinal subsequence also has trivial $\varprojlim^1$. (Note that this is used in our proof of 11.3.2 and so should be done without using 11.3.2.)

4. Prove 11.3.6.

5. Establish $\varprojlim^1$: towers-Groups $\to$ Pointed Sets and $\varprojlim^1$: towers-R-modules $\to$ R-modules as functors.

6. Give a version of Remark 11.3.9 for $\bar{a}$ and $\bar{b}$.

7. Prove 11.3.15.

8. Prove: $\varprojlim_n^1\left\{\bigoplus_n^{\infty}\mathbb{Z}\right\}$ is isomorphic to $\left(\prod_1^{\infty}\mathbb{Z}\right)\Big/\bigoplus_1^{\infty}\mathbb{Z}$.

9. Prove that if $\{G_n\}$ and $\{H_n\}$ are inverse sequences of countable groups, then $\{G_n \times H_n\}$ is semistable iff $\{G_n\}$ and $\{H_n\}$ are semistable.

10. An inverse sequence $\{G_n\}$ in Groups is *pro-trivial* if for each m there exists $n \geq m$ such that the image of f_m^n is trivial. Prove that when each G_n is countable, $\{G_n\}$ is pro-trivial iff $\varprojlim\{G_n\}$ and $\varprojlim^1\{G_n\}$ are trivial. Prove that $\{G_n\}$ is pro-trivial iff it is isomorphic in pro-Groups to the trivial inverse system $\{1\}$.

11. In Sect. 7.4 we defined a direct sequence $\{G_n\}$ in Groups to be *ind-trivial* if for each m there exists $n \geq m$ such that the image of f_n^m is trivial. Prove that when each G_n is finitely generated, $\{G_n\}$ is ind-trivial iff $\varinjlim_n\{G_n\} = \{1\}$ iff $\{G_n\}$ is isomorphic in ind-Groups to the trivial direct system $\{1\}$.

12. Let $\{G_{m,n}\}$ be an inverse system of groups indexed by the directed set $(\mathbb{N}\times\mathbb{N}, <)$ where the partial ordering $<$ is generated by $(m, n) < (m + 1, n)$ and $(m, n) < (m, n + 1)$. Construct a short exact sequence of pointed sets[20]

$$\varprojlim_n^1 \varprojlim_m\{G_{m,n}\} \rightarrowtail \varprojlim_n^1\{G_{n,n}\} \twoheadrightarrow \varprojlim_n \varprojlim_m^1\{G_{m,n}\}.$$

11.4 Homology of ends

Let X be an oriented CW complex having locally finite type. So far, we have made little attempt to compute $H_*^{\infty}(X; R)$. We will be interested in doing so only when X is countably infinite, but even in this case it may happen that $H_*^{\infty}(X; R)$ is uncountable. We will see examples; in the meantime we point out that $C_n^{\infty}(X; R)$ is uncountable when X has infinitely many n-cells.

It is convenient to define related groups $H_*^e(X; R)$. Consider the short exact sequence of chain complexes:

---

[20] See [21] and [103] for more on this.

$$\begin{CD}
@. @VVV @VVV @VVV @. \\
0 @>>> C_n(X;R) @>i>> C_n^\infty(X;R) @>p>> C_n^\infty(X;R)/C_n(X;R) @>>> 0 \\
@. @VV\partial V @VV\partial V @VV\partial V @. \\
0 @>>> C_{n-1}(X;R) @>i>> C_{n-1}^\infty(X;R) @>p>> C_{n-1}^\infty(X;R)/C_{n-1}(X;R) @>>> 0 \\
@. @VVV @VVV @VVV @.
\end{CD}$$

where i is inclusion and p is projection. We define

$$H_{n-1}^e(X;R) := H_n(C_*^\infty(X;R)/C_*(X;R)).$$

We call $H_*^e(X;R)$ the *homology of the end*[21] of X.

Applying the usual algebra (see Sect. 2.1) we get:

Proposition 11.4.1. *There is an exact sequence*

$$\cdots \longrightarrow H_n(X;R) \overset{i_*}{\longrightarrow} H_n^\infty(X;R) \overset{p_*}{\longrightarrow} H_{n-1}^e(X;R)$$

$$\overset{\partial_*}{\longrightarrow} H_{n-1}(X;R) \longrightarrow \cdots$$

$\square$

If X has finite type, i_* is an isomorphism for all n, so $H_*^e(X;R) = 0$. If X is n-connected, $n \geq 1$, and X^1 is infinite, we can apply Exercise 5 of Sect. 7.1 to get $H_i^\infty(X;R) \cong H_{i-1}^e(X;R)$ for $2 \leq i \leq n$ and, using 11.1.3, $H_0^e(X;R) \cong H_1^\infty(X;R) \oplus R$. More generally, 11.4.1 shows that if one knows two of $H_*(X;R)$, $H_*^\infty(X;R)$ and $H_*^e(X;R)$, the exact sequence gives information about the third.

Clearly, $H_n^e(X;R) \cong Z_n^e(X;R)/B_n^e(X;R)$ where

$$Z_n^e(X;R) = \{c \in C_{n+1}^\infty(X;R) \mid \partial c \in C_n(X;R)\}$$
$$B_n^e(X;R) = B_{n+1}^\infty(X;R) + C_{n+1}(X;R) \subset C_{n+1}^\infty(X;R).$$

The elements of $Z_n^e(X;R)$ and $B_n^e(X;R)$ are called *n-cycles of the end* of X and *n-boundaries of the end* of X, respectively.

Proposition 11.4.2. *Let X be path connected. Then $H_n^e(X;R) = 0$ for all $n < 0$.*

---

[21] The reason for the shift of dimension is explained after 11.4.13.

Proof. For $n < -1$, this is trivial. For $n = -1$, it follows from 11.1.3 together with the above exact sequence when X^1 is infinite. When X^1 is finite then X^0 is finite, so $Z^e_{-1} = C^\infty_0 = C_0 = B^e_{-1}$.   □

Just as in Sect. 11.1, CW-proper maps induce homomorphisms on H^e_* possessing the usual functorial properties. We leave their statements to the reader. Note in particular that H^e_* is a proper homotopy invariant of strongly locally finite CW complexes.

Countability will be important to us, so we note:

Proposition 11.4.3. *A path connected CW complex X having locally finite type is countable.*

Proof. We begin with X^1. If X^1 is infinite then, by 11.1.4, X^1 is the union of countably many finite subcomplexes. Assume, inductively, that X^n is countable. Since X^{n+1} is locally finite, each cell of X^n meets only finitely many $(n + 1)$-cells. Hence X^{n+1} is countable. So X is countable.   □

A *filtration* of a CW complex[22] X is a collection of subcomplexes $\emptyset = K_{-1} \subset K_0 \subset K_1 \subset \dots$ such that $X = \bigcup_{i=0}^{\infty} K_i$, and we say that $\{K_i\}$ *filters*[23] X. When X has locally finite type, we wish to filter X by subcomplexes K_i having finite type, and to relate the homology modules[24] $H_*(X \stackrel{c}{-} K_i; R)$ to the modules $H^e_*(X; R)$. We begin with:

Proposition 11.4.4. *If X is strongly locally finite and A is finite then the CW neighborhood $N(A)$ is finite.*

Proof. The collection $\{C(e) \mid e$ is a cell of $X\}$ is locally finite, and A is compact.   □

It is not true that if X has locally finite type and A has finite type then $N(A)$ has finite type. Nor is 11.4.4 true when the word "strongly" is dropped.

Now, let X be a countable CW complex of locally finite type. A *finite type filtration* of X is a filtration $K_0 \subset K_1 \subset \dots$ by full subcomplexes of finite type. If X is strongly locally finite, each K_i must then be finite, and we call $\{K_i\}$ a *finite filtration*. In particular, when X is finite-dimensional every finite type filtration is a finite filtration.

A subset U of a locally compact space Y is a *neighborhood of the end* if $\mathrm{cl}(Y - U)$ is compact. A sequence $\{U_i \mid i \in \mathbb{N}\}$ of neighborhoods of the end of Y is a *basis for the neighborhoods of the end* of Y if for each neighborhood U of the end, there is some i such that $U_i \subset U$. Note that Y is compact iff every such collection $\{U_i\}$ contains $\emptyset$.

---

[22] This is the case of "G-filtration" in Sect. 7.4 in which G is the trivial group.

[23] Later we will also need filtrations $\dots \subset K_n \subset K_{n+1} \subset \dots$ indexed by $\mathbb{Z}$ which will also be called "filtrations."

[24] Recall the discussion of CW complements in Sect. 1.5.

Proposition 11.4.5. *With X as above, let $\{K_i\}$ be a finite type filtration of X. Then, for each n, $\{(X \overset{c}{-} K_i) \cap X^n\}$ is a basis for the neighborhoods of the end of X^n.*

Proof. First, we show that $U_i := (X \overset{c}{-} K_i) \cap X^n = X^n \overset{c}{-} K_i^n$ is a neighborhood of the end of X^n. We have $X^n - U_i \subset N_{X^n}(K_i^n)$, by 1.5.5. So $\mathrm{cl}(X^n - U_i) \subset N_{X^n}(K_i^n)$. Thus $\mathrm{cl}(X^n - U_i)$ is compact, by 11.4.4.

Let U be a neighborhood of the end of X^n. Then $\mathrm{cl}(X^n - U)$ is compact. So $\mathrm{cl}(X^n - U) \subset K_i^n$ for some i, by 1.2.13. In other words, $X^n - \mathrm{int}\, U \subset K_i^n$, so $\mathrm{int}\, U \supset X^n - K_i^n \supset (X \overset{c}{-} K_i) \cap X^n = U_i$. $\square$

As for the existence of finite type filtrations, we know by 11.4.3 that every path connected locally finite CW complex is countable and we have:

Proposition 11.4.6. *Let X be a countable CW complex of locally finite type. Well-order the set of vertices of X, and let K_i be the full subcomplex of X generated by the first i vertices in the well-ordering. Then $\{K_i\}$ gives a finite type filtration of X. If X is strongly locally finite, each K_i is finite.* $\square$

When X has only finitely many vertices, all but finitely many of the K_i in 11.4.6 are equal to X. This happens if X is finite, or in cases such as the one described in 10.1.13.

We use finite type filtrations to compute $H_*^\infty(X; R)$ and $H_*^e(X; R)$ (when X is countable of locally finite type). Let $\{K_i\}$ be such a filtration. Define $C_*(X, X \overset{c}{-} K_i; R)$ to be $C_*(X; R)/C_*(X \overset{c}{-} K_i; R)$. The chain complex $(\varprojlim_i \{C_*(X, X \overset{c}{-} K_i; R)\}, \varprojlim \partial)$ is isomorphic to the chain complex

$$\cdots \longrightarrow C_n^\infty(X; R) \overset{\partial}{\longrightarrow} C_{n-1}^\infty(X; R) \longrightarrow \cdots$$

and, since each induced bond $C_n(X, X \overset{c}{-} K_{i+1}; R) \to C_n(X, X \overset{c}{-} K_i; R)$ is an epimorphism, 11.3.1 and 11.3.2 imply that $\varprojlim_i^1 \{C_n(X, X \overset{c}{-} K_i; R)\}$ is trivial.

Therefore, 11.3.7 gives us:

Theorem 11.4.7. *Let X be a countable oriented CW complex having locally finite type. Let $\{K_i\}$ be a finite type filtration of X. For each n, there is a natural short exact sequence of R-modules*

$$0 \longrightarrow \varprojlim_i^1 \{H_{n+1}(X, X \overset{c}{-} K_i; R)\} \overset{\bar{a}}{\longrightarrow} H_n^\infty(X; R) \overset{\bar{b}}{\longrightarrow} \varprojlim_i \{H_n(X, X \overset{c}{-} K_i; R)\} \longrightarrow 0. \quad \square$$

Using the companion Theorem 11.3.8, we get a similar short exact sequence for $H_*^e(X; R)$ as follows. Consider the short exact sequences

$$0 \longrightarrow C_n(X \overset{c}{-} K_i; R) \longrightarrow C_n(X; R) \longrightarrow C_n(X, X \overset{c}{-} K_i; R) \longrightarrow 0$$

We identify $\varprojlim_i \{C_n(X, X -^c K_i; R)\}$ with $C_n^\infty(X; R)$ as before. Clearly $\varprojlim_i \{C_n(X \stackrel{c}{-} K_i; R)\}$ is trivial. By 11.3.4, we get a short exact sequence of chain complexes:

$$0 \longrightarrow C_*(X; R) \longrightarrow C_*^\infty(X; R) \longrightarrow \varprojlim_i{}^1\{C_*(X \stackrel{c}{-} K_i; R)\} \longrightarrow 0.$$

(Here, we are using the fact that $\lim^1$ of the "constant" inverse sequence $\{C_n(X; R), \mathrm{id}\}$ is trivial, by 11.3.1 and 11.3.2.) Comparing this with the exact sequence at the start of this section, we see that the chain complexes $C_*^\infty(X; R)/C_*(X; R)$ and $\varprojlim^1\{C_*(X \stackrel{c}{-} K_i; R)\}$ are isomorphic. Applying 11.3.8 to the chain complex $C_*(X \stackrel{c}{-} K_i; R)$ we get:

Theorem 11.4.8. *With hypotheses as in 11.4.7, we have, for each n, a natural short exact sequence of R-modules:*

$$0 \longrightarrow \varprojlim_i{}^1\{H_{n+1}(X \stackrel{c}{-} K_i; R)\} \stackrel{a}{\longrightarrow} H_n^e(X; R) \stackrel{b}{\longrightarrow} \varprojlim_i\{H_n(X \stackrel{c}{-} K_i; R)\} \longrightarrow 0.$$

$\square$

Remark 11.4.9. Theorems 11.4.7 and 11.4.8 are the fundamental tools for computing $H_n^\infty(X; R)$ and $H_n^e(X; R)$. In Remark 11.3.9, we stated the algebraic meaning of the homomorphisms a and b which occur in 11.3.8. These translate into geometric interpretations of the homomorphisms a and b in 11.4.8, as we now explain; see Fig. 11.1. Let z be an n-cycle of the end of X. Then z is an infinite $(n+1)$-chain with finite boundary. By 11.4.5, we can write

$$z = \sum_{j=0}^\infty x_j \quad \text{where } x_j \text{ is a finite } (n+1)\text{-chain in } X \stackrel{c}{-} K_j. \text{ Let } c_i = \partial\left(\sum_{j=i}^\infty x_j\right).$$

Then c_i is a finite n-cycle in $X \stackrel{c}{-} K_i$, and $\partial x_i = c_i - c_{i+1}$. So $(\{c_i\})$ is an element of $\varprojlim_i\{H_n(X \stackrel{c}{-} K_i; R)\}$. The formula for b is $b(\{z\}) = (\{c_i\})$. The monomorphism a takes the equivalence class of $(\{d_i\})$ to $\left\{\sum_{i=1}^\infty d_i\right\}$, where $d_i \in Z_{n+1}(X \stackrel{c}{-} K_i)$. We leave the corresponding interpretation of $\bar{a}$ and $\bar{b}$, in 11.4.7, as an exercise.

By 11.3.2, we get (see Fig. 11.2):

Corollary 11.4.10. *Let X and $\{K_i\}$ be as in 11.4.8. The homomorphism $b : H_n^e(X; R) \to \varprojlim_i\{H_n(X \stackrel{c}{-} K_i; R)\}$ is an isomorphism iff for each i, there exists $j \geq i$ such that for each $k \geq j$, each finite cellular $(n+1)$-cycle in $X \stackrel{c}{-} K_j$ is homologous in $X \stackrel{c}{-} K_i$ to a finite cellular $(n+1)$-cycle in $X \stackrel{c}{-} K_k$.*
$\square$

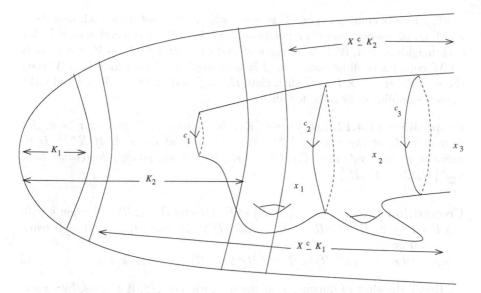

Fig. 11.1.

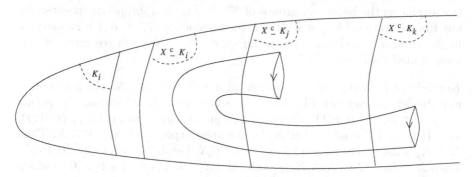

Fig. 11.2.

A locally compact space Y is *movable at the end* if for each neighborhood U of the end there is a neighborhood $V \subset U$ of the end such that for each neighborhood $W \subset U$ of the end there is a map $g : V \to W$ making the following diagram commute up to homotopy:

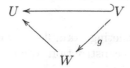

(i.e., there exists $H : V \times I \to U$ such that $H_0 =$ inclusion and $H_1 = g$.)

Example 11.4.11. $\mathbb{R}^n$ is movable at the end.

The locally compact space Y is *n-movable at the end* if for each neighborhood, U, of the end there is a neighborhood, $V \subset U$, of the end such that for each neighborhood, $W \subset U$, of the end and each map $f : K \to V$, where K is a CW complex of dimension $\leq n$, f is homotopic in U to a map into W (i.e., there exists $H : K \times I \to U$ such that $H_0 = f$ and $H_1(K) \subset W$). Obviously, movable implies n-movable for all n.

Proposition 11.4.12. *Let X and $\{K_i\}$ be as in 11.4.7, and let $r > n$. X^r is n-movable at the end iff X^{n+1} is n-movable at the end. If X^{n+1} is n-movable at the end, then $\{H_n(X \overset{c}{-} K_i; R)\}$ is semistable; hence we have $\underleftarrow{\lim}^1_i \{H_n(X \overset{c}{-} K_i; R)\} = 0$.* $\qquad\square$

Proposition 11.4.13. (i) $H_0^e(\mathbb{R}; R) \cong R \oplus R$ and $H_k^e(\mathbb{R}; R) = 0$ when $k \neq 0$.
(ii) *For $n \geq 2$, $H_0^e(\mathbb{R}^n; R) \cong H_{n-1}^e(\mathbb{R}^n; R) \cong R$, and $H_k^e(\mathbb{R}^n; R) = 0$ when $k \neq 0$ or $n - 1$.*
(iii) *For all n, $H_n^\infty(\mathbb{R}^n; R) \cong R$ and $H_k^\infty(\mathbb{R}^n; R) = 0$ when $k \neq n$.* $\qquad\square$

Recall the shift of dimension in the definition of H_*^e. It results, for example, in the above statement that the homology groups of the end of $\mathbb{R}^n$ are isomorphic to the homology groups of S^{n-1}. This is appropriate, because for any finite filtration $\{K_i\}$ of $\mathbb{R}^n$, the inverse sequence $\{\mathbb{R}^n \overset{c}{-} K_i\}$ is isomorphic in the category pro-Homotopy to the space S^{n-1}. We will see more of this kind of analysis in Sect. 17.5.

Example 11.4.14. Here is an example of a CW complex X which is not 1-movable at the end (see Fig. 11.3). X is the graph in $\mathbb{R}^2$ having the points $\{(i, j) \mid i \in \mathbb{N}, j = 0 \text{ or } 1\}$ as vertices, and having the segments $[(i, j), (i+1, j)]$ and $[(i, 0), (i, 1)]$ as edges. Let K_i be the subcomplex $[0, i] \times [0, 1] \cap X$. Then $X \overset{c}{-} K_i$ is the subcomplex $[i+1, \infty) \times [0, 1] \cap X$. Let $J_i = [i+1, i+2] \times [0, 1] \cap X$. Then $X \overset{c}{-} K_i = J_i \cup (X \overset{c}{-} K_{i+1})$ and $J_i \cap (X \overset{c}{-} K_{i+1}) = \{i+2\} \times [0, 1]$ which is contractible. The reduced Mayer-Vietoris sequence gives an exact sequence

$$0 \longrightarrow H_1(X \overset{c}{-} K_{i+1}; R) \oplus H_1(J_i; R) \overset{j_*}{\longrightarrow} H_1(X \overset{c}{-} K_i; R) \longrightarrow 0$$

The restriction of j_* to $H_1(X \overset{c}{-} K_{i+1}; R)$ is the homomorphism induced by $X \overset{c}{-} K_{i+1} \hookrightarrow X \overset{c}{-} K_i$. J_i is homeomorphic to S^1 so that $H_1(J_i; R) \cong R$. Thus, all the bonds in the inverse sequence $\{H_1(X \overset{c}{-} K_i; R)\}$ are monomorphisms while none is an epimorphism. Hence this inverse sequence is not semistable.

A useful device for constructing examples X with interesting $H_*^e(X)$ is the "inverse mapping telescope" construction, dual to the mapping telescope of Sect. 4.3. Let

$$X_1 \overset{f_1}{\longleftarrow} X_2 \overset{f_2}{\longleftarrow} X_3 \longleftarrow \cdots$$

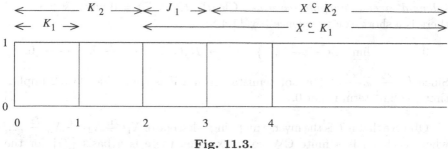

Fig. 11.3.

be an inverse sequence of CW complexes and cellular maps. Let $T = \coprod_{i=1}^{\infty} M(f_i)/\sim$ where $M(f_i)$ is the mapping cylinder of f_i and $\sim$ is the equivalence relation generated by identifying, for all i, the canonical copies of X_{i+1} in $M(f_{i+1})$ and $M(f_i)$ – see Sect. 4.1. If $q : \coprod_{i=1}^{\infty} M(f_i) \to T$ is the quotient map, it is clear that T admits a CW complex structure whose cells are $\{q(e) \mid e$ is a cell of $M(f_i)$ for some $i\}$. Compare the proof of 4.1.1. T is the *inverse mapping telescope* of the given inverse sequence. We abuse notation, writing $M(f_i)$ for $q(M(f_i)) \subset T$. If each X_i is path connected and locally finite, so is T. Assume each X_i is finite and path connected. Let $K_i = \bigcup_{j=1}^{i-1} M(f_j)$. Then $\{K_i\}$ is a finite filtration of T. With the obvious identifications we get homotopy commutative diagrams

$$X \stackrel{c}{\,\underline{\ }\,} K_i \hookleftarrow X \stackrel{c}{\,\underline{\ }\,} K_{i+1}$$

$$\uparrow \qquad\qquad \uparrow$$

$$X_{i+1} \stackrel{f_{i+1}}{\longleftarrow} X_{i+2}$$

whose vertical arrows are homotopy equivalences. Hence in the commutative diagram

$$\cdots \longleftarrow H_k(X \stackrel{c}{\,\underline{\ }\,} K_i; R) \longleftarrow H_k(X \stackrel{c}{\,\underline{\ }\,} K_{i+1}; R) \longleftarrow \cdots$$

$$\uparrow \cong \qquad\qquad\qquad \uparrow \cong$$

$$\cdots \longleftarrow H_k(X_{i+1}; R) \stackrel{f_{i+1*}}{\longleftarrow} H_k(X_{i+2}; R) \longleftarrow \cdots$$

the unmarked arrows are induced by inclusion and the vertical arrows are isomorphisms.

Example 11.4.15. To get an example of the exact sequence of 11.4.8 in which the $\varprojlim^1$ term does not vanish, apply the above construction with $X_i = S^1$ for all i, and f_i the map of degree 2, $e^{2\pi it} \mapsto e^{4\pi it}$. The resulting T is the *dyadic solenoid inverse mapping telescope.* Clearly $H_k^e(T; \mathbb{Z}) = 0$ when $k > 0$, and there is a short exact sequence (by 11.4.8):

$$0 \longrightarrow \varprojlim{}^1\{\mathbb{Z} \xleftarrow{\times 2} \mathbb{Z} \xleftarrow{\times 2} \ldots\} \longrightarrow H_0^e(T; \mathbb{Z}) \longrightarrow \mathbb{Z} \longrightarrow 0.$$

Since $\{\mathbb{Z} \xleftarrow{\times 2} \mathbb{Z} \xleftarrow{\times 2} \ldots\}$ is not semistable, and $\mathbb{Z}$ is countable, 11.3.2 implies that the $\varprojlim^1$ term is not 0. $\square$

Observe that if T is the inverse mapping telescope of $X_1 \xleftarrow{f_1} X_2 \xleftarrow{f_2} X_3 \xleftarrow{f_3} \ldots$, where each X_i is a finite CW complex, then there is a basis $\{T_i\}$ for the neighborhoods of the end of T such that each T_i has the homotopy type of a finite CW complex. This is because T_i is the inverse mapping telescope of $X_i \xleftarrow{f_i} X_{i+1} \xleftarrow{f_{i+1}} X_{i+2} \longleftarrow \ldots$, so T_i has the homotopy type of X_i. Hence $H_1(T_i; \mathbb{Z})$ is finitely generated. However, if X is the graph in Example 11.4.14, the discussion shows that for every neighborhood U of the end of X, $H_1(U; \mathbb{Z})$ is infinitely generated. Hence, X is quite different from an inverse mapping telescope.

Source Notes: The general viewpoint of this section is similar to that of [110], though that book is set in a different category. See also [60].

Exercises

1. Prove that when K_i is a finite filtration of the strongly locally finite CW complex X, then for each i there exists j with $K_i \subset \mathrm{int}_X K_j$.
2. Give an example of X of locally finite type having a vertex v such that $N(v)$ does not have finite type.
3. Let A be a subcomplex of a finite-dimensional locally finite CW complex. If A^0 is finite prove that A is finite.
4. If X is strongly locally finite and path connected and if A is a finite non-empty subcomplex of X, prove that $\{N^n(A)\}$ is a finite filtration of X, where $N^1(A) := N(A)$ and for $n \geq 1$ $N^{n+1}(A) := N(N^n A)$).
5. Give an example where X is locally finite and A is finite but $N(A)$ is not finite.
6. Prove that any countable CW complex is homotopy equivalent to a countable strongly locally finite CW complex. Prove that any CW complex is homotopy equivalent to a locally finite-dimensional CW complex (i.e., every point has a neighborhood which is a finite-dimensional CW complex). *Hint*: form a "direct mapping telescope."
7. What is meant by saying that the exact sequences in 11.4.7 and 11.4.8 are natural? Prove naturality.
8. Interpret the homomorphisms $\bar{a}$ and $\bar{b}$ of 11.4.7 in the spirit of 11.4.9.
9. Prove 11.4.12.
10. Prove 11.4.13.
11. Discover a Mayer-Vietoris sequence for $H_*^e(\cdot, R)$.
12. Consider the short exact sequence in 11.4.7. The middle and right terms only depend on the $(n + 1)$-skeleton, so the same is true of the left term. Yet the modules $H_{n+1}(X, X \xrightarrow{c} K_i)$ depend on the $(n + 2)$-skeleton. Explain why their $\lim^1$ does not.

13. In Sect. 11.1 we defined $B_n(y)$ to be the ball of radius n about y where y is a vertex of an infinite locally finite path connected graph Y. Give an example to show that $B_n(y)$ need not be a full subgraph of Y.

14. Use Example 11.4.14 to construct another example (in addition to 11.4.15) where the $\lim^1$ term in 11.4.8 does not vanish.

Cohomology of CW Complexes

Here we introduce cellular cohomology, and cellular cohomology based on finite chains. The first of these is popularly called "ordinary cohomology'. There is an intriguing double duality in this. From one point of view ordinary cohomology is considered to be the "dual" of homology as defined in Chap. 2. From another point of view, which will be made precise when we discuss Poincaré Duality in Chap. 15, cohomology based on finite chains is "dual" to homology, while ordinary cohomology is "dual" to the infinite cellular homology theory of Sect. 11.1.

As in Chap. 11, having the two cohomology theories enables us to define cohomology at the end of a CW complex.

12.1 Cohomology based on infinite and finite (co)chains

Just as the cellular boundary homomorphism algebraically sums up the faces of a cell of an oriented CW complex, there is a "dual" homomorphism called the "coboundary" which algebraically sums up the cells of which a given cell is a face. This feature of the "coboundary" is somewhat obscured in many books, because the authors define the "coboundary" on "cochains" rather than on chains. So our treatment, although it goes back to the beginnings of cohomology, may appear slightly eccentric.[1] To minimize the eccentricity, we will give both versions of the coboundary, and we will stay rather close to standard forms.

Let X be an oriented CW complex. The *coboundary homomorphism* $\delta : C_n^\infty(X; R) \to C_{n+1}^\infty(X; R)$ is defined by

$$\delta\left(\sum_\alpha m_\alpha e_\alpha^n\right) = \sum_\beta \left(\sum_\alpha m_\alpha [e_\beta^{n+1} : e_\alpha^n]\right) e_\beta^{n+1}.$$

---

[1] Expressing cohomology in terms of chains rather than cochains makes Poincaré Duality more obvious: see Sect. 15.2.

This is well defined by 2.5.8 and 1.2.13. The reader can check that δ is a homomorphism.

Clearly, the following diagram commutes:[2]

$$
\begin{array}{ccc}
C_n^\infty(X;R) & \xrightarrow{\ \delta\ } & C_{n+1}^\infty(X;R) \\
\cong \Big\uparrow \phi_n & & \cong \Big\uparrow \phi_{n+1} \\
\operatorname{Hom}_R(C_n(X;R),R) & \xrightarrow{\ \partial^*\ } & \operatorname{Hom}_R(C_{n+1}(X;R),R).
\end{array}
$$

Here, $\phi_k(g) = \sum_\alpha g(e_\alpha^k).e_\alpha^k$ and $\partial^*(g) = g \circ \partial$. Moreover, each ϕ_k is clearly an isomorphism. It is customary to call the elements of $\operatorname{Hom}_R(C_n(X;R),R)$ *cellular n-cochains* in X with *coefficients* in R and to abbreviate $\operatorname{Hom}_R(C_n(X;R),R)$ to $C^n(X;R)$, a notation we will sometimes refer back to for purposes of clarification. In view of the above equivalence, we will also refer to ∂^* as the *coboundary homomorphism*.

Proposition 12.1.1. *The composition*

$$
C_n^\infty(X;R) \xrightarrow{\ \delta\ } C_{n+1}^\infty(X;R) \xrightarrow{\ \delta\ } C_{n+2}^\infty(X;R)
$$

is zero for all n.

Proof. This follows from 2.3.3 and the corresponding statement for ∂^*.   □

A *cochain complex* over R is a pair (C,δ) where C is a graded R-module and $\delta : C \to C$ is a homomorphism of degree 1, called the *coboundary*, such that $\delta \circ \delta = 0$. If $(\{C_n\},\delta)$ is a cochain complex and if one defines $C_n' = C_{-n}$, then $(\{C_n'\},\delta)$ is a chain complex. In this way one obtains from Sect. 2.1 properties of cochain complexes analogous to the stated properties of chain complexes. Elements of ker δ [resp. image δ] are called *cocycles* [resp. *coboundaries*]. In dimension n, ker δ/im δ is the nth *cohomology module*, denoted $H^n(C)$.

By Proposition 12.1.1, $(C_*^\infty(X;R),\delta)$ is a cochain complex, the *cellular cochain complex* of X; as explained, this term is also used for the isomorphic cochain complex $(C^*(X;R),\partial^*)$. The resulting *cellular cohomology modules* are denoted[3] by $H^*(X;R)$.

A cellular map $f : X \to Y$ induces a homomorphism $f^\# : C_n^\infty(Y;R) \to C_n^\infty(X;R)$ defined by:

$$
f^\#\left(\sum_\beta m_\beta \tilde{e}_\beta^n\right) = \sum_\alpha \left(\sum_\beta m_\beta [e_\alpha^n : \tilde{e}_\beta^n : f]\right) e_\alpha^n.
$$

---

[2] Some authors (e.g., [29]) use $(-1)^{n+1}\partial^*$ instead of ∂^* in this diagram, a convention which would force a corresponding change in our definition of δ.

[3] For consistency, we should write $H_\infty^*(X;R)$ because these cohomology modules are based on infinite chains; but tradition forbids this.

In the alternative language, $f^\# : C^n(Y; R) \to C^n(X; R)$ (i.e.,
$f^\# : \mathrm{Hom}_R(C_n(Y; R), R) \to \mathrm{Hom}_R(C_n(X; R), R))$ is the dual of
$f_\# : C_n(X; R) \to C_n(Y; R)$. This $f^\#$ is a *cochain map*, meaning it commutes
with δ. Clearly, $(g \circ f)^\# = f^\# \circ g^\#$ whenever this makes sense for cellular
maps f and g, and of course $\mathrm{id}^\# = \mathrm{id}$.

We now turn to cohomology based on finite chains. We have $C_*(X; R) \subset C_*^\infty(X; R)$. In general, it is not the case that δ maps $C_n(X; R)$ into $C_{n+1}(X; R)$,
but this is the case when X^{n+1} is locally finite, and then we will also write
$\delta : C_n(X; R) \to C_{n+1}(X; R)$ for the restriction of the previous δ. So when
X has locally finite type, $(C_*(X; R), \delta)$ is a cochain complex called the *cellular cochain complex of X based on finite chains*. Let $\mathrm{Hom}_R^f(C_n(X; R), R)$ or
$C_f^n(X; R)$ denote[4] the subset of $\mathrm{Hom}_R(C_n(X; R), R)(= C^n(X; R))$ consisting
of homomorphisms which take the value 0 on all but finitely many generators
(oriented n-cells of X). Then we have a commutative diagram

$$
\begin{array}{ccc}
C_n(X; R) & \xrightarrow{\;\delta\;} & C_{n+1}(X; R) \\
\Big\uparrow{\scriptstyle \phi_n} & & \Big\uparrow{\scriptstyle \phi_{n+1}} \\
C_f^n(X; R) & \xrightarrow{\;\partial^*\;} & C_f^{n+1}(X; R)
\end{array}
$$

where each ϕ_k [resp. ∂^*] is the restriction of the previous ϕ_k [resp. ∂^*]. Again,
these ϕ_k's are isomorphisms.

The cohomology modules coming from the cochain complex $(C_*(X; R), \delta)$,
or, equivalently, from $(C_f^*(X; R), \partial^*)$ (when X has locally finite type) are the
cellular cohomology modules based on finite chains with coefficients in R. They
are denoted $H_f^*(X; R)$.

A cellular CW-proper map $f : X \to Y$ between CW complexes of locally
finite type induces a cochain map

$$ f^\# : C_n(Y; R) \to C_n(X; R) $$

where $f^\#$ is the restriction of the previous $f^\#$ (one checks that the previous
formula remains well defined). Just as δ can be interpreted as ∂^*, this $f^\#$ can
be interpreted as $f^\# : C_f^n(Y; R) \to C_f^n(X; R)$. We write f^* for both $H^n(f^\#)$
and $H_f^n(f^\#)$.

Here are some basic properties of H^* and H_f^* whose proofs are similar to
those of the corresponding statements for H_* and H_*^∞ in Sect. 2.4 and Sect.
10.2. WHENEVER A STATEMENT INVOLVES H_f^n THE CW COMPLEX
IS ASSUMED TO HAVE LOCALLY FINITE $(n + 1)$-SKELETON.

Proposition 12.1.2. *If X is path connected, $H^0(X; R) \cong R$. When X^1 is
infinite and path connected, $H_f^0(X; R) = 0$.* $\square$

---

[4] Here "f" stands for "finite'.

Proposition 12.1.3. *If $\{X_\alpha \mid \alpha \in \mathcal{A}\}$ is the set of path components of X then $H^n(X;R)$ is isomorphic to $\prod_\alpha H^n(X_\alpha;R)$, and $H^n_f(X;R)$ is isomorphic to $\bigoplus_\alpha H^n_f(X_\alpha;R)$.* $\quad\square$

Proposition 12.1.4. *When X is finite, $H^*(X;R) = H^*_f(X;R)$.* $\quad\square$

Proposition 12.1.5. *The cohomology of the one-point CW complex $\{p\}$ is: $H^0(\{p\};R) \cong H^0_f(\{p\};R) \cong R$, and, for $n \neq 0$, $H^n(\{p\};R) = H^n_f(\{p\};R) = 0$.* $\quad\square$

Proposition 12.1.6. *With the usual CW complex structure on $[0,\infty)$, $H^n_f([0,\infty);R) = 0$ for all n.* $\quad\square$

Proposition 12.1.7. *If X has dimension d, then $H^n(X;R) = H^n_f(X;R) = 0$ for all $n > d$.* $\quad\square$

Proposition 12.1.8. *If $i : X^{n+1} \hookrightarrow X$, then $i^* : H^j(X;R) \to H^j(X^{n+1};R)$ and $i^* : H^j_f(X;R) \to H^j_f(X^{n+1};R)$ are isomorphisms for $j \leq n$.* $\quad\square$

We now turn to homotopy invariance of cohomology.

Theorem 12.1.9. *[Homotopy Invariance] If $f,g : X \to Y$ are homotopic cellular maps, then $f^* = g^* : H^*(Y;R) \to H^*(X;R)$. In particular, cellular cohomology is a topological invariant.*

Proof. Although this theorem is analogous to 2.7.10, we prove it using 2.7.14 instead. The reason is explained in Remark 12.1.11 below.

By 1.4.3 there is a cellular homotopy $F : X \times I \to Y$ from f to g. Let D be as in 2.7.14. Consider the diagram:

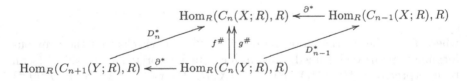

Since $f^\# = (f_\#)^*$, etc., it is merely the dual of 2.7.14 to say $D^*_n\partial^* + \partial^* D^*_{n-1} = f^\# - g^\#$; or equivalently $D^*\delta + \delta D^* = f^\# - g^\#$. Applying both sides to a cocycle c in Y we find that $f^\#(c)$ and $g^\#(c)$ differ by a coboundary.[5] $\quad\square$

---

[5] If the cocycle c is $\sum_\beta m_\beta \tilde{e}^n_\beta$ (in the proof of 12.1.9) then $f^\#(c)$ and $g^\#(c)$ differ by the coboundary of $\sum_{\alpha,\beta} m_\beta [e^{n-1}_\alpha \times I : \tilde{e}^n_\beta : F]e^{n-1}_\alpha$. The compact support of this $(n-1)$-dimensional cochain can be found from this formula using 2.5.10. In applications, one sometimes needs to know a finite subcomplex of X in which $f^\#(c)$ and $g^\#(c)$ are cohomologous.

By a similar proof (compare 11.1.9) we get:

Theorem 12.1.10. *Let $f, g : X \to Y$ be CW-cellular maps where X and Y have locally finite type. Assume either* (a) *f and g are CW-proper homotopic or* (b) *X is locally finite, Y is strongly locally finite, and f is properly homotopic to g. Then $f^* = g^* : H_f^*(Y; R) \to H_f^*(X; R)$.* □

It follows that on strongly locally finite CW complexes, H_f^* is a topological invariant.

Remark 12.1.11. We wish to treat homology and cohomology on an equal footing. But there are places where that is impossible, and we have just met one. Our proof of 2.7.10 (homotopy invariance of homology) was direct and geometrical in the sense that we avoided using a chain homotopy. It was based on the idea that if $c \in C_n(X; R)$ there is a sensible notion of $c \times I \in C_n(X \times I; R)$. If we were to imitate this proof of 2.7.10 in proving 12.1.9, we would need a method of associating with each infinite n-chain $c \in C_n^\infty(Y; R)$ an infinite $(n-1)$-chain (call it $c \div I$) in some space (call it $Y \div I$); we would need $F : X \times I \to Y$ to induce a map $\tilde{F} : X \to Y \div I$; and the proof would consist of showing that, when c is a cocycle, $\delta\tilde{F}^\#(c \div I) = \pm(f^\#(c) - g^\#(c))$. The candidate for $Y \div I$ is clear: it is Y^I, the space of all maps $I \to Y$ with the compact-open topology; indeed, the adjoint map $\tilde{F} : X \to Y^I$ is defined by $\tilde{F}(x)(t) = F(x, t)$. So it would only remain to define $c \div I \in C_n^\infty(Y^I; R)$. However, the space Y^I does not admit the structure of a CW complex, so the notion of a cellular chain in Y^I is meaningless. It is possible to repair this defect by working throughout with singular chains or cubical singular chains, for then there is a canonical way of associating with each singular n-simplex [resp. singular n-cube] in Y a singular $(n-1)$-simplex [resp. singular $(n-1)$-cube] in Y^I. Since our aim is to work with cellular chains (and since we have given an easy proof of 12.1.9 anyway), this "repair" will not be pursued here. And besides, it is not clear that a similar repair would be possible in the analogous "proof" of 12.1.10. In summary, this lack of symmetry between the cellular chain complexes and the cellular cochain complexes, finite and infinite, is forced on us by the fact that the category of CW complexes and cellular maps does not possess path-space objects dual to cartesian products with I.

As in Sect. 2.5 and Sect. 10.2, the various cellular approximation theorems 1.4.3, 10.1.14 and 10.2.3 allow us to regard $H^*(\cdot; R)$ as a contravariant functor from the category of oriented CW complexes and homotopy classes of maps to the category, $\mathcal{D}$, of graded R-modules and homomorphisms of degree 0; and to regard $H_f^*(\cdot; R)$ as a contravariant functor from $\mathcal{C}$ to $\mathcal{D}$, where $\mathcal{C}$ is either (a) the category of oriented CW complexes of locally finite type and CW-proper homotopy classes of CW-proper maps or (b) the category of oriented strongly locally finite CW complexes and proper homotopy classes of proper maps. Thus $H^*(\cdot; R)$ is a homotopy invariant, and $H_f^*(\cdot; R)$ is a CW-proper homotopy invariant in case (a) or a proper homotopy invariant in case (b).

For oriented CW pairs (X, A) let

$$C_n^\infty(X, A; R) = \{\textstyle\sum m_\alpha e_\alpha^n \in C_n^\infty(X; R) \mid m_\alpha = 0 \text{ whenever } e_\alpha^n \text{ is a cell of } A\},$$

and[6]

$$C_n(X, A; R) = C_n^\infty(X, A; R) \cap C_n(X; R).$$

We have short exact sequences of cochain complexes

$$0 \longleftarrow C_*^\infty(A; R) \overset{i^\#}{\longleftarrow} C_*^\infty(X; R) \longleftarrow C_*^\infty(X, A; R) \longleftarrow 0$$

and (if X has locally finite type)

$$0 \longleftarrow C_*(A; R) \overset{i^\#}{\longleftarrow} C_*(X; R) \longleftarrow C_*(X, A; R) \longleftarrow 0.$$

The cohomology modules of the cochain complexes $(C_*^\infty(X, A; R), \delta)$ and $(C_*(X, A; R), \delta)$ are denoted by $H^*(X, A; R)$ and $H_f^*(X, A; R)$ respectively and called *relative cohomology modules*. For the usual algebraic reasons (see Sect. 2.1) we get a commutative diagram whose horizontal rows are exact:

$$\cdots \longleftarrow H^n(A; R) \longleftarrow H^n(X; R) \longleftarrow H^n(X, A; R) \longleftarrow H^{n-1}(A; R) \longleftarrow \cdots$$
$$\uparrow \qquad\qquad \uparrow \qquad\qquad \uparrow \qquad\qquad \uparrow$$
$$\cdots \longleftarrow H_f^n(A; R) \longleftarrow H_f^n(X; R) \longleftarrow H_f^n(X, A; R) \longleftarrow H_f^{n-1}(A; R) \longleftarrow \cdots$$

Of course $H^n(X, A; R) = Z^n(X, A; R)/B^n(X, A; R)$. Here $Z^n(X, A; R) = \{c \in C_n^\infty(X; R) \mid \delta c = 0$ and the coefficient in c corresponding to each n-cell of A is $0\}$; and $B^n(X, A; R) = \{c \in C_n^\infty(X; R) \mid c = \delta d$ where $d \in C_{n-1}^\infty(X; R)$ and the coefficient in d corresponding to each $(n-1)$-cell of A is $0\}$. These are the *relative n-cocycles* and *relative n-coboundaries* respectively. $Z_f^n(X, A; R)$ and $B_f^n(X, A; R)$ are defined similarly. Thus $Z^n(X, A; R) \subset Z^n(X; R)$, $B^n(X, A; R) \subset B^n(X; R)$, etc.

The relative cohomology modules $H^*(X, A; R)$ and $H_f^*(X, A; R)$ have properties analogous to the corresponding properties of relative homology modules as described in Sects. 2.6, 2.7 and 11.1.

Remark 12.1.12. Reduced cohomology based on infinite chains makes sense, and is entirely dual to the theory described in Sect. 2.9. Reduced cohomology based on finite chains does not always make sense because the reduced coboundary $\tilde{\delta} : R \to C_0(X; R)$ maps 1 to the sum of all the vertices – an infinite chain when X is infinite.

---

[6] This is abuse of notation: in Sect. 11.3, $C_n(X, A; R)$ meant $C_n(X; R)/C_n(A; R)$, whereas here $C_n(X, A; R)$ denotes the module of chains whose A-coefficients are all zero; these are really two ways of describing the same idea.

Exercises

1. Establish the *Milnor exact sequence* for the cohomology of X where $X = \bigcup_i K_i$ each K_i being finite, $K_1 \subset K_2 \subset \cdots$, namely,

$$0 \to \varprojlim_i{}^1 H^{n-1}(K_i; R) \to H^n(X; R) \to \varprojlim_i H^n(K_i; R) \to 0.$$

2. Prove 12.1.2–12.1.8.
3. Discover Mayer-Vietoris sequences for $H^*(\cdot; R)$ and $H_f^*(\cdot; R)$.
4. Guided by Sect. 2.9, write out the theory of reduced cohomology based on infinite chains.
5. Prove that an n-equivalence induces an isomorphism on $H^k(\cdot; R)$ for all $k \leq n - 1$.
6. Prove that a CW-proper n-equivalence induces an isomorphism on $H_f^k(\cdot; R)$ for all $k \leq n - 1$.
7. Show $\delta(C_n(X, A)) \subset C_{n+1}(X, A)$. Show $Z^n(X, A) = \ker(\delta : C_n(X, A) \to C_{n+1}(X, A))$ and $B^n(X, A) = \text{image } (\delta : C_{n-1}(X, A) \to C_n(X, A))$.

12.2 Cohomology of ends

Let X be an oriented CW complex having locally finite type. We define cohomology modules of the end of X in much the same way as we defined homology of the end in Sect. 11.4.

Consider the short exact sequence of cochain complexes and cochain maps:

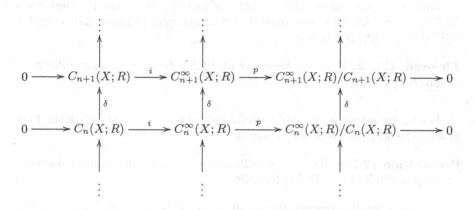

Let $H_e^n(X; R) = H^n(C_*^\infty(X; R)/C_*(X; R))$; we call $H_e^*(X; R)$ the *cohomology of the end* of X. As usual there is a long exact sequence[7]

---

[7] Since i is a cochain map, an argument could be made for writing i_* rather than i^* for the induced homomorphism; similarly for p. However, our notation is customary.

$$\cdots \longrightarrow H^{n+1}(X;R) \xrightarrow{\ i^*\ } H_f^{n+1}(X;R) \xrightarrow{\ \delta^*\ } H_e^n(X;R) \xrightarrow{\ p^*\ } H^n(X;R) \longrightarrow \cdots$$

Clearly, $H_e^n(X;R) = Z_e^n(X;R)/B_e^n(X;R)$ where

$$Z_e^n(X;R) = \{c \in C_n^\infty(X;R) \mid \delta c \in C_{n+1}(X;R)\}$$

and

$$B_e^n(X;R) = B^n(X;R) + C_n(X;R) \subset C_n^\infty(X;R).$$

The elements of $Z_e^n(X;R)$ and $B_e^n(X;R)$ are called *n-cocycles of the end* of X and *n-coboundaries of the end* of X, respectively. Obviously $H_e^n(X;R) = 0$ when $n < 0$.

Just as in Sect. 11.4, CW-proper maps or proper maps induce homomorphisms on H_e^* under sensible hypotheses, and these possess the usual functorial properties. We omit the details.

We now give the cohomology analogs of 11.4.7 and 11.4.8.

Let $\{K_i\}$ be a finite type filtration of X. We have a direct sequence of cochain complexes $\{(C_*(X, X \xrightarrow{c} K_i; R), \delta)\}$ with bonds (inclusion)$^\#$ whose direct limit is isomorphic to the cochain complex $(C_*(X;R), \delta)$. From 11.3.16, we get:

Theorem 12.2.1. *Let X be a countable oriented CW complex having locally finite type. Let $\{K_i\}$ be a finite type filtration of X. There is an isomorphism*
$$\bar{c} : \varinjlim_i \{H^n(X, X \xrightarrow{c} K_i; R)\} \to H_f^n(X;R). \qquad \square$$

Similarly, we have the direct sequence of cochain complexes $\{(C_*^\infty(X \xrightarrow{c} K_i; R), \delta)\}$ whose direct limit is isomorphic to the cochain complex $C_*^\infty(X;R)/C_*(X;R)$, giving:

Theorem 12.2.2. *With hypotheses as in 12.2.1, there is an isomorphism*
$$c : \varinjlim_i \{H^n(X \xrightarrow{c} K_i; R)\} \to H_e^n(X;R). \qquad \square$$

For future reference we note a useful fact about the "bottom" of the long exact sequence; the proof is an exercise:

Proposition 12.2.3. *When X is path connected and infinite the monomorphism $p^* : H^0(X;R) \to H_e^0(X;R)$ splits.* $\qquad \square$

If Y is a locally compact Hausdorff space, it is customary to define the *singular cohomology modules of Y with compact supports* to be $H_c^*(Y;R) = \varinjlim_K \{H_\Delta^*(Y, Y - K; R)\}$ where H_Δ^* denotes singular cohomology (see Exercise 5) and the directed set is the set of all compact subsets K of Y directed by inclusion. The reader should compare this with Theorem 12.2.1.

Remark 12.2.4. Many theorems in this chapter and elsewhere in this book are stated for a CW complex X of locally finite type. In such a context a statement about X which only involves H_n^∞, H_{n-1}^e, H_f^n or H_e^n holds provided X^{n+1} is locally finite because these invariants only depend on X^{n+1}.

Convention 12.2.5. *Throughout this book it is understood that whenever H_n^∞, H_{n-1}^e, H_f^n or H_e^n appears in a statement, the relevant CW complex has locally finite $(n+1)$-skeleton.*

Exercises

1. Write down the isomorphisms $\bar{c}$ in 12.2.1 and c in 12.2.2.
2. Explain why there is no shift of dimension in the definition of $H_e^*(X;R)$ while there is such a shift in the definition of $H_*^e(X;R)$.
3. Discover a Mayer-Vietoris sequence for $H_*^e(\cdot;R)$.
4. Up to what value of k does a CW-proper n-equivalence induce an isomorphism on (a) $H_k^e(\cdot;R)$ and (b) $H_e^k(\cdot;R)$?
5. Discover "singular cohomology" by applying the methods of Sect. 12.1 to the singular chain complex of Sect. 2.2.

12.3 A special case: Orientation of pseudomanifolds and manifolds

Before continuing with the general theory we pause to discuss infinite chains and cochains in an important case.

A CW *n-pseudomanifold* is a locally finite regular CW complex X having the following properties: (i) every cell of X is a face of an n-cell; (ii) every $(n-1)$-cell of X is a face of at most two n-cells; (iii) if $e_\alpha^n \neq e_\beta^n$ are cells in the same path component of X, there is a finite sequence $e_\alpha^n = e_{\alpha_0}^n, e_{\alpha_1}^n, \cdots, e_{\alpha_k}^n = e_\beta^n$ of n-cells of X such that each $e_{\alpha_i}^n \cap e_{\alpha_{i+1}}^n$ is an $(n-1)$-cell of X. Such a sequence is called a *gallery*. The subcomplex of X consisting of those $(n-1)$-cells which are faces of exactly one n-cell, together with their faces, is called the *boundary* of X and is denoted ∂X. Note that X is not required to be path connected.

Proposition 12.3.1. *If K is a combinatorial n-manifold then $|K|$ is a CW n-pseudomanifold, and the two notions of boundary coincide.*   □

The cone on a path connected finite n-dimensional pseudomanifold X is a finite $(n+1)$-dimensional pseudomanifold but is not a topological manifold if X is not a sphere or a ball.[8]

---

[8] If X is an n-pseudomanifold, it is clear that $X - X^{n-2}$ is a topological n-manifold. For that reason a pseudomanifold is sometimes described as a "manifold with a codimension two singularity;" however, if X satisfies (i) and (ii) and $X - X^{n-2}$ is a manifold, it does not follow that X is a pseudomanifold.

Remark 12.3.2. In the older literature, an *n-pseudomanifold* is a path connected CW complex $X = |K|$ for some abstract simplicial complex K, which satisfies (i)–(iii) above. We have dropped the requirement of path connectedness, but have added the requirement of local finiteness. In the literature of buildings, a simplicial complex $X = |K|$ (not necessarily locally finite) satisfying (i)–(iii) and with $\partial X = \emptyset$ is called a *thin chamber complex*.

Let X be a CW n-pseudomanifold. Orient the cells of X in the sense of Sect. 2.5. Let R be an integral domain.[9] A relative cycle $c \in Z_n^\infty(X, \partial X; R)$ is called a *fundamental cycle* (over R) if the coefficient in c of every n-cell is a unit of R. We say that X is *R-orientable* if X has a fundamental cycle over R. Of course, $B_n^\infty(X, \partial X; R) = 0$, so there is only a formal difference between $Z_n^\infty(X, \partial X; R)$ and $H_n^\infty(X, \partial X; R)$. If c is a fundamental cycle, the corresponding homology class in $H_n^\infty(X, \partial X; R)$ is called a *fundamental class* (over R). Clearly, X is R-orientable iff each path component of X is R-orientable.

Proposition 12.3.3. *Let $c \in Z_n^\infty(X, \partial X; R)$, let e_α^n and e_β^n be n-cells in the same path component of X, and let r_α, r_β be the coefficients of e_α^n and e_β^n in c. Then $r_\alpha = \pm r_\beta$.*

Proof. First assume $e_\alpha^n \cap e_\beta^n = e_\gamma^{n-1}$. Then e_γ^{n-1} is not a face of any other n-cell. Since $\partial c = 0$ we get $r_\alpha [e_\alpha^n : e_\gamma^{n-1}] + r_\beta [e_\beta^n : e_\gamma^{n-1}] = 0$. By 5.3.10 this implies $r_\alpha = \pm r_\beta$. The general case now follows by induction, using Axiom (iii) in the definition of a pseudomanifold. □

Proposition 12.3.4. *Let X be path connected. If c and c' are fundamental cycles over R then, for some unit $u \in R$, $c = uc'$.*

Proof. Let the coefficient of e_α^n in c [resp. c'] be the unit u_α [resp. u_α']. Then c and $u_\alpha(u_\alpha')^{-1}c'$ have the same e_α^n-coefficient, so the coefficient of e_α^n in $c - u_\alpha(u_\alpha')^{-1}c'$ is zero. By 12.3.3, this implies $c = u_\alpha(u_\alpha')^{-1}c'$. □

Obviously, if $u \neq 1$ and c is a fundamental cycle, then $uc \neq c$. Hence the last proposition implies that when X is path connected and R-orientable the fundamental cycles on X are in bijective correspondence with the units of R.

A choice of fundamental cycle c on X (if X is R-orientable) is called an *R-orientation* of X, and X is *R-oriented* by c. Summarizing:

Proposition 12.3.5. *Let X, as above, be path connected and R-orientable. If two fundamental cycles have the same e_α^n-coefficient for some α, then they are identical. Hence, one fully specifies an R-orientation for X by specifying a unit of R as the coefficient of a single n-cell.* □

---

[9] Recall that an *integral domain* is a commutative ring with the property that $rs = 0$ implies $r = 0$ or $s = 0$.

Every CW n-pseudomanifold X is $\mathbb{Z}_2$-orientable and has a unique $\mathbb{Z}_2$-orientation. The terms "$\mathbb{Z}$-orientable,' "$\mathbb{Z}$-orientation," and "$\mathbb{Z}$-oriented" are usually abbreviated to *orientable, orientation,* and *oriented*[10]. If X is orientable, the underlying CW complex X can be oriented (in the sense of Sect. 2.5) so that $\sum\limits_{\alpha \in \mathcal{A}} e_\alpha^n$ is a fundamental class, where $\mathcal{A}$ indexes the n-cells of X. Whenever X is path connected and orientable, X has two orientations, since $\mathbb{Z}$ has two units.

Example 12.3.6. The CW manifolds $\overset{\bullet}{I}{}^{n+1}$ (homeomorphic to S^n) and $\mathbb{R}^n$ are orientable. The projective plane and the Möbius band (equipped with regular CW complex structures) are not orientable.

Now we use the fact that R is an integral domain:

Proposition 12.3.7. *Let X be path connected. If $H_n^\infty(X, \partial X; R) \neq 0$ then X has a fundamental cycle over R; i.e., X is R-orientable.*

Proof. Let $z = \sum\limits_\alpha r_\alpha e_\alpha^n$ be a non-zero relative cycle in X. Since $\partial z = 0$, $r_\alpha = \pm r_\beta$ when e_α^n and e_β^n share an $(n-1)$-face. Since X is path connected, the gallery property then implies $r_\alpha = \pm r$ for all α, where $r \neq 0 \in R$. Thus $z = rz_0$ where every coefficient in z_0 is ± 1. Now $0 = \partial z = r\partial z_0$. If $\partial z_0 \neq 0$ then some e_β^{n-1} has non-zero coefficient s_β in ∂z_0, so $rs_\beta = 0$, contradicting the fact that R is an integral domain. So z_0 is a fundamental cycle. $\qquad\square$

Corollary 12.3.8. *Let X be path connected. The following are equivalent:*

 (i) *X is R-orientable;*
 (ii) *$H_n^\infty(X, \partial X; R)$ is isomorphic to R;*
 (iii) *$H_n^\infty(X, \partial X; R) \neq 0$.* $\qquad\square$

The hypothesis that R be an integral domain is only needed for the (iii) $\Rightarrow$ (i) part of 12.3.8.

Proposition 12.3.9. *Let X be path connected. If $\partial X \neq \emptyset$, $H_n(X; R) = 0$ and $H_n^\infty(X; R) = 0$. If X is non-compact, $H_n(X, \partial X; R) = 0$.* $\qquad\square$

The property of R-orientability of n-pseudomanifolds is a proper homotopy invariant of pairs in the following sense.

Proposition 12.3.10. *Let X and Y be CW n-pseudomanifolds, and let $f : (X, \partial X) \to (Y, \partial Y)$ be a cellular proper homotopy equivalence (of pairs). If $c \in Z_n^\infty(X, \partial X; R)$ is a fundamental cycle for X then $f_\#(c) \in Z_n^\infty(Y, \partial Y; R)$ is a fundamental cycle for Y.*

---

[10] The two meanings of "oriented" and "orientation" – here and in Sect. 2.5 – should be noted.

Proof. We may assume that X is path connected: otherwise work with each path component separately. Clearly, $f_\#(c) \in Z_n^\infty(Y, \partial Y; R)$. Hence, by 12.3.3, $f_\#(c) = r.d$ where $r \in R$, $d \in C_n^\infty(Y; R)$ and every coefficient in d is ± 1. Let $g : (Y, \partial Y) \to (X, \partial X)$ be a proper homotopy inverse for f. We know that $g_\# \circ f_\#(c)$ is homologous to c, and hence equals c. So $r.g_\#(d) = c$. Thus r is a unit of R, so $f_\#(c)$ is a fundamental cycle. $\square$

As a special case, suppose X' is a regular oriented CW complex which is a subdivision of X. Let c be a fundamental cycle (over R) for X. The identity map[11] $i : X \to X'$ is cellular. By 12.3.10, $i_\#(c)$ is a fundamental cycle (over R) for X'. The cycle c defines an R-orientation on X; the *inherited orientation* on X' is that defined by $i_\#(c)$.

As another special case, let $X = Y$ in 12.3.10, and let X be path connected and orientable. We say that f (in 12.3.10) is *orientation preserving* [resp. *orientation reversing*] if for some fundamental class $c \in Z_n^\infty(X, \partial X; \mathbb{Z})$ $f_\#(c) = c$ [resp. $f_\#(c) = -c$].

Proposition 12.3.11. *If $H_{n-1}^\infty(X, \partial X; \mathbb{Z}_2) = 0$ then X is orientable.*

The *support* of a k-chain c is the set of k-cells appearing in c with non-zero coefficient.

Proof (of 12.3.11). Let $c = \sum_\alpha e_\alpha^n$ where α ranges over all the n-cells of X. Then $\partial c = 2d + e$, where d has support outside ∂X, e has support in ∂X, and every coefficient in d and in e is ± 1. So $0 = 2(\partial d) + \partial e$, implying that $2(\partial d)$, hence also ∂d, has support in ∂X. So $d \in Z_{n-1}^\infty(X, \partial X; \mathbb{Z})$. Since $H_{n-1}^\infty(X, \partial X; \mathbb{Z}_2) = 0$, there exist $b \in C_n^\infty(X; \mathbb{Z})$, $a \in C_{n-1}^\infty(X; \mathbb{Z})$, and $f \in C_{n-1}^\infty(\partial X; \mathbb{Z})$ such that $\partial b = d + 2a + f$, and a has support outside ∂X. Now $\partial(b \pm 2e_\alpha^n) = d + 2a + f \pm 2.\partial e_\alpha^n$ which has the form $d + 2a' + f'$; so we may assume that every non-zero coefficient in b is 1. Collecting terms, we have $\partial c = 2.\partial b - 4a + e - 2f$, so $\partial(2b - c) = 4a - e + 2f$. All coefficients in $2b - c$ are ± 1 so each coefficient in $\partial(2b - c)$ is 0, ± 1 or ± 2. Hence $a = 0$. Thus $2b - c \in Z_n^\infty(X, \partial X; \mathbb{Z})$. If X were non-orientable, we would have $2b = c$, by 12.3.9. But $2b \neq c$, so X is orientable. $\square$

A special case of Poincaré Duality says that if $X = |K|$, where K is a combinatorial n-manifold, then $H_{n-1}^\infty(X, \partial X; \mathbb{Z}_2)$ is isomorphic to $H^1(X; \mathbb{Z}_2)$; this is proved in 15.1.9. We use this here to get an important consequence of 12.3.1 and 12.3.11.

Corollary 12.3.12. *If $X = |K|$, as above, and if $H^1(X; \mathbb{Z}_2) = 0$, then X is orientable. In particular, every simply connected combinatorial manifold is orientable.* $\square$

Note that 12.3.12 does not necessarily hold for pseudomanifolds.

Again, let X be a CW n-pseudomanifold whose cells are oriented.

---

[11] As sets of points $X = X'$, so there is an identity map from X to X'.

Proposition 12.3.13. *If X is non-orientable then $H_{n-1}^{\infty}(X, \partial X; \mathbb{Z})$ contains an element of order 2.*

Proof. We refer to the proof of 12.3.11. We have $2d \in B_{n-1}^{\infty}(X, \partial X; \mathbb{Z})$ and $d \in Z_{n-1}^{\infty}(X, \partial X; \mathbb{Z})$. If d were itself a boundary, we would have b and f with $\partial b = d + f$. So $\partial(2b - c) = 2f - e$. Since there is no fundamental cycle, $c = 2b$, contradicting the definition of c. So d defines an element of order 2. □

Proposition 12.3.14. *If X is path connected and orientable, $H_f^n(X, \partial X; \mathbb{Z}) \cong \mathbb{Z}$. If X is non-orientable, $H_f^n(X, \partial X; \mathbb{Z}) \cong \mathbb{Z}_2$. In both cases, if e_{α}^n is an oriented n-cell then the cohomology class of the n-cocycle e_{α}^n generates.*

Proof. If e_{α}^n and e_{β}^n share an $(n-1)$-face d then $\delta d = \pm e_{\alpha} \pm e_{\beta}$, so e_{α} is cohomologous to $\pm e_{\beta}$. Since X is path connected, it follows that every element of $Z_f^n(X, \partial X; \mathbb{Z})$ is cohomologous to $k e_{\alpha}^n$ for some $k \in \mathbb{Z}$. Thus $H_f^n(X, \partial X; \mathbb{Z})$ is cyclic. If X is orientable then the n-cells can be oriented so that $\sum_{\alpha} e_{\alpha}^n$ is a fundamental cycle, and in that case the coefficient sum in a (relative) coboundary is zero. So $k e_{\alpha}^n$ is not a coboundary when $k \neq 0$. Thus, in the orientable case $H_f^n(X, \partial X; \mathbb{Z})$ is freely generated by e_{α}^n.

Let X be non-orientable and let P be the set of infinite n-chains c having the following properties: (i) every coefficient in ∂c is 0, 1 or -1; and (ii) when e_{α}^n and e_{β}^n are in the support of c, they can be joined by a gallery consisting of n-cells in the support of c. The set P is partially ordered by $c_1 \leq c_2$ if the support of c_1 lies in the support of c_2. The hypotheses of Zorn's Lemma are satisfied, so there is a maximal element $c_{\max} \in P$. Since X is non-orientable, there is an n-cell of X which is not in the support of $c_{\max}$. Hence, using a gallery, there exist an $(n-1)$-cell d and n-cells e_{α}^n, e_{β}^n such that $e_{\alpha}^n \cap e_{\beta}^n = d$, e_{α}^n is in the support of $c_{\max}$, and e_{β}^n is not in the support of $c_{\max}$. If $[e_{\beta}^n : d] = -[e_{\alpha}^n : d]$ then there will be another $(n-1)$-face $\bar{d}$ of e_{β}^n which is also a face of some e_{γ}^n in the support of $c_{\max}$, so that $[e_{\beta}^n : \bar{d}] = [e_{\gamma}^n : \bar{d}]$; otherwise $c_{\max} + e_{\beta}^n$ would be in P, contradicting maximality. Thus in this case e_{α}^n is cohomologous to e_{β}^n, and e_{γ}^n is cohomologous to $-e_{\beta}^n$, so e_{α}^n is cohomologous to $-e_{\alpha}^n$. On the other hand, if $[e_{\beta}^n : d] = [e_{\alpha}^n : d]$ one could change the orientation on e_{β}^n, and return to the previous case. So when X is non-orientable the cohomology class of e_{α}^n has order ≤ 2. The sum of the coefficients in a relative coboundary is even, so $H_f^n(X, \partial X; \mathbb{Z}) \cong \mathbb{Z}_2$. □

Now we turn to covering spaces. A covering space of a combinatorial n-manifold is a combinatorial n-manifold, but a covering space of a CW n-pseudomanifold need not be a CW n-pseudomanifold. On the other hand, if the universal cover, $\tilde{X}$, is a CW n-pseudomanifold, then so is every covering space of X, including X itself.

For the rest of this section X is a path connected oriented CW complex such that $\tilde{X}$ is an n-pseudomanifold. Pick a base vertex v. Let $G = \pi_1(X, v)$. Orient the cells of $(\tilde{X}, \tilde{v})$ as in Sect. 8.1 so that each covering transformation preserves the orientations of the n-cells of X. Let $p : (\tilde{X}, \tilde{v}) \to (X, v)$ be the universal covering projection. When $\tilde{X}$ is orientable, let H be the subgroup of G consisting of orientation preserving covering transformations. Clearly H has index ≤ 2 in G.

Proposition 12.3.15. *Let the n-pseudomanifold $\tilde{X}$ be orientable. Then X is orientable iff $H = G$.*

Proof. Let $c = \sum\limits_{g,\alpha} r_{\alpha,g} g \tilde{e}^n_\alpha$ be a fundamental cycle (over $\mathbb{Z}$) in $\tilde{X}$. Then $r_{\alpha,g} = \pm 1$. First, let $H = G$. Then $r_{\alpha,g} = r_{\alpha,1}$ for all $g \in G$. Let $d = \sum\limits_{\alpha} r_{\alpha,1} e^n_\alpha \in C^\infty_n(X; \mathbb{Z})$. We claim d is a fundamental cycle in X. To see this, let $\tilde{e}^{n-1}_\beta$ be a face of $g_1 \tilde{e}^n_{\alpha_1}$ and $g_2 \tilde{e}^n_{\alpha_2}$. Then $r_{\alpha_1, g_1} [g_1 \tilde{e}^n_{\alpha_1} : \tilde{e}^{n-1}_\beta] + r_{\alpha_2, g_2} [g_2 \tilde{e}^n_{\alpha_2} : \tilde{e}^{n-1}_\beta] = 0$. Hence, $r_{\alpha_1, 1} [e^n_{\alpha_1} : e^{n-1}_\beta] + r_{\alpha_2, 1} [e^n_{\alpha_2} : e^{n-1}_\beta] = 0$. Since this holds for any e^{n-1}_β in $\overset{\circ}{X}$, the claim follows.

Now, let X be orientable and suppose $H \neq G$. Then H is a (normal) subgroup of index 2. By the previous paragraph, the covering space $\bar{X}(H)$ is orientable. We have a two-to-one covering projection $q_{G^+} : \bar{X}(H) \to X$. Each cell e^n_α of X is covered by two cells $\bar{e}^n_{\alpha,1}$ and $\bar{e}^n_{\alpha,2}$ of $\bar{X}(H)$. If we orient these via q_{G^+}, we see that a fundamental class $c = \sum\limits_{\alpha} r_\alpha e^n_\alpha$ in X yields a fundamental class $d = \sum\limits_{\alpha} r_\alpha (\bar{e}^n_{\alpha,1} + \bar{e}^n_{\alpha,2})$ in $\bar{X}(H)$, and $q_{G^+\#}(d) = 2c$. Since $H \neq G$, there is an orientation reversing covering transformation $\gamma : \bar{X}(H) \to \bar{X}(H)$. Thus $2c = q_{G^+\#}(d) = q_{G^+\#} \circ \gamma(d) = -q_{G^+\#}(d) = -2c$, a contradiction. $\square$

Corollary 12.3.16. *If $\tilde{X}$ is orientable and X is non-orientable, X has an orientable path connected double cover $q_{G^+} : \bar{X}(H) \to X$.* $\square$

The awkward feature of 12.3.15 and 12.3.16 is the hypothesis that the universal cover be orientable. Unlike manifolds – see 12.3.12 – simply connected pseudomanifolds are not always orientable; e.g., the cone on the projective plane. However, by 12.3.12 and 12.3.16 we get:

Corollary 12.3.17. *Every non-orientable path connected CW n-manifold has an orientable path connected double cover.* $\square$

Exercises

1. Give an example of a 2-pseudomanifold which is not a manifold, and of a 2-pseudomanifold whose boundary is not a pseudomanifold.
2. Prove that an orientable pseudomanifold is R-orientable for any commutative ring R.

3. Prove that the cone on a path connected finite pseudomanifold is a pseudoman-ifold.

4. Give a counterexample to the converse of 12.3.11.

5. Give an example of a non-pseudomanifold covering space of a pseudomanifold.

6. Show that if K is a combinatorial manifold triangulating the surface $T_{g,d}$ then K is orientable, and if K triangulates the surface $U_{h,d}$ then K is non-orientable.

7. Give a counterexample to 12.3.7 when 2 is a 0-divisor.

8. In a pseudomanifold X, let $c = \sum_\alpha u_\alpha e_\alpha^n$ where each $u_\alpha = \pm 1$. Then $\partial c = 2d + e$ where d is supported outside ∂X and e is supported in ∂X. Show that $d \in Z_{n-1}^\infty(X, \partial X; \mathbb{Z})$ and that $\{d\} = 0$ or has order 2 in $H_{n-1}(X, \partial X; \mathbb{Z})$ depending on whether or not X is orientable. Show that if X is non-orientable and path connected then the torsion subgroup of $H_{n-1}^\infty(X, \partial X; \mathbb{Z})$ has order 2. Describe explicitly a cycle whose homology class is non-zero.

12.4 Review of more homological algebra

For details of the algebra reviewed here see, for example, [83].

Let R be a (not necessarily commutative) ring[12] with $1 \neq 0$. The *tensor product* $B \otimes_R A$ of a right R-module B and a left R-module A has the structure of an abelian group; it is generated by elements of the form $b \otimes a$, where $b \in B$ and $a \in A$, subject to bilinearity and relations of the form $br \otimes a = b \otimes ra$ where $r \in R$. If R is commutative, the left action of R on $B \otimes_R A$ defined by $r(b \otimes a) = br \otimes a$ makes $B \otimes_R A$ into an R-module, and $B \otimes_R A$ is understood to carry this left R-module structure. If R is not commutative, $B \otimes_R A$ is understood to be an abelian group only, unless an R-action is specified.[13]

If $(\{C_n\}, \partial)$ is an R-chain complex and B is a right R-module, then $(\{B \otimes_R C_n\}, \mathrm{id} \otimes \partial)$ is a $\mathbb{Z}$-chain complex whose homology groups are denoted by $H_*(C; B)$ and are called *the homology groups of C with coefficients in B*. Of course, if R is commutative we get an R-chain complex and homology R-modules.

Dually, if C and A are left R-modules, then $\mathrm{Hom}_R(C, A)$ has the structure of an abelian group. If R is commutative, the left action of R on $\mathrm{Hom}_R(C, A)$ defined by $(r.f)(c) = r.f(c)$ makes $\mathrm{Hom}_R(C, A)$ into an R-module (since $r.f \in \mathrm{Hom}_R(C, A)$). If R is not commutative, $\mathrm{Hom}_R(C, A)$ is understood to be an abelian group only, unless an R-action is specified.

If $(\{C_n\}, \partial)$ is an R-chain complex and A is a left R-module, then $(\{\mathrm{Hom}_R(C_n, A)\}, \partial^*)$ is a $\mathbb{Z}$-cochain complex whose cohomology groups are denoted by $H^*(C; A)$; they are the *cohomology groups of C with coefficients in A*. Again, if R is commutative we get an R-cochain complex and cohomology R-modules.

---

[12] For this section only we suspend our standing convention (Sect. 2.1) that R denotes a commutative ring. In this section the status of R will change several times.

[13] For group rings RG we elaborate on this convention in Sect. 8.1.

FROM HERE UNTIL AFTER REMARK 12.4.6, R IS UNDERSTOOD TO BE A PID. In particular, R is a commutative ring without zero divisors, and every submodule of a free R-module is free. When R is commutative, the distinction between left and right R-modules is not worth making, since a left module M becomes a right module under the R-action $m.r = rm$, and vice versa.

For any R-module A, there exist short exact sequences $0 \to F_1 \xrightarrow{i} F_0 \xrightarrow{j} A \to 0$ in which F_0 and F_1 are free R-modules. For any R-module B, let $\mathrm{Tor}_R(B, A)$ be defined to make an exact sequence

$$0 \longrightarrow \mathrm{Tor}_R(B,A) \longrightarrow B \otimes_R F_1 \xrightarrow{\mathrm{id}\,\otimes i} B \otimes_R F_0 \xrightarrow{\mathrm{id}\,\otimes j} B \otimes_R A \longrightarrow 0.$$

Up to canonical isomorphism, $\mathrm{Tor}_R(B, A)$ is independent of the choice of F_0 and of the epimorphism $F_0 \to A$. Both $\cdot \otimes_R \cdot$ and $\mathrm{Tor}_R(\cdot, \cdot)$ are covariant functors of two variables, which commute with direct sums and direct limits. A useful fact is:

Proposition 12.4.1. *If A or B is torsion free, then $\mathrm{Tor}_R(B, A) = 0$.*

Theorem 12.4.2. (Universal Coefficient Theorem in homology) *With $(\{C_n\}, \partial)$ and B as above, and each C_n free, there is a natural short exact sequence of R-modules*

$$0 \longrightarrow B \otimes_R H_n(C) \xrightarrow{\ \beta\ } H_n(C;B) \longrightarrow \mathrm{Tor}_R(B, H_{n-1}(C)) \longrightarrow 0.$$

This sequence splits, naturally in B but unnaturally in C. ☐

Dually, if B and A are R-modules, let $\mathrm{Ext}_R(A, B)$ be defined to make an exact sequence

$$0 \longrightarrow \mathrm{Hom}_R(A, B) \xrightarrow{\ j^*\ } \mathrm{Hom}_R(F_0, B) \xrightarrow{\ i^*\ } \mathrm{Hom}_R(F_1, B)$$

$$\longrightarrow \mathrm{Ext}_R(A, B) \longrightarrow 0.$$

As with Tor_R, $\mathrm{Ext}_R(A, B)$ is independent of the choice of F_0 and of the epimorphism $F_0 \to A$. Both $\mathrm{Hom}_R(\cdot, \cdot)$ and $\mathrm{Ext}_R(\cdot, \cdot)$ are functors of two variables, contravariant in the first and covariant in the second. They convert direct sums into direct products, so they commute with finite direct sums. Since R is a domain, $\mathrm{Hom}_R(A, R)$ is torsion free for all A.

A useful fact is:

Proposition 12.4.3. *If A is free then $\mathrm{Ext}_R(A, B) = 0$.* ☐

The next statement is proved in [146, Sect. 5.5.2]:

Proposition 12.4.4. $\text{Ext}_R(R/(r), R) \cong R/(r)$; $\text{Ext}_R(R/(r), R/(q)) \cong R/(s)$ *where* $s = (r, q)$, *the greatest common divisor of* r *and* q. $\qquad\square$

Theorem 12.4.5. (Universal Coefficient Theorem in cohomology) *With* $(\{C_n\}, \partial)$ *and* B *as above, and each* C_n *free, there is a natural short exact sequence of* R-*modules*

$$0 \longrightarrow \text{Ext}_R(H_{n-1}(C), B) \longrightarrow H^n(C; B) \xrightarrow{\ \alpha\ } \text{Hom}_R(H_n(C), B) \longrightarrow 0.$$

This sequence splits, naturally in B *but unnaturally in* C. $\qquad\square$

Remark 12.4.6. By examining the proofs of 12.4.2 and 12.4.5, one sees that the monomorphism β in 12.4.2 is given by $\beta(b \otimes \{z\}) = \{z \otimes b\}$, where $\{\cdot\}$ denotes homology class; and the epimorphism α in 12.2.5 is given by $\alpha(\{f\})(\{z\}) = f(z)$, where $f \in \text{Hom}_R(C_n, B)$, $z \in Z_n(C)$, and $\{\cdot\}$ denotes cohomology or homology as appropriate.

Let X be a CW complex. Let R be a commutative ring and let M be an R-module. We define the *homology and cohomology modules of* X *with coefficients in* M.

(i) $H_n(X; M)$ is the homology of the chain complex

$$(\{M \otimes_R C_n(X; R)\}, \text{id} \otimes \partial);$$

(ii) $H^n(X; M)$ is the cohomology of the cochain complex

$$(\{\text{Hom}_R(C_n(X; R), M)\}, \partial^*).$$

When X has locally finite type

(iii) $H_f^n(X; M)$ is the cohomology of the cochain complex

$$(\{M \otimes_R C_n(X; R)\}, \text{id} \otimes \delta);$$

(iv) $H_n^\infty(X; M)$ is the homology of the chain complex

$$(\{\text{Hom}_R(C_n(X; R), M)\}, \delta^*).$$

Similar definitions are made for pairs.

Remark 12.4.7. We can make the $\mathbb{Z}$-module $R \otimes_\mathbb{Z} C_n(X; \mathbb{Z})$ into an R-module by defining $r.(r' \otimes c) = (rr') \otimes c$. This construction is called *extension of scalars*. It gives a "canonical" isomorphism of R-chain complexes $R \otimes_\mathbb{Z} C_n(X; \mathbb{Z}) \to C_n(X; R)$ (Exercise 1). The resulting homology R-modules are therefore also "canonically" isomorphic. We have denoted both homology constructions by $H_*(X; R)$. However, there is a delicate point here. The derivation from $\{C_n(X; R), \partial\}$ makes $H_*(X; R)$ into an R-module naturally, as described in Chap. 2. The derivation of $H_*(X; R)$ from $\{R \otimes_{\mathbb{Z}} C_n(X; \mathbb{Z}), \mathrm{id} \otimes \partial\}$ would only give the homology groups ($= \mathbb{Z}$-modules) with coefficients in the (right R-module) R, were it not for the fact that extension of scalars induces an R-module structure on these homology groups. This becomes important when using the Universal Coefficient Theorem. The extension of scalars also gives an R-module structure to $\mathrm{Tor}_{\mathbb{Z}}(R, B)$, where R is a PID and B is any R-module. Thus the exact sequence of $\mathbb{Z}$-modules

$$0 \longrightarrow R \otimes_{\mathbb{Z}} H_n(X; \mathbb{Z}) \longrightarrow H_n(X; R) \longrightarrow \mathrm{Tor}_{\mathbb{Z}}(R, H_{n-1}(X; \mathbb{Z})) \longrightarrow 0$$

is in fact an exact sequence of R-modules. For example, one deduces from this that $H_1(X; \mathbb{Q})$ is isomorphic to $\mathbb{Q} \otimes_{\mathbb{Z}} H_1(X; \mathbb{Z})$ as a $\mathbb{Q}$-vector space and not merely as an abelian group. Similarly, we make the $\mathbb{Z}$-module $\mathrm{Hom}_{\mathbb{Z}}(C_n(X; \mathbb{Z}), R)$ into an R-module by the rule $(r.f)(c) = r.f(c)$. A parallel discussion can then be given concerning cohomology, ending with the conclusion that we have an exact sequence of R-modules

$$0 \longrightarrow \mathrm{Ext}_{\mathbb{Z}}(H_{n-1}(X; \mathbb{Z}); R) \longrightarrow H^n(X; R) \longrightarrow \mathrm{Hom}_{\mathbb{Z}}(H_n(X; \mathbb{Z}); R) \longrightarrow 0.$$

Similar remarks hold for H_f^* and H_*^∞.

Another version of the Universal Coefficient Theorem will be useful. A proof is given in [146, Sect. 5.5.10]:

Theorem 12.4.8. *Let R be a PID, and let B be a finitely generated R-module. Let $(\{C_n\}, \partial)$ be a free R-chain complex. There is a natural short exact sequence of R-modules*

$$0 \longrightarrow B \otimes_R H^n(C) \xrightarrow{\beta} H^n(C; B) \longrightarrow \mathrm{Tor}_R(B, H^{n+1}(C)) \longrightarrow 0.$$

This sequence splits naturally in B but unnaturally in C. $\qquad\square$

If $(\{C_m'\}, \partial')$ and $(\{C_n''\}, \partial'')$ are R-chain complexes, their *tensor product* is the $\mathbb{Z}$-chain complex $(\{C_p\}, \partial)$ where $C_p = \bigoplus_{m+n=p} C_m' \otimes_R C_n''$ and $\partial : C_p \to C_{p-1}$ is defined by $\partial(c' \otimes c'') = (\partial c') \otimes c'' + (-1)^{\deg c'} c' \otimes (\partial c'')$. This is easily seen to be a chain complex. The homology of this tensor product is given by the *Künneth Formula*:

Theorem 12.4.9. *Let R be a PID. Assume either that each C_m' is free or that each C_n'' is free. There is a natural short exact sequence of R-modules*

$$0 \to \bigoplus_{m+n=p} H_m(C') \otimes_R H_n(C'') \to H_p(C) \to \bigoplus_{m+n=p-1} \mathrm{Tor}_R(H_m(C'), H_n(C'')) \to 0.$$

This sequences splits. $\qquad\square$

Exercises

1. Write down the "canonical" isomorphism $R \otimes_{\mathbb{Z}} C_n(X; \mathbb{Z}) \to C_n(X; R)$ in Remark 12.4.7.
2. When R is a PID and M is an R-module, apply 12.4.2 to get two short exact sequences whose middle terms are $H_f^n(X; M)$ and $H_n(X; M)$ respectively.
3. Prove 12.3.14 using Universal Coefficient Theorems.

12.5 Comparison of the various homology and cohomology theories

All our homology and cohomology theories have turned out to be independent of how the CW complex was oriented. FROM NOW ON, WE WILL ONLY REFER TO A CHOICE OF ORIENTATION OF CELLS WHEN DISCUSSING CHAINS.

We begin with two immediate consequences of the Universal Coefficient Theorem in cohomology:

Theorem 12.5.1. *Let X be a CW complex, let R be a PID, and let M be an R-module. There is a short exact sequence of R-modules*

$$0 \longrightarrow \operatorname{Ext}_R(H_{n-1}(X; R), M) \longrightarrow H^n(X; M) \longrightarrow \operatorname{Hom}_R(H_n(X; R), M) \longrightarrow 0.$$

This is natural with respect to homotopy classes of maps. It splits, but unnaturally, with respect to M. □

Theorem 12.5.2. *Let X be a CW complex having locally finite type, and let R and M be as in 12.5.1. There is a short exact sequence of R-modules*

$$0 \longrightarrow \operatorname{Ext}_R(H_f^{n+1}(X; R), M) \longrightarrow H_n^{\infty}(X; M) \longrightarrow \operatorname{Hom}_R(H_f^n(X; R), M) \longrightarrow 0.$$

This is natural with respect to CW-proper homotopy classes of CW-proper maps. For strongly locally finite CW complexes, it is natural with respect to proper homotopy classes of proper maps. It splits, but unnaturally, with respect to M. □

We wish to compare properties of $H_e^*(X; R)$ with those of the inverse sequence $\{H_*(X \xrightarrow{c} K_i; R)\}$, where $\{K_i\}$ is a finite type filtration. This will require some algebraic preliminaries.

An inverse sequence $\{M_n\}$ of R-modules is (i) *pro-finitely generated*, (ii) *semistable*, (iii) *stable*, (iv) *pro-trivial*, (v) *pro-torsion free*, (vi) *pro-torsion* if it is pro-isomorphic[14] to an inverse sequence (i) whose modules are finitely generated, (ii) whose bonds are epimorphisms, (iii) whose bonds are isomorphisms,[15] (iv) whose modules are trivial, (v) whose modules are torsion free,

---

[14] Recall the convenient characterization of pro-isomorphism in 11.2.10.

[15] Note that (iii) is compatible with "stable" as defined in Sect. 11.2.

(vi) whose modules are torsion, respectively. We leave the proof of the following as an exercise:

Proposition 12.5.3. *The above definitions are equivalent to the following intrinsic definitions:*

(i) $\forall m$, $\exists n \geq m$ *such that image* $(M_n \to M_m)$ *is finitely generated;*

(ii) $\forall m$, $\exists n$ *such that* $\forall k \geq n$, *image* $(M_k \to M_m) = $ *image* $(M_n \to M_m)$;

(iii) *there is a cofinal subsequence* $\{M_{n_i}\}$ *such that image* $(M_{n_{i+2}} \to M_{n_{i+1}})$ *is mapped by the bond isomorphically onto image* $(M_{n_{i+1}} \to M_{n_i})$ *for all i;*

(iv) $\forall m$, $\exists n$ *such that* $M_n \to M_m$ *is zero;*

(v) $\forall m$, $\exists n$ *such that* tor $M_n \subset \ker(M_n \to M_m)$;

(vi) $\forall m$, $\exists n$ *such that image* $(M_n \to M_m)$ *is torsion.*   □

Let $\bar{M}$ denote $M/\text{tor } M$. Then $M \mapsto \bar{M}$ defines a covariant functor from R-modules to R-modules. When convenient, we will use the alternative name M *mod torsion* for $\bar{M}$. By 11.3.14 we have:

Proposition 12.5.4. *For any direct sequence $\{M_i\}$, the canonical homomorphism* $(\varinjlim_n \{M_n\})^- \to \varinjlim_n \{\bar{M}_n\}$ *is an isomorphism.*   □

Proposition 12.5.5. *Let R be a PID. Let $\{M_n\}$ be a pro-finitely generated inverse sequence of R-modules. Then $\{M_n\}$ is semistable iff $\{\bar{M}_n\}$ is semistable.*

Proof. We may assume that each M_n is finitely generated. By 11.3.11, $\{M_n, f_n^m\}$ can be replaced by $\{\bar{M}_n \oplus \text{tor } M_n, \bar{f}_n^m \oplus (f_n^m \mid \text{tor } M_n)\}$. By 11.3.12, $\{\text{tor } M_n\}$ is semistable. The desired conclusion is immediate.   □

Proposition 12.5.6. *Let R be a PID. Let $\{M_n, f_n^m\}$ be a pro-finitely generated inverse sequence of R-modules. Then $\{M_n\}$ is semistable iff the direct limit of the dual direct sequence $\{\text{Hom}_R(M_n, R), (f_n^m)^*\}$ is a countably generated free R-module.*

Proof. Let $\{M_n\}$ be semistable. Let $N_n = \bigcap_{m > n}$ image f_n^m. Then N_n is finitely generated. Let $g_n^{n+1} : N_{n+1} \to N_n$ be the restriction of f_n^{n+1}. Then each g_n^{n+1} is onto and $\{N_n, g_n^m\}$ is pro-isomorphic to $\{M_n, f_n^m\}$. In the inverse sequence $\{\bar{N}_n, \bar{g}_n^m\}$, each $\bar{N}_n$ is free and finitely generated and each $\bar{g}_n^{n+1}$ is a splittable epimorphism. Thus $\{\text{Hom}_R(\bar{N}_n, R), (\bar{g}_n^m)^*\}$ is a direct sequence of finitely generated free modules whose bonds are splittable monomorphisms (see Exercise 6). The direct limit of such a sequence is clearly countably generated and free. But $\text{Hom}_R(\bar{N}_n, R)$ is isomorphic to $\text{Hom}_R(N_n, R)$, for all n, by means of isomorphisms which identify $(\bar{g}_n^{n+1})^*$ with $(g_n^{n+1})^*$. And $\{\text{Hom}_R(N_n, R), (g_n^m)^*\}$ is ind-isomorphic to $\{\text{Hom}_R(M_n, R), (f_n^m)^*\}$, a fact obtained by dualizing the corresponding pro-isomorphism statement above. By 11.2.6, the direct limit of the last sequence must also be countably generated and free.

Conversely, let $\varinjlim\{\mathrm{Hom}_R(M_n, R), (f_n^m)^*\} =: L$ be free. By hypothesis and 11.2.6, we may assume each M_n is finitely generated. By 11.3.11, each $\mathrm{Hom}_R(M_n, R)$ is finitely generated and free. Clearly, L is free on a countable set of generators, so L is the direct limit of a sequence $\{L_n, h_n^m\}$ in which each L_n is finitely generated and free, and each h_n^m is a splittable monomorphism – just take more and more R-summands of L. By 11.3.15, $\{\mathrm{Hom}_R(M_n, R)\}$ and $\{F_n\}$ are ind-isomorphic. Hence, dualizing again, the inverse sequences $\{\mathrm{Hom}_R(\mathrm{Hom}_R(M_n, R), R)\}$ and $\{\mathrm{Hom}_R(L_n, R)\}$ are pro-isomorphic. The first of these is obviously isomorphic to $\{\bar{M}_n\}$, while the second is an inverse sequence whose bonds are (splittable) epimorphisms, hence semistable. By 12.5.5, $\{M_n\}$ is semistable. $\qquad\square$

Proposition 12.5.7. *Let R be a PID. A pro-finitely generated inverse sequence of R-modules $\{M_n\}$ is pro-torsion free iff the direct limit of the corresponding direct sequence $\{\mathrm{Ext}_R(M_n, R)\}$ is trivial.*

Proof. This proof uses homological algebra not needed elsewhere in this book. We may assume that each M_n is finitely generated. By 11.3.11 and 12.4.3, $\varinjlim\{\mathrm{Ext}_R(M_n, R)\} = 0$ iff $\varinjlim\{\mathrm{Ext}_R(\mathrm{tor}\ M_n, R)\} = 0$. The PID R has a field of quotients F obtained by inverting all non-zero elements of R (see page 69 of [101].) Clearly $\mathrm{Hom}_R(\mathrm{tor}\ M_n, F) = 0$. The underlying abelian group of F is divisible, so F is an injective R-module (see [83, Chap. 1, Sect. 7.1]), hence $\mathrm{Ext}_R(\mathrm{tor}\ M_n, F) = 0$ [83, Chap. 3, Sect. 2.6]. Thus $\mathrm{Hom}_R(\mathrm{tor}\ M_n, F/R)$ is isomorphic to $\mathrm{Ext}_R(\mathrm{tor}\ M_n, R)$; this follows by applying [83, Chap. 3, Sect. 5.2] to the short exact sequence $0 \to R \to F \to F/R \to 0$. By 11.3.15, $\varinjlim \mathrm{Hom}_R(\mathrm{tor}\ M_n, F/R) = 0$ iff for each m there exists $n \geq m$ such that the composition $\mathrm{tor}\ M_n \to \mathrm{tor}\ M_m \to F/R$ is zero. By 11.3.11, this is equivalent to saying that the inverse sequence $\{\mathrm{tor}\ M_n\}$ is pro-trivial. This in turn is equivalent to saying that $\{M_n\}$ is pro-torsion free. $\qquad\square$

Proposition 12.5.8. *Let R be a PID. Let the inverse sequence of R-modules $\{M_n\}$ be pro-finitely generated.*

(a) *$\{M_n\}$ is pro-torsion iff $\varinjlim\{\mathrm{Hom}_R(M_n, R)\} = 0$.*
(b) *$\{\bar{M}_n\}$ is stable and R-rank $(\varinjlim\{M_n\}) = \rho < \infty$ iff $\varinjlim\{\mathrm{Hom}_R(M_n, R)\}$ is a free R-module of rank $\rho < \infty$.*

Proof. The proofs are similar to that of 12.5.6. $\qquad\square$

Let X be a countable CW complex having locally finite type. Let $\{K_i\}$ be a finite type filtration of X.

Proposition 12.5.9. *If $H_n(X; R)$ is finitely generated and R is a PID, then $H_n(X \stackrel{c}{-} K_i; R)$ is finitely generated for all i. Hence $\{H_n(X \stackrel{c}{-} K_i; R)\}$ is pro-finitely generated.*

Proof. Replacing X by X^{n+1}, we may assume X is strongly locally finite. Then, by 11.4.4, $N(K_i)$ is finite. By 1.5.5, $X = N(K_i) \cup (X \overset{c}{-} K_i)$. And $N(K_i) \cap (X \overset{c}{-} K_i)$ is finite. The Mayer-Vietoris sequence gives exactness in

$$H_n(N(K_i) \cap (X \overset{c}{-} K_i); R) \to H_n(N(K_i); R) \oplus H_n(X \overset{c}{-} K_i; R) \to H_n(X; R).$$

By 2.7.7, the left term is finitely generated. Hence $H_n(X \overset{c}{-} K_i; R)$ is finitely generated.   □

Theorem 12.5.10. *Let X be a countable CW complex having locally finite type, let R be a PID, let $H_n(X; R)$ and $H_{n-1}(X; R)$ be finitely generated, and let $\{K_i\}$ be a finite type filtration of X.*

(i) $H_e^n(X; R)$ *mod torsion is free iff* $\{H_n(X \overset{c}{-} K_i; R)\}$ *is semistable;*

(ii) $H_e^n(X; R)$ *is torsion free iff* $\{H_{n-1}(X \overset{c}{-} K_i; R)\}$ *is pro-torsion free;*[16]

(iii) $H_e^n(X; R)$ *is torsion iff* $\{H_n(X \overset{c}{-} K_i; R)\}$ *is pro-torsion;*

(iv) $H_e^n(X; R)$ *mod torsion is free with finite rank ρ iff* $\{H_n(X \overset{c}{-} K_i; R)$ *mod torsion$\}$ is stable with free inverse limit of finite rank ρ.*

Proof. By 12.5.9, $\{H_k(X \overset{c}{-} K_i; R)\}$ is pro-finitely generated for $k = n - 1$ and n.

(i) By the previous propositions and Sect. 12.2, $\{H_n(X \overset{c}{-} K_i; R)\}$ is semistable iff $\varinjlim_i \{\mathrm{Hom}_R(H_n(X \overset{c}{-} K_i), R)\}$ is free iff $\varinjlim_i \{H^n(X \overset{c}{-} K_i; R)\}$ mod torsion is free iff $H_e^n(X; R)$ mod torsion is free. (ii) By 12.5.7, the following are equivalent:

1. $\{H_{n-1}(X \overset{c}{-} K_i; R)\}$ is pro-torsion free
2. $\varinjlim_i \mathrm{Ext}_R(H_{n-1}(X \overset{c}{-} K_i; R), R) = 0$

3. $H_e^n(X; R)$ is torsion free.

(iii) and (iv) are proved similarly using 12.5.8.   □

Corollary 12.5.11. *Let X be a path connected CW complex having locally finite type, and let R be a PID. Then $H_e^0(X; R)$ is countably generated and free. If $H_1(X; R)$ is finitely generated, then $H_e^1(X; R)$ is torsion free.*

Proof. By 11.4.3, X is countable. Applying 12.5.10 with $n = 0$ we find that $H_e^0(X; R)$ is torsion free. To show that $H_e^0(X; R)$ is free, we apply 12.5.6 so we must show that $\{H_0(X \overset{c}{-} K_i; R)\}$ is semistable. By 2.7.2, each $H_0(X \overset{c}{-} K_i; R)$ is free with finite rank equal to the number of path components of $X \overset{c}{-} K_i$. Identifying generators with path components, one sees (by considering things at the chain level) that, for $j > i$, the image of $H_0(X \overset{c}{-} K_j; R) \to H_0(X \overset{c}{-} K_i; R)$ is freely generated by those path components of $X \overset{c}{-} K_i$ which meet $X \overset{c}{-} K_j$. The proof of 12.5.9 shows that there are only finitely many path components. Thus $\{H_0(X \overset{c}{-} K_i; R)\}$ is semistable.

For the second part, we apply 12.5.10 with $n = 1$. In particular, we have observed that $\{H_0(X \overset{c}{-} K_i; R)\}$ is pro-torsion free. Hence $H_e^1(X; R)$ is torsion free.   □

[16] This holds even if $H_n(X; R)$ is infinitely generated.

Exercises

1. Prove the naturality statements in 12.5.1 and prove 12.5.2.
2. Prove 12.5.3, 12.5.4, 12.5.8, 12.5.10 (iii) and (iv).
3. Compute $H_e^*(T; \mathbb{Z})$ where T is the dyadic solenoid inverse mapping telescope (Example 11.4.15).
4. Establish the following exact sequence: $0 \to B \otimes_R H_e^n(X; R) \to H_e^n(X; B) \to \mathrm{Tor}_R(B, H_e^{n+1}(X; R)) \to 0$ using 12.4.8 (where B is finitely generated over the PID R).
5. Prove that the inverse sequence of modules $\{M_n\}$ is stable iff it is pro-isomorphic to a module.
6. Prove that the dual of an inverse sequence of splittable epimorphisms is a direct sequence of splittable monomorphisms, and vice versa.
7. Let $X = |K|$ where K is a combinatorial n-manifold with empty boundary. Assume X is path connected and non-compact and that $H^n(X; \mathbb{Z}) \cong H^{n-1}(X; \mathbb{Z}) = 0$. Using 12.5.10, what can be concluded about $\{H_{n-1}(X \overset{c}{-} K_i; \mathbb{Z})\}$ and $\{H_{n-2}(X \overset{c}{-} K_i; \mathbb{Z})\}$ when X is orientable? non-orientable?

12.6 Homology and cohomology of products

Let X and Y be oriented CW complexes of locally finite type. We give $X \times Y$ the product orientation. Let R be a PID. The R-module $C_p(X \times Y; R)$ is freely generated by the cells $e_\alpha^i \times \tilde{e}_\beta^j$ such that e_α^i and $\tilde{e}_\beta^j$ are cells of X and Y respectively and $i + j = p$. Letting E_p be the set of all (oriented) p-cells of $X \times Y$, the function $E_p \to \bigoplus_{i+j=p} C_i(X; R) \otimes_R C_j(Y; R)$ defined by $e_\alpha^i \times \tilde{e}_\beta^j \mapsto e_\alpha^i \otimes \tilde{e}_\beta^j$ extends uniquely to a homomorphism

$$\zeta_p : C_p(X \times Y; R) \to \bigoplus_{i+j=p} C_i(X; R) \otimes_R C_j(Y; R).$$

And the R-bilinear map

$$C_i(X; R) \times C_j(Y; R) \to C_{i+j}(X \times Y; R)$$

defined by $(e_\alpha^i, \tilde{e}_\beta^j) \mapsto e_\alpha^i \times \tilde{e}_\beta^j$ induces a homomorphism

$$C_i(X; R) \otimes_R C_j(Y; R) \to C_{i+j}(X \times Y; R).$$

As i and j vary with $i + j = p$, the latter homomorphisms fit together to provide an inverse for ζ_p. Thus ζ_p is an isomorphism of R-modules. Moreover, by the formulas for incidence numbers in 2.5.17, one obtains the coboundary formula

$$\delta(e_\alpha^i \times \tilde{e}_\beta^j) = (\delta e_\alpha^i) \times \tilde{e}_\beta^j + (-1)^i e_\alpha^i \times (\delta \tilde{e}_\beta^j)$$

and one checks that $\delta\zeta = \zeta\delta$. Hence ζ defines an isomorphism from the cochain complex $(C_*(X \times Y; R), \delta)$ to the tensor product of the cochain complexes $(C_*(X; R), \delta)$ and $C_*(Y; R), \delta)$ as defined in Sect. 12.4. There is a similar boundary formula satisfying $\partial\zeta = \zeta\partial$, so ζ also defines an isomorphism of chain complexes from $(C_*(X \times Y; R), \partial)$ to the tensor product[17] of $(C_*(X; R), \partial)$ and $(C_*(Y; R), \partial)$. Applying the Künneth Formula 12.4.9, we get:

Proposition 12.6.1. *There are natural short exact sequences of R-modules*

$$0 \to \bigoplus_{i+j=p} H_f^i(X; R) \otimes_R H_f^j(Y; R) \to H_f^p(X \times Y; R) \to \bigoplus_{i+j=p+1} \mathrm{Tor}_R(H_f^i(X; R), H_f^j(Y; R)) \to 0$$

and

$$0 \to \bigoplus_{i+j=p} H_i(X; R) \otimes H_j(Y; R) \to H_p(X \times Y; R) \to \bigoplus_{i+j=p-1} \mathrm{Tor}_R(H_i(X; R), H_j(Y; R)) \to 0.$$

These sequences split. $\square$

Exercises

1. Prove that $\partial(e_\alpha^i \times \tilde{e}_\beta^j) = (\partial e_\alpha^i) \times e_\beta^j + (-1)^i e_\alpha^i \times (\partial \tilde{e}_\beta^j)$.
2. Prove that $\partial\zeta = \zeta\partial$.

---

[17] In our context all this is straightforward, but in more abstract versions the existence of these isomorphisms follows from the Eilenberg-Zilber Theorem.

PART IV: TOPICS IN THE COHOMOLOGY OF INFINITE GROUPS

The emphasis here is on understanding the topological content of certain kinds of cohomological statements. In particular, the cohomology of a group with group-ring coefficients really can be seen as encoding facts about homology-at-infinity of the universal cover of a suitable $K(G,1)$-complex. This is the theme of Chapter 13. We also treat ends of pairs of groups in some detail; there are two interesting definitions, giving different results, and we explain both. Finally, we give an "old-fashioned" highly geometric treatment of Poincaré Duality, to bring out the notion of "dual cell", which is often lost in more sophisticated treatments of the subject.

13

Cohomology of Groups and Ends Of Covering Spaces

This chapter is about the cohomology modules $H^*(G, RG)$ and their connection with asymptotic homological invariants of the group G. The first interesting case involves the classical subject of ends of spaces and ends of groups. Our treatment of homology and cohomology of ends in Part III enables us to begin building a theory of "higher ends" of groups which will occupy much of the rest of the book.

13.1 Cohomology of groups

We carry over notation and conventions from Sect. 8.1.

Let M be a left RG-module and let $\{F_n\}$ be a free RG-resolution of R. The *cohomology R-modules of G with coefficients in M* are computed from the R-cochain complex

$$\cdots \xleftarrow{\ \partial_2^*\ } \operatorname{Hom}_{RG}(F_1, M) \xleftarrow{\ \partial_1^*\ } \operatorname{Hom}_{RG}(F_0, M) \longleftarrow 0$$

where $\partial_n^*(f)(x) := f\partial_n(x)$; they are denoted $H^*(G, M)$. By 8.1.1 they are well defined.[1]

Analogous to 8.1.4 we have

Proposition 13.1.1. *Let (X, v) be a $K(G, 1)$-complex. Then $H^*(G, R) \cong H^*(X; R)$.* $\qquad\square$

The group-cohomology of most interest in this book is $H^n(G, RG)$, where the action of G on RG is induced by left translation in G. When the group G is of type FP_n we will explain how the R-module $H^n(G, RG)$ encodes asymptotic homological information about G in dimensions $< n$.

---

[1] As remarked in Sect.12.1, some authors define $\partial_n^*(f)(x) = (-1)^{n+1} f\partial_n(x)$. The cocycles and coboundaries are not altered by this.

13.2 Homology and cohomology of highly connected covering spaces

In this section we relate the homology and cohomology of covering spaces of a $K(G,1)$-complex to the homology and cohomology of G with appropriate coefficient modules.

Let (X, v) be a pointed path connected CW complex.[2] We write G for $\pi_1(X, v)$. We saw in Sect. 3.4 that path connected pointed covering spaces of (X, v) correspond to subgroups of G. We will need special RG-modules associated with subgroups.

Let $H \leq G$. We write $H\backslash G = \{Hg \mid g \in G\}$. When A is a set, RA or $R(A)$ denotes the free R-module generated by A. Taking $A = H\backslash G$, we make $R(H\backslash G)$ into a right RG-module via the action $(Hg).\bar{g} = Hg\bar{g}$. We write $R\hat{A} = \prod_{a \in A} R(a)$ where $R(a)$ is the free left R-module generated by the one-element set $\{a\}$; this is a cartesian product of R-modules; $R\hat{A}$ is the *completion* of RA. We make $R(H\backslash G)\hat{}$ into a right RG-module via the obvious extension of the right G-action on $R(H\backslash G)$. As before we will use infinite summation notation for elements of $R(H\backslash G)\hat{}$.

Consider the short exact sequence of right RG-modules:

$$0 \longrightarrow R(H\backslash G) \overset{i}{\longrightarrow} R(H\backslash G)\hat{} \overset{p}{\longrightarrow} R(H\backslash G)^e \longrightarrow 0$$

where $R(H\backslash G)^e := R(H\backslash G)\hat{}/R(H\backslash G)$.

If $H \leq G$ and $p_H : (\tilde{X}, \tilde{v}) \to (\bar{X}(H), \bar{v})$ is the corresponding covering projection (see Sect. 3.2), we orient the cells of $\bar{X}(H)$ so as to make p_H and $q_H : \bar{X}(H) \to X$ orientation preserving on cells. As in Sect. 8.1, we choose a cell $\tilde{e}^n_\alpha$ "over" each e^n_α. A typical n-cell of the covering space $\bar{X}(H)$ is $Hg\tilde{e}^n_\alpha = p_H(g\tilde{e}^n_\alpha)$. We will be discussing infinite chains in $\bar{X}(H)$, so we note that if X has locally finite type or locally finite n-skeleton then the same is true of every $\bar{X}(H)$ (see 10.1.24).

For each k we have a commutative diagram of R-modules

$$
\begin{array}{ccccc}
R(H\backslash G) \otimes_G C_k(\tilde{X}; R) & \overset{i \otimes \mathrm{id}}{\longrightarrow} & R(H\backslash G)\hat{} \otimes_G C_k(\tilde{X}; R) & \overset{p \otimes \mathrm{id}}{\longrightarrow} & R(H\backslash G)^e \otimes_G C_k(\tilde{X}; R) \\
\downarrow{\scriptstyle \phi_k} & & \downarrow{\scriptstyle \phi_k^\infty} & & \downarrow{\scriptstyle \phi_k^e} \\
C_k(\bar{X}(H); R) & \longrightarrow & C_k^\infty(\bar{X}(H); R) & \longrightarrow & C_k^\infty(\bar{X}(H); R)/C_k(\bar{X}(H); R)
\end{array}
$$

---

[2] Note that Convention 12.2.5 applies throughout; i.e., X is assumed to have locally finite skeleta where appropriate.

in which ϕ_k^∞ is defined[3] on generators by: $\phi_k^\infty \left(\left(\displaystyle\sum_{Hg} m_{Hg,\alpha} Hg \right) \otimes \tilde{e}_\alpha^k \right) =$

$\displaystyle\sum_{Hg} m_{Hg,\alpha} Hg \tilde{e}_\alpha^k$. The top line is exact because of 12.4.1 and the exact sequence

preceding it, since $C_k(\tilde{X}; R)$ is free.[4] The bottom line is obviously exact. Clearly, ϕ_k, the restriction of ϕ_k^∞, is well defined, hence also ϕ_k^e.

Proposition 13.2.1. $\{\phi_k\}$ *is a chain isomorphism. Each ϕ_k^∞ and ϕ_k^e is a well defined monomorphism. If X has locally finite type then $\{\phi_k^\infty\}$ and $\{\phi_k^e\}$ are chain maps. If X has finite type then $\{\phi_k^\infty\}$ and $\{\phi_k^e\}$ are chain isomorphisms.*

For the proof of 13.2.1, we need:

Lemma 13.2.2. *Let e_α^k and e_β^{k-1} be cells of X, and let $g, \bar{g} \in G$. Then $[H g \tilde{e}_\alpha^k : H \bar{g} \tilde{e}_\beta^{k-1}] = \displaystyle\sum_{h \in H} [g \tilde{e}_\alpha^k : h \bar{g} \tilde{e}_\beta^{k-1}]$.*

Proof.

$$\partial(g \tilde{e}_\alpha^k) = \sum_{h \in H} [g \tilde{e}_\alpha^k : h \bar{g} \tilde{e}_\beta^{k-1}] h \bar{g} \tilde{e}_\beta^{k-1} + \text{ other terms}.$$

$$(p_H)_\# \partial(g \tilde{e}_\alpha^k) = \sum_{h \in H} [g \tilde{e}_\alpha^k : h \bar{g} \tilde{e}_\beta^{k-1}] H \bar{g} \tilde{e}_\beta^{k-1} + \text{ other terms}.$$

In the last line, "other terms" is a chain independent of $H \bar{g} \tilde{e}_\beta^{k-1}$. But $(p_H)_\#(g \tilde{e}_\alpha^k) = H g \tilde{e}_\alpha^k$, so

$$\partial(p_H)_\#(g \tilde{e}_\alpha^k) = [H g \tilde{e}_\alpha^k : H \bar{g} \tilde{e}_\beta^{k-1}] H \bar{g} \tilde{e}_\beta^{k-1} + \text{ other terms},$$

where, again, "other terms" is independent of $H \bar{g} \tilde{e}_\beta^{k-1}$. The result follows. $\square$

Proof (of 13.2.1). A calculation gives:

$$\partial \phi_k^\infty \left(\sum_{Hg} m_{Hg,\alpha} Hg \otimes \tilde{e}_\alpha^k \right) = \sum_\beta \sum_{Hg} \sum_{H\bar{g}} m_{Hg,\alpha} [H g \tilde{e}_\alpha^k : H \bar{g} \tilde{e}_\beta^{k-1}] H \bar{g} \tilde{e}_\beta^{k-1}$$

$$\phi_{k-1}^\infty \partial \left(\sum_{Hg} m_{Hg,\alpha} Hg \otimes \tilde{e}_\alpha^k \right) = \sum_\beta \sum_{Hg} \sum_{\bar{g} \in G} m_{Hg,\alpha} [\tilde{e}_\alpha^k : \bar{g} \tilde{e}_\beta^{k-1}] H \bar{g} \tilde{e}_\beta^{k-1}.$$

---

[3] Recall that $C_n^\infty(Z; R)$ makes sense for any CW complex Z. But Z must have locally finite n-skeleton for the boundary homomorphism to make sense on $C_n^\infty(Z; R)$.

[4] When M is a right RG-module, the abelian group $M \otimes_G \mathbb{Z}G$ is canonically isomorphic to M. Thus the operations $\cdot \otimes_G C_k(\tilde{X}; R)$ and $\cdot \otimes_R C_k(X; R)$ have essentially the same effect. That is why 12.4.1 can be used here.

Lemma 13.2.2 implies that these are the same. So $\{\phi_k^\infty\}$ and $\{\phi_k\}$ are chain maps as claimed; hence also $\{\phi_k^e\}$. Obviously ϕ_k^∞ and ϕ_k are monomorphisms; ϕ_k^∞ is onto if (and only if) α varies over a finite set, whereas ϕ_k is always onto. □

Corollary 13.2.3. *Let X be a $K(G,1)$-complex. Then, for all k,*

$$H_k(\bar{X}(H); R) \cong H_k(G, R(H\backslash G)).$$

If X has finite type then

$$H_k^\infty(\bar{X}(H); R) \cong H_k(G, R(H\backslash G)^\wedge)$$

and

$$H_{k-1}^e(\bar{X}(H); R) \cong H_k(G, R(H\backslash G)^e).$$

Proof. $C_*(\tilde{X}; R)$ gives a free RG-resolution of R. Apply 13.2.1. □

Proposition 13.2.4. *Let X be $(n-1)$-aspherical. Then for all $k \leq n-1$, $H_k(\bar{X}(H); R) \cong H_k(G, R(H\backslash G))$. If X^n is finite, the other conclusions of 13.2.3 hold for $k \leq n-1$.*

Proof. By 7.1.5 there is a $K(G,1)$-complex Y with $Y^n = X^n$. The first claim follows from 13.2.1. If X^n is finite, $\partial : C_k^\infty(\bar{Y}(H); R) \to C_{k-1}^\infty(\bar{Y}(H); R)$ is well defined for $k \leq n$. The rest of the proof of 13.2.1 therefore works for $k \leq n-1$. Similarly for $C_k^e(\bar{Y}(H); R)$. □

Note in particular the special cases of 13.2.3 and 13.2.4 in which $H = \{1\}$ or $H = G$. We conclude that when X is a $K(G,1)$-complex, $H_k(G, R) \cong H_k(X; R)$ for all k (see 8.1.4), and $H_k(G, RG) = 0$ for all $k > 0$, while $H_0(G, RG) \cong R$. Similar conclusions follow from the weaker hypothesis of 13.2.4 when $k \leq n-1$.

The next proposition is (a special case of) *Shapiro's Lemma*:

Proposition 13.2.5. *Let $H \leq G$. Then, for all k,*

$$H_k(G, R(H\backslash G)) \cong H_k(H, R).$$

Proof. Let X be a $K(G,1)$-complex. By 13.2.1, we have $H_k(\bar{X}(H); R) \cong H_k(G, R(H\backslash G))$ and (applying 13.2.1 with H replacing G) $H_k(\bar{X}(H); R) \cong H_k(H, R)$. □

In the commutative diagram preceding Proposition 13.2.1, the fact that the top line is exact gives:

Proposition 13.2.6. *There is an exact sequence*

$$\cdots \to H_n(G, R(H\backslash G)) \to H_n(G, R(H\backslash G)^\wedge) \to H_n(G, R(H\backslash G)^e) \to H_{n-1}(G, R(H\backslash G)) \to \cdots.$$

□

Corollary 13.2.7. *For $k > 1$, $H_k(G, RG\hat{\ }) \cong H_k(G, RG^e)$. If G is infinite and finitely generated then $H_1(G, RG^e) \cong H_1(G, RG\hat{\ }) \oplus R$.*

Proof. For $k > 1$, the claim follows from 13.2.6 and the preceding remarks. The claim for $k = 1$ also follows from 13.2.6, because, by 13.2.3 (with $n = 0$) and 11.1.3, $H_0(G, RG\hat{\ }) = 0$ when G is infinite and finitely generated.   □

We now turn to cohomology. The details are similar to the homology case, so we will leave them as exercises.

For each k, there is a commutative diagram of R-modules

$$
\begin{array}{ccccc}
\mathrm{Hom}_G(C_k(\tilde{X}; R), R(H\backslash G)) & \rightarrowtail & \mathrm{Hom}_G(C_k(\tilde{X}; R), R(H\backslash G)\hat{\ }) & \twoheadrightarrow & \mathrm{Hom}_G(C_k(\tilde{X}; R), R(H\backslash G)^e) \\
\uparrow{\scriptstyle\psi_k} & & \uparrow{\scriptstyle\psi_k^\infty} & & \uparrow{\scriptstyle\psi_k^e} \\
C_k(\bar{X}(H); R) & \rightarrowtail & C_k^\infty(\bar{X}(H); R) & \longrightarrow & C_k^\infty(\bar{X}(H); R)/C_k(\bar{X}(H); R)
\end{array}
$$

in which ψ_k^∞ is defined by: $\psi_k^\infty \left(\displaystyle\sum_{\alpha, Hg} m_{\alpha, Hg} Hg\tilde{e}_\alpha^k \right) (\tilde{e}_\beta^k) = \displaystyle\sum_{Hg} m_{\beta, Hg} Hg$. The top line is exact (by 12.4.3, for reasons similar to those given in the homology case), since $C_k(\tilde{X}; R)$ is free. The bottom line is obviously exact. Clearly ψ_k, the restriction of ψ_k^∞, is well defined, hence also ψ_k^e. Note that we are converting $R(H\backslash G)\hat{\ }$, $R(H\backslash G)$ and $R(H\backslash G)^e$ into left RG-modules by the usual rule $g.x := x.g^{-1}$.

Proposition 13.2.8. $\{\psi_k^\infty\}$ *is a cochain isomorphism. Each ψ_k and ψ_k^e is a well defined monomorphism. If X has locally finite type, $\{\psi_k\}$ and $\{\psi_k^e\}$ are cochain maps. If X has finite type, $\{\psi_k\}$ and $\{\psi_k^e\}$ are cochain isomorphisms.*
□

Corollary 13.2.9. *Let X be $(n - 1)$-aspherical. Then for all $k \leq n - 1$, $H^k(\bar{X}(H); R) \cong H^k(G, R(H\backslash G)\hat{\ })$. If X^n is finite then $H_f^k(\bar{X}(H); R) \cong H^k(G, R(H\backslash G))$ and $H_e^k(\bar{X}(H); R) \cong H^k(G, R(H\backslash G)^e)$ for $k \leq n - 1$.*   □

In particular, when X is a $K(G, 1)$-complex, $H^k(G, R) \cong H^k(X; R)$ for all k (see 8.1.4), and $H^k(G, RG\hat{\ }) = 0$ for all $k > 0$, while $H^0(G, RG\hat{\ }) \cong R$. The same holds for $k \leq n - 1$ when X is merely $(n - 1)$-aspherical.

Another case of Shapiro's Lemma with proof similar to that of 13.2.5 is:

Proposition 13.2.10. *Let $H \leq G$. For all k, $H^k(G, R(H\backslash G)\hat{\ }) \cong H^k(H, R)$.*
□

Proposition 13.2.11. *Let $H \leq G$. $H^0(G, R(H\backslash G)) = 0$ [resp. $\cong R$] iff H has infinite [resp. finite] index in G. In particular, $H^0(G, RG) = 0$ [resp. $\cong R$] iff G is infinite [resp. finite].*

Proof. Let (X, v) be a $K(G, 1)$-complex having one vertex. By 8.1.2 we see that $H^0(G, R(H\backslash G))$ is the module of cocycles in $\operatorname{Hom}_G(C_0(\tilde{X}; R), R(H\backslash G))$. Since $\tilde{X}^1$ is path connected, any such cocycle f must satisfy $f(g\tilde{v}) = gf(\tilde{v}) = f(\tilde{v})$ for all $g \in G$ because (see Sect. 8.1) g preserves orientation on 0-cells. If $H\backslash G$ is infinite, this implies $f(\tilde{v}) = 0$, hence $f = 0$. If $H\backslash G$ is finite, any such $f(\tilde{v})$ is an R-multiple of the sum of the generators of $R(H\backslash G)$. $\square$

Just as in the homology case, we have:

Proposition 13.2.12. *There is an exact sequence*

$$\cdots \leftarrow H^n(G, R(H\backslash G)\hat{\ }) \leftarrow H^n(G, R(H\backslash G)) \leftarrow H^{n-1}(G, R(H\backslash G)^e) \leftarrow H^{n-1}(G, R(H\backslash G)\hat{\ }) \leftarrow \cdots$$

$\square$

Corollary 13.2.13. *For $k > 1$, $H^k(G, RG) \cong H^{k-1}(G, RG^e)$. If G is infinite, $H^0(G, RG^e) \cong H^1(G, RG) \oplus R$.* $\square$

Proof. The first part is clear. For the last part, the short exact sequence splits. $\square$

Proposition 13.2.14. *Let $H \leq G$ where H has type F_n. If H has finite index in G then for $k \leq n$, $H^k(G, RG) \cong H^k(H, RH)$.*

Proof. By 7.2.4, G has type F_n iff H has type F_n. Let X be as in 13.2.9 with X^n finite and $\pi_1(X, v) \cong G$. By 13.2.9, $H^k(G, (RG)^e) \cong H_e^k(\tilde{X}; R) \cong H^k(H, (RH)^e)$ for $k \leq n - 1$. Now apply 13.2.13. $\square$

Remark 13.2.15. A defect of the cellular cochain complexes is that they do not exhibit at the level of chains the important *cup product*, a homomorphism $H^k \otimes H^l \to H^{k+l}$ which makes H^* into a graded ring. We are presenting topology for group theory. In so far as the reader needs cohomology of groups to be equipped with a cup product the exposition given here is inadequate. However, our cohomology modules are canonically isomorphic to corresponding cohomology modules defined in terms of singular cochains, as is expounded in most advanced books on algebraic topology, so the reader who needs cup products can proceed using those sources. A treatment more directly applicable to group theory can be found for example in [29].

Next, we consider $H^n(G, RG)$ as a right RG-module. The R-module $H^n(G, RG)$ is calculated from the cochain complex $\{\operatorname{Hom}_G(F_n, RG), \partial^*\}$. The right module RG is to be changed into a left RG-module via the action $\bar{g}.(\Sigma n_g g) = \Sigma n_g g(\bar{g})^{-1}$. With this understood, there is a right RG-module structure on $\operatorname{Hom}_G(F_n, RG)$ given by $(f.\bar{g})(y) = (\bar{g})^{-1}(f(y))$. This structure is preserved by ∂^* and hence induces a right RG-module structure on $H^*(G, RG)$.

On the other hand, when X is an oriented $K(G, 1)$-complex the free left action of G on $\tilde{X}$ (oriented as in Sect. 8.1) induces a left action on $C_*(\tilde{X}; R)$

which can be converted as usual into a right action by the formula $c.\bar{g} = (\bar{g})^{-1}c$. When X has locally finite type, this action is preserved by δ. Taking $F_n = C_n(\tilde{X}; R)$, a straightforward check yields:

Proposition 13.2.16. *If G has type F_∞, then ψ_* is an isomorphism of right RG-cochain complexes. Hence the two right RG-module structures on $H^*(G, RG)$ (coming from the resolution and from the covering transformations) coincide.* $\square$

We end this section with two propositions which will be needed later. They show what information is retained at the "outer limit" of finiteness properties of groups.

Let X be a $K(G, 1)$-complex with finite n-skeleton. By 13.2.9, $H^k(G, RG) \cong H^k_f(\tilde{X}^n; R)$ for all $k \le n - 1$. For $k = n$ we have:

Proposition 13.2.17. *If R is a PID there is a free R-module F such that $H^n_f(\tilde{X}^n; R) \cong H^n(G, RG) \oplus F$. Thus $H^n(G, RG)$ is free over R iff $H^n_f(\tilde{X}^n; R)$ is free over R.*

Proof. We will find F such that $H^{n-1}(G, RG^e) \oplus F \cong H^n_f(\tilde{X}^n; R)$; by 13.2.13 this is enough. For any PID R, $H^{n-1}(G, RG^e) \cong H^{n-1}_e(\tilde{X}^n; R)$ by 13.2.9. From Sect. 12.2 we get an exact sequence

$$H^n(\tilde{X}^n; R) \xleftarrow{i^*} H^n_f(\tilde{X}^n; R) \xleftarrow{\delta^*} H^{n-1}_e(\tilde{X}^n; R) \xleftarrow{p^*} H^{n-1}(\tilde{X}^n; R).$$

We have $H^{n-1}(\tilde{X}^n; R) = 0$ and $\tilde{H}_{n-1}(\tilde{X}^n; R) = 0$ (exercise). Thus p^* is a monomorphism and, by 12.5.1, $H^n(\tilde{X}^n; R) \cong \operatorname{Hom}(Z_n(\tilde{X}^n; R), R)$. The latter is a countable product of copies of R since $Z_n(\tilde{X}^n; R)$ is a free R-module. Thus $F := \operatorname{image} i^*$ is a countably generated submodule of a product of copies of R, and is therefore free by Lemma 13.2.18 below. Thus $H^n_f(\tilde{X}^n; R)$ decomposes as claimed. $\square$

Lemma 13.2.18. *If R is a PID and M is a countably generated submodule of a product of copies of R, then M is a free R-module.*

Proof. We may assume $M \le P = \prod_1^\infty R$. We take R to be $\mathbb{Z}$: the case of a general PID is done similarly and is an exercise. Enumerate a countable set of generators of M: $m_1, m_2, \ldots$. Let m_1 have height α_1; in other words, $m_1 = \alpha_1 h_1$ where α_1 is a non-zero integer and $h_1(h_{11}, h_{12}, \cdots) \in P$ with the integers h_{1i} relatively prime. It follows that there are finitely many integers t_i and n_i such that $t_1 h_{1n_1} + \cdots + t_r h_{1n_r} = 1$. Define $p_1 : P \to \mathbb{Z}$ by $p_1(x_1, x_2, \cdots) = \sum_{i=1}^r t_i x_{n_i}$. Then $p_1(h_1) = 1$, so $P \cong \langle h_1 \rangle \oplus \ker p_1$ and $m_1 \in \langle h_1 \rangle$. If $M \subset \langle h_1 \rangle$ we are done. Otherwise, let m_k be the first member of M not in $\langle h_1 \rangle$: we assume

$m_k = m_2$ for simplicity of notation. Let $\beta_2 = p_1(m_2)$. Then $m_2 - \beta_2 h_1 \neq 0$ and it lies in ker p_1. Let α_2 be the height of $m_2 - \beta_2 h_1$. We have $m_2 = (m_2 - y_2 h_1) + y_2 h_1 = \alpha_2 h_2 + y_2 h_1$ where h_2 lies in ker p_1 and has height 1. So $m_2 \in \langle h_1, h_2 \rangle$, and since $h_2 \in$ ker p_1, h_1 and h_2 freely generated $\langle h_1, h_2 \rangle$. Thus $\langle h_1 \rangle \hookrightarrow \langle h_1, h_2 \rangle$ is a split monomorphism. We have $h_2 = (h_{21}, h_{22}, \cdots) \in$ ker p_1 where the integers h_{2i} are relatively prime. As before, we find $s_1 h_{2,m_1} +$

$$\cdots + s_q h_{2,m_q} = 1 \text{ and we define } p_2 : P \to \mathbb{Z} \text{ by } p_2(x_1, x_2, \cdots) = \sum_{i=1}^{q} s_i x_{m_i}.$$

Then $p_2(h_2) = 1$, so $p_2 \mid:$ ker $p_1 \to \mathbb{Z}$ is surjective and ker $p_1 \cong \langle h_2 \rangle \oplus$ ker p_2. Proceeding as before, either $M \subset \langle h_1, h_2 \rangle$ or we find h_3 with $\langle h_1, h_2 \rangle \hookrightarrow \langle h_1, h_2, h_3 \rangle$ a split monomorphism. In this way we find a copy of $\bigoplus_1^{\infty} \mathbb{Z}$ lying in P and containing M. Hence M is free. $\square$

Proposition 13.2.19. *Let X be a $K(G,1)$-complex with finite n-skeleton. As R-modules, $H_e^n(\tilde{X}^n; R)$ is torsion free iff $H^{n+1}(G, RG)$ is torsion free.*

Proof. By 13.2.13, $H^{n+1}(G, RG)$ is torsion free iff $H^n(G, RG^e)$ is torsion free. By 13.2.8 and 13.2.9, there is a short exact sequence of cochain complexes

$$0 \longrightarrow A \longrightarrow C_n^{\infty}(\tilde{X}; R)/C_n(\tilde{X}; R) \longrightarrow C_*^{\infty}(\tilde{X}^n; R)/C_*(\tilde{X}^n; R) \longrightarrow 0$$

where $A_k = 0$ when $k \leq n$ and $A_k = \mathrm{Hom}_G(C_k(\tilde{X}; R), RG^e)$ when $k > n$. The corresponding long exact sequence is in part:

$$0 \longrightarrow H^n(G, RG^e) \longrightarrow H_e^n(\tilde{X}^n; R) \longrightarrow C_{n+1}^{\infty}(\tilde{X}; R)/C_{n+1}(\tilde{X}; R)$$

from which the result follows. $\square$

Exercises

1. Prove that G is countable if and only if there is a $K(G,1)$ of locally finite type.
2. Prove 13.2.14 without the F_n hypothesis.
3. Starting with $C_*(\tilde{X}; \mathbb{Z})$, write down a functorial isomorphism between $H^1(G, \mathbb{Z})$ and the group of all homomorphisms $G \to \mathbb{Z}$.
4. Give a second solution to the previous exercise by using the Universal Coefficient Theorem 12.2.5 and Theorem 3.1.20.
5. Prove the following extension of Corollary 13.2.9: if X^n is finite then the module $H^n(G, R(H \backslash G))$ can be computed as kernel/image in

$$C_{n+1}^{\infty}(\bar{X}(H); R) \xleftarrow{\delta} C_n(\bar{X}(H); R) \xleftarrow{\delta} C_{n-1}(\bar{X}(H); R).$$

6. State and prove the homological analog of the previous exercise, i.e., an extension of Proposition 13.2.4.
7. Prove 13.2.18 when R is a PID (as distinct from the case $R = \mathbb{Z}$ covered in the proof).

8. Prove that if there is a finite n-dimensional $K(G, 1)$-complex then $H^n(G, \mathbb{Z}G)$ is finitely generated as a $\mathbb{Z}G$-module.
9. Let the finitely presented group act freely and cocompactly on the CW complex X. Prove that $H_1(X; R)$ is finitely generated as an RG-module. (*Hint*: prove that $\pi_1(X, x)$ is finitely generated as a G-group.)
10. The proofs given for 13.2.6 and 13.2.12 are topological. Derive these results from the Bockstein sequences.

13.3 Topological interpretation of $H^*(G, RG)$

Here we discuss the R-modules $H^n(G, RG)$, especially the cases $R = \mathbb{Z}$ and $R = \mathbb{Q}$. In the latter case, $H^n(G, \mathbb{Q}G)$ is a $\mathbb{Q}$-vector space, and thus the only invariant is its $\mathbb{Q}$-dimension. In the case $R = \mathbb{Z}$, $H^n(G, \mathbb{Z}G)$ is an abelian group, and when G has type F_n the structure of that abelian group is of topological interest. In particular, we want to relate this group to the homology of the end of the universal cover of a suitable $K(G, 1)$-complex. In Chaps. 16 and 17 we will give a parallel study of homotopy properties of the end, thus exhibiting important invariants of G which cannot be expressed in terms of homological algebra.

We saw in 13.2.11 that $H^0(G, RG)$ is trivial if G is infinite, and is isomorphic to R if G is finite. We dispose of the finite case:

Proposition 13.3.1. *If G is finite, $H^n(G, RG) = 0$ for all $n > 0$.*

Proof. By 7.2.5, there is a $K(G, 1)$-complex X of finite type. Since G is finite, 3.2.13 implies that $\tilde{X}$ also has finite type, so $H_f^n(\tilde{X}; R) = H^n(\tilde{X}; R) = 0$ for all $n > 0$. Apply 13.2.9. $\square$

Next, we give a theorem relating the structure of the R-module $H^n(G, RG)$, where G has type F_n, to homological properties of the end of the universal cover of a $K(G, 1)$-complex.[5]

Theorem 13.3.2. *Let $n \geq 0$ and let G be a group of type F_n; when $n = 0$ assume G is countable. Let X be an $(n-1)$-aspherical CW complex with finite n-skeleton whose fundamental group is isomorphic to G. Let $\{K_i\}$ be a finite filtration of $\tilde{X}^n$. Let R be a PID.*

(i) *For $k \leq n$, $H^k(G, RG)$ mod torsion is a countably generated free R-module iff $\{H_{k-1}(\tilde{X}^n \overset{c}{-} K_i; R)\}$ is semistable.*
(ii) *For $k \leq n + 1$, $H^k(G, RG)$ is a torsion free R-module iff the module $\{H_{k-2}(\tilde{X}^n \overset{c}{-} K_i; R)\}$ is pro-torsion free.*
(iii) *For $k \leq n$, $H^k(G, RG)$ is a torsion R-module iff $\{\tilde{H}_{k-1}(\tilde{X}^n \overset{c}{-} K_i; R)\}$ is pro-torsion.*

---

[5] Recall from Sect. 2.9 that $\tilde{H}_*$ denotes reduced homology.

(iv) *For $k \le n$, $H^k(G, RG)$ mod torsion is a free R-module with finite rank ρ iff $\{\tilde{H}_{k-1}(\tilde{X}^n \stackrel{c}{-} K_i; R)$ mod torsion$\}$ is stable with free (over R) inverse limit of finite rank ρ.*

Moreover, if G has type F_∞, if X is a $K(G, 1)$-complex of finite type, and if $\{K_i\}$ is a finite type filtration of $\tilde{X}$, the conclusions (i)–(iv) hold, with $\tilde{X}$ replacing $\tilde{X}^n$, for all k.

Proof. Combine 12.5.11 (applied to the appropriate skeleton), 13.2.9 and 13.2.13. For the case $k = n + 1$ in (ii), use 13.2.19. There are a few details to be worked out in (iii) and (iv) when $k = 1$ (details concerning the behavior of $\tilde{H}_0$); we leave these as an exercise.   □

The hypothesis that G be countable when $n = 0$ is merely to ensure that there is a finite filtration of $(\tilde{X})^0$. Parts (iii) and (iv) hold for finite groups because $H_{-1}(\emptyset; R) \cong R$.

For later use, we reorganize 13.3.2 in a different way; there is no new content:

Theorem 13.3.3. *Let n, G, X, $\{K_i\}$ and R be as in 13.3.2.*

(i) $H^k(G, RG) = 0$ *for all $k < n$ and $H^n(G, RG)$ is a countably generated free R-module iff $\{\tilde{H}_k(\tilde{X}^n \stackrel{c}{-} K_i; R)\}$ is pro-trivial for all $k \le n - 2$ and $H_{n-1}(\tilde{X}^n \stackrel{c}{-} K_i; R)$ is semistable.*

(ii) $H^k(G, RG) = 0$ *for all $k \le n$ and $H^{n+1}(G, RG)$ is a torsion free R-module iff $\{\tilde{H}_k(\tilde{X}^n \stackrel{c}{-} K_i; R)\}$ is pro-trivial for all $k \le n - 1$.*

(iii) $H^k(G, RG) = 0$ *for all $k \le n$ iff $\{\tilde{H}_k(\tilde{X}^n \stackrel{c}{-} K_i; R)\}$ is pro-trivial for all $k \le n - 2$ and $\{\tilde{H}_{n-1}(\tilde{X}^n \stackrel{c}{-} K_i; R)\}$ is pro-torsion.*

(iv) $H^k(G, RG) = 0$ *for all $k \le n - 1$ and $H^n(G, RG)$ is a free R-module of finite rank ρ iff $\{\tilde{H}_k(\tilde{X}^n \stackrel{c}{-} K_i; R)\}$ is pro-trivial for all $k \le n - 2$ and $\{\tilde{H}_{n-1}(\tilde{X}^n \stackrel{c}{-} K_i; R)$ mod torsion$\}$ is stable with free (over R) inverse limit of finite rank ρ.*

Moreover, if G has type F_∞, if X is a $K(G, 1)$-complex of finite type, and if $\{K_i\}$ is a finite type filtration of $\tilde{X}$, the conclusions (i)–(iv) hold, with $\tilde{X}$ replacing $\tilde{X}^n$, for all k.   □

Source Notes: 13.3.2 and 13.3.3 appear in [70] and [71].

Exercise

What does 13.3.3 say when $\tilde{X}^n = \tilde{X} = \mathbb{R}^n$?

13.4 Ends of spaces

Throughout this section Y denotes a path connected, strongly locally finite CW complex. A *proper ray* in Y is a proper map $\omega : [0, \infty) \to Y$. Two proper rays ω_1 and ω_2 in Y *define the same end* of Y if $\omega_1|\mathbb{N}$ and $\omega_2|\mathbb{N}$ are properly homotopic; here, $\mathbb{N} = \{0, 1, 2, \cdots\}$ considered as a discrete subspace of $[0, \infty)$. This is an equivalence relation; an *end* of Y is an equivalence class of proper rays. The set of ends of Y is denoted by $\mathcal{E}(Y)$.

One should think of a proper ray as a way of reaching out towards infinity in Y, and one should think of an end as a sort of component at infinity, so that "having one end" means "connected at infinity." However, as we shall see in Sect. 16.1, there are other (different but also useful) notions of "connected at infinity."

A proper map $f : Y \to Z$ between path connected, strongly locally finite CW complexes induces a function $f_\# : \mathcal{E}(Y) \to \mathcal{E}(Z)$. Properly homotopic proper maps induce the same function,[6] and one obtains a covariant functor from the category of path connected, strongly locally finite CW complexes and proper homotopy classes to the category Sets (of sets and functions). A proper 1-equivalence $Y \to Z$ induces a bijection $\mathcal{E}(Y) \to \mathcal{E}(Z)$.

From 10.1.14 we obtain:

Proposition 13.4.1. *The inclusion $Y^1 \hookrightarrow Y$ induces a bijection $\mathcal{E}(Y^1) \to \mathcal{E}(Y)$.* □

Combining this with 11.1.4 we get:

Corollary 13.4.2. *The set of ends of Y is non-empty iff Y is infinite. More precisely, if Y is infinite and $y \in Y$ there is a proper ray in Y whose initial point is y.* □

Corollary 13.4.2 would have to be modified if we were discussing ends of spaces Y which are not path connected. See Exercise 3.

The *number of ends* of Y is ∞ [resp. $m \geq 0$] if $\mathcal{E}(Y)$ is infinite [resp. has m members]. We often say that Y *has m ends* or Y *has infinitely many ends*.

Example 13.4.3. The quotient CW complex obtained by identifying the points "0" in m copies of $[0, \infty)$ has m ends. The space $\{(x, y) \in \mathbb{R}^2 \mid y = 0 \text{ or } x \in \mathbb{Z}\}$ with the obvious structure of a graph has ∞ ends. The space illustrated in Fig. 11.3 has one end. By 13.4.2 a path connected CW complex Y has 0 ends iff Y is finite.

From 1.5.5 and 11.4.4 we get:

---

[6] Indeed, weakly properly homotopic maps (see Sect. 17.7) induce the same function on the set of ends; in other words $\mathcal{E}(Y)$ is a shape invariant, not just a strong shape invariant.

Proposition 13.4.4. *If Y is infinite and K is a finite subcomplex then $Y \overset{c}{-} K$ is non-empty and has finitely many path components.* ☐

Following tradition we say that a path component of $Y \overset{c}{-} K$ is *bounded* if it is compact, and *unbounded* otherwise.

The next two propositions are exercises:

Proposition 13.4.5. *If Y has one end then given any finite subcomplex K of Y there is a finite subcomplex L such that any two points of $Y \overset{c}{-} L$ lie in the same path component of $Y \overset{c}{-} K$.* ☐

Proposition 13.4.6. *Two proper rays ω_1 and ω_2 in Y define the same end iff for every finite subcomplex K of Y there exists $k \in \mathbb{N}$ such that $\omega_1([k, \infty))$ and $\omega_2([k, \infty))$ lie in the same path component of $Y \overset{c}{-} K$.* ☐

Pick a finite filtration $\{L_i\}$ of Y. A proper ray ω in Y picks out an element (Z_i) of[7] $\varprojlim_i \{\pi_0(Y \overset{c}{-} L_i)\}$; in the notation of 13.4.6, Z_i is the path component of $Y \overset{c}{-} L_i$ containing $\omega(k(i))$ and $Z_i \supset Z_{i+1}$ for all i. From 13.4.6 we deduce:

Proposition 13.4.7. *This correspondence induces a bijection $\mathcal{E}(Y) \to \varprojlim_i \{\pi_0(Y \overset{c}{-} L_i)\}$.* ☐

By 13.4.4, $\{\pi_0(Y \overset{c}{-} L_i)\}$ is an inverse sequence of finite sets. If each is regarded as a discrete space and the inverse limit is interpreted in the category Spaces then, via the bijection of 13.4.7, $\mathcal{E}(Y)$ becomes a compact totally disconnected metrizable space called the *space of ends* of Y. A proper 1-equivalence $Y \to Z$ induces a homeomorphism $\mathcal{E}(Y) \to \mathcal{E}(Z)$ between spaces of ends. This is developed in the Appendix at the end of the section.

Proposition 13.4.8. *Let Y have m ends where $0 \le m \le \infty$. For all i, $Y \overset{c}{-} L_i$ has finitely many path components. When $m < \infty$, there exists i_0 such that whenever a finite subcomplex K contains L_{i_0}, $Y \overset{c}{-} K$ has exactly m unbounded path components. When $m = \infty$, the number of unbounded path components of $Y \overset{c}{-} L_i$ is a weakly monotonic unbounded function of i.*

Proof. The first sentence follows from 13.4.4 (or from the discussion preceding 11.1.4). The rest follows from 13.4.7; to see this, observe that a compact path component of $Y \overset{c}{-} L_i$ lies in L_j for some j. ☐

Addendum 13.4.9. *When m is finite and $i \ge i_0$, there is a path connected finite subcomplex K of Y containing L_i such that $Y \overset{c}{-} K$ has exactly m path components. When m is infinite K can still be chosen so that all the path components of $Y \overset{c}{-} K$ are unbounded.*

---

[7] Recall that the set of path components of a space W is denoted by $\pi_0(W)$. In the obvious way, π_0 is a functor from Spaces to Sets.

Proof. For m finite, the subcomplex $Y \stackrel{c}{-} L_i$ has m unbounded path components and finitely many bounded path components. Enlarge L_i (by the addition of a finite graph, say) to get a finite path connected subcomplex K'. Then $Y \stackrel{c}{-} K'$ has m unbounded path components, as well as, perhaps, bounded path components $C_1, \cdots, C_r$. The required K is $K' \cup \left[\bigcup_{j=1}^{r} N(C_j) \right]$. This proof also gives the second part. $\qquad\square$

The next proposition will be used in the proof of Theorem 13.5.9.

Proposition 13.4.10. *Let Y have m ends, where $m < \infty$. Let K and L be finite subcomplexes of Y such that each of $Y \stackrel{c}{-} K$ and $Y \stackrel{c}{-} L$ has m unbounded path components $C_1, \cdots, C_m$ and $D_1, \cdots, D_m$, respectively, indexed so that there is a proper ray in each $C_i \cap D_i$. Then each $C_i \stackrel{c}{-} (C_i \cap D_i)$ is compact.*

Proof. By 13.4.9, we may assume $Y \stackrel{c}{-} K$ and $Y \stackrel{c}{-} L$ have no bounded path components. If $m = 1$, $Y \stackrel{c}{-} (C_1 \cap D_1) = K \cup L$ which is compact. The subcomplex $C_1 \stackrel{c}{-} (C_1 \cap D_1)$ is therefore a closed subset of a compact set. Now let $m \geq 2$. First, consider the special case $K \subset L$. Suppose D_i fails to cover all but a compact subset of C_i. Then there is a point x in the set $C_i \stackrel{c}{-} [(C_i \cap D_i) \cup N(L)]$. This x must lie in D_j for some $j \neq i$. But $D_j \cap C_j \neq \emptyset$. So the path connected set D_j meets C_i and C_j and misses $N(L)$; hence it misses $N(K)$, a contradiction. For the general case, let J be finite with $K \cup L \cup J$ and $Y \stackrel{c}{-} J$ having no bounded path components. If E_i is the appropriate unbounded path component of $Y \stackrel{c}{-} J$, then $C_i \stackrel{c}{-} (C_i \cap E_i)$ and $D_i \stackrel{c}{-} (D_i \cap E_i)$ are compact, so $C_i \stackrel{c}{-} (C_i \cap D_i)$ is compact. $\qquad\square$

We saw in 2.7.2 that when Z is an oriented CW complex we may regard $H_0(Z; R)$ as the free R-module generated by the set $\pi_0(Z)$. We need more precision. Suppose the vertex v has orientation $\epsilon = \pm 1$. This v determines an element of $\pi_0(Z)$, namely, the path component $C(v)$ containing v; and it also determines an element of $H_0(Z; R)$, namely, the homology class represented by the 0-cycle ϵv. The function $h_Z : \pi_0(Z) \to H_0(Z; R)$, $C(v) \mapsto \epsilon\{v\}$, is well defined, injective, and satisfies the universal property for free R-modules generated by sets.[8] Moreover, h_Z is natural in the sense that if $f : Z_1 \to Z_2$ is a cellular map then the following diagram commutes:

$$
\begin{array}{ccc}
\pi_0(Z_1) & \xrightarrow{\ h_{Z_1}\ } & H_0(Z_1; R) \\
{\scriptstyle f_\#}\downarrow & & \downarrow{\scriptstyle f_*} \\
\pi_0(Z_2) & \xrightarrow{\ h_{Z_2}\ } & H_0(Z_2; R).
\end{array}
$$

These remarks are used in the proof of:

---

[8] When $R = \mathbb{Z}$, h_Z can be regarded as the 0-dimensional case of the Hurewicz homomorphism; compare Sect. 4.5.

Proposition 13.4.11. *Let R be a PID. Y has m ends iff $H_e^0(Y; R)$ is a free R-module of rank m. Y has infinitely many ends iff $H_e^0(Y; R)$ is free of countably infinite rank.*

Proof. We saw in 12.5.11 that $H_e^0(Y; R)$ is free. First, let m be finite. By 12.5.10 (iv), $H_e^0(Y; R)$ has rank m iff $\{H_0(Y \overset{c}{-} L_i; R)\}$ is stable with free inverse limit of rank m, iff $\{\pi_0(Y \overset{c}{-} L_i)\}$ is stable[9] with inverse limit having m elements. Apply 13.4.7. The case $m = \infty$ is now immediate.   □

The above commutative diagram and 13.4.7 suggest natural functions

$$\mathcal{E}(Y) \rightarrowtail \varprojlim\{\pi_0(Y \overset{c}{-} L_i)\} \overset{h_{\mathcal{E}(Y)}}{\rightarrowtail} \varprojlim\{H_0(Y \overset{c}{-} L_i; R)\}.$$

We will take these up in Sect. 16.1.

Appendix: Topology of the space of ends

A space is *totally disconnected* if each of its components consists of one point. A space is 0-*dimensional* if there is a basis for the neighborhoods of each point consisting of sets which are both closed and open (equivalently: whose frontiers are empty). We set as an exercise:[10]

Proposition 13.4.12. *A compact metrizable space is 0-dimensional iff it is totally disconnected.*   □

In using 13.4.12, note that, by first countability, bases for the neighborhoods of points can always be taken to be countable.

Every compact metrizable totally disconnected space has cardinality $\leq$ cardinality of $\mathbb{R}$. Here are some examples: a finite discrete space, a "convergent sequence with limit point" $\{0\} \cup \{\frac{1}{n} \mid n \geq 1\}$, the "middle-third" Cantor set.[11]

Proposition 13.4.13. *If Z is a compact totally disconnected metrizable space, there is an inverse sequence $\{Z_n\}$ of finite discrete spaces whose inverse limit is homeomorphic to Z.*

Proof. Pick a metric for Z. By 13.4.12, for each $n \geq 1$ there is a finite cover $\mathcal{Z}_n$ of Z by pairwise disjoint closed-and-open sets of diameter $\leq \frac{1}{n}$, and these $\mathcal{Z}_n$'s can be chosen inductively so that each member of $\mathcal{Z}_{n+1}$ lies in a member of $\mathcal{Z}_n$. Choose any such set of inclusions to define the bond $\mathcal{Z}_{n+1} \to \mathcal{Z}_n$. Giving each $\mathcal{Z}_n$ the discrete topology, we get the desired inverse sequence.   □

We leave as an exercise:

---

[9] In the category Sets: see Sect. 11.2.

[10] Or see [92, pp. 20–22].

[11] i.e., the set of numbers in I having triadic expansion each of whose numerators is 0 or 2.

Proposition 13.4.14. (a) *The inverse limit space of an inverse sequence of finite discrete spaces is compact totally disconnected and metrizable.*

(b) *If two such sequences have homeomorphic inverse limit spaces then they are pro-isomorphic.*[12]   □

The compact totally disconnected metrizable space Z is *perfect* if no one-point subset is open[13] in Z. The "middle-third" Cantor set is an example. In fact, as a further exercise the reader should prove:

Proposition 13.4.15. (a) *Such a space Z is perfect iff Z is homeomorphic to the inverse limit of an inverse sequence $\{Z_n\}$ of finite discrete spaces such that for every n and every $z \in Z_n$ there exists k such that the pre-image of z in Z_{n+k} contains more than one point.*

(b) *Any two inverse sequences $\{Z_n\}$ and $\{Z'_n\}$ as in (a) are pro-isomorphic.*

An immediate corollary of 13.4.15 is the following classical theorem:

Theorem 13.4.16. *Every perfect compact totally disconnected metrizable space is homeomorphic to the "middle third" Cantor Set.*   □

In view of 13.4.16, we will call any such space a *Cantor Set*.

The space $\mathcal{E}(Y)$ of ends of Y is a compact totally disconnected metrizable space. One asks: which spaces can occur as $\mathcal{E}(Y)$? We leave the following answer as an exercise:

Proposition 13.4.17. *For any compact totally disconnected metrizable space Z there is a tree T such that $\mathcal{E}(T)$ is homeomorphic to Z.*   □

Exercises

1. Prove that Y has one end iff for some ring R $\{H_0(Y \overset{c}{-} L_i; R)\}$ is stably R.
2. Prove that if Y_1 and Y_2 are infinite and path connected then $Y_1 \times Y_2$ has one end.
3. When Y is not path connected what becomes of 13.4.2? Distinguish the cases of finitely many and infinitely many path components.
4. Write out the proof that $h : \pi_0(Y) \to H_0(Y; R)$ satisfies the universal defining property of a free R-module.
5. Prove 13.4.12. Extend it to the locally compact case.
6. Prove 13.4.14–13.4.16.
7. Prove 13.4.17. *Hint*: Compare 17.5.6.
8. The bijection in 13.4.7 is natural (or functorial). What does this mean? Prove it.

---

[12] In pro-Sets or, equivalently, in pro-Spaces.

[13] i.e., Every point is a "limit point."

13.5 Ends of groups and the structure of $H^1(G, RG)$

Let G be a finitely generated group. Let X be a path connected CW complex with fundamental group isomorphic to G and having finite 1-skeleton. The *number of ends* of G is the number of ends of the (strongly) locally finite graph $\tilde{X}^1$. That this is well defined follows from 13.5.5, below.

By the remark preceding 3.2.5, we have:

Proposition 13.5.1. *If G is generated by $\{g_1, \cdots, g_k\}$, the number of ends of G is the number of ends of the corresponding Cayley graph of G.*   □

Proposition 13.5.2. *If H has finite index in G, then H and G have the same number of ends.*

Proof. X and its finite covering space $\bar{X}(H)$ have the same universal cover.
□

Let R be a PID. Next we show in 13.5.5 that the number of ends of G determines and is determined by the R-module $H^1(G, RG)$, and therefore does not depend on the choice of X or R.

Proposition 13.5.3. *The R-module $H^1(G, RG)$ is countably generated and free.*

Proof. Apply the first two parts of 13.3.2 and the following lemma.   □

Lemma 13.5.4. *Let Y be a path connected countable CW complex of locally finite type, and let $\{L_i\}$ be a finite type filtration of Y. Then $\{H_0(Y \stackrel{c}{-} L_i; R)\}$ is semistable.*

Proof. By 12.5.9, each $H_0(Y \stackrel{c}{-} L_i; R)$ is finitely generated and free. The image of each bond $H_0(Y \stackrel{c}{-} L_j; R) \to H_0(Y \stackrel{c}{-} L_i; R)$ is a direct summand generated by the finite number of path components of $Y \stackrel{c}{-} L_i$ which meet $Y \stackrel{c}{-} L_j$. Fixing i, this finite number decreases with j until it eventually stabilizes.   □

Theorem 13.5.5. *The number of ends of a finite group is 0. The number of ends of an infinite finitely generated group G is well defined and equals $1 + \mathrm{rank}_R(H^1(G, RG))$.*

Here, "rank_R" means rank as a free R-module (see 13.5.3); the value ∞ is permitted. This is well defined.[14]

Proof. For a finite group, the result follows from 3.2.13 and 13.4.2. Assume G is infinite. Let ρ be a non-negative integer. We use both the result and the notation of 13.3.2: $H^1(G, RG)$ has rank ρ iff $\{\tilde{H}_0(\tilde{X}^1 \stackrel{c}{-} K_i; R)\}$ is stable with free inverse limit of rank ρ, iff $\{H_0(\tilde{X}^1 \stackrel{c}{-} K_i; R)\}$ is stable with free

---

[14] Rank is certainly well defined when R is a PID, as is the case here. More general conditions, which include the case $R = \mathbb{Z}G$, are given in [42, p. 36].

inverse limit of rank $1 + \rho$ (by 2.9.1), iff $\varprojlim_i \{\pi_0(\tilde{X}^1 \stackrel{c}{-} K_i)\}$ has $1 + \rho$ elements.

Apply 13.4.7. Similarly $H^1(G, RG)$ has infinite rank iff $\varprojlim_i \{\pi_0(\tilde{X}^1 \stackrel{c}{-} K_i)\}$ is infinite,[15] and again 13.4.7 gives the result. $\qquad\square$

Example 13.5.6. The groups $\mathbb{Z}^n$ $(n > 1)$ have one end. The group $\mathbb{Z}$ has two ends. A free group of rank $n > 1$ has infinitely many ends. To see this, consider the universal covers of T^n, S^1 and the n-fold wedge of circles; the last is an infinite tree every vertex of which has valence $2n$. These are examples of every case which can arise; more precisely, we have:

Theorem 13.5.7. *The number of ends of a finitely generated group is 0, 1, 2 or ∞. Hence for G infinite and finitely generated, the R-module $H^1(G, RG)$ is isomorphic to 0 or R or $\bigoplus_1^\infty R$.*

Proof. Suppose G has m ends where $3 \le m < \infty$. Let Γ be the Cayley graph of G with respect to a finite set of generators. By 13.5.5, G is infinite, hence Γ is infinite. By 13.4.9, there is a finite path connected subgraph K of Γ such that[16] $\Gamma \stackrel{c}{-} K$ has exactly m path components $Z_1, \ldots, Z_m$, all unbounded, and for any finite subgraph L of Γ containing K the graph $\Gamma \stackrel{c}{-} L$ has exactly m unbounded path components, one lying in each Z_i. Since G is infinite and acts freely, and $N(K)$ is finite (by 11.4.4), there exists $g \in G$ such that $N(g(K)) = g(N(K)) \subset Z_1$. Let $Z'_1, \ldots, Z'_m$ be the unbounded path components of $\Gamma \stackrel{c}{-} (K \cup g(K))$, where $Z'_i \subset Z_i$. Then $Z'_i = Z_i$ when $i > 1$. Now, $Z_2 \cup Z_3 \cup N(K)$ is path connected. So $Z'_2 \cup Z'_3 \cup N(K)$ is a path connected subset of $\Gamma \stackrel{c}{-} g(K)$, implying that Z'_2 and Z'_3 lie in the same unbounded path component of $\Gamma \stackrel{c}{-} g(K)$. But each unbounded component of $\Gamma \stackrel{c}{-} g(K)$ contains an unbounded component of $\Gamma \stackrel{c}{-} (K \cup g(K))$. This is a contradiction, since $m < \infty$. $\qquad\square$

Just as we define the "number of ends" of G rather than the "set of ends" of G, so we define the *homeomorphism type of the ends* of G to be that of the space Z if $\mathcal{E}(Z)$ is homeomorphic to $\mathcal{E}(\Gamma)$ with Γ as above. The proof of 13.5.7 clearly gives:

Addendum 13.5.8. *The homeomorphism type of the ends of G is either that of a discrete space having at most two points or that of a Cantor set.* $\qquad\square$

Perhaps it is because these spaces have different cardinal numbers that it is usual to speak only of the "number of ends" of G, and we will do so when the homeomorphism type is irrelevant.

---

[15] By 12.2.1 and 13.2.9, $H^1(G, RG)$ is countable when G is finitely generated; however, $\varprojlim_i \{\pi_0(\tilde{X}^1 \stackrel{c}{-} K_i)\}$ is either finite or uncountable.

[16] Here and elsewhere in this chapter we use CW complements in order to stay in the world of CW complexes. Ordinary complements would do just as well.

A finitely generated group has 0 ends iff it is finite. One asks: which finitely generated groups have 1 end, 2 ends, ∞ ends? Equivalently, when is $H^1(G, RG)$ 0, R or $\bigoplus_1^\infty R$?

Theorem 13.5.9. *A finitely generated group G has two ends iff G has an infinite cyclic subgroup of finite index.*

Proof. Since the space $\mathbb{R}$ has two ends, the group $\mathbb{Z}$ has two ends. By 13.5.2, the same holds for any group G in which $\mathbb{Z}$ has finite index.

Conversely, let G have two ends. Let Γ be the Cayley graph with respect to a finite set of generators. By 13.4.9, there is a path connected finite subgraph K such that $\Gamma \stackrel{c}{-} K$ has exactly two path components, Z^+ and Z^-, both of which are unbounded. The group G acts on the (two-element) set of ends of Γ, so some subgroup of index ≤ 2 fixes the two ends. Thus, we may assume that G fixes the ends.

We first show that G has an element of infinite order. Since G is infinite, there is some $g \in G$ such that $g(N(K)) \cap N(K) = \emptyset$. So $g(N(K)) \subset Z^+ \cup Z^-$. Path connectedness implies that $g(N(K))$ lies in Z^+ or Z^-; say $g(N(K)) \subset Z^+$. We claim $g(N(Z^+)) \subset Z^+$. Suppose not. Then for some $x \in N(Z^+)$, $g(x) \in Z^- \cup N(K)$; this x is in Z^+ because $g(N(K)) \subset Z^+$. There is a proper ray ω in Z^+ with $\omega(0) = x$. Since the image of ω misses K, the image of the proper ray $g \circ \omega$ misses $g(K)$; and $g \circ \omega(0) = g(x)$. There is a proper ray τ in $Z^- \cup N(K)$ with $\tau(0) = g(x)$. The proper rays $g \circ \omega$ and τ both miss $g(K)$, and they define opposite ends (since g fixes ends). Hence, $g(K)$ does not separate Γ. This is a contradiction. Thus $g(N(Z^+)) \subset Z^+$. Indeed, $g(Z^+)$ is a proper subcomplex of Z^+ since $g(K) \subset Z^+$. It follows that $g^n(Z^+) \neq Z^+$ for all $n \geq 1$, and hence that g has infinite order.

Let G_0 be the infinite cyclic subgroup of G generated by g. Claim: There is a finite subgraph L of Γ such that $\bigcup_{n=-\infty}^{\infty} g^n(L) = \Gamma$. The Theorem follows from this, for if $g_1, \cdots, g_k$ are the elements of G whose corresponding vertices in Γ lie in L, then every vertex of Γ corresponds to $g^n g_i$ for some n and i, implying that G_0 has index $\leq k$ in G.

To prove the Claim let $L = \Gamma \stackrel{c}{-} (Z^- \cup g(Z^+))$. By 13.4.10, $g(Z^+)$ covers all but a compact piece of Z^+. Hence L is finite. One sees that $x \in L$ iff $x \in K \cup N(Z^+)$ and $x \notin N(gZ^+)$. For any $y \in \Gamma$ there is a greatest integer n such that $y \in g^n(K \cup N(Z^+))$. Then $y \in g^n(L)$. So $\Gamma = \bigcup_n g^n(L)$ as claimed. $\square$

Another characterization of finitely generated groups with two ends is given in 13.5.16 below.

The classification of groups with infinitely many ends is more difficult:

Theorem 13.5.10. (Stallings' Theorem) *A finitely generated group G has ∞ ends iff either*[17] *(i) $G = A \underset{C}{*} B$ where C is finite having index ≥ 2 in A and in B, with one of these indices being ≥ 3; or (ii) $G = A \underset{\phi}{*}$ where ϕ is an isomorphism between finite subgroups having index ≥ 2 in A.*

The proof is given in the next section.

Next, we consider pairs of groups. Apart from its own interest this extension of the theory will give us some more information[18] about ends of groups (in 13.5.12, 13.5.15 and 13.5.16).

Let H be a subgroup of G. The *number of ends* of the pair (G, H) is the number of ends of the locally finite graph $\bar{X}(H)^1$. By 13.4.11, this number ($\leq \infty$) is the rank of the free R-module $H^0_e(\bar{X}(H); R)$ which, by 13.2.9, is isomorphic to $H^0(G, R(H\backslash G)^e)$. Thus the number of ends of (G, H) does not depend on the choice of X or R. By 3.2.13, the pair (G, H) has 0 ends iff $[G : H] < \infty$.

Proposition 13.5.11. *If N is a normal subgroup of G then the number of ends of the pair (G, N) is equal to the number of ends of the group G/N.*

Proof. Let $f : (X, x) \to (Y, y)$ be a cellular map between 2-dimensional complexes having the following properties: each of X and Y has just one vertex, the base point; $f| : X^1 \to Y^1$ is an isomorphism of finite graphs (i.e., a homeomorphism taking each 1-cell of X onto a 1-cell of Y); there are isomorphisms $\alpha : G \to \pi_1(X, x)$ and $\beta : G/N \to \pi_1(Y, y)$ so that $\beta^{-1} \circ f_\# \circ \alpha$ is the quotient homomorphism[19] $G \to G/N$. We described in Sect. 3.2 how to build $\tilde{Y}^1$, $\tilde{X}^1$ and, hence, $\bar{X}(N)^1$. Inspection of that description shows that $\bar{X}(N)^1$ and $\tilde{Y}^1$ are isomorphic graphs; indeed the pointed map $\bar{X}(N) \to \hat{Y}$ covering f restricts to an isomorphism of 1-skeleta. Hence $\bar{X}(N)$ and $\tilde{Y}$ have the same number of ends. $\square$

Corollary 13.5.12. *The number of ends of a finitely generated group G is the number of ends of any path connected free[20] G-CW complex whose quotient by the action of G is a finite CW complex.*

Proof. Let $\bar{Z}$ be the free G-CW complex and Z the finite quotient. Write $H = \pi_1(Z, z)$. By covering space theory (see Sect. 3.4), there is a short exact sequence of groups $N \rightarrowtail H \twoheadrightarrow G$ such that $\bar{Z} = \bar{Z}(N)$. By 13.5.11, the number of ends of G is the number of ends of $\bar{Z}$. $\square$

It follows from 13.5.11 and 13.5.7 that if N is normal in G then the number of ends of the pair (G, N) is 0, 1, 2 or ∞. For arbitrary subgroups H this is not true:

---

[17] This notation is explained in the footnotes to Sect. 6.2.
[18] For another useful fact about ends of groups see Exercise 3 of Sect. 16.8.
[19] Using the methods of Sect. 3.1, it is not hard to produce such.
[20] One may even allow finite cell stabilizers here. See Exercise 2 in Sect. 17.2.

Example 13.5.13. For each $n \geq 3$ we describe a pair of groups (G, H) having n ends. Let X be a closed path connected orientable surface of genus g and let Y be a compact subsurface (i.e., $\mathrm{cl}_X(X - Y)$ is also a surface) whose boundary consists of n circles; Y is to be placed in X so that no boundary circle of Y is homotopically trivial in $\mathrm{cl}_X(X - Y)$, something which can be achieved for given n by choosing g large enough. (Clearly, there is a triangulation of X in which Y and $\mathrm{cl}_X(X - Y)$ are subcomplexes, but we will leave such matters implicit here.) These conditions ensure that the inclusion of any of the n boundary circles into Y or into the appropriate path component of $\mathrm{cl}_X(X-Y)$ induces a monomorphism on fundamental group. Let $G = \pi_1(X, x)$ and let $H = \pi_1(Y, x)$. This decomposition of X into path connected subsurfaces expresses G as the fundamental group of a graph of groups in which H is a vertex group. By 6.2.1 the obvious homomorphism $H \to G$ is a monomorphism. We identify H with its image, regarding (G, H) as a pair of groups.

Consider the covering projection $q_H : \bar{X}(H) \to X$, with base point $\bar{x}$ over x. By 3.4.10 the path component $Y_{\bar{x}}$ of $q_H^{-1}(Y)$ is a copy of Y which "carries" the entire fundamental group $H = \pi_1(\bar{X}(H), \bar{x})$. Let $C_1, \ldots, C_n$ be the boundary circles in $Y_{\bar{x}}$, and let U_i be the path component of $\mathrm{cl}_{\bar{X}(H)}(\bar{X}(H) - Y_{\bar{x}})$ containing C_i. The inclusion $C_i \hookrightarrow U_i$ induces a monomorphism on fundamental group by construction, and the same is true of $C_i \hookrightarrow Y_{\bar{x}}$. Thus the decomposition of $\bar{X}(H)$ into path connected surfaces expresses H as the fundamental group of a graph of groups in which H itself is a vertex group. It follows that $C_i \hookrightarrow U_i$ induces an isomorphism on fundamental group, and $U_i \neq U_j$ when $i \neq j$.

If we attach a disk to U_i along C_i by a homeomorphic attaching map, we get a simply connected surface with empty boundary. If U_i were compact this would be a 2-sphere, implying U_i is a disk, which is impossible since C_i is not homotopically trivial in U_i. The only remaining possibility is that U_i is a simply connected open surface. By 12.3.8, 4.5.2 and 4.1.4 it follows that U_i is contractible. It is a classical theorem that every contractible open surface is homeomorphic to $\mathbb{R}^2$. Therefore, the pairs (U_i, C_i) and $(S^1 \times [0, \infty), S^1 \times \{0\})$ are homeomorphic. Hence, $\bar{X}(H)$ has n ends; that is, (G, H) has n ends.

We remarked that the number of ends of (G, H) is 0 iff H has finite index in G. At the other extreme we may ask what happens when H is finite:

Example 13.5.14. Consider the infinite dihedral group $G = \langle a, b \mid a^2, b^2 \rangle$, and let H be the subgroup of order 2 generated by a. The graphs $\tilde{X}^1$ and $\bar{X}(H)^1$ are shown in Fig. 13.1: clearly $(G, \{1\})$ has two ends while (G, H) has one end even though $\{1\}$ has finite index in H.

However, finite normal subgroups are better behaved:

Proposition 13.5.15. *If N is a finite normal subgroup of G, then the number of ends of (G, N) is equal to the number of ends of G. Hence G and G/N have the same number of ends.*

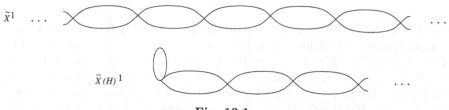

$\tilde{X}^1 \quad \cdots$

$\bar{X}_{(H)}{}^1$

Fig. 13.1.

Proof. If G is finite, both numbers are 0. Assume G is infinite. Then p_N : $\tilde{X}^1 \to \bar{X}(N)^1$ is a finite-to-one covering projection, hence a proper map. The induced function[21] $p_{N\#} : \mathcal{E}(\tilde{X}^1) \to \mathcal{E}(\bar{X}(N)^1)$ is clearly surjective. We must show it is injective. Let ω and τ be proper rays in $\tilde{X}^1$ with initial point such that the proper rays $p_N \circ \omega$ and $p_N \circ \tau$ define the same end of $\bar{X}(N)^1$. It is easy to see that $p_N \circ \tau$ is properly homotopic (in $\bar{X}(N)^1$) to a proper ray σ of the form $\nu_0.\beta_1.\nu_1.\beta_2.\cdots$; here, "." means path product in the sense of Sect. 3.4, ν_k is a loop at $p_N \circ \omega(k)$, and β_k is the path $p_N \circ \omega \mid [k, k+1]$. Some details of definition should be filled in here, but the idea is adequately captured by the pictures in Sect. 16.1. By 2.4.6 and 2.4.7, we may assume, by properly homotoping τ if necessary, that $p_N \circ \tau = \sigma$. Then the proper rays ω and τ are such that for all $k \in \mathbb{N}$, $\tau(k)$ is a translate of $\omega(k)$ by an element of N. It is convenient to assume, as we may, that X has only one vertex. Then using the isomorphism χ of Sect. 3.2, we may identify G with the 0-skeleton of $\tilde{X}^0$ (having chosen a base point). Since N is finite, there is some $m \in \mathbb{N}$ such that every element of N can be joined to the vertex 1 by a path of length $\leq m$ (i.e., involving $\leq m$ edges). Hence the vertices g and gh (where $g \in G$ and $h \in N$) can be joined by a path of length $\leq m$. Since N is normal, the vertices g and hg can be joined by a path of length $\leq m$. In particular this is true of $\omega(k)$ and $\tau(k)$ for each k, so ω and τ define the same end of $\tilde{X}$. $\square$

With 13.5.9 this gives:

Corollary 13.5.16. *If there is an exact sequence of groups $N \rightarrowtail G \twoheadrightarrow Q$ with N finite and Q isomorphic to $\mathbb{Z}$ or $\mathbb{Z}_2 * \mathbb{Z}_2$ then G has two ends.* $\square$

Again, let H be a subgroup of the (finitely generated) group G. By 13.2.11, $H^0(G, R(H\backslash G))$ is trivial if H has infinite index in G, and is isomorphic to R if H has finite index. Analogous to 13.5.3 is:

Proposition 13.5.17. *The R-module $H^1(G, R(H\backslash G))$ is free.*

Proof. By 13.2.12 we have an exact sequence

$$H^1(G, R(H\backslash G)\check{}) \xleftarrow{i} H^1(G, R(H\backslash G)) \xleftarrow{j} H^0(G, R(H\backslash G)^e) \xleftarrow{k} H^0(G, R(H\backslash G)\check{}) \xleftarrow{r} H^0(G, R(H\backslash G))$$

---

[21] Recall that $\mathcal{E}(Y)$ denotes the set of ends of Y.

If $[G : H] = \infty$, $H^0(G, R(H\backslash G)) = 0$ and $k : R \to H^0(G, R(H\backslash G)^e))$ is a split monomorphism: this follows from 13.2.9 and 12.2.3. By 13.2.9 and 13.4.11, $H^0(G, R(H\backslash G)^e)$ is free. Hence its direct summand image $j = \ker i$ is free. If $[G : H] < \infty$, r is an isomorphism $R \to R$, so j is a monomorphism, hence in this case too image $j = \ker i$ is free. Using 13.2.9, 12.4.3 and 12.5.1, one sees that $H^1(G, R(H\backslash G)\hat{\ }) \cong H^1(\bar{X}(H); R) \cong \mathrm{Hom}_R(H_1(\bar{X}(H); R), R) =: M$. Since $H_1(\bar{X}(H); R)$ is a quotient of some free R-module F_0, M embeds in $\mathrm{Hom}_R(F_0, R)$; see Sect. 12.4. The latter is a product of copies of R, and image i is a countably generated submodule. So, by 13.2.18, image i is free. Thus $H^1(G, R(H\backslash G)) \cong \ker i \oplus \mathrm{image}\ i$ is free. $\qquad\square$

Addendum 13.5.18. *Let H have infinite index in G. The number of ends of (G, H) is $\leq 1 + \mathrm{rank}_R(H^1(G, R(H\backslash G)))$ with equality iff the homomorphism i is trivial.* $\qquad\square$

This should be compared with 13.5.3; by 13.2.10, when H is trivial so is i. We now consider other cases where i is trivial.

Proposition 13.5.19. *Let H, generated by h, be an infinite cyclic subgroup of the one-ended finitely generated group G. Then the number of ends of (G, H) equals $1 + \mathrm{rank}_{\mathbb{Z}}(H^1(G, \mathbb{Z}(H\backslash G)))$.*

Proof. First, assume G is finitely presented. Then X can be taken to have finite 2-skeleton and, by 13.2.9 and 13.5.18, we are to show that

$$i : H_f^1(\bar{X}(H); \mathbb{Z}) \to H^1(\bar{X}(H); \mathbb{Z})$$

is zero. By 13.2.10 and Exercise 3 in Sect. 13.2, there is an obvious identification of $H^1(\bar{X}(H); \mathbb{Z})$ with $\mathrm{hom}(H, \mathbb{Z})$. Suppose $i \neq 0$. Then there is $\theta \in Z_f^1(\bar{X}(H); \mathbb{Z})$ such that $i(\{\theta\}) = \phi : H \to \mathbb{Z}$ with ϕ non-trivial. Then ϕ is a monomorphism. The subgroup H has infinite index in G by 13.5.9, so $\bar{X}(H)$ is non-compact. Let L be a finite subcomplex of $\bar{X}(H)^1$ supporting θ, let C be an unbounded path component of $\bar{X}(H)^1 \overset{c}{-} L$, and let v be a vertex of C. Write $j_C : C \hookrightarrow \bar{X}(H)$. We have a commutative diagram

$$H_f^1(C; \mathbb{Z}) \xleftarrow{\quad j_C^* \quad} H_f^1(\bar{X}(H); \mathbb{Z})$$

$$\downarrow \qquad\qquad\qquad\qquad \downarrow{\scriptstyle i}$$

$$\mathrm{hom}(\pi_1(C, v), \mathbb{Z}) \cong H^1(C; \mathbb{Z}) \xleftarrow{\quad} H^1(\bar{X}(H); \mathbb{Z}) \cong \mathrm{hom}(H, \mathbb{Z}).$$

Since $j_C^*(\{\theta\}) = 0$, ϕ is mapped to 0 so $\phi \circ j_{C\#} : \pi_1(C) \to \mathbb{Z}$ is trivial and therefore $j_{C\#}$ is trivial. Hence, writing $p_H : \tilde{X} \to \bar{X}(H)$ as usual for the covering projection, we see that $p_H^{-1}(C)$ consists of a pairwise disjoint collection of copies of C indexed by H; this follows from 3.4.9. Let B be the frontier of C in $\bar{X}(H)^1$. Then B is a subcomplex of $N(L)$ and is therefore finite. Thus there is a copy of B in $\tilde{X}^1$ separating one of the copies of C in $\tilde{X}^1$ from another. But $\tilde{X}^1$ has one end, so it cannot be separated by a finite subcomplex in such a way that more than one complementary path component is infinite – a contradiction.

When G is merely finitely generated, the argument is essentially the same, but $H^1_f(\bar{X}(H); \mathbb{Z})$ must be replaced by the cohomology group described in Exercise 5 of Sect. 13.2, since now X^1 is finite but X^2 is not. $\qquad \square$

More generally, we have:

Proposition 13.5.20. *Let H be a subgroup of infinite index in the finitely generated group G. Assume that whenever there is a short exact sequence $N \rightarrowtail H \twoheadrightarrow \mathbb{Z}$ then the pair (G, N) has one end. Then the number of ends of (G, H) equals $1 + \mathrm{rank}_{\mathbb{Z}}(H^1(G, \mathbb{Z}(H \backslash G)))$.*

Proof. The proof is the same as that of 13.5.19. In that proof, we used only the facts that H had infinite index in G and that the covering space $\bar{X}(\ker \phi)$ had one end. The details are an exercise. $\qquad \square$

Finally, we state a generalization of 13.5.7 whose (similar) proof is left as an exercise:

Theorem 13.5.21. *If H has infinite index in its normalizer in G, then the number of ends of (G, H) is 1, 2 or ∞.* $\qquad \square$

Analogs of Stallings' Theorem 13.5.10 for pairs of groups can be found in [139].

In Sect. 14.5, we will return to ends of pairs of groups, introducing another approach.

Source Notes. The first systematic study of ends of pairs of groups seems to be [86]. The material on ends of pairs of groups in this section is mostly in [139] and [140], except for 13.5.18–13.5.20, which is in [152]. Example 13.5.13 is in [140]. Martin Roller showed me Example 13.5.14. The theory of ends of groups is extended to (discrete) groups which are not finitely generated in [49], and to compactly generated locally compact topological groups in [130] and, earlier, in [86].

Exercises

1. Prove the "if" part of 13.5.10.
2. Prove that a finitely generated torsion free group is accessible.
3. How many ends does $G_1 \times G_2$ have?
4. Prove that if the finitely generated group G has a finitely generated normal subgroup of infinite index, then G has one end.
5. Prove 13.5.20.

6. Prove: If an infinite group G acts freely on a path connected strongly locally finite CW complex Y, then Y has 1, 2 or ∞ ends.

7. Prove that if an infinite group G acts freely on a path connected strongly locally finite CW complex Y which has two ends, then Y is finite mod G, and G has two ends.

8. Let G be the Klein Bottle group $\langle a, b \mid bab^{-1} = a^{-1} \rangle$ and let H be the infinite cyclic subgroup generated by b. Show that 13.5.19 is false for this case when the ring R is $\mathbb{Z}_2$ rather than $\mathbb{Z}$. (This is easier if a simple special case of Poincaré Duality given in 15.1.9 is used.)

9. Give another proof of the $\mathbb{Z}$-part of Corollary 13.5.16 by expressing G as an HNN extension.

10. Give an example to show that $H^1(G, \mathbb{Z}G)$ is not necessarily free when G is not finitely generated. *Hint*: consider a free resolution when G is free of countably infinite rank.

11. Show that Thompson's group F has one end.

12. Prove 13.5.21.

13.6 Proof of Stallings' Theorem

The easy part of Stallings' Theorem 13.5.10 is the "if" part: that the indicated decompositions imply infinitely many ends; that part is an exercise. In this section we deal with the "only if" part. We are given G with infinitely many ends. We will show that G acts rigidly on a tree with finite edge stabilizers as expected from the statement of 13.5.10. Then 6.2.7 will give the required decomposition of G.

A. Trees and Posets

There is a useful correspondence between trees and partially ordered sets of a particular kind. A *tree poset* is a poset $(E, \leq)$ equipped with an involution $\tau \mapsto \bar{\tau}$ (i.e., $\bar{\bar{\tau}} = \tau$) satisfying the following axioms:

(i) $\forall \tau, \tau \neq \bar{\tau}$;
(ii) $\forall \sigma \leq \tau, \bar{\tau} \leq \bar{\sigma}$;
(iii) $\forall \sigma, \tau, \sigma \leq \mu \leq \tau$ for only finitely many μ;
(iv) $\forall \sigma, \tau$ at least one of the following holds: $\sigma \leq \tau, \sigma \leq \bar{\tau}, \bar{\sigma} \leq \tau, \bar{\sigma} \leq \bar{\tau}$;
(v) $\forall \sigma, \tau$, it is not the case that $\sigma \leq \tau$ and $\sigma \leq \bar{\tau}$.

The example to have in mind involves a tree T. Let E be the set of all non-degenerate edges of T. We write $\tau_1 \leq \tau_k$ if there is a reduced edge[22] path $(\tau_1, \ldots, \tau_k)$. The set $(E, \leq)$ with involution $\tau \mapsto \tau^{-1}$ is a tree poset.

Indeed, this relation $\tau_1 \leq \tau_k$ makes sense on any graph Γ. Then (i) and (ii) hold and one has:

(a) Γ is path connected iff for two edges σ and τ at least one of the following is true: $\sigma \leq \tau, \sigma \leq \bar{\tau}, \bar{\sigma} \leq \tau, \bar{\sigma} \leq \bar{\tau}$.

---

[22] Recall that an edge is an oriented 1-cell.

(b) Γ, being path connected, is simply connected iff $\sigma \leq \tau$ and $\tau \leq \sigma$ imply $\sigma = \tau$.

Having just defined a function Φ : Trees $\longrightarrow$ Tree Posets we now define its inverse. Let $(E, \leq)$ be a tree poset. We write $\sigma < \tau$ if $\sigma \leq \tau$ with $\sigma \neq \tau$, and we write $\sigma \vdash \tau$ if $\sigma \leq \mu \leq \tau$ implies $\mu = \sigma$ or $\mu = \tau$. We define $\sigma \sim \tau$ by: either $\sigma = \tau$ or $\sigma \vdash \bar{\tau}$. We leave the next proposition as an exercise:

Proposition 13.6.1. *The relation $\sim$ is an equivalence relation on E.* $\square$

Let $t(\tau)$ denote the equivalence class of τ under[23] $\sim$. The classes $t(\tau)$ are the vertices of a simplicial complex K. Two vertices form a 1-simplex of K if they are $t(\tau)$ and $t(\bar{\tau})$ for some $\tau \in E$. This K is a 1-dimensional abstract simplicial complex, and we write $T := |K|$.

Theorem 13.6.2. *The graph T is a tree. The function $(E, \leq) \mapsto T$ from Tree Posets to Trees is inverse to Φ.*

Proof. For distinct edges σ and τ of T we have $\sigma \vdash \tau$ iff there is a two-edge edge path (σ, τ). By 13.6.1 it follows that, for general σ and τ, $\sigma \leq \tau$ iff there is a reduced edge path in T starting with σ and ending with τ. So the two partial orderings are equivalent.

To see that T is a tree one checks the conditions (a) and (b) above. $\square$

For any group G the function Φ provides a bijection between G-trees and G-(tree posets).

The plan of the proof of Stallings' Theorem is to use Bass-Serre theory to exhibit the splitting of an infinite-ended finitely generated group G. For this we need a suitable G-tree. We will find, instead, a G-(tree poset), whose pre-image under Φ will be the required G-tree.

B. Building a tree poset

Let G be a finitely generated group having more than one end. Let Γ be the Cayley graph of G with respect to some finite set of generators. By 12.4.5, Γ is an infinite locally finite left G-graph. When restricted to $G = \Gamma^0$, this G-action is by left multiplication. There is also the action of G on G by right multiplication (which does not extend to an action on Γ). We will express things in terms of subsets of G, but the reader should bear in mind that each such subset U defines the full subgraph of Γ generated by U.

In this section the letters U, V, W, X always label subsets of G. We write $U \overset{a}{\subseteq} V$ if $U - V$ is finite; in words, U is *almost a subset* of V. We write $U \overset{a}{=} V$ if $U \overset{a}{\subseteq} V$ and $V \overset{a}{\subseteq} U$; in words, U is *almost equal* to V. The set U is *almost invariant* if, for all $g \in G$, $U \overset{a}{=} Ug$, and in that case we write $[U]$ for the set of all almost invariant subsets V which are almost equal to U. We write $[V] \leq [W]$ if $V \overset{a}{\subseteq} W$.

---

[23] Think of this as the "terminus" of τ; compare Sect. 6.2.

There is an obvious bijection between the subsets of G and the $\mathbb{Z}_2$-vector space, $C_0^\infty(\Gamma; \mathbb{Z}_2)$, of 0-chains in Γ. We identify each set U with the corresponding infinite 0-chain in Γ. Thus the coboundary δU picks out the (unoriented) 1-simplexes of Γ which have one vertex in U and the other in $G - U$. We leave as an exercise:

Lemma 13.6.3. *The set U is almost invariant iff δU is a finite 1-chain in Γ.*   □

It is convenient to write $U^c := G - U$.

Corollary 13.6.4. *There are almost invariant sets U such that both U and U^c are infinite.*

Proof. Since Γ has more than one end, we can pick a finite path connected subgraph Δ of Γ and let $Z_1, \ldots, Z_m$ be the unbounded path components of $\Gamma \stackrel{c}{-} \Delta$ with $m \geq 2$. Let $U = Z_1^0$, the set of vertices of Z_1. By 13.6.3, U is almost invariant, and U^c contains $Z_2^0 \cup \cdots \cup Z_m^0$.   □

We now make a first attempt at constructing a tree poset. Using 13.6.4, choose an almost invariant set U such that U and U^c are infinite. Let $E = \{[gU], [gU^c] \mid g \in G\}$. Then $(E, \leq)$ is a G-poset which admits the well-defined involution $[V] \rightarrow [V^c]$ (here V stands for some gU or gU^c). We want to choose U so that $(E, \leq)$, equipped with this involution, is a tree poset. Clearly (i), (ii) and (v) in that definition hold. In the next two subsections we verify (iii) when G has infinitely many ends, and (iv) when U is carefully chosen.

C. Verification of (iii)

We denote by $|U|$ the full subgraph of Γ generated by $U \subseteq G$. Thus $|U^c| = \Gamma \stackrel{c}{-} |U|$. By *almost all* $g \in U$ we mean all but finitely many $g \in U$.

Proposition 13.6.5. *Let U_0 and U_1 be almost invariant. Then for almost all $g \in G$ at least one of the following is true: $gU_1 \subseteq U_0$, $gU_1^c \subseteq U_0$, $gU_1 \subseteq U_0^c$, $gU_1^c \subseteq U_0^c$.*

Proof. If U_0 or U_0^c is empty the Proposition is trivial, so we assume they are non-empty. For $i = 0$ or 1 let Δ_i be a finite path connected full subgraph of Γ containing the support of δU_i. Then for almost all $g \in U_0$ we have (i) $\Delta_0 \cap g\Delta_1 = \emptyset$ and (ii) $g\Delta_1^0 \subseteq U_0$. Fix such a $g \in U_0$. By (i) there is a path component $E(g)$ of $|U_1|$ or of $|U_1^c|$ such that $\Delta_0 \subseteq gE(g)$.
Let E be any path component of $|U_1|$ or of $|U_1^c|$. Then $E \cap \Delta_1 \neq \emptyset$, so $gE \cap g\Delta_1 \neq \emptyset$, so $gE \cap |U_0| \neq \emptyset$ by (ii). If $gE \cap |U_0^c| \neq \emptyset$ then by path connectedness $gE \cap \Delta_0 \neq \emptyset$, so $gE \cap gE(g) \neq \emptyset$, so $E = E(g)$. Thus, when $E \neq E(g)$ we have $gE \subseteq |U_0|$. It follows that $|U_0^c|$ meets just one gE, namely, $gE(g)$, and this implies $|U_0^c| \subseteq g|U_1|$ or $|U_0^c| \subseteq g|U_1^c|$. Thus for almost all $g \in U_0$, either $gU_1 \subseteq U_0$ or $gU_1^c \subseteq U_0$. Applying the same argument to U_0^c we get the required conclusion.   □

Proposition 13.6.6. *Let* $H = \{g \in G \mid gU \stackrel{a}{=} U\}$. *If* G *has infinitely many ends then* H *is finite.*

Proof. Recall that both U and U^c are infinite. We will show that if H is infinite then G has an infinite cyclic subgroup of finite index, implying G has two ends by 13.5.9.

We may assume that $H \cap U$ is infinite and that $1 \in U$. By (the proof of) 13.6.5 we have $gU \subseteq U - \{1\}$ or $gU^c \subseteq U - \{1\}$ for almost all $g \in U$. Fix such an element $g \in H \cap U$. Since $gU \stackrel{a}{=} U$ we have $gU \subseteq U - \{1\}$. Thus for all $n > 0$ we have $g^n U \subseteq U - \{1\}$, so g has infinite order. We will show that $\langle g \rangle$ has finite index in G.

Since $1 \in U$, $g^n \in U$ for all $n > 0$, and since $1 \notin g^n U$, $g^{-n} \in U^c$. We next observe that $\cap \{g^n U \mid n > 0\} = \emptyset$, for if $h \in g^n U$ for all $n > 0$ then $g^{-n} \in Uh^{-1}$, contradicting the fact that $U \stackrel{a}{=} Uh^{-1}$ (since all the elements g^{-n} are different). From this we get $U = \cup\{g^n U - g^{n+1}U \mid n \geq 0\} = \cup\{g^n(U - gU) \mid n \geq 0\}$. But $U - gU$ is finite, so U lies in the union of finitely many cosets $\langle g \rangle h$. Similarly for U^c. So $\langle g \rangle$ has finite index in G. $\square$

Proposition 13.6.7. *Let* G *have infinitely many ends. If* V, W *and* X *are infinite almost invariant sets with infinite complements, then* $\{g \in G \mid V \stackrel{a}{\subseteq} gW \stackrel{a}{\subseteq} X\}$ *is finite.*

Proof. We may assume $V \stackrel{a}{\subseteq} X$ and, enlarging V if necessary, that $V \not\subseteq X$. Let $V \stackrel{a}{\subseteq} gW \stackrel{a}{\subseteq} X$. Then either $gW \not\subseteq X$ or $V \not\subseteq gW$. It is enough to show that $\{g \in G \mid gW \stackrel{a}{\subseteq} X \text{ and } gW \not\subseteq X\}$ and $\{g \in G \mid V \stackrel{a}{\subseteq} gW \text{ and } V \not\subseteq gW\}$ are finite.

By 13.6.5 one of the following is true for almost all $g : gW \subseteq X$, $gW \subseteq X^c$, $gW^c \subseteq X$, $gW^c \subseteq X^c$. If $gW \stackrel{a}{\subseteq} X$ and $gW \not\subseteq X$, the only possibility is $gW^c \subseteq X^c$. In that case $gW \stackrel{a}{=} X$. By 13.6.6, this only holds for finitely many g. The case $V \stackrel{a}{\subseteq} gW$ and $V \not\subseteq gW$ is treated similarly. $\square$

The content of 13.6.7 is that (iii) in the definition of a tree poset holds for $(E, \leq)$ when G has infinitely many ends.

D: Verification of (iv)

The previous arguments involve a chosen infinite almost invariant set U such that U^c is also infinite. By 13.6.3, δU has finite support in Γ. We call such a set U *narrow* if (among all such U) the number, k, of 1-simplexes in the support of δU is as small as possible.

Proposition 13.6.8. *If* $U_1 \supseteq U_2 \supseteq \cdots$ *are narrow sets, and if* $V := \bigcap_n U_n$ *is non-empty, then the sequence stabilizes, i.e., for some* N, $U_n = U_N$ *for all* $n \geq N$.

Proof. If an edge e lies in the support of δV then one vertex of e lies in every U_n, while there is an integer N such that the other vertex of e does not lie in U_n when $n \geq N$. Thus e is in the support of δU_n when $n \geq N$. The support of δV lies in the support of such δU_n. In particular, V is almost invariant by 13.6.3.

From now on $n \geq N$. The $\mathbb{Z}_2$-0-chain U_n can be written as $(U_n + V) + V$, hence $\delta U_n = \delta(U_n + V) + \delta V$. No edge appears in both $\delta(U_n + V)$ and δV because if it did then one vertex would be in the set V and the other in the set $U_n - V$, hence that edge could not be in the support of δU_n, contradicting the previous paragraph.

Let W be an infinite set which is either $U_n - V$ or V; one of those must be infinite since U_n is infinite and, as chains, $U_n = (U_n + V) + V$. Then W is almost invariant, and one easily shows that W^c is infinite too. It follows that the number of edges in the support of δW is $\leq k$, because $\delta U_n = \delta(U_n + V) + \delta V$. By minimality that number is k and W is narrow. Since δV and $\delta(U_n + V)$ have no edge in common, and since $V \neq \emptyset$, it must be the case that $W = V$ and the chain $U_n + V = 0$. Thus $U_n = V$. □

The stabilizing set U_N in 13.6.8 will be called a *minimal* narrow set.

If U is narrow, so is U^c, so any $g \in G$ lies in a narrow set, hence, by 13.6.8, there is a minimal narrow set containing g.

Proposition 13.6.9. *Fix $g_0 \in G$. Let U be a minimal narrow set containing g_0. For any narrow set V at least one of the following holds: $U \overset{a}{\subseteq} V$, $U \overset{a}{\subseteq} V^c$, $U^c \overset{a}{\subseteq} V$, $U^c \overset{a}{\subseteq} V^c$.*

Proof. Write $W_1 = U \cap V$, $W_2 = U \cap V^c$, $W_3 = U^c \cap V$, and $W_4 = U^c \cap V^c$. We are to prove that one of the W_i is finite. The support of δW_i lies in the union of the supports of δU and δV. Since the sets W_i are pairwise disjoint, any edge appearing in δU or in δV meets exactly two of the sets W_i. So the number of edges lying in the support of at least one δW_i is $\leq 4k$. Suppose all the W_i's are infinite. Then there are at least k edges in the support of each δW_i, hence precisely k in each support. Thus one of the W_i, say W_j, is a minimal narrow set containing g_0. Hence $W_j = U$. If $j = 1$ then $U \subseteq V$, so $W_2 = \emptyset$, a contradiction; and there is a similar contradiction if $j = 2, 3$ or 4. It follows that some W_i is finite. □

The content of 13.6.9 is that if we choose U to be a minimal narrow set containing some chosen element $g_0 \in G$, then for any $g \in G$ one of the following holds: $[U] \leq [gU]$, $[U] \leq [gU^c]$, $[U^c] \leq [gU]$, $[U^c] \leq [gU^c]$. It follows that (iv) holds for $(E, \leq)$.

We summarize:

Theorem 13.6.10. *Let G be a finitely generated group with infinitely many ends. There is an almost invariant set $U \subseteq G$ with respect to which the G-poset $(E, \leq)$ is a G-(tree poset).* □

E. Completion of the proof of Stallings' Theorem

Let T be the G-tree determined by $(E, \leq)$ in 13.6.10 via the function Φ. The stabilizer of $[U] \in E$ is $\{g \in G \mid gU \stackrel{a}{=} U\}$; this is finite by 13.6.6. So edge stabilizers in T are finite.

From the definition of E it is clear that there are at most two G-orbits. These are represented by the two possible orientations of a 1-cell of T. If no member of G takes one to the other then the G-action on T is rigid. Otherwise we can replace the tree T by its barycentric subdivision sd T, which is also a tree, and the G-action on sd T is rigid. Thus there is no loss of generality in assuming that T is a rigid G-tree with just one orbit of 1-cells. This expresses G as either a free product with amalgamation $A \underset{C}{*} B$ or an HNN extension $A\underset{\phi}{*}$ where the subgroups involved are finite.

If $G \cong A\underset{\phi}{*}$ with $A \geq C_1 \underset{\phi}{\overset{\cong}{\longrightarrow}} C_2$ where C_1 is finite, the case $A = C_1$, and hence also $A = C_2$, can be ruled out since, by 13.5.16, either would imply that G has two ends. .

If $G \cong A\underset{C}{*}B$, the case $[A : C] = [B : C] = 2$ can be ruled out also, since 13.5.16 would imply that G has two ends.

It remains to rule out the case $C = A$, the case in which G fixes a vertex of T. Suppose this happens. Then T consists of a wedge of copies of I whose 0-points are identified. That wedge point is fixed by G. It follows that T is a rigid G-tree; no barycentric subdivision is needed.

In this tree, the relation $\sigma < \tau$ implies $\sigma \vdash \tau$. By (the proof of) 13.6.5, we have $gU \subseteq U$ or $gU^c \subseteq U$ for almost all $g \in U$. Since $\{g \mid gU \stackrel{a}{=} U\}$ is finite there exists $g \in U$ such that gU is not almost equal to U or U^c, and $gU \subseteq U$ or $gU^c \subseteq U$; and there exists $h \in U^c$ such that hU is not almost equal to U or U^c, and $hU \subseteq U$ or $hU^c \subseteq U$. Consider the case $gU \subseteq U$. Then $g^2U \subseteq gU \subseteq U$, so $[g^2U] \leq [gU] \leq [U]$, hence $gU \stackrel{a}{=} U$ or $g^2U \stackrel{a}{=} gU$: both give contradictions. The other three cases give similar contradictions.

Thus the proof of 13.5.10 is complete. $\Box$

Remark 13.6.11. There is an important companion theorem to 13.5.10. A finitely generated group G is *accessible* if G can be decomposed as the fundamental group of a finite graph of groups whose edge groups are finite and whose vertex groups have at most one end. A theorem of Dunwoody [54] says that every finitely presented group is accessible. Examples of finitely generated groups which are not accessible are found in [56]. When G is accessible the process of decomposition described in 13.5.10 can be done iteratively in such a way that it terminates.

A group is *slender* if all its subgroups are finitely generated. For example, finitely generated abelian groups are slender. This notion occurs in connection with JSJ decompositions. In more detail:

Remark 13.6.12. Let G be a finitely presented group. Stallings' Theorem is about decomposing G as the fundamental group of a graph of groups where the edge groups are finite. This is commonly described as *splitting* G over finite subgroups. When G has one end, no such splitting is possible, but one can seek splittings of G whose edge groups are infinite. Typically, the edge groups in such a decomposition of G are required to belong to a named class of slender edge groups; for example (and this is an important special case) all the edge groups might be required to be infinite cyclic. Such splittings are called *JSJ decompositions*[24] of G provided stringent conditions involving permissible vertex groups and permissible relations between the edge groups and the vertex groups are met – see, for example, the papers cited below for details. The point is that, for various classes of slender groups, JSJ decompositions of G exist, and are essentially unique. The first theorem of this kind (for arbitrary finitely presented groups) was given by Rips-Sela [133], with subsequent versions by various authors, e.g., in [55] and in [66].

Source Notes: This proof of Stallings' Theorem is due to Scott and Wall [140], based in part on previous proofs by Dunwoody [53] and D. Cohen [41]. Stallings' original proof appeared in [147].

Exercises

1. Prove the "if" part of 13.5.10.
2. Prove 13.6.1.
3. Let U be as in 13.6.9. Define $U_1 := \{g \in G \mid U \overset{a}{\subseteq} gU \text{ or } U^c \overset{a}{\subseteq} gU\}$. Prove that for all $g \in G$ one of the following holds: $U_1 \subseteq gU_1$, $U_1 \subseteq gU_1^c$, $U_1^c \subseteq gU_1$, $U_1^c \subseteq gU_1^c$.

13.7 The structure of $H^2(G, RG)$

In 13.2.11, we saw that the R-module $H^0(G, RG)$ is R or 0, depending on whether G is finite or infinite, and that when G is finite $H^n(G, RG) = 0$ for all $n > 0$. We saw in Sect. 13.5 that when G is infinite and finitely generated, the R-module $H^1(G, RG)$ is 0 or R or $\overset{\infty}{\underset{1}{\bigoplus}} R$, and we were able to describe the groups G for which the latter two cases occur – hence, by default, the other case too. Throughout this section G is a finitely presented group and R is a PID. We will consider $H^2(G, RG)$. By 13.3.2 and 2.7.2 we have:

Proposition 13.7.1. *The R-module $H^2(G, RG)$ is torsion free.* $\square$

We will study $H^2(G, RG)$ by studying $H^1_e(\tilde{X}; R)$ where $\tilde{X}$ is a simply connected free G-CW complex whose quotient, X, is finite. The method will

---

[24] So called because the idea was inspired by results in 3-manifold topology due independently to W. Jaco and P. Shalen and to K. Johannson in which the slender groups are free abelian groups of rank 2 – the fundamental groups of suitable tori.

require that G have an infinite cyclic subgroup, which we will call J. Therefore, we first consider free J-actions in their own right, not assuming the quotients are compact.

With J an infinite cyclic group, let Y be a simply connected free J-CW complex of locally finite type; then Y is countable, by 11.4.3. The group J is generated by an automorphism $j : Y \to Y$. Now J also acts freely on $\mathbb{R}$, with the generator j acting as the translation $x \mapsto x + 1$. Applying the Borel Construction (Sect. 6.1), we get a commutative diagram

$$
\begin{array}{ccccc}
\mathbb{R} & \longleftarrow & \mathbb{R} \times Y & \longrightarrow & Y \\
\downarrow{\scriptstyle p_1} & & \downarrow{\scriptstyle r} & & \downarrow{\scriptstyle p_2} \\
S^1 & \xleftarrow{\ q_1\ } & Z & \xrightarrow{\ q_2\ } & J\backslash Y
\end{array}
$$

in which Z is the quotient of $\mathbb{R} \times Y$ by the diagonal action of J. This gives us two ways of looking at Z, when we regard q_1 and q_2 as stacks of CW complexes.

First, we consider the resulting decomposition of Z over S^1, thought of as a CW complex with one vertex, v, and one 1-cell, e.

Proposition 13.7.2. *Z is the mapping torus of $j : Y \to Y$.*

Proof. Apply 6.1.3. The space $q_1^{-1}(v)$ is a copy of Y, and Z is obtained by attaching $Y \times B^1$ via the attaching maps id on $Y \times \{-1\}$ and j on $Y \times \{1\}$. $\square$

Next, consider the following commutative diagram:

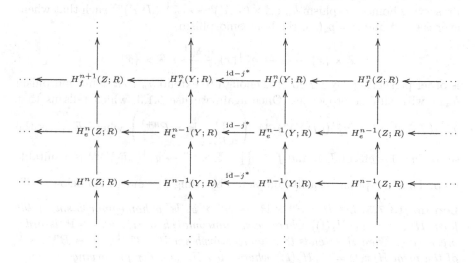

Here, the vertical exact sequences are as in Sect. 12.2. By 13.7.2, Z is the adjunction complex $Y \underset{f}{\cup} (Y \times I)$ where $f : Y \times \{0, 1\} \to Y$ is defined by

$f(y, 0) = y$ and $f(y, 1) = j(y)$. There is a short exact sequence of cochain complexes

$$0 \longleftarrow C_f^*(Y; R) \xleftarrow{i_1^{\#} - (i_0 \circ j)^{\#}} C_f^*(Y \times I; R) \xleftarrow{q^{\#}} C_f^*(Z; R) \longleftarrow 0$$

where $i_0 : y \mapsto (y, 0)$, $i_1 : y \mapsto (y, 1)$, and $q : Y \times I \to Z$ is the quotient map. Identifying $H_f^*(Y \times I; R)$ with $H_f^*(Y; R)$ via the isomorphism i_0^*, we get the top horizontal exact sequence. The other horizontal sequences[25] are formed similarly. One checks:

Proposition 13.7.3. *The above diagram commutes.* $\square$

Next, we examine the decomposition of Z over $J \backslash Y$.

Proposition 13.7.4. *There is a homeomorphism $h : \mathbb{R} \times (J \backslash Y) \to Z$ which is fiber preserving; i.e., $q_2 \circ h$ is projection on the $J \backslash Y$ factor.*

For readers familiar with bundles, this says that the fiber bundle q_2 with fiber $\mathbb{R}$ is trivial. A short proof would read: this is a principal bundle which has a section. What follows is an elementary proof.

Proof (of 13.7.4). There is certainly a fiber preserving homeomorphism $h_0 : \mathbb{R} \times (J \backslash Y)^0 \to q_2^{-1}((J \backslash Y)^0)$. Moreover, we can choose h_0 to be order preserving on each fiber ($\mathbb{R} \times \{point\}$), in the sense that whenever $\tilde{v}$ is a vertex of Y and $v = p_2(\tilde{v})$, the homeomorphism

$$\mathbb{R} \times \{\tilde{v}\} \xrightarrow{r|} q_2^{-1}(v) \xrightarrow{h_0^{-1}} \mathbb{R} \times \{v\}$$

is order preserving. As an induction hypothesis, assume we are given a fiber preserving homeomorphism $h_n : \mathbb{R} \times (J \backslash Y)^n \to q_2^{-1}((J \backslash Y)^n)$ such that whenever $\tilde{x} \in Y^n$ and $x = p_2(\tilde{x})$, the homeomorphism

$$\mathbb{R} \times \{\tilde{x}\} \xrightarrow{r|} q_2^{-1}(x) \xrightarrow{h_n^{-1}} \mathbb{R} \times \{x\}$$

is order preserving. By 1.2.23, it is enough to extend h_n to a homeomorphism h_{n+1} with similar properties. Once again we use 6.1.3, which tells us that

$$q_2^{-1}((J \backslash Y)^{n+1}) = q_2^{-1}((J \backslash Y)^n) \underset{f}{\cup} \left(\coprod_{e \in E_{n+1}} \mathbb{R} \times B^{n+1} \right) \text{ where } E_{n+1} \text{ is the}$$

set of $(n+1)$-cells of $J \backslash Y$ and $f : \coprod_{e \in E_{n+1}} \mathbb{R} \times S^n \to q_2^{-1}((J \backslash Y)^n)$ is a suitable map. Thus the proof is easily completed by using Lemma 13.7.5 below. $\square$

Lemma 13.7.5. *Let $H : S^n \times \mathbb{R} \to S^n \times \mathbb{R}$ be a homeomorphism of the form $H(x, t) = (x, H_x(t))$ where each homeomorphism $H_x : \mathbb{R} \to \mathbb{R}$ is order preserving. Then H extends to a homeomorphism $\hat{H} : B^{n+1} \times \mathbb{R} \to B^{n+1} \times \mathbb{R}$ of the form $\hat{H}(x, t) = (x, \hat{H}_x(t))$ where each $\hat{H}_x$ is order preserving.*

---

[25] The horizontal exact sequences in the diagram are special cases of what is called the *Wang sequence*.

Proof. Note that when $n > 0$ the hypothesis on each H_x holds iff it holds on one H_x; but we also need this lemma for $n = 0$. The required $\hat{H}$ is given by the formula $\hat{H}(x, t) = (x, (1 - |x|)t + |x|H_{\frac{x}{|x|}}(t))$ if $x \neq 0$, and by $\hat{H}(0, t) = t$. $\square$

The idea behind the above proof is simple: any convex linear combination of order-preserving homeomorphisms of $\mathbb{R}$ is an order-preserving homeomorphism. Therefore, we are "coning" each H_x, gradually deforming it to the identity as we approach the point $0 \in B^{n+1}$. This process is continuous in x.

Proposition 13.7.6. *If W is a countable CW complex of locally finite type, then for all n, $H_f^n(W; R) \cong H_f^{n+1}(W \times \mathbb{R}; R)$.*

Proof. This follows from 12.6.1. Here is a short direct proof. Let $\{K_i\}$ be a finite type filtration of W. Then $\{K_i \times [-i, i]\}$ is a finite type filtration of $W \times \mathbb{R}$. Fix i, and let $X_i = (W \times \mathbb{R}) \stackrel{c}{-} (K_i \times [-i, i])$. We can write $X_i = A_i \cup B_i$ where $A_i W \times [0, \infty) \stackrel{c}{-} (K_i \times [0, i])$ and $B_i W \times (-\infty, 0] \stackrel{c}{-} (K_i \times [-i, 0])$. We have $A_i \cap B_i = (W \stackrel{c}{-} K_i) \times \{0\}$. The pair $(W \times [i+1, \infty), W \times [i+1, \infty))$ is a strong deformation retract of the pair $(W \times [0, \infty), A_i)$; so $H^*(W \times [0, \infty), A_i; R) = 0$. Similarly, $H^*(W \times (-\infty, 0], B_i; R) = 0$. By the appropriate Mayer-Vietoris sequence, there is a natural isomorphism

$$\delta^* : H^n(W, W \stackrel{c}{-} K_i; R) \to H^{n+1}(W \times \mathbb{R}, A_i \cup B_i; R).$$

By taking the direct limit and applying 12.2.1, we get the desired result. $\square$

By 13.7.4 and 13.7.6 we get:

Corollary 13.7.7. *For each n, $H_f^n(J \backslash Y; R) \cong H_f^{n+1}(Z; R)$.* $\square$

Proposition 13.7.8. *Assume Y^2 does not have two ends. If $j^* : H_e^1(Y; R) \to H_e^1(Y; R)$ agrees with the identity on a non-trivial R-submodule, A, then either $A \cong R$ or $J \backslash Y$ has more than one end.*

Proof. Recall that Y is assumed to be simply connected. The space $J \backslash Y$ is not compact (i.e., does not have 0 ends) because Y does not have two ends; see 13.5.9. Assuming $J \backslash Y$ has one end, we must show $A \cong R$. We have an exact sequence

$$H^1(J \backslash Y; R) \longleftarrow H_f^1(J \backslash Y; R) \longleftarrow H_e^0(J \backslash Y; R) \stackrel{\alpha}{\longleftarrow} H^0(J \backslash Y; R) \longleftarrow 0.$$

By 13.4.11, $H_e^0(J \backslash Y; R) \cong R$, so 12.2.3 implies α is an isomorphism. The R-module $H^1(J \backslash Y; R)$ is isomorphic to R, by 3.1.19 and 12.5.1. So $H_f^1(J \backslash Y; R)$ embeds in R, which implies, by 11.3.11, that $H_f^1(J \backslash Y; R) \cong R$ or 0. Consider the diagram

$$H_f^2(Y;R) \xleftarrow{\text{id}-j^*} H_f^2(Y;R) \longleftarrow H_f^1(J\backslash Y;R) \longleftarrow H_f^1(Y;R)$$

$$\uparrow \qquad\qquad\qquad \uparrow$$

$$H_e^1(Y;R) \xleftarrow{\text{id}-j^*} H_e^1(Y;R) \geq A$$

$$\uparrow$$

$$0 = H^1(Y;R)$$

Here, the horizontal exact sequences come from 13.7.3, 13.7.4 and 13.7.6. From the exactness, we conclude that A is a non-trivial quotient of a submodule of $H_f^1(J\backslash Y;R)$. Hence $H_f^1(J\backslash Y;R) \cong R$, and A must be generated by one element. By 12.5.11, A is torsion free, so by 11.3.11, $A \cong R$. $\square$

A similar proof gives:

Proposition 13.7.9. *Assume* Y^2 *has one end. If* $j^*: H_e^1(Y;R) \to H_e^1(Y;R)$ *agrees with the identity on a non-trivial* R*-submodule* A*, then* A *is a free* R*-module.*

Proof. Consider the first exact sequence given in the proof of 13.7.8. The monomorphism α splits by 12.2.3, and we have seen that $H^1(J\backslash Y;R)$ is isomorphic to R. Moreover, $H_e^0(J\backslash Y;R)$ is free, by 12.5.11. Thus $H_f^1(J\backslash Y;R)$ is R-free. From the diagram of 13.7.3 we see that $H_f^2(Y;R) = 0$; this is where we use the hypothesis that Y is one-ended. It follows from the second diagram in the proof of 13.7.8 that A embeds in a free R-module and hence is R-free. $\square$

Now we return to the finitely presented group G acting freely on $\tilde{X}$ with finite quotient X. The covering transformations give $H_e^1(\tilde{X};R)$ the structure of a right RG-module; compare 13.2.16.

Proposition 13.7.10. *Let there exist* $j \in G$ *of infinite order, and a sub-RG-module* A *of* $H_e^1(\tilde{X};R)$ *on which* $j^*: H_e^1(\tilde{X};R) \to H_e^1(\tilde{X};R)$ *agrees with the identity. Then, as an* R*-module,* A *is either trivial or is isomorphic to* R *or is infinitely generated. If, in addition,* $\tilde{X}$ *has one end then, as an* R*-module,* A *is trivial or is isomorphic to* R *or to* $\bigoplus_1^\infty R$.

Proof. By hypothesis, $\tilde{X}$ is non-compact. If $\tilde{X}$ has two ends, then, by 13.5.9 and 12.2.2, $H_e^1(\tilde{X};R) \cong H_e^1(\mathbb{R};R) = 0$, implying A is trivial. Assume $\tilde{X}$ does not have two ends, and suppose A is a finitely generated R-module which is neither 0 nor R. By 12.5.11, A is free of finite rank > 1. By 13.7.8, $J\backslash\tilde{X} = \bar{X}(J)$ has more than one end. Let K be a finite path connected subcomplex of $\bar{X}(J)$ which separates $\bar{X}(J)$. Certainly, K can be chosen so that the inclusion induces an epimorphism $i_\#: \pi_1(K,\bar{v}) \to \pi_1(\bar{X}(J),\bar{v})$. Temporarily, we will assume that K can be chosen so that $i_\#$ is an isomorphism. Then 3.4.9 implies

$p_J^{-1}(K)$ is the universal cover of K, where $p_J : \tilde{X} \to \bar{X}(J)$ is the covering projection. Since $\tilde{K}$ has two ends, $H_f^1(\tilde{K}; R) \cong R$.

Let $L_1, \cdots, L_r$ be the path components of $\bar{X}(J) \overset{c}{-} K$. At least two of these are unbounded, say L_1 and L_2. Write $L^- = (N(L_1)) \cup K$ and $L^+ = N(L_2 \cup \cdots \cup L_r) \cup K$. Then $L^+ \cup L^- = \bar{X}(J)$ and $L^+ \cap L^- = K$. Writing $M^{\pm} = p_J^{-1}(L^{\pm})$, we have $M^+ \cup M^- = \tilde{X}$ and $M^+ \cap M^- = \tilde{K}$. See Fig. 13.2.

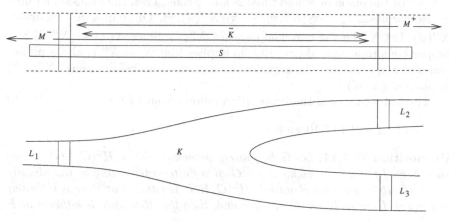

Fig. 13.2.

In the following diagram, the row is the Mayer Vietoris sequence, and the column is exact:

$$H_f^2(M^+; R) \oplus H_f^2(M^-; R) \overset{\alpha}{\longleftarrow} H_f^2(\tilde{X}; R) \longleftarrow H_f^1(\tilde{K}; R) \cong R$$

$$\uparrow \beta$$

$$A \leq H_e^1(\tilde{X}; R)$$

$$\uparrow$$

$$0 = H^1(\tilde{X}; R)$$

The key point is that since A is finitely generated over R and is invariant under the action of G, $\beta(A) \subset \ker \alpha$. [Proof: Let finite cocycles $c_1, \cdots, c_k$ represent R-generators of $\beta(A)$ and let a finite subcomplex S of $\tilde{X}$ support $c_1, \cdots, c_k$. Since X is compact and J has infinite index in G, there exist g_+ and g_- in G such that $g_+(S) \cap M^- = \emptyset$ and $g_-(S) \cap M^+ = \emptyset$. Thus (inclusion)* : $H_f^2(\tilde{X}; R) \to H_f^2(M^{\pm}; R)$ maps $\beta(A)$ to 0, so $\alpha\beta(A) = 0$. It follows that there is an epimorphism $R \to A$, a contradiction.]

If K cannot be chosen so that $i_{\#}$ is an isomorphism, we alter the argument slightly. Since $i_{\#}$ is a split epimorphism, the theory of Tietze transformations (Sect. 3.1) can be used to attach finitely many 2-cells $e_1^2, \cdots, e_s^2$ to $K \subset \bar{X}(J)$

to form a new pair of complexes $(\bar{X}'(J), K')$, where the inclusion induces an isomorphism $i'_\# : \pi_1(K', \bar{v}) \to \pi_1(\bar{X}'(J), \bar{v})$. These 2-cells are attached to $\bar{X}(J)$ along loops in K which are homotopically trivial in $\bar{X}(J)$. So, just as in 4.1.10, $\bar{X}'(J)$ is homotopy equivalent to $\bar{X}(J) \vee \left(\bigvee_{i=1}^{s} S_i^2 \right)$. Thus, using 4.1.8, we can attach s 3-cells to $\bar{X}'(J)$ to produce $\bar{X}''(J)$ homotopy equivalent to $\bar{X}(J)$. Let K'' be the union of K and these 2- and 3-cells. Then the inclusion induces an isomorphism $i''_\# : \pi_1(K'', \bar{v}) \to \pi_1(\bar{X}''(J), \bar{v})$. Of course, K'' separates $\bar{X}''(J)$. Let $\tilde{X}''$ be the universal cover of $\bar{X}''(J)$. Since $\bar{X}(J) \hookrightarrow \bar{X}''(J)$ is a proper homotopy equivalence, 10.1.23 implies that $\tilde{X} \hookrightarrow \tilde{X}''$ is also a proper homotopy equivalence. The argument now proceeds as before, using $(\tilde{X}'', K'')$ in place of $(\tilde{X}, K)$.

The last sentence of the Proposition follows from 13.7.9.   □

By 13.7.1 and 13.7.10 we get:

Proposition 13.7.11. *Let G be finitely presented. Then $H^2(G, RG)$ is torsion free. If there is an element $j \in G$ of infinite order acting as the identity on $H^2(G, RG)$, then the R-module $H^2(G, RG)$ is either 0 or R or is infinitely generated. If, in addition, G has one end, then this R-module is either 0 or R or $\bigoplus_1^{\infty} R$.*   □

When $R = \mathbb{Z}$ we can improve the first part of 13.7.11, getting rid of the hypothesis on j:

Theorem 13.7.12. (Farrell's Theorem) *Let the finitely presented group G act freely on $\tilde{X}$ with compact quotient. Assume G has an element of infinite order.[26] Then the abelian group $H_e^1(\tilde{X}; \mathbb{Z})$ is trivial or is isomorphic to $\mathbb{Z}$ or is infinitely generated. Hence the abelian group $H^2(G, \mathbb{Z}G)$ is trivial or is isomorphic to $\mathbb{Z}$ or is infinitely generated.*

Proof. By 12.4.8 and 12.2.2, there is a monomorphism of $\mathbb{Z}_2$-modules $\beta : \mathbb{Z}_2 \otimes_\mathbb{Z} H_e^1(\tilde{X}; \mathbb{Z}) \to H_e^1(\tilde{X}; \mathbb{Z}_2)$. In fact, by naturality of the universal coefficient sequence, β is a monomorphism of $\mathbb{Z}_2 G$-modules. Assume $H_e^1(\tilde{X}; \mathbb{Z})$ is a finitely generated abelian group. Then it is $\mathbb{Z}$-free (see 12.5.10) of finite rank, say, r, and we are to show $r = 0$ or 1. Let $A = \text{image}(\beta)$. Then A is a finite $\mathbb{Z}_2 G$-submodule. If $j \in G$ is an element of infinite order, then some power of j agrees with the identity on A. By 13.7.10, it follows that A is 0 or $\mathbb{Z}_2$. Hence $r = 0$ or 1. The last sentence now follows from 13.2.9 and 13.2.13.   □

---

[26] At time of writing, the question of whether every infinite finitely presented group contains an element of infinite order is open. M. Kapovich and B. Kleiner have announced a new proof of 13.7.12 without the hypothesis that G contains an element of infinite order.

Theorem 13.7.12 leaves open the question of whether $H^2(G, \mathbb{Z}G)$ is always a free abelian group. For finitely presented groups G the answer is unknown in general. We will give a positive answer in 16.5.1 under the hypothesis that G is semistable at each end. The corresponding question when R is a field k has been fully answered for torsion free finitely presented groups G in [64]: the k-vector space $H^2(G, kG)$ has dimension 0 or 1 or ∞.

Source Notes: Farrell's Theorem appeared in [63].

Exercises

1. Give an example of a finitely generated (but not finitely presented) group G such that $H^2(G, \mathbb{Z}G)$ is not free. *Hint*: compare Exercise 10 in Sect. 13.5.
2. Prove that the countably infinite product of copies of $\mathbb{Z}$ is not a free abelian group.

13.8 Asphericalization and an example of $H^3(G, \mathbb{Z}G)$

In this section we will construct a finite aspherical 3-pseudomanifold whose universal cover is a non-orientable 3-pseudomanifold. Its fundamental group G will thus have type F and geometric dimension 3. We will show that $H^3(G, \mathbb{Z}G) \cong \mathbb{Z}_2$. (Until now we have not seen a group G of type F_n for which $H^n(G, \mathbb{Z}G)$ is not free abelian.)

For this we require a general procedure, "asphericalization," for making complexes aspherical. It is of independent interest.[27]

Recall from Sect. 5.2 the abstract simplicial complex **n** with $|\mathbf{n}| = \Delta^n$. A pair (K, π) is *a simplicial complex over* **n** if K is an abstract non-empty simplicial complex, and $\pi : K \to \mathbf{n}$ is a simplicial map which is injective on each simplex of K. Here is a source of examples:

Proposition 13.8.1. *Let Y be a regular CW complex of dimension $\leq n$ and let $\pi : \mathrm{sd}\, Y \to \mathbf{n}$ map the vertex e^k_α of $\mathrm{sd}\, Y$ (e^k_α being a cell of Y) to $p_k \in V_\mathbf{n}$. Then $(\mathrm{sd}\, Y, \pi)$ is a simplicial complex over* **n**. $\square$

Note that if $K = \overset{\bullet}{\Delta}{}^2$ there is no π such that (K, π) is a simplicial complex over **1**. In general, the existence of π depends on the simplicial structure of K rather than on the topology of $|K|$.

An *aspherical model over* Δ^n is a map $f : X \to \Delta^n = |\mathbf{n}|$ where

(i) X is an aspherical CW complex;
(ii) for every non-empty subcomplex J of **n**, $f^{-1}(|J|)$ is a non-empty subcomplex of X each of whose path components is aspherical;

---

[27] A more refined procedure called "hyperbolization" turns suitable complexes into hyperbolic complexes. See [47].

(iii) for every base vertex v, the inclusion map induces a monomorphism $\pi_1(f^{-1}(|J|), v) \to \pi_1(X, v)$.

Given (K, π) and (X, f) as above, the *asphericalization of $|K|$ by (X, f)* is the pull-back $X \triangle |K| := f^*(|K|)$ in the diagram

$$
\begin{array}{ccc}
X \triangle |K| & \longrightarrow & |K| \\
\downarrow & & \downarrow {\scriptstyle |\pi|} \\
X & \xrightarrow{\ f\ } & \Delta^n.
\end{array}
$$

Example 13.8.2. Take $|K|$ to be $\mathbb{R}$ where the integers are vertices. Take $n = 1$, Δ^1 having vertices p_0 and p_1. Define π to take evens to p_0 and odds to p_1. Take X to be a compact orientable surface with three boundary components B_0, B_1 and B_2. Let $f^{-1}(p_0) = B_0$ and $f^{-1}(p_1) = B_1 \cup B_2$. The space $X \triangle |K|$ is illustrated in Fig. 13.3.

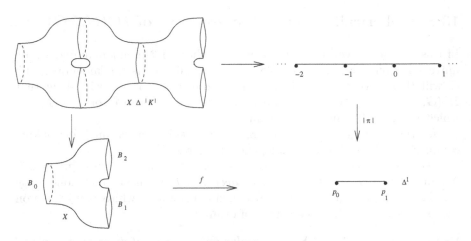

Fig. 13.3.

We can also regard $X \triangle |K|$ as the pull-back $|\pi|^*(X)$. As such it has a natural CW complex structure, being assembled out of copies of the various $f^{-1}(|\sigma|)$ for $\sigma \in K$ in order of increasing dimension.

Proposition 13.8.3. *Let $f : X \to \Delta^n$ be an aspherical model over Δ^n and let (K, π) be a simplicial complex over $\mathbf{n}$. Then $X \triangle |K|$ is non-empty and each of its path components is aspherical. Moreover, for any non-empty subcomplex L of K and any base vertex w of $X \triangle |L|$, the inclusion map induces a monomorphism $\pi_1(X \triangle |L|, w) \to \pi_1(X \triangle |K|, w)$, where $(L, \pi|)$ is considered as a simplicial complex over $\mathbf{n}$.*

For $\sigma \in K$ let $\bar{\sigma}$ [resp. $\overset{\bullet}{\sigma}$] denote the subcomplex of K consisting of σ and its faces [resp. the proper faces of σ]. A simplex of K is *principal* if it is not a proper face of another simplex of K.

Proof (of 13.8.3). First we handle the special case in which K is a simplex, i.e., $K = \bar{\sigma}$ for some $\sigma \in K$. Then for any non-empty subcomplex L of $\bar{\sigma}$, $X\triangle|L| = f^{-1}(|\pi|(|L|))$. By Properties (i)–(iii), above, the Proposition holds.

For the case where K is finite we do induction on N, the number of simplexes of K. If K is a point, we are done, by the special case. For larger N, let K_0 be the subcomplex of K obtained by removing a principal simplex of K. Then $X\triangle|K|$ is the adjunction complex obtained from $(X\triangle|K_0|) \amalg (X\triangle|\bar{\sigma}|)$ by gluing along the obvious copies of $X\triangle|\overset{\bullet}{\sigma}|$ in both. If $X\triangle|\overset{\bullet}{\sigma}|$ is empty (which happens iff σ is a 0-simplex) then by induction and/or the special case, $X\triangle|K|$ is non-empty and every path component is aspherical. If $X\triangle|\overset{\bullet}{\sigma}|$ is non-empty, $\pi_1(X\triangle|\overset{\bullet}{\sigma}|, w) \to \pi_1(X\triangle|K_0|, w)$ and $\pi_1(X\triangle|\overset{\bullet}{\sigma}|, w) \to \pi_1(X\triangle|\bar{\sigma}|, w)$ are monomorphisms by induction and/or the special case; so 7.1.9 implies that every path component of $X\triangle|K|$ is aspherical.

Now let L be a non-empty proper subcomplex of K. There is a principal simplex τ of K such that L is a subcomplex of the complex K_1 obtained by removing τ from K. For any suitable vertex w, the following diagram commutes, where all arrows are induced by inclusion:

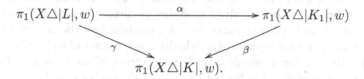

Now, α is a monomorphism by induction, and β is a monomorphism by 6.2.1 (since $X\triangle|K|$ is the adjunction space of $X\triangle|K_1|$ and $X\triangle|\bar{\tau}|$ by gluing across $X\triangle|\overset{\bullet}{\tau}|$), so γ is a monomorphism. The extension of this argument to the case where K is infinite is left as an exercise.   □

Addendum 13.8.4. *If K is path connected and n-dimensional then $X\triangle K$ is path connected.*

Proof. If $K = \bar{\sigma}$ then σ is n-dimensional, so $X\triangle|K| = X$. If $K \neq \bar{\sigma}$ and K is finite, one shows easily that K is the union of two connected proper subcomplexes K_1 and K_2. By induction on the number of principal simplexes of K, the subcomplexes $X\triangle K_1$ and $X\triangle K_2$ are path connected, and their intersection is non-empty by Property (ii), above. So $X\triangle K$ is path connected. Again, the infinite case is left as an exercise.   □

We can now construct the promised 3-pseudomanifold. Let X be a compact 3-manifold whose boundary is U_4, the surface obtained from S^2 by attaching four crosscaps (see 5.1.8); in Exercise 1 the reader is asked to construct such a space X. Identifying S^2 with $\overset{\bullet}{\Delta}{}^3$, we think of ∂X as being obtained by attaching one crosscap to the interior of each of the four 2-simplexes of $\overset{\bullet}{\Delta}{}^3$. Then ∂X contains the 1-skeleton of Δ^3. Form an aspherical model over Δ^3, $f : X \to \Delta^3$; f is the "identity" on the 1-skeleton of Δ^3, f maps the interior of each 2-cell-with-crosscap onto the interior of the corresponding 2-cell of Δ^3, and f maps $\overset{\circ}{X}$ onto $\overset{\circ}{\Delta}{}^3$.

Let J be a triangulation of some compact 3-manifold with connected non-orientable boundary. Form a simplicial complex L from J by coning over ∂J; i.e., $v * \partial J$ is identified with J along ∂J to define L. Then $|L| = \{v\} \cup_c |J|$, where $c : |\partial J| \to \{v\}$ is the constant map. This L is a 3-pseudomanifold without boundary. Let $K = \mathrm{sd}\, L$, and let $\pi : K \to \mathbf{3}$ be as in Proposition 13.8.1 (reading $Y = |L|$). Then (K, π) is a simplicial complex over $\mathbf{3}$. The key point is that the base of the dual cone of v in L (the link of v in K) is a non-orientable surface.

By 13.8.3 and 13.8.4, $W := X \triangle K$ is a compact path-connected aspherical CW 3-pseudomanifold without boundary, containing a vertex the base of whose dual cone is a non-orientable surface. So the universal cover $\tilde{W}$ has vertices whose dual cones have the same property. Hence $\tilde{W}$ is a non-orientable pseudomanifold (exercise). Writing $G = \pi_1(W, w)$, we see that G has type F and geometric dimension 3. By 12.3.14 and 13.2.9, $H^3(G, \mathbb{Z}G) \cong \mathbb{Z}_2$.

Remark 13.8.5. There is a construction called "relative asphericalization" in which a subcomplex of the given complex is already aspherical and is not to be changed during the construction. An unpublished theorem of M. Davis says that any compact aspherical polyhedron is a retract of a closed orientable manifold. Hence, any group of type F is a retract of an orientable Poincaré Duality group as defined in Sect. 15.3. See [87] for more details.

Source Notes: The asphericalization construction goes back to Gromov. See also [47]. The example here is from [12].

Exercises

1. Describe a compact 3-manifold with connected non-orientable boundary.
2. Why is id : $\Delta^2 \to \Delta^2$ not an aspherical model over Δ^2?
3. Prove 13.8.3 and 13.8.4 when K is infinite.
4. Prove that $\tilde{W}$ is non-orientable. State a more general theorem implied by your proof (about pseudomanifolds having non-orientable links).
5. Show that $\pi_1(W, w)$ has cohomological dimension 3 over $\mathbb{Z}$ and cohomological dimension 2 over $\mathbb{Q}$.

13.9 Coxeter group examples of $H^n(G, \mathbb{Z}G)$

We have seen that for an infinite finitely presented group G the cohomology $H^k(G, \mathbb{Z}G)$ is torsion free when $k \leq 2$, and in Sect. 13.8 we saw an example where $H^3(G, \mathbb{Z}G)$ is of order 2. Now we show that the methods used in Sect. 9.1 can be refined to give Coxeter groups G with $H^*(G, \mathbb{Z}G)$ quite varied.

A Coxeter system (G, S) is *right angled*[28] if whenever $s_1 \neq s_2$, $m(s_1, s_2) = 2$ or ∞. The *corresponding abstract graph* $\Gamma(G, S)$ has S as its vertex set and, for $s_1 \neq s_2$, $\{s_1, s_2\}$ is a 1-simplex iff $m(s_1, s_2) = 2$. This in turn defines the *corresponding flag complex* $L(G, S)$, and the function $(G, S) \mapsto L(G, S)$ is a bijection between right angled Coxeter systems (for which S is finite) and finite flag complexes. By Exercise 2 of §9.1 we have

Proposition 13.9.1. *Let (G, S) be a right angled Coxeter system and let $T \subset S$. $\langle T \rangle$ is finite iff T spans a simplex of $L(G, S)$. In fact this correspondence gives an isomorphism of abstract simplicial complexes between the poset of non-trivial finite special subgroups of G and the abstract first derived sd $L(G, S)$.* □

For the rest of this section $d \geq 2$ is an integer, L is a finite connected closed combinatorial $(d-1)$-dimensional manifold which is also a flag complex, and (G, S) is the corresponding right angled Coxeter system, so that $L = L(G, S)$. Define $K = \text{sd } L$, the first barycentric subdivision. By 13.9.1, we may identify this K with K in Sect. 9.1, the base of the cone F. When $\langle T \rangle$ is a non-trivial finite standard subgroup of G, T is a simplex of L. Let $K(T)$ denote the corresponding subcomplex of K, an abstract first derived of this simplex. Recall that $N_K(K(T))$ denotes the simplicial neighborhood of $K(T)$ in K. We use the notation of Sect. 9.1.

Lemma 13.9.2. $F_{\sigma(T)} = N_K(K(T))$.

Proof. Let τ be a simplex of $F_{\sigma(T)}$. Then τ is a simplex of $F_{\{s\}}$ for some $s \in T$, so τ is a face of a simplex μ whose initial vertex is $\langle\{s\}\rangle$. Thus τ is a face of a simplex which shares a vertex with $K(T)$, so τ is a simplex of $N(K(T))$.

Conversely, let τ be a simplex of $N(K(T))$. Then τ is a face of a simplex ν having a vertex in $K(T)$. Write $\nu = \{\langle T_0 \rangle, \cdots, \langle T_k \rangle\}$ and let $\langle T_i \rangle$ be a vertex of $K(T)$. Then $T_i \subset T$ by 9.1.2, so ν is a face of a simplex μ whose initial vertex is $\langle\{s\}\rangle$ for some $s \in T$. Thus τ is a face of a simplex of $F_{\{s\}}$, so τ is a simplex of $F_{\{s\}}$, and $F_{\{s\}}$ is a subcomplex of $F_{\sigma(T)}$. □

Proposition 13.9.3. *When $\langle T \rangle$ is a non-trivial finite standard subgroup of G, $|F_{\sigma(T)}|$ is a PL $(d-1)$-ball.*

Proof (Sketch). This requires knowledge of piecewise linear topology, in particular, of regular neighborhoods in PL manifolds; we have set things up so that references are easily given. By 13.9.2, $F_{\sigma(T)}$ is the simplicial neighborhood of $K(T)$ in the closed combinatorial $(d-1)$-manifold $K = \text{sd } L$, and $K(T)$ is an abstract first derived of a simplex of L. Thus (see [136, Chap. 3]),

---

[28] A *right angled Coxeter group* is a group G for which there exists a right angled Coxeter system (G, S).

$|F_{\sigma(T)}|$ is a regular neighborhood of $|K(T)|$. Since $|K(T)|$ is collapsible, any regular neighborhood of it in the PL $(d-1)$-manifold $|L|$ is a PL $(d-1)$-ball ([136, Chap. 3, Sect. 27]).   □

Recall from 9.1.6 that $A_n = \bigcup\limits_{i=0}^{n} g_i F$ and $A_n \cap g_{n+1}F = g_{n+1}F_{\sigma(B(g_{n+1}))}$.

Define $\overset{\bullet}{A}_n$ to be the subcomplex of A_n satisfying $|\overset{\bullet}{A}_n| = \mathrm{fr}_{|D|}|A_n|$.

Proposition 13.9.4. *Each $\overset{\bullet}{A}_n$ is a finite connected closed combinatorial manifold of dimension $d-1$. $\overset{\bullet}{A}_n$ is orientable iff $|L|$ is orientable.*

Proof. The proof is by induction on n, starting with $\overset{\bullet}{A}_0 = K = \mathrm{sd}\ L$. Assume the Proposition for $\overset{\bullet}{A}_n$. Let $B_{n+1} = g_{n+1}F_{\sigma(B(g_{n+1}))}$. The full subcomplex of D generated by $\overset{\bullet}{A}_n$ and $g_{n+1}F$ contains $\overset{\bullet}{A}_{n+1}$. By 13.9.3, $|B_{n+1}|$ is a PL $(d-1)$-ball. Denoting its interior by $|\overset{\circ}{B}_{n+1}|$, we have

$$|\overset{\bullet}{A}_{n+1}| = \mathrm{cl}_{|D|}(|\overset{\bullet}{A}_n| - |\overset{\circ}{B}_{n+1}|) \cup \mathrm{cl}_{|D|}(g_{n+1}|K| - |\overset{\circ}{B}_{n+1}|).$$

Because $|B_{n+1}|$ is a regular neighborhood in the closed PL manifolds $|\overset{\bullet}{A}_n|$ and $g_{n+1}|K|$, the two closures in this union are PL $(d-1)$-manifolds with a PL $(d-2)$-sphere as their common boundary; this follows from piecewise linear topology as in [136, Chap. 3]. The desired conclusions follow for $\overset{\bullet}{A}_{n+1}$.   □

Of course we have seen a simpler way of describing the expression of $|\overset{\bullet}{A}_{n+1}|$ as the union of two closures in the last proof: $|\overset{\bullet}{A}_{n+1}|$ is the connected sum of the manifolds $|\overset{\bullet}{A}_n|$ and $g_{n+1}|K|$, i.e., $|\overset{\bullet}{A}_{n+1}| = |\overset{\bullet}{A}_n| \# g_{n+1}|K|$. See Fig. 13.4.

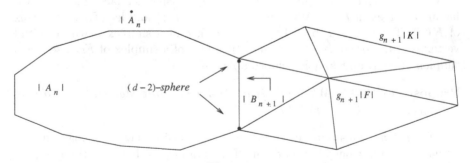

Fig. 13.4.

It is clear that $|\overset{\bullet}{A}_n|$ is a strong deformation retract of $|\overset{\bullet}{A}_n| \cup g_{n+1}|F|$. Thus there is a homotopy equivalence r making the following diagram commute up to homotopy:

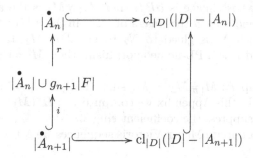

Write $f = r \circ i : |\dot{A}_{n+1}| = |\dot{A}_n| \# g_{n+1}|K| \to |\dot{A}_n|$; see Fig. 13.4. This is just the kind of map f appearing in the Appendix to this section. The following theorem combines a number of previous results: the fact that the last diagram commutes up to homotopy, 12.2.2, 13.2.13, 13.2.14, 9.1.10, as well as 13.9.7 in the Appendix below:

Theorem 13.9.5. *Let $d \geq 2$, let L be a non-empty finite connected closed combinatorial manifold of dimension $d-1$ which is also a flag complex, and let (G, S) be the corresponding right angled Coxeter system (i.e., $L = L(G, S)$). If $|L|$ is orientable then*[29] *we have isomorphisms of abelian groups:*

$$H^k(G, \mathbb{Z}G) \cong \begin{cases} \bigoplus_G \tilde{H}^{k-1}(|L|; \mathbb{Z}) & \text{if } k \neq d \\ \mathbb{Z} & \text{if } k = d \end{cases}$$

If $|L|$ is non-orientable (so that $d \geq 3$), the same holds except in dimensions $d-1$ and d, where $H^d(G, \mathbb{Z}G) \cong \mathbb{Z}_2$ and $H^{d-1}(G, \mathbb{Z}G) \cong \bigoplus_G H^{d-2}(|L|; \mathbb{Z}) \oplus F$

where F is a free abelian group of countably infinite rank. □

Corollary 13.9.6. *The group G has one end, $H^2(G, \mathbb{Z}G)$ is free abelian, and every torsion free subgroup of finite index in G has geometric dimension d.*

Proof. By 13.5.2 and 13.5.5, G has one end. We have $H^2(G, \mathbb{Z}G)$ free abelian because $H^1(|L|; \mathbb{Z})$ is (finitely generated and) free abelian. Let H be a torsion free subgroup of finite index (see 9.1.10). By 9.1.11, H has geometric dimension $\leq d$; and by 13.9.5, H has geometric dimension $\geq d$. □

Appendix: Homology of connected sums

Let M_1 and M_2 be path connected n-manifolds where $n \geq 1$. Recall that $M_1 \# M_2$ is obtained by removing the interior of an unknotted n-ball B_i^n in

---

[29] $\tilde{H}^*$ denotes reduced cohomology. Note that $\tilde{H}^{-1}(|L|) = 0$ because $|L|$ is non-empty, and $\tilde{H}^0(|L|) = 0$ because L is path connected.

each $\overset{\circ}{M}_i$ (thus $N_i := M_i - \text{int } B_i^n$ is a manifold), and picking an embedding $h : \partial B_1^n \to N_2$ whose image is ∂B_2^n; then $M_1 \# M_2$ is the adjunction space $N_2 \cup_h N_1$. We denote by S the image of ∂B_1^n in $M_1 \# M_2 : S$ is the $(n-1)$-sphere along which N_1 is glued to N_2 to get $M_1 \# M_2$. If M_1 and M_2 are PL manifolds and h is a PL homeomorphism, then $M_1 \# M_2$ is clearly a PL manifold.

There is a map $f : M_1 \# M_2 \to M_1$ such that $f \mid N_1 = \text{id}_{N_1}$ and $f(N_2) = B_1^n$; see Fig. 5.1. In this Appendix we compute $f^* : H^*(M_1) \to H^*(M_1 \# M_2)$. Throughout, we suppress the coefficient ring $\mathbb{Z}$.

From the appropriate Mayer-Vietoris sequences we get a commutative diagram:[30]

$$
\begin{array}{ccccccccc}
0 = H^i(S,S) & \longleftarrow & H^i(N_1,S) \oplus H^i(N_2,S) & \overset{\cong}{\longleftarrow} & H^i(M_1 \# M_2, S) & \longleftarrow & H^{i-1}(S,S) = 0 \\
& & \big\uparrow{\scriptstyle \text{id} \oplus f^*} & & \big\uparrow{\scriptstyle f^*} & & \\
0 = H^i(S,S) & \longleftarrow & H^i(N_1,S) \oplus H^i(B_2^n,S) & \overset{\cong}{\longleftarrow} & H^i(M_1,S) & \longleftarrow & H^{i-1}(S,S) = 0
\end{array}
$$

We also have isomorphisms:

$$
H^i(N_j,S) \underset{\text{excision}}{\overset{\cong}{\longleftarrow}} H^i(M_j, B_j^n) \overset{\cong}{\longrightarrow} \tilde{H}^i(M_j)
$$

which are sufficiently canonical to give a commutative diagram:

$$
\begin{array}{ccc}
H^i(M_1 \# M_2, S) & \overset{\cong}{\longrightarrow} & \tilde{H}^i(M_1) \oplus \tilde{H}^i(M_2) \\
\big\uparrow{\scriptstyle f^*} & & \big\uparrow{\scriptstyle \text{id} \oplus f^*} \\
H^i(M_1,S) & \overset{\cong}{\longrightarrow} & \tilde{H}^i(M_1) \oplus H^i(B,S).
\end{array}
$$

Substituting this into the Mayer-Vietoris sequences for $(M_1 \# M_2, S)$ and (M_1, S), we get a commutative diagram, exact in the horizontal directions:

$$
\begin{array}{ccccccccc}
\tilde{H}^i(S) & \longleftarrow & \tilde{H}^i(M_1 \# M_2) & \longleftarrow & \tilde{H}^i(M_1) \oplus \tilde{H}^i(M_2) & \overset{\alpha}{\longleftarrow} & \tilde{H}^{i-1}(S) \\
\big\uparrow{\scriptstyle \text{id}} & & \big\uparrow{\scriptstyle f^*} & & \big\uparrow{\scriptstyle \text{id} \oplus f^*} & & \big\uparrow{\scriptstyle \text{id}} \\
\tilde{H}^i(S) & \longleftarrow & \tilde{H}^i(M_1) & \longleftarrow & \tilde{H}^i(M_1) \oplus H^i(B_2^n, S) & \overset{\beta}{\longleftarrow} & \tilde{H}^{i-1}(S).
\end{array}
$$

From this we get:

Proposition 13.9.7. (a) *If M_1 and M_2 are orientable then $M_1 \# M_2$ is orientable; $f^* : H^n(M_1) \to H^n(M_1 \# M_2)$ is an isomorphism, and for $i \leq n-1$ the following diagram commutes*

---

[30] To simplify notation we reuse the symbol f for various restrictions/corestrictions of the map f.

$$\tilde{H}^i(M_1 \# M_2) \xleftarrow{\cong} \tilde{H}^i(M_1) \oplus \tilde{H}^i(M_2)$$

$$\uparrow{f^*} \qquad\qquad \uparrow{i_1}$$

$$\tilde{H}^i(M_1) \xleftarrow{\cong} \tilde{H}^i(M_1)$$

Here, i_1 is inclusion as the first coordinate.

(b) If M_1 and M_2 are both non-orientable (implying $n \geq 2$), then $M_1 \# M_2$ is non-orientable; $f^* : H^n(M_1) \to H^n(M_1 \# M_2)$ is an isomorphism, and for $i \leq n - 2$ the diagram in Part (a) commutes. For $i = n - 1$, the following diagram commutes and the horizontal lines are exact:

$$H^{n-1}(S) \xleftarrow{j} H^{n-1}(M_1 \# M_2) \xleftarrow{\cong} H^{n-1}(M_1) \oplus H^{n-1}(M_2) \longleftarrow 0$$

$$\uparrow{f^*} \qquad\qquad\qquad \uparrow{i_1}$$

$$0 \longleftarrow H^{n-1}(M_1) \xleftarrow{\cong} H^{n-1}(M_1) \longleftarrow 0$$

Here, $H^{n-1}(S) \cong \mathbb{Z}$ and the image of j is $2H^{n-1}(S)$.   □

Proof. We make some comments, leaving the rest to the reader.

(i) $H^n(M_2)$ is infinite cyclic if M_2 is orientable, and has order 2 if M_2 is non-orientable. Indeed, the map $f^* : H^n(B_2^n, S) \to H^n(M_2)$ can be regarded as id $: \mathbb{Z} \to \mathbb{Z}$ or the epimorphism $\mathbb{Z} \to \mathbb{Z}_2$ in these two cases.

(ii) The homomorphism $\tilde{H}^{n-1}(S) \to H^n(M_1) \oplus H^n(M_2)$ has trivial kernel in the orientable case, and has kernel of index 2 when both M_1 and M_2 are non-orientable.

(iii) There is a cell e_α^n in M_1 such that $f^{-1}f(e_\alpha^n) = e_\alpha^n$. Treating e_α^n as a cocycle the statement about isomorphisms in dimension n becomes clear.

□

Exercises

1. Fill in the details of the proof of 13.9.7.
2. Work through the (suppressed) case (c) of 13.9.7: M_1 orientable and M_2 non-orientable.
3. Which of the hypotheses on L rules out G being finite?
4. Rewrite this section for the cases $d = 1$ and $d = 0$.
5. Prove that if $|L|$ is a $(d-1)$-dimensional homology sphere, then G is a Poincaré Duality group of dimension d (whether or not $|L|$ bounds a compact contractible manifold).
6. Is the example in Sect. 13.8 included in the construction in Sect. 13.9?
7. Give a counterexample to the converse of 16.5.2.

13.10 The case $H^*(G, RG) = 0$

Here we consider what it means for all the cohomology with group ring coefficients to be trivial. An example will be discussed in the next section.

Theorem 13.10.1. *Let G be of type FP_∞ over R and let $H^n(G, RG) = 0$ for all n. Then $\mathrm{cd}_R G = \infty$.*

The proof uses two lemmas from homological algebra; for their proofs the reader is referred to Chapter 8 of [29]:

Lemma 13.10.2. *The cohomological dimension, $\mathrm{cd}_R G$, is the infimum of numbers n such that, for all $i > n$ and all RG-modules M, $H^i(G, M) = 0$. If there is no such number, $\mathrm{cd}_R G = \infty$.* □

Lemma 13.10.3. *If G has type FP_n over R and if $H^n(G, RG) = 0$ then $H^n(G, \Phi) = 0$ for any free RG-module Φ.* □

For the proof of 13.10.1 we also need a long exact sequence in cohomology. Let

$$0 \to M' \to M \to M'' \to 0$$

be an exact sequence of left RG-modules. It is a well-known fact of homological algebra (compare 12.4.3) that when Φ is a free RG-module the sequence

$$0 \to \mathrm{Hom}_{\mathbb{Z}G}(\Phi, M') \to \mathrm{Hom}_{\mathbb{Z}G}(\Phi, M) \to \mathrm{Hom}_{\mathbb{Z}G}(\Phi, M'') \to 0$$

is also exact. In this way the free resolution $\{F_n\}$ gives a short exact sequence of cochain complexes, and hence (see Sect. 2.1) a long exact *Bockstein sequence*

$$0 \to H^0(G, M') \to H^0(G, M) \to H^0(G, M'') \to H^1(G, M') \to \cdots$$

Proof (of 13.10.1). Suppose $\mathrm{cd}_R G = d < \infty$. Then, by 13.10.2, there is a module M such that $H^d(G, M) \neq 0$ while $H^{d+1}(G, K) = 0$ for all modules K. There is a free module Φ mapping onto M; call its kernel K. Applying the Bockstein sequence to

$$0 \to K \to \Phi \to M \to 0$$

we get exactness in:

$$H^d(G, \Phi) \to H^d(G, M) \to H^{d+1}(G, K) \to H^{d+1}(G, \Phi).$$

By 13.10.2 and 13.10.3 this gives $H^{d+1}(G, K) \neq 0$, a contradiction. □

Another consequence of $H^*(G, RG) = 0$ follows from Theorem 13.3.3:

Theorem 13.10.4. *Let G be a group of type F_∞, let X be a $K(G, 1)$-complex of finite type, let $\{K_i\}$ be a finite type filtration of $\tilde{X}$, and let R be a PID. Then $H^*(G, \mathbb{Z}G) = 0$ iff $\{\tilde{H}_k(\tilde{X}^n \stackrel{c}{-} K_i; R)\}$ is pro-trivial for all k.* □

Loosely, this is saying that G is "acyclic at infinity" and that, in view of Theorem 13.10.1, this can only happen when G (of type F_∞) is infinite-dimensional. □

Exercises

1. When G is of type F_∞ and has finite cohomological dimension, prove that the cohomological dimension of $G = \sup\{n \mid H^n(G, \mathbb{Z}G) \neq 0\}$.
2. The Bockstein sequence is given here for cohomology. Write down the details of a homology version.

13.11 An example of $H^*(G, RG) = 0$

We saw in 9.3.19 that Thompson's group F has type F_∞, hence also type FP_∞ over any commutative ring R.

Theorem 13.11.1. $H^*(F, RF) = 0$.

Proof. The proof requires a return to the notation of Sect. 9.3. We sketch it, leaving the reader to fill in the details.[31]

For $b \in B$ let $\lambda(b)$ be the smallest integer such that an expansion of b of length $\lambda(b)$ has the form $[1_F, T]$. Let $\nu(b)$ be the largest integer such that a chain $b_0 < b_1 < \cdots < b_{\nu(b)} = b$ exists in B. Let

$$B(p) = \{b \in B \mid \lambda(b) + \nu(b) \geq p\}.$$

Then $B(p)$ is a directed sub-poset of B, so $|B(p)|$ is contractible. The set $B - B(p)$ is finite, for if $b \notin B(p)$ then, for some b', we have $[1, y_0] < b'$ and $b' > b$ where $\nu(b') < p$; there are only finitely many such b'. For large n, $H^i(F, RF) \cong H^i_c(|B_n|; R) \cong \varinjlim_p H^i(|B_n|, |B_n| \cap |B(p)|; R)$; this follows from 9.3.18 and 12.2.1. By appropriate analogs (for $B(p)$) of 9.3.21 and 9.3.22, given k there is $m(k)$, the same for all p, such that, when $n \geq m(k)$, $|B_n| \cap |B(p)|$ is k-connected.   □

Remark. We saw in 9.2.6 that F has infinite geometric dimension. We get another proof of this by 13.11.1 and 13.10.1.

Remark. The group T of Sect. 9.4 also satisfies $H^*(T, \mathbb{Z}T) = 0$. See [30].

Source Note. This proof of 13.11.1 appears in [30]. The first proof in [33] is quite different.

---

[31] Or see §4F of [30].

Filtered Ends of Pairs of Groups

Proper and CW-proper homotopy theory as described in Chap. 10 can be regarded as the homotopy theory of maps which preserve a finite (or finite type) filtration. In this chapter we introduce a generalization in which the filtration is by complexes which are not necessarily of finite type. Although all the main ideas have already been seen in our discussion of proper homotopy, there is need for an exposition of the foundations of the generalized theory. The corresponding homology and cohomology theories of filtered ends are discussed. The immediate occasion is our discussion of an alternative way of counting ends of pairs of groups. That appears in Sect. 14.5.

14.1 Filtered homotopy theory

A *topological filtration* of a space X is a family $\{K_i\}_{i \in \mathbb{Z}}$ of subsets of X satisfying $K_i \subset K_{i+1}$ for all i, $\bigcup_i K_i = X$, and $\bigcap_i K_i = \emptyset$. Because the indexing directed set is $\mathbb{Z}$, the filtering family $\{K_i\}$ and the family of complements $\{X - K_i\}$ can both be viewed as either inverse sequences or direct sequences. The pair $(X, \{K_i\})$ is a *filtered space*. A *filtered map* is a map $f : X \to Y$ which induces morphisms of pro-Spaces and of ind-Spaces with respect to both of these families. In detail, f is a filtered map iff

1. (pro-map with respect to $\{K_i\}$): $\forall i$, $\exists j$ such that $f(K_j) \subset L_i$;
2. (pro-map with respect to $\{X - K_i\}$): $\forall i$, $\exists j$ such that $f(X - K_j) \subset Y - L_i$;
3. (ind-map with respect to $\{K_i\}$): $\forall i$, $\exists j$ such that $f(K_i) \subset L_j$;
4. (ind-map with respect to $\{X - K_i\}$): $\forall i$, $\exists j$ such that $f(X - K_i) \subset Y - L_j$.

A filtered space $(X, \mathcal{K})$ is *topologically well filtered* if for each i there exist j and k such that $\mathrm{cl}_X K_k \subset \mathrm{int}_X K_i \subset \mathrm{cl}_X K_i \subset \mathrm{int}_X K_j$. In that case $\bigcap_i \mathrm{cl}\, K_i = \emptyset$ and $\bigcup_i \mathrm{int}\, K_i = X$.

Lemma 14.1.1. *Let $(X, \mathcal{K})$ be a topologically well filtered space and let C be a compact subset of X. There exist i and j such that $C \subset \operatorname{int} K_i$ and $C \subset X - \operatorname{cl} K_j$.*   □

If X is Hausdorff and $\mathcal{K}$ consists of compact sets then some K is empty (see Exercises). Using this we see that filtered maps generalize proper maps:

Proposition 14.1.2. *Let $(X, \mathcal{K})$ and $(Y, \mathcal{L})$ be topologically well filtered Hausdorff spaces where each K_i and each L_j is compact. A map $f : X \to Y$ is filtered iff it is proper.*

Proof. Use Lemma 14.1.1 and the fact that compact subsets of Hausdorff spaces are closed. Note that two of the four conditions in the definition of filtered map hold trivially.   □

If $(X, \mathcal{K})$ is a topologically well filtered space, $(X \times I, \mathcal{K} \times I)$ is a topologically well filtered space where $\mathcal{K} \times I := \{K_i \times I\}$. A *filtered homotopy* is a filtered map $(X \times I, \mathcal{K} \times I) \to (Y, \mathcal{L})$. The definitions of *filtered homotopic relative to a subspace, filtered homotopy equivalence, filtered homotopy inverse,* etc., are analogous to the proper case.

We now turn to CW complexes, WHICH IN THIS SECTION ARE ALWAYS ASSUMED TO HAVE LOCALLY FINITE TYPE. Recall from Sect. 11.4 that a "filtration" on a CW complex X is a topological filtration of X by subcomplexes[1] and that if A is a subcomplex of X its CW neighborhood $N(A)$ or $N_X(A)$ is the union of all cell carriers which meet A. A *well filtered CW complex* is a filtered CW complex $(X, \mathcal{K})$ with the property that for each n and i there exist j and k such that $N_{X^n}(K_k^n) \subset K_i$ and $N_{X^n}(K_i^n) \subset K_j$. It follows that for each n the filtered space $(X^n, \{K_i^n\})$ is topologically well filtered. Note that if $\mathcal{K}$ is a finite type filtration then, by 1.5.5 and 11.4.4, $(X, \mathcal{K})$ is a well filtered CW complex.

If $(X, \mathcal{K})$ and $(Y, \mathcal{L})$ are well filtered CW complexes, a *CW-filtered map* $f : (X, \mathcal{K}) \to (Y, \mathcal{L})$ is a map $f : X \to Y$ such that for each n there exists k such that $f(X^n) \subset Y^k$ and $f| : (X^n, \mathcal{K}^n) \to (Y, \mathcal{L})$ is a filtered map, where $\mathcal{K}^n := \{K_i^n \mid K_i \in \mathcal{K}\}$. Referring back to the topological definitions, we find a condition $f(X - K_i) \subset Y - L_j$, whereas in the context of CW complexes it would be more natural to encounter the condition $f(X \stackrel{c}{-} K_i) \subset Y \stackrel{c}{-} L_j$. The reader can check that for well filtered CW complexes the definitions are not changed by substituting CW complement for complement.

By analogy with 10.2.3 we have:

Theorem 14.1.3. (CW-Filtered Cellular Approximation Theorem)
Let $f : (X, \mathcal{K}) \to (Y, \mathcal{L})$ be a filtered map between well filtered CW complexes, and let A be a subcomplex of X on which f is cellular. Then f is filtered homotopic, rel A, to a CW-filtered cellular map.   □

---

[1] Though, as promised in a footnote in Sect. 11.4, the filtrations here are indexed by $\mathbb{Z}$.

The terms *CW-filtered homotopy*, etc., are defined in the obvious way.

Two topological filtrations $\mathcal{K}$ and $\mathcal{L}$ of a space X are *equivalent* if each member of $\mathcal{K}$ lies in a member of $\mathcal{L}$ and vice versa. It then follows that the "identity map" $\mathrm{id}_X : (X, \mathcal{K}) \to (X, \mathcal{L})$ is filtered and is therefore a filtered homotopy equivalence.

Given a topological filtration $\mathcal{K}$ of a CW complex X, we sometimes need an equivalent filtration of X by subcomplexes. There are two reasonable candidates. Define the *CW envelope* of a subset A of X to be the smallest subcomplex $E(A)$ containing A, and define the *CW envelope* of $\mathcal{K}$ to be $E(\mathcal{K}) := \{E(K_i) \mid K_i \in \mathcal{K}\}$. Define the CW *spine* of $A \subset X$ to be the largest subcomplex $S(A)$ lying in A and define the CW *spine* of $\mathcal{K}$ to be $S(\mathcal{K}) := \{S(K_i) \mid K_i \in \mathcal{K}\}$.

Proposition 14.1.4. *If X is a strongly locally finite CW complex and $(X, \mathcal{K})$ is topologically well filtered then $E(\mathcal{K})$ and $S(\mathcal{K})$ are filtrations of X.*

Proof. The only non-trivial part in the envelope case involves showing that $\bigcap_i E(K_i) = \emptyset$. Suppose otherwise. Since this intersection is a subcomplex of X it contains a vertex v. It is not hard to see that $E(K_i) = \bigcup\{C(e) \mid e \cap K_i \neq \emptyset\}$. Thus for every i there is a cell e_i such that $e_i \cap K_i \neq \emptyset$ and v is a vertex of $C(e_i)$. Since X is strongly locally finite there can only be finitely many different carriers containing v, so $D \bigcup_i C(e_i)$ is compact. By 14.1.1, $D \subset X - \mathrm{cl}\, K_j$ for some j. But $D \cap K_i \neq \emptyset$ for all i, which is a contradiction. The proof for $S(\mathcal{K})$ is similar. $\square$

We are assuming that our CW-complexes have locally finite type, so usually 14.1.4 will be applied to a skeleton (which is strongly locally finite by 10.1.12).

When $(X, \mathcal{K})$ is as in 14.1.4, i.e., X is strongly locally finite and $(X, \mathcal{K})$ topologically well filtered, we say that $\mathcal{K}$ is *CW-compatible* if $(X, E(\mathcal{K}))$ is a well filtered CW complex and $E(\mathcal{K})$ is equivalent to $\mathcal{K}$. Clearly, this happens iff $S(\mathcal{K})$ is equivalent to $\mathcal{K}$. To illustrate this we give two examples involving the following general set-up. We are given a map $h : X \to M$ where $\mathcal{L}$ is a topological filtration of the space M; then $h^{-1}\mathcal{L} := \{h^{-1}(L_i) \mid L_i \in \mathcal{L}\}$ is a topological filtration of X, and $(X, h^{-1}\mathcal{L})$ is topologically well filtered whenever $(M, \mathcal{L})$ is well filtered.[2]

Example 14.1.5. Let X be a rigid G-CW complex such that $G\backslash X$ is finite, let $M = \mathbb{R}$ where a left action of G on $\mathbb{R}$ by translations is given, let $\mathcal{L} = \{L_i\}$ where $i \in \mathbb{Z}$ and $L_i := (-\infty, i)$, and let $h : X \to \mathbb{R}$ be a G-map. Then $h^{-1}\mathcal{L}$ is CW-compatible. A particular case is this. Assume G has type F_n, X is the n-skeleton of the universal cover of a $K(G, 1)$-complex which has finite

---

[2] Note that even if $(M, \mathcal{L})$ is a well filtered CW complex and h is a cellular map, the spaces $h^{-1}(L_i)$ might not be subcomplexes of X.

n-skeleton, $\chi : G \to \mathbb{R}$ is a *character* (i.e., a homomorphism into the additive group of real numbers), and the G-action[3] on $\mathbb{R}$ is $g.r = r + \chi(g)$. Pick a vertex v_i in each G-orbit of vertices, define $h(v_i) = 0$, and extend h equivariantly to map X^0 into $\mathbb{R}$; then (since the action on X is free and $\mathbb{R}$ is contractible) one can proceed skeleton by skeleton to define a G-map $h : X \to \mathbb{R}$. The filtered homotopy theory in this case leads to the Sigma invariants of G; this will be developed in Sect. 18.3.

Example 14.1.6. Let $p : X \to Y$ be a covering projection with $(Y, \mathcal{L})$ a well filtered CW complex. Assume X has the CW structure such that p maps each open cell homeomorphically onto an open cell. Then $(X, p^{-1}\mathcal{L})$ is a well filtered CW complex, so, trivially, $p^{-1}\mathcal{L}$ is CW-compatible. For example, given a pair of groups (G, H) and a free action of G on a path connected graph X such that $G \backslash X$ is finite, choose a finite filtration $\mathcal{L}$ of $H \backslash X$. The well filtered graphs $(H \backslash X, \mathcal{L})$ and $(X, p^{-1}\mathcal{L})$ yield two notions of the "number of ends of the pair of groups (G, H)," where $p : X \to H \backslash X$ is the quotient covering projection. The first of these was discussed in Sect. 13.5; the second will be discussed in Sect. 14.5.

Next, we give a practical way of recognizing filtered homotopy equivalences (Theorem 14.1.8). For this we need a new concept.

A cellular map $f : X \to Y$ between CW complexes is *CW-Lipschitz* if for each n there exists m such that for every cell e of X^n the carrier $C(f(e))$ has at most m cells. Two cellular maps $X \to Y$ are *CW-Lipschitz homotopic* if there is a *CW*-Lipschitz homotopy $X \times I \to Y$ between them. A *CW-Lipschitz homotopy equivalence* is a cellular map $f : X \to Y$ for which there exists a cellular CW-Lipschitz map $g : Y \to X$ so that $g \circ f$ and $f \circ g$ are CW-Lipschitz homotopic to the appropriate identity maps. Here is an important source of examples:

Example 14.1.7. If X and Y are finite connected CW complexes and $f : X \to Y$ is a cellular map, then any lift to universal covers $\tilde{f} : \tilde{X} \to \tilde{Y}$ is CW-Lipschitz. Hence if f is a homotopy equivalence, $\tilde{f}$ is a CW-Lipschitz homotopy equivalence.

Theorem 14.1.8. *Let $f : (X, \mathcal{K}) \to (Y, \mathcal{L})$ be a CW-filtered map between well filtered CW complexes. If f is a CW-Lipschitz homotopy equivalence then f is a CW-filtered homotopy equivalence.*

In particular, applying 14.1.2, this gives a useful condition for a map to be a proper homotopy equivalence.

For the proof we need two lemmas.

Lemma 14.1.9. *Let $(X, \mathcal{K})$ be a well filtered CW complex and let $F : X \times I \to X$ be a (cellular) CW-Lipschitz homotopy with $F_0 = \mathrm{id}_X$. Then F is CW-filtered.*

---

[3] i.e., identify $\mathbb{R}$ with the group $\mathrm{Transl}(\mathbb{R})$ of translations of $\mathbb{R}$, so that $\chi : G \to \mathrm{Transl}(\mathbb{R})$ is an action of G on $\mathbb{R}$ by translations.

Proof. We need only check that $F \mid X^n \times I$ is filtered for given n. Let m be such that for every cell e of X^n the carrier $C(F(e \times I))$ contains at most m cells. Since $F(e \times I)$ is path connected and e is a cell of this carrier, $F(e \times I) \subset N_{X^n}^m(C(e))$. Thus for all i we have[4] $F(K_i \times I) \subset N_{X^n}^m(K_i)$ and

$$F((X^n \stackrel{c}{-} K_i) \times I) \subset N_{X^n}^m(X^m \stackrel{c}{-} K_i).$$

Since $(X, \mathcal{K})$ is well-filtered, there exists j such that $N_{X^n}^m(K_i) \subset K_j$. Moreover, given i, this j satisfies

$$F((X^n \stackrel{c}{-} K_j) \times I) \subset N_{X^n}^m(X \stackrel{c}{-} K_j) \subset X \stackrel{c}{-} K_i.$$

The verification of the remaining conditions for a map to be filtered is similar.
$\square$

Lemma 14.1.10. *Let $f : (X, \mathcal{K}) \to (Y, \mathcal{L})$ be a (cellular) CW-filtered map and let $g : Y \to X$ be a cellular map such that $g \circ f$ and $f \circ g$ are CW-filtered maps which are CW-filtered homotopic to the appropriate identity maps. Then g is a CW-filtered map.*

Proof. Given L_i we seek K_j such that $g(L_i) \subset K_j$. There exists k such that $fg(L_i) \subset L_k$ and there exists j such that $f(X - K_j) \subset Y - L_k$. So $g(L_i) \subset K_j$. Similarly, given K_i we seek L_j such that $g(Y \stackrel{c}{-} L_j) \subset X - K_i$. There exists k such that $f(K_i) \subset L_k$ and there exists j such that $fg(Y - L_j) \subset Y - L_k$. So $g(Y - L_j) \subset X - K_i$. The rest of the proof is similar. $\square$

Proof (of Theorem 14.1.8). Let g be a CW-Lipschitz homotopy inverse for f. By 14.1.8 $g \circ f$ and $f \circ g$ are CW-filtered homotopic to the appropriate identity maps. Hence, by 14.1.10, g is a CW-filtered map. $\square$

Recall from Sect. 2.7 the definition of "n-equivalence." There are analogs of "CW-filtered homotopy equivalence" and "CW-Lipschitz homotopy equivalence" in which the phrase "homotopy equivalence" is replaced by "n-equivalence." The definitions are obvious.

Addendum 14.1.11. *Theorem 14.1.8 remains true if "homotopy equivalence" is everywhere replaced by "n-equivalence."* $\square$

Here is an application of 14.1.11. Consider the "particular case" discussed in Example 14.1.5. An action of G on $\mathbb{R}$ by translations is given, i.e., a homomorphism $\chi : G \to \mathbb{R}$ (identified with $\text{Transl}(\mathbb{R})$) and G is known to have type F_n. We choose an n-dimensional $(n-1)$-connected rigid G-CW complex X which is finite mod G, and we have a G-map $h : X \to \mathbb{R}$ (either given to us or constructed as in 14.1.5). The space $\mathbb{R}$ is filtered by $\mathcal{L} = \{L_i := (-\infty, i)\}_{i \in \mathbb{Z}}$.

Proposition 14.1.12. *If $h_1 : X_1 \to \mathbb{R}$ and $h_2 : X_2 \to \mathbb{R}$ both satisfy these conditions (for X and h above), there is a cellular map $f : X_1 \to X_2$ which is a CW-filtered $(n-1)$-equivalence $(X_1, h_1^{-1}\mathcal{L}) \to (X_2, h_2^{-1}\mathcal{L})$.*

---

[4] Recall from Sect. 1.5 that $N^m(A)$ denotes $N(N^{m-1}(A))$ and $N^1(A) = N(A)$.

Proof. By 7.1.8 and the n-equivalence version of Example 14.1.7, there is a CW-Lipschitz (cellular) $(n - 1)$-equivalence $f : X_2 \to X_1$, and f is CW-filtered as a map $(X_2, f^{-1}h_1^{-1}\mathcal{L}) \to (X_1, h_1^{-1}\mathcal{L})$. As explained in Example 14.1.5, these are CW-compatible topological filtrations, so we can proceed as if we had well-filtered CW complexes and conclude that f is a CW-filtered $(n - 1)$-equivalence. It only remains to show that $f^{-1}h_1^{-1}\mathcal{L}$ and $h_2^{-1}\mathcal{L}$ are equivalent filtrations. Since both h_2 and $h_1 \circ f$ are G-maps from $X_2 \to \mathbb{R}$, and X_2 is finite mod G, it follows that there is a finite upper bound to the set $\{|h_2(x) - h_1 \circ f(x)| \mid x \in X\}$. This implies the two filtrations are equivalent. $\square$

The last proposition shows that those algebraic topology invariants of $(X, h^{-1}\mathcal{L})$ which only depend on the n-skeleton are invariants of the character $\chi : G \to \mathbb{R}$. This discussion is continued in Sect. 18.3.

Source Note. The idea of filtered homotopy theory as a useful generalization of proper homotopy theory appears in [60].

Exercises

1. Let $(X, \mathcal{K})$ be a filtered space where each K_i is compact. Show that if X is Hausdorff then some K_i is empty.
2. Give an example of subcomplexes $K \subset L$ of X where $K \subset \text{int}_X L$ but $N_X(K) \not\subset L$.
3. Prove that if the filtration $\{K_i\}$ on a CW-complex X of locally finite type makes $(X^n, \{K_i^n\})$ into a topologically well filtered space for each n, then $(X, \mathcal{K})$ is a well filtered CW complex. (*Hint:* Use 1.5.1).
4. Show that if $f : (X, \mathcal{K}) \to (Y, \mathcal{L})$ is a filtered map between topologically well filtered Hausdorff spaces and if each $K \in \mathcal{K}$ is compact, then f is a proper map.
5. Let A be a subset of the CW complex X. Show that CW envelope and CW spine are related by: $E(A) = S(A) \cup \bigcup\{C(e) \mid e \text{ is a cell of } X, \mathring{e} \cap A \neq \emptyset \text{ and } C(e) \not\subset A\}$.

14.2 Filtered chains

Let X be an oriented CW complex of locally finite type, let $\mathcal{K}$ be a filtration of X so that $(X, \mathcal{K})$ is a well filtered CW-complex, and let $\mathcal{A}$ index the n-cells of X. An infinite cellular n-chain $c = \sum_{\alpha \in \mathcal{A}} r_\alpha e_\alpha^n$ is *locally finite with respect to* $\mathcal{K}$ if for each i, the coefficient $r_\alpha \in R$ is non-zero for only finitely many cells e_α^n in K_i. Since $(X, \mathcal{K})$ is well filtered, ∂c is also locally finite with respect to $\mathcal{K}$. We define the R-module $C_n^{\mathcal{K}}(X; R)$ to be the submodule of $C_n^\infty(X; R)$ consisting of chains which are locally finite with respect to $\mathcal{K}$. Elements of $C_n^{\mathcal{K}}(X; R)$ are *filtered locally finite n-chains* in X (with respect to $\mathcal{K}$). They form a chain complex whose homology modules are denoted $H_*^{\mathcal{K}}(X; R)$, the *filtered locally finite homology* modules of $(X, \mathcal{K})$.

Filtered maps induce chain maps on filtered locally finite chains. There is a theory of $H_*^{\mathcal{K}}(X;R)$, analogous to that of Sect. 11.1 for $H_*^{\infty}(X;R)$, giving a covariant functor $H_*^{\cdot}(\cdot;R)$ on the filtered homotopy category.

If A is a subcomplex of X there is a short exact sequence

$$0 \to C_*^{\mathcal{K}|}(A;R) \to C_*^{\mathcal{K}}(X;R) \to C_*^{\mathcal{K}}(X;R)/C_*^{\mathcal{K}|}(A;R) \to 0$$

as in Sect. 11.1, from which the usual homology exact sequence for $H_*^{\cdot}(\cdot;R)$ follows directly. And, just as in Sect. 11.4, there is a short exact sequence

$$0 \to C_*(X;R) \to C_*^{\mathcal{K}}(X;R) \to C_*^{\mathcal{K}}(X;R)/C_*(X;R) \to 0$$

which leads us to define the *homology modules of the $\mathcal{K}$-end of X*: $H_{n-1}^{\mathcal{K},e}(X;R)$ denotes the nth homology of $C_*^{\mathcal{K}}/C_*$. There is a corresponding homology exact sequence analogous to that in 11.4.1:

$$\cdots \longrightarrow H_n(X;R) \xrightarrow{i_*} H_n^{\mathcal{K}}(X;R) \xrightarrow{p_*} H_{n-1}^{\mathcal{K},e}(X;R) \xrightarrow{\partial_*} H_{n-1}(X;R) \longrightarrow \cdots$$

Note that $H_{-1}^{\mathcal{K},e}(X;R)$ can fail to be zero.

We saw in Sect. 14.1 that the inclusions $X^n \stackrel{c}{\cap} K_{i-1} \hookrightarrow X^n \stackrel{c}{\cap} K_i$ define an inverse sequence and a direct sequence. The inverse sequence plays the role previously played by a basis for the neighborhoods of the end in Sect. 11.4. By analogy with 11.4.7 and 11.4.8 we have:

Proposition 14.2.1. *There are short exact sequences*

$$0 \to \varprojlim_{i \geq j}{}^1 \{H_{n+1}(X, X \stackrel{c}{\cap} K_i;R)\} \xrightarrow{\bar{a}} H_n^{\mathcal{K}}(X \stackrel{c}{\cap} K_j;R) \xrightarrow{\bar{b}} \varprojlim_{i \geq j}\{H_n(X, X \stackrel{c}{\cap} K_i;R) \to 0$$

$$0 \to \varinjlim_{j \to -\infty} \varprojlim_{i \geq j}{}^1 H_{n+1}(X, X \stackrel{c}{\cap} K_i;R) \to H_n^{\mathcal{K}}(X;R) \to \varinjlim_{j \to -\infty} \varprojlim_{i \geq j} H_n(X, X \stackrel{c}{\cap} K_i;R) \to 0$$

and, for any j:

$$0 \to \varprojlim_{i \geq j}{}^1 \{H_{n+1}(X \stackrel{c}{\cap} K_i;R)\} \xrightarrow{a} H_n^{\mathcal{K},e}(X;R) \xrightarrow{b} \varprojlim_{i \geq j}\{H_n(X \stackrel{c}{\cap} K_i;R) \to 0.$$

$\square$

If $\mathcal{K}$ is a finite type filtration then $H_*^{\mathcal{K}}(X;R)$ and $H_*^{\mathcal{K},e}(X;R)$ reduce to $H_*^{\infty}(X;R)$ and $H_*^e(X;R)$ respectively.

Turning to cohomology, an infinite cellular n-chain $c = \sum_{\alpha \in \mathcal{A}} r_\alpha e_\alpha^n$ is *locally cofinite with respect to $\mathcal{K}$* if c is supported by some K_i (i.e., $r_\alpha = 0$ when e_α^n is not a cell of K_i). Since $(X, \mathcal{K})$ is well filtered, δc is also locally cofinite with respect to $\mathcal{K}$. We define the R-module $C_n^{\mathcal{K},\mathrm{cof}}(X;R)$ to be the submodule of $C_n^\infty(X;R)$ consisting of chains which are locally cofinite with respect to $\mathcal{K}$. Elements of $C_n^{\mathcal{K},\mathrm{cof}}(X;R)$ are *filtered locally cofinite n-chains* in X (with respect to $\mathcal{K}$). They form a cochain complex whose cohomology modules are denoted $H_{\mathcal{K}}^*(X;R)$, the *filtered locally cofinite cohomology* modules of $(X,\mathcal{K})$.

Filtered maps induce cochain maps on $C_*^{\cdot,\mathrm{cof}}$. There is a theory of $H_{\mathcal{K}}^*(X;R)$ analogous to that of Sect. 12.1 for $H_f^*(X;R)$. Just as in Sect. 12.2, the short exact sequence

$$0 \to C_*^{\mathcal{K},\mathrm{cof}}(X;R) \to C_*^\infty(X;R) \to C_*^\infty(X;R)/C_*^{\mathcal{K},\mathrm{cof}}(X;R) \to 0$$

leads us to define the *cohomology of the $\mathcal{K}$-end* of X: $H_{\mathcal{K},e}^n(X;R)$ denotes the nth cohomology of $C_*^\infty/C_*^{\mathcal{K},\mathrm{cof}}$, from which follows an exact sequence analogous to that in Sect. 12.2:

$$\leftarrow H^{n+1}(X;R) \xleftarrow{i^*} H_{\mathcal{K}}^{n+1}(X;R) \xleftarrow{\delta^*} H_{\mathcal{K},e}^n(X;R) \xleftarrow{p^*} H^n(X;R) \leftarrow \cdots .$$

By analogy with 12.2.1 and 12.2.2 we have:

Proposition 14.2.2. *There are isomorphisms* $\varinjlim_{i\geq 0}\{H^n(X, X \xrightarrow{c} K_i;R)\} \to$ $H_{\mathcal{K}}^n(X;R)$ *and* $\varinjlim_{i\geq 0}\{H^n(X \xrightarrow{c} K_i;R)\} \to H_{\mathcal{K},e}^n(X;R).$ $\square$

If $\mathcal{K}$ is a finite type filtration then $H_{\mathcal{K}}^*(X;R)$ and $H_{\mathcal{K},e}^*(X;R)$ reduce to $H_f^*(X;R)$ and $H_e^*(X;R)$ respectively.

For reference, we summarize the invariance properties:

Proposition 14.2.3. *The homology and cohomology theories* $H_*^{\mathcal{K}}(X;R)$, $H_*^{\mathcal{K},e}(X;R)$, $H_{\mathcal{K}}^*(X;R)$ *and* $H_{\mathcal{K},e}^*(X;R)$ *are filtered homotopy invariants.* $\square$

The more complete version of 14.2.3 in which the categories and functors are described explicitly is left to the reader.

Remark 14.2.4. A filtered map $f : (X,\mathcal{K}) \to (Y,\mathcal{L})$ between well filtered CW complexes is *properly filtered* if for each n and each $K_i \in \mathcal{K}$ the restriction $f \mid K_i : K_i^n \to Y$ is a proper map. All the definitions in filtered homotopy theory have properly filtered analogs, and there is a Properly Filtered Cellular Approximation Theorem analogous to 14.1.3. One can define *filtered locally finite cohomology* [resp. *filtered locally cofinite homology*] by using the coboundary δ on filtered locally finite chains [resp. the boundary ∂ on filtered locally cofinite chains] and the resulting modules are invariants of properly filtered homotopy theory rather than of filtered homotopy theory. When X is an oriented manifold these arise as the Poincaré duals of $H_*^{\mathcal{K}}(X;R)$ and $H_{\mathcal{K}}^*(X;R)$ – see Exercise 5 of Sect. 15.2.

Example 14.2.5. We continue the discussion of Example 14.1.5. In Sect. 13.2 we met the right RG-module $(RG)\hat{}$. Elements of $(RG)\hat{}$ are written as $\sum_{g \in G} r_g g$ with each $r_g \in R$. Those where all but finitely many r_g are zero form the ring RG, but the ring multiplication given in Sect. 8.1 does not extend to make the module $(RG)\hat{}$ into a ring. However, any non-zero character $\chi : G \to \mathbb{R}$ defines

$$(RG)\hat{}_\chi := \left\{ \sum_{g \in G} r_g g \mid \text{for any } t \geq 0, r_q \neq 0 \text{ for only} \right.$$

$$\left. \text{finitely many } g \text{ with } \chi(g) \leq t \right\}.$$

One checks that the multiplication in RG does extend to $(RG)\hat{}_\chi$, making the latter a ring called the *Novikov ring* defined by χ. Regarding this ring as a right G-module (via right multiplication by elements of G) and using the notation of 14.1.5, it is clear that the homology of the chain complex $\{(RG)\hat{}_\chi \otimes_G C_*(\tilde{X}; R), \mathrm{id} \otimes \tilde{\partial}\}$ is canonically isomorphic to $H_*^{h^{-1}\mathcal{L}}(\tilde{X}; R)$. This is often written as $H_*(X; (RG)\hat{}_\chi)$.

Exercises

1. Work through Sects. 14.1 and 14.2 for the case where $X = K_i$ for some i.
2. What is the correct analog of 11.1.3 for filtered locally finite homology?
3. Give an example where $H_{-1}^{\mathcal{K},e}(X; R)$ is non-zero.
4. Give an example of a non-well-filtered $(X, \mathcal{K})$ with $\mathcal{K}$ a filtration of X, and a chain c which is locally finite with respect to $\mathcal{K}$ while ∂c is not locally finite with respect to $\mathcal{K}$.
5. Prove 14.2.1 and 14.2.2.

14.3 Filtered ends of spaces

The theory of ends of CW complexes described in Sect. 13.4 generalizes to well filtered CW complexes. We explain this here, and will apply it to group theory in the next sections.

Let $(Y, \mathcal{L})$ be a well filtered path connected CW complex of locally finite type. A *filtered ray* in $(Y, \mathcal{L})$ is a filtered map $\omega : [0, \infty) \to Y$ where $[0, \infty)$ has the filtration[5] $\{[0, i]\}$. Two filtered rays *define the same filtered end* of $(Y, \mathcal{L})$ if their restrictions to $\mathbb{N} \subset [0, \infty)$ are filtered homotopic. This is an equivalence relation, and an equivalence class is called a *filtered end* of $(Y, \mathcal{L})$. We indicate how the propositions in Sect. 13.4 generalize to this setting.

---

[5] for $i \in \mathbb{Z}$, $[0, i]$ is empty when $i < 0$.

A filtered map induces a function between the corresponding sets of filtered ends. Indeed there is an obvious covariant functor from the category of well filtered $(Y, \mathcal{L})$'s and filtered homotopy classes to the category Sets which sends $(Y, \mathcal{L})$ to the set of filtered ends of $(Y, \mathcal{L})$. A filtered 1-equivalence induces a bijection on filtered ends. In particular, by 14.1.3, the inclusion $Y^1 \hookrightarrow Y$ induces a bijection of filtered ends.

If ω is a filtered ray there is a corresponding point of $\varprojlim_i \{\pi_0(Y \overset{c}{-} L_i)\}$

whose i^{th} entry is the path component of $Y \overset{c}{-} L_i$ containing all but a compact subset of the image of ω. This depends only on the filtered end defined by ω and establishes:

Proposition 14.3.1. *This association defines a functorial (with respect to filtered homotopy classes) bijection between the set of filtered ends of $(Y, \mathcal{L})$ and the set $\varprojlim_i \{\pi_0(Y \overset{c}{-} L_i)\}$. In particular, the set of filtered ends of $(Y, \mathcal{L})$ is non-empty iff there is a nested sequence of non-empty subcomplexes $U_1 \supset U_2 \supset \cdots$ with U_i a path component of $Y \overset{c}{-} L_i$.*   □

By analogy with the group theoretic case we say that an inverse sequence $\{S_i\}$ of sets is *semistable* if for each i there exists $j \geq i$ such that for all $k \geq j$ the image of S_k in S_i equals the image of S_j in S_i. As with 11.2.1, $\{S_i\}$ is semistable iff it is isomorphic in pro-(Sets) to an inverse sequence whose bonds are surjections. In the situation of 13.4.7 the inverse sequence $\{\pi_0(Y \overset{c}{-} L_i)\}$ consisted of finite sets and hence was semistable. But that need not be the case here, so we must discuss the condition under which semistability holds.

A subset A of Y is $\mathcal{L}$-*bounded* if $A \subset L_i$ for some i, and is $\mathcal{L}$-*unbounded* otherwise. To a certain extent $\mathcal{L}$-bounded subcomplexes of Y play the same role that finite subcomplexes play in the proper case (by 14.1.1, finite subcomplexes are $\mathcal{L}$-bounded) but with the important difference that the CW complement of an $\mathcal{L}$-bounded subcomplex may be $\mathcal{L}$-unbounded even though all its path components are $\mathcal{L}$-bounded; in that case, although Y is $\mathcal{L}$-unbounded, $(Y, \mathcal{L})$ has no filtered ends. We say that $(Y, \mathcal{L})$ is *regular* if for each i the union of all $\mathcal{L}$-bounded path components of $Y \overset{c}{-} L_i$ is $\mathcal{L}$-bounded.

Proposition 14.3.2. $(Y, \mathcal{L})$ *is regular iff the inverse sequence of sets* $\{\pi_0(Y \overset{c}{-} L_i)\}$ *is semistable.*

Proof. For each i, we write $\pi_0(Y \overset{c}{-} L_i) = B_i \cup U_i$ where B_i is the set of $\mathcal{L}$-bounded path components and U_i is the set of $\mathcal{L}$-unbounded path components. If $(Y, \mathcal{L})$ is regular then for each i there exists j such that the elements of B_i are not in the image of $B_j \cup U_j \to B_i \cup U_i$, while every element of U_i is in the image of $B_k \cup U_k$ for all $k \geq I$; then U_i is the image of $B_k \cup U_k$ for all $k \geq j$, and that implies semistability. Conversely, assume semistability. Then, given i, there exists j such that for all $k \geq j$ the image of $B_k \cup U_k \to B_i \cup U_i$ is the same. We claim that this image is disjoint from B_i. To see this, note that, for $k \geq i$, U_k is mapped into U_i, and each member of B_i (i.e., each $\mathcal{L}$-bounded path component of $Y \stackrel{c}{-} L_i$) is disjoint from $Y \stackrel{c}{-} L_m$ for some $m \geq j$ dependent on i; so it is not in the image of $B_m \cup U_m \to B_i \cup U_i$, which is the same as the image of $B_j \cup U_j \to B_i \cup U_i$. The claim implies each $\mathcal{L}$-bounded path component of $Y \stackrel{c}{-} L_i$ lies in L_j, and that is the meaning of regularity. $\square$

It follows that regularity is preserved by filtered 1-equivalences. Here is an example which is not regular:

Example 14.3.3. Let Y be the subset of $\mathbb{R}^2$ consisting of all points (x, y) such that either $y = 0$ and $x \geq 0$, or $x \in \mathbb{N}$ and $0 \leq y \leq x$. If we take points (x, y) with integer entries as vertices and the obvious closed intervals of length 1 as edges, Y becomes a locally finite graph. Let $L_i = \{(x, y) \in Y \mid y \leq i\}$, and write $\mathcal{L} = \{L_i\}$. Then $(Y, \mathcal{L})$ is well filtered, $Y \stackrel{c}{-} L_i$ is $\mathcal{L}$-unbounded for all i, and each path component of $Y \stackrel{c}{-} L_i$ is $\mathcal{L}$-bounded.

The *number of filtered ends* of $(Y, \mathcal{L})$ is $m \geq 0$ [resp. is ∞] if the set of filtered ends has m members [resp. is infinite]. For $0 \leq m \leq \infty$, we also say that $(Y, \mathcal{L})$ *has m filtered ends*. By 14.3.1 this number is preserved by filtered 0-equivalences.

An inverse sequence $\{S_i\}$ of sets is *stable* if there is a cofinal subsequence $\{S_{n_i}\}$ such that image $(S_{n_{i+2}} \to S_{n_{i+1}})$ is mapped bijectively onto image $(S_{n_{i+1}} \to S_{n_i})$ for all i. "Stable" implies "semistable." If $\{S_i\}$ is stable then $\varprojlim_i \{S_i\}$ has the same cardinal number as image $(S_{n_{i+2}} \to S_{n_{i+1}})$ for any i. If $\{S_i\}$ is semistable but not stable then $\varprojlim_i \{S_i\}$ is infinite.

By 14.3.1 and 14.3.2 we have:

Proposition 14.3.4. *Let $(Y, \mathcal{L})$ be regular and let $m < \infty$. The number of filtered ends of $(Y, \mathcal{L})$ is m iff $\{\pi_0(Y \stackrel{c}{-} L_i)\}$ is stable with inverse limit of cardinality m.* $\square$

Analogous to 13.4.8 is:

Corollary 14.3.5. *Let $(Y, \mathcal{L})$ be regular with m filtered ends where $0 \leq m \leq \infty$. Then $m = 0$ iff Y is $\mathcal{L}$-bounded. When m is finite, there exists i_0 such that whenever an $\mathcal{L}$-bounded subcomplex K contains L_{i_0} then $Y \stackrel{c}{-} K$ has exactly m $\mathcal{L}$-unbounded path components. When $m = \infty$, the number of $\mathcal{L}$-unbounded path components of $Y \stackrel{c}{-} L_i$ is a weakly monotonic unbounded function of i, whose value may be ∞ for some (finite) i.* $\square$

Addendum 14.3.6. *Let m be finite and let $i \geq i_0$. If $(Y, \mathcal{L})$ is essentially 0-connected[6] there is a path connected $\mathcal{L}$-bounded subcomplex K of Y containing L_i such that $Y \stackrel{c}{-} K$ has exactly m path components.*

---

[6] See Sect. 7.4.

Proof. Similar to that of 13.4.9; essential 0-connectedness ensures we can enlarge L_i to make an $\mathcal{L}$-bounded path connected subcomplex K'. $\square$

The analog of 13.4.11 is:

Proposition 14.3.7. *Let R be a PID and let $\mathcal{L}$ be regular. For $m < \infty$, $(Y, \mathcal{L})$ has m filtered ends iff $H^0_{\mathcal{L},e}(Y; R)$ is a free R-module of rank m; $(Y, \mathcal{L})$ has ∞ filtered ends iff $H^0_{\mathcal{L},e}(Y; R)$ is not finitely generated.*

Proof. First, assume $m < \infty$. By 14.3.4, the following are equivalent:

(i). $(Y, \mathcal{L})$ has m filtered ends;
(ii). $\{\pi_0(Y \overset{c}{-} L_i)\}$ is stable with inverse limit having m elements;
(iii). $\{H_0(Y \overset{c}{-} L_i; R)\}$ is stable with free inverse limit of rank m;
(iv). $H^0_{\mathcal{L},e}(Y; R)$ is free of rank m.[7]

If $(Y, \mathcal{L})$ has ∞ filtered ends then $\{H_0(Y \overset{c}{-} L_i; R)\}$ is pro-isomorphic to an inverse sequence of splittable epimorphisms and $\operatorname{rank}_R(H_0(Y \overset{c}{-} L_i; R))$ is either infinite for sufficiently large i, or is finite for all i but goes to ∞ with i. In both cases the direct sequence $\{\operatorname{Hom}_R(H_0(Y \overset{c}{-} L_i), R)\}$ is ind-isomorphic to a direct sequence of splittable monomorphisms whose direct limit is not finitely generated. By 14.2.2, $H^0_{\mathcal{L},e}(Y; R)$ is not finitely generated. Conversely, if $H^0_{\mathcal{L},e}(Y; R)$ is not finitely generated then $(Y, \mathcal{L})$ has ∞ ends, by the first part of this proof. $\square$

Exercises

1. Construct a well filtered graph $(Y, \mathcal{L})$ such that for each i, $Y \overset{c}{-} L_i$ has an $\mathcal{L}$-unbounded path component whose intersection with $Y \overset{c}{-} L_{i+1}$ is the union of $\mathcal{L}$-bounded path components.
2. State and prove filtered analogs of 13.4.9 and 13.4.10.
3. Give an example for 14.3.5 where some $Y \overset{c}{-} L_i$ has infinitely many $\mathcal{L}$-unbounded path components.
4. Show that $(X, h^{-1}\mathcal{L})$ in Example 14.1.5 is regular.

14.4 Filtered cohomology of pairs of groups

This section is the filtered analog of Sect. 13.2. We only discuss cohomology, leaving the corresponding treatment of homology as an exercise.

Let H be a subgroup of the group G. The set of (right) cosets of H in G is $H \backslash G := \{Hg \mid g \in G\}$. Define $R^H(G)\hat{\ }$ to be the submodule of $RG\hat{\ }$ consisting of all $\sum_{g \in G} r_g.g$ such that the set of all g with $r_g \neq 0$ lies in finitely many cosets. In particular, $R^{\{1\}}(G)\hat{\ } = RG$. We have a short exact sequence

---

[7] By 12.5.8, 12.5.1 and 14.2.2.

$$0 \to R^H(G)\hat{\ } \xrightarrow{\ i\ } RG\hat{\ } \longrightarrow R^H(G)^e \longrightarrow 0$$

where i is inclusion and the right term is by definition the quotient (right) RG-module.

To keep things simple, let (X, v) be a $K(G, 1)$-complex of finite type (hence G has type[8] F_∞). Let $\mathcal{K} := \{K_i\}$ be a finite type filtration of the covering space $\bar{X}(H)$. The covering projection is $p_H : \tilde{X} \to \bar{X}(H)$: let $L_i = p_H^{-1}(K_i)$. Then $\mathcal{L} = \{L_i\}$ is a filtration of $\tilde{X}$, and both $(\bar{X}(H), \mathcal{K})$ and $(\tilde{X}, \mathcal{L})$ are regular well filtered CW complexes. By analogy with Sect. 13.2, and with notation as in Sect. 14.2, we have a commutative diagram of R-modules, where R is a PID:

$$
\begin{array}{ccccc}
\mathrm{Hom}_G(C_k(\tilde{X}; R), R^H(G)\hat{\ }) & \rightarrowtail & \mathrm{Hom}_G(C_k(\tilde{X}; R), RG\hat{\ }) & \twoheadrightarrow & \mathrm{Hom}_G(C_k(\tilde{X}; R), R^H(G)^e) \\
\uparrow{\psi_k^H} & & \uparrow{\psi_k^\infty} & & \uparrow{\psi_k^{H,e}} \\
C_k^{\mathcal{L},\mathrm{cof}}(\tilde{X}; R) & \rightarrowtail & C_k^\infty(\tilde{X}; R) & \twoheadrightarrow & C_k^\infty(\tilde{X}; R)/C_k^{\mathcal{L},\mathrm{cof}}(\tilde{X}; R)
\end{array}
$$

in which both lines are exact and all three vertical arrows are cochain isomorphisms; compare Sect. 13.2. Hence:

Proposition 14.4.1. ψ_*^H and $\psi_*^{H,e}$ induce isomorphisms

$$H_\mathcal{L}^*(\tilde{X}; R) \to H^*(G, R^H(G)\hat{\ })$$

and

$$H_{\mathcal{L},e}^*(\tilde{X}; R) \to H^*(G, R^H(G)^e)$$

respectively.   □

By a proof similar to that of 13.2.11 one shows:

Proposition 14.4.2. Let $H \leq G$. Then $H^0(G, R^H(G)\hat{\ }) = 0$ [resp. $\cong R$] iff H has infinite [resp. finite] index in G.   □

From the top line of the above commutative diagram we get a long exact sequence

$$\cdots \leftarrow H^n(G, RG\hat{\ }) \leftarrow H^n(G, R^H(G)\hat{\ }) \leftarrow H^{n-1}(G, R^H(G)^e) \leftarrow H^{n-1}(G, RG\hat{\ }) \leftarrow \cdots$$

We know that $H^n(G, RG\hat{\ }) = 0$ for $n > 0$ and $H^0(G, RG\hat{\ }) \cong R$. Assuming H has infinite index in G, 14.4.2 implies that the sequence starts with a monomorphism $H^0(G, RG\hat{\ }) \rightarrowtail H^0(G, R^H(G)^e)$ which splits using 14.4.1; compare 12.2.3. Hence:

Proposition 14.4.3. For $k > 1$, $H^k(G, R^H(G)\hat{\ }) \cong H^{k-1}(G, R^H(G)^e)$. If H has infinite index in G, $H^0(G, R^H(G)^e) \cong H^1(G, R^H(G)\hat{\ }) \oplus R$.   □

---

[8] By 12.2.4 and 12.2.5, appropriate modifications of this section apply to groups of type F_n.

Exercise

1. Write out the analogous homology. What plays the role of $R^H(G)\hat{}$?

14.5 Filtered ends of pairs of groups

Let G be a finitely generated group, let H be a subgroup and let R be a PID. We discussed the number of ends of the pair (G, H) in Sect. 13.5, but there is a competing definition of equal interest: "filtered ends of pairs."

Let X be a path connected CW complex with fundamental group isomorphic to G and having finite 1-skeleton. As in Sect. 13.5, we will only be concerned with the 1-skeleton of X and of its covering spaces, all locally finite graphs. We will use the filtered cohomology of Sects. 14.3 and 14.4 only in the lowest dimensions (see the footnote in that section – here it applies for groups of type F_1).

Choose a finite filtration $\mathcal{K} := \{K_i\}$ for the locally finite graph $\bar{X}(H)^1$. We have the covering projection $p_H : \tilde{X}^1 \to \bar{X}(H)^1$. Let $L_i = p_H^{-1}(K_i)$, and let $\mathcal{L} = \{L_i\}$. Then $(\tilde{X}^1, \mathcal{L})$ is a regular well filtered graph. The *number of filtered ends* of (G, H) is the number of filtered ends of $(\tilde{X}^1, \mathcal{L})$. By 14.3.7, this number is either ∞ or, if finite, is the rank of the free R-module $H^0_{\mathcal{L},e}(\tilde{X}^1; R)$, which, by 14.4.1, is isomorphic to $H^0(G, R^H(G)^e)$. Thus the number of filtered ends of (G, H) does not depend on the choice of X or R.

Let $\tilde{e}(G, H)$ and $e(G, H)$ denote the number of filtered ends and the number of ends of (G, H), respectively. For comparison with 13.5.18, we note a consequence of 14.4.3, 14.4.1 and 14.3.7:

Proposition 14.5.1. *Let H have infinite index in G. If $\tilde{e}(G, H) < \infty$ then $\tilde{e}(G, H) = 1 + \mathrm{rank}_R(H^1(G, R^H(G)\hat{}))$; $\tilde{e}(G, H) = \infty$ iff $H^1(G, R^H(G)\hat{})$ is not finitely generated.*[9] $\square$

By 14.3.5 we have:

Proposition 14.5.2. $\tilde{e}(G, H) = 0$ *iff* $e(G, H) = 0$ *iff* H *has finite index in* G. $\square$

Proposition 14.5.3. $\tilde{e}(G, H) \geq e(G, H)$; *hence, if* $\tilde{e}(G, H) = 1$ *then we have* $e(G, H) = 1$.

Proof. By the (proof of the) Homotopy Lifting Property 2.4.6 the obvious function from filtered ends of (G, H) to ends of (G, H) is surjective. For the last part apply 14.5.2. $\square$

---

[9] In [100] the number of ends of the pair (G, H) is defined to be 0 when H has finite index in G and, by the formula in 14.5.1, with $R = \mathbb{Z}_2$, when H has infinite index in G. It follows that $\tilde{e}(G, H)$ equals the number of ends of (G, H) in that sense. In [23] the number $\tilde{e}(G, H)$ is called "the number of coends" of H with respect to G. Yet another definition of "the number of ends of (G, H)" is studied in [2].

Proposition 14.5.4. *If $K \leq H \leq G$, where $[H : K] < \infty$, then $\tilde{e}(G, H) = \tilde{e}(G, K)$. In particular, if H is finite, $\tilde{e}(G, H)$ is the number of ends of G.*

Proof. In this case the covering projection $\bar{X}(K)^1 \to \bar{X}(H)^1$ is proper, so one gets the same filtration of $\tilde{X}^1$ from $\bar{X}(K)^1$ as from $\bar{X}(H)^1$. □

Example 14.5.5. In Example 13.5.14, $e(G, H) = 1$ while $\tilde{e}(G, H) = 2$.

Example 14.5.6. If $G = \mathbb{Z} \times \mathbb{Z}$ and $H = \mathbb{Z} \times \{0\}$, then $\tilde{e}(G, H) = e(G, H) = 2$.

Example 14.5.7. For the pair (G, H) in Example 13.5.13 with $e(G, H) = n \geq 3$, we have $\tilde{e}(G, H) = \infty$. To see this, note that in 13.5.13 we showed that for each i the inclusion $U_i \hookrightarrow \bar{X}(H)$ induces a monomorphism on fundamental group whose image has infinite index in H. By 3.4.9, $\tilde{e}(G, H) = \infty$.

If N is normal in G it can happen that $\tilde{e}(G, N)$ is greater than $e(G, N)$, which, by 13.5.11, is the number of ends of the group G/N:

Example 14.5.8. Let $\phi : H \to H$ be a monomorphism which is not an epimorphism. Let $G = H *_\phi$ be the resulting ascending HNN extension. Then G has a presentation $\langle H, t \mid t^{-1}xt\phi(x)^{-1}, \forall x \in H \rangle$. The standard homomorphism $G \to \mathbb{Z}$ taking H to 0 and t to 1 has kernel N, and H is a proper subgroup of N. Now assume that H is finitely presented, and let Z be a finite CW complex such that $\pi_1(Z, v) \cong H$. Let $h : (Z, v) \to (Z, v)$ be a map inducing ϕ on fundamental group. Then G is the fundamental group of $T(h)$, the mapping torus of h, and N is the fundamental group of $\mathrm{Tel}(h)$, the mapping telescope of h. The latter has two ends, so $e(G, N) = 2$. Inspection of the picture of $\mathrm{Tel}(h)$ in Sect. 4.3 shows that there is a basis for the neighborhoods of the end whose path components U_i "on the left side" have the property that for all i the map $U_{i+1} \hookrightarrow U_i$ induces $\phi : H \to H$ on fundamental group. It follows, by 3.4.9, that $\tilde{e}(G, N) = \infty$. (For example, take G to be the Baumslag-Solitar group $B(1, 2)$ and $H \cong \mathbb{Z}$; then $N \cong$ the dyadic rational numbers.)

In this example N is not finitely generated. However, when N is finitely generated and normal in G, $\tilde{e}(G, N) = e(G, N)$:

Proposition 14.5.9. *If N is a finitely generated normal subgroup of G then $\tilde{e}(G, N)$ is the number of ends of G/N. Hence $\tilde{e}(G, N) = e(G, N)$.*

Proof. We know that $\tilde{e}(G, N) = 0$ iff G/N is finite, so we may assume N has infinite index in G. Let (X, x) be a pointed 2-dimensional CW complex whose 1-skeleton is finite and whose fundamental group is (identified with) G. A finite subcomplex C of the 1-skeleton $\bar{X}(N)^1$ "carries" N in the sense that $C \hookrightarrow \bar{X}(N)$ induces an epimorphism of fundamental groups. Since G/N is infinite there are covering transformations moving C into any path component of $\bar{X}(N)^1 \stackrel{c}{-} C$. This implies (by 3.4.9) that $p_{N\#} : \pi_0(\bar{X} \stackrel{c}{-} p_N^{-1}(C)) \to \pi_0(\bar{X}(N)^1 \stackrel{c}{-} C)$ is a bijection. The same holds when C is replaced by any larger compact set. Note that the 2-cells play no role in this statement. So $\tilde{e}(G, N) = e(G, N)$. □

Two subgroups H_1 and H_2 of G are *commensurable* if $H_1 \cap H_2$ has finite index in both. The *commensurator* of a subgroup H of G is $\text{Comm}_G(H) :=$ $\{g \in G \mid gHg^{-1} \text{ and } H \text{ are commensurable}\}$. This is a subgroup of G which contains H. In general, the role of the normalizer for $e(G, H)$ is played instead by the commensurator for $\tilde{e}(G, H)$. The analog of 13.5.21 is:

Theorem 14.5.10. *If H is finitely generated and has infinite index in the commensurator $\text{Comm}_G(H)$, then $\tilde{e}(G, H) = 1, 2$ or ∞.*

Before proving 14.5.10 we explain the role of the commensurator. With notation as above we take $\tilde{X}^1$ to be Γ, the Cayley graph of G with respect to a finite set of generators. The vertices of Γ are labeled by elements g of G, and the vertices of $H\backslash\Gamma$ by cosets Hg. As before, we choose a finite filtration $\{K_i\}$ of $H\backslash\Gamma$ and we filter Γ by $\mathcal{L} = \{p_H^{-1}(K_i)\}$ where $p_H : \Gamma \to H\backslash\Gamma$ is the covering projection.

Proposition 14.5.11. *If $g \in \text{Comm}_G(H)$ then there is a finite set $F \subset gH$ such that $gH \subset HF$.*

Proof. The subgroup $K = gHg^{-1} \cap H$ has finite index in gHg^{-1}, so there is a finite set $F_0 \subset H$ such that $gHg^{-1} = KgF_0g^{-1} \subset HgF_0g^{-1}$. The required F is gF_0. $\square$

Corollary 14.5.12. *If L is an $\mathcal{L}$-bounded subgraph of Γ and $g \in \text{Comm}_G(H)$, then gL is also $\mathcal{L}$-bounded.*

Proof. By 14.5.11, $gH \subset HF$, so for any $\bar{g} \in G$ we have $gH\bar{g} \subset HF\bar{g}$. This implies that g takes the vertices of L into an $\mathcal{L}$-bounded set. It follows easily that gL is $\mathcal{L}$-bounded. $\square$

Proposition 14.5.13. *If H has infinite index in $\text{Comm}_G(H)$, then for any finite set $F \subset G$ there exists $g \in \text{Comm}_G(H)$ such that $gHF \cap HF = \emptyset$.*

Proof. Suppose F exists such that for every $g \in \text{Comm}_G(H)$ $gHF \cap HF \neq \emptyset$. Then, for each such g, there exists $f, \bar{f} \in F$ and $h, \bar{h} \in H$ such that $g = \bar{h}(\bar{f}f^{-1})h^{-1}$. For pairs $(f, \bar{f})$ arising in this way we then have $f\bar{f}^{-1} \in \text{Comm}_G(H)$, hence, by 14.5.11, there exists a finite set $F_0(f, \bar{f}) \subset H$ such that $\bar{f}f^{-1}H \subset H\bar{f}f^{-1}F_0(f, \bar{f})$. This implies $H\bar{f}f^{-1}H \subset H\bar{f}f^{-1}F_0(f, \bar{f})$, so every g in $\text{Comm}_G(H)$ lies in $H\bar{f}f^{-1}F_0(f, \bar{f})$ for some such f and $\bar{f}$. But this contradicts our infinite index hypothesis. $\square$

As with 14.5.12 we have a corollary whose proof is left as an exercise:

Corollary 14.5.14. *If H has infinite index in $\text{Comm}_G(H)$ and if L is an $\mathcal{L}$-bounded subgraph of Γ then there exists $g \in \text{Comm}_G(H)$ such that $gL \cap L = \emptyset$.* $\square$

Proof (of Theorem 14.5.10). Suppose (G, H) has m filtered ends where $3 \leq m < \infty$. Since H is finitely generated, Brown's Criterion 7.4.1 makes 14.3.6 applicable, so there is an $\mathcal{L}$-bounded path connected subgraph L of Γ such that for any $\mathcal{L}$-bounded subgraph L' of Γ containing L the graph $\Gamma \stackrel{c}{-} L'$ has exactly m $\mathcal{L}$-unbounded path components; moreover, each path component $Z_1, \cdots, Z_m$ of $\Gamma \stackrel{c}{-} L$ is $\mathcal{L}$-unbounded and contains an $\mathcal{L}$-unbounded path component of $\Gamma \stackrel{c}{-} L'$. By 11.4.4 (applied to $H \backslash \Gamma$), $N(L)$ is $\mathcal{L}$-bounded so, by 14.5.14, there exists $g \in \mathrm{Comm}_G(H)$ such that $N(gL) = g(N(L))$ lies in some Z_i, say Z_1. By 14.5.12, $N(gL)$ is $\mathcal{L}$-bounded. Let $Z'_1, \cdots, Z'_m$ be the $\mathcal{L}$-unbounded path components of $\Gamma \stackrel{c}{-} (L \cup g(L))$, where $Z'_i \subset Z_i$. Then $Z'_i = Z_i$ when $i > 1$. Now $Z_2 \cup Z_3 \cup N(L)$ is path connected. So $Z'_2 \cup Z'_3 \cup N(L)$ is a path connected subset of $\Gamma \stackrel{c}{-} g(L)$, implying that Z'_2 and Z'_3 lie in the same $\mathcal{L}$-unbounded path component of $\Gamma \stackrel{c}{-} g(L)$. But each $\mathcal{L}$-unbounded path component of $\Gamma \stackrel{c}{-} g(L)$ contains exactly one $\mathcal{L}$-unbounded path component of $\Gamma \stackrel{c}{-} (L \cup g(L))$. This is a contradiction, since $m < \infty$. $\qquad \square$

The classification of pairs of groups having two filtered ends, analogous to 13.5.9, is:

Theorem 14.5.15. *Let H be a finitely generated subgroup of G having infinite index in $\mathrm{Comm}_G(H)$. Then $\tilde{e}(G, H) = 2$ iff there are subgroups G_1 and H_1 of finite index in G and H respectively such that H_1 is normal in G_1 and G_1/H_1 is infinite cyclic.*

For the proof of 14.5.15 we need a variation on 14.5.13:

Lemma 14.5.16. *For each finite set $F \subset G$ there is a finite set $F_0 \subset \mathrm{Comm}_G(H)$ such that whenever $g \in \mathrm{Comm}_G(H) - HF_0H$ then $gHF \cap HF = \emptyset$.*

Proof. Suppose F exists such that for every finite set $F_0 \subset \mathrm{Comm}_G(H)$ there is some $g \in \mathrm{Comm}_G(H) - HF_0H$ with $gHF \cap HF \neq \emptyset$. Then $ghf = \bar{h}\bar{f}$ for some $h, \bar{h} \in H$ and $f, \bar{f} \in F$, so $g = \bar{h}\bar{f}f^{-1}h^{-1}$ with $\bar{f}f^{-1} \in \mathrm{Comm}_G(H) - HF_0H$. By 14.5.11 there is a finite set $F_1(f, \bar{f}) \subset H$ such that $\bar{f}f^{-1}H \subset HF_1(f, \bar{f})$, hence $g \in H\bar{f}f^{-1}H \subset HF_1(f, \bar{f})$. Let F_2 be the union of all the (finitely many) sets $F_1(f, \bar{f})$. Then $g \in HF_2 \subset HF_2H$. In summary, there is a finite set $F_2 \subset H$ such that for every finite set $F_0 \subset \mathrm{Comm}_G(H)$ the set $(\mathrm{Comm}_G(H) - HF_0H) \cap HF_2H$ is non-empty. Taking $F_0 = F_2$, this is a contradiction. $\qquad \square$

With Γ as above we have:

Corollary 14.5.17. *If L is an $\mathcal{L}$-bounded subgraph of Γ there is a finite set $F_0 \subset \mathrm{Comm}_G(H)$ such that whenever $g \in \mathrm{Comm}_G(H) - HF_0H$ then $gL \cap L = \emptyset$.* $\qquad \square$

Proof (of 14.5.15). Let $\tilde{e}(G, H) = 2$. The proof is somewhat analogous to that of 13.5.9. We begin with $(\Gamma, \mathcal{L})$ as in the proof of 14.5.10. As in that proof, the fact that H is finitely generated implies that there is an $\mathcal{L}$-bounded path connected subgraph L of Γ such that $\Gamma \overset{c}{=} L$ has exactly two $\mathcal{L}$-unbounded path components Z^+ and Z^-. The group $\mathrm{Comm}_G(H)$ acts on the (two-element) set of filtered ends, so a subgroup $\overline{\mathrm{Comm}}_G(H)$ of index ≤ 2 fixes the two ends. Passing to a subgroup of index 2 if necessary, we assume $H \leq \overline{\mathrm{Comm}}_G(H)$. For this L, let $F_0 \subset \mathrm{Comm}_G(H)$ be as in 14.5.17. The infinite index hypothesis ensures (by 14.5.11) that there exists $g \in \overline{\mathrm{Comm}}_G(H) - HF_0H$, i.e., that this set is non-empty: we pick one such g. As in the proof of 13.5.9, one shows that there is a finite set $P \subset G$ such that $G = \langle g \rangle HP$.

Let $H_1 = \bigcap\limits_{n \in \mathbb{Z}} g^n H g^{-n}$. *Claim* : H_1 has finite index in H. Assuming the Claim, there is a finite set $Q \subset H$ such that $G = \langle g \rangle H_1 Q P$. Let $G_1 = \langle g \rangle H_1$. Clearly G_1 is a subgroup of G, and H_1 is normal in G_1 with G_1/H_1 infinite cyclic.

It remains to prove the Claim. We may assume L is H-invariant (replacing it by HL if necessary), hence Z^+ is also H-invariant. For any $n \in \mathbb{Z}$ and $h \in H$ there is an integer k such that $g^n h^k g^{-n} \in H$; this is because $g^n \in \mathrm{Comm}_G(H)$. Hence $g^n h^k g^{-n} Z^+ = Z^+$. Now $g^n h g^{-n} \in \mathrm{Comm}_G(H)$, so, by 14.5.17, $g^n h g^{-n} \in HF_0H$; otherwise $g^n h g^{-n}(Z^+)$ would either properly contain or be properly contained in Z^+, both of which are incompatible with the fact that $g^n h^k g^{-n} \in H$. Thus for all $n \in \mathbb{Z}$ and all $h \in H$, $g^n h g^{-n} \in HF_0H$ which, by 14.5.11, lies in HF_1 for some finite set $F_1 \subset F_0 H$. Let $F^{(n)}$ be the smallest subset of F_1 such that $g^n H g^{-n} \subset HF^{(n)}$. Then, for every $f \in F^{(n)}$, $Hf \cap g^n H g^{-n} \neq \emptyset$. We may alter $F^{(n)}$ so that $F^{(n)} \subset g^n H g^{-n}$. Writing $K_n = H \cap g^n H g^{-n}$, it follows easily that $g^n H g^{-n} = K_n F^{(n)}$. So $K_{-n} = g^{-n} K_n g^n$ has index $\leq |F^{(n)}| \leq |F_1|$ in H. Thus there is an upper bound $|F_1|$ to the indices of the subgroups K_n in H. By Exercise 4 in Sect. 3.4, there are only finitely many subgroups of a finitely generated group having a given index. So there are only finitely many distinct subgroups K_n, and $H_1 := \bigcap\limits_{n \in \mathbb{Z}} K_n$ is a finite intersection. By Exercise 5 in Sect. 3.4, H_1 has finite index in H. The Claim is proved.

The proof of the converse is similar to that of 14.5.10. $\square$

Remark 14.5.18. Analogous to 13.5.10, one might expect a splitting theorem saying that if G and H are finitely generated and $\tilde{e}(G, H) \geq 2$ then G splits as $A \underset{C}{*} B$ or $A \underset{C}{*}$ where C is commensurable with H; compare our sketch of the proof of 13.5.10. Without further hypotheses this is false. Using the language of manifolds, let M be a closed orientable aspherical 3-manifold, N a closed orientable surface and $f : N \to M$ a "good" map inducing a monomorphism $\phi : \pi_1(N, v) \to \pi_1(M, v)$. Let $G = \pi_1(M, v)$ and $H = \mathrm{image}(\phi)$. Then $\tilde{e}(G, H) = e(G, H) = 2$. In [137] there are examples where f does not lift to an embedding into any compact covering space of M. This implies G does not split as above.

Example 14.5.19. Let $n \geq 3$ be an integer. In 13.5.13 we saw a pair (G, H) for which $e(G, H) = n$ and $\tilde{e}(G, H) = \infty$. Here we discuss a pair (G, H) such that $e(G, H) = \tilde{e}(G, H) = n$. Let M be a compact 3-manifold whose boundary $T = \partial M$ is a torus. Let $f : S^1 \to S^1$ be a map of degree $n - 1$, let $F = \mathrm{id} \times f : T \to T$ (where T is identified with $S^1 \times S^1$) and let $N = M(F)$ be the mapping cylinder of F. Let $X = M \cup_T N$ in which ∂M is identified with $i(T) \subseteq M(F)$. With base point $x \in T$, we write $A := \pi_1(M, x)$, $B := \pi_1(N, x)$, $H : \pi_1(T, x)$, and $G := \pi_1(X, x)$. Then G is isomorphic to $A *_H B$ in a natural way. Now assume H is malnormal[10] in A, as happens, for example, when H is a prime hyperbolic knot group. Then [100] $e(G, H) = \tilde{e}(G, H) = n$. The proof is sketched as an exercise.

Source Notes. Filtered ends of pairs, under another name and presented in an algebraic setting, are due to P. Kropholler and M. Roller [100]. Remark 14.5.18 was made to me by G. Swarup.

Exercises

1. Use Addendum 14.1.11 to give an alternative proof that the number of filtered ends of (G, H) does not depend on the choice of X.
2. Prove that $\mathrm{Comm}_G H$ is a subgroup of G.
3. Let $H_1, \ldots, H_n$ be subgroups of G such that for each i either H_i has finite index in H_{i+1} or H_{i+1} has finite index in H_i. Prove that H_1 and H_n are commensurable.
4. Prove Corollary 14.5.14.
5. Fill in the missing steps in the proof of 14.5.15.
6. Find a counterexample to 14.5.9, when N is not finitely generated, where $e(G, N) = 1$ and $\tilde{e}(G, N) = \infty$.
7. Give an example of (G, H) and (G, K) where $e(G, H) = e(G, K)$ and $\tilde{e}(G, H) \neq \tilde{e}(G, K)$.
8. Fill in the proof that (G, H) in Example 14.5.19 has the properties claimed as follows: First, use the Normal Form Theorem (stated in the proof of 18.3.19) to prove

   Lemma. *Let H be malnormal in A, let B be abelian and let $c \in G = A *_H B$.*
   (i) *If $c \in B$ then $c^{-1} H c = H$;*
   (ii) *If $c \notin B$ then $(c^{-1} H c) \cap H = \{1\}$;*
   (iii) *If $c \in A \cup B$ or $c = c_1 c_2$ where $c_1 \in A - H$ and $c_2 \in B - H$ then $(c^{-1} A c) \cap H = H$;*
   (iv) *If $c \in G$ is not covered by (iii) then $(c^{-1} A c) \cap H = \{1\}$.*

   Then, using 3.4.9 and 3.4.10, study the covering space $q_H : (\bar{X}(H), \bar{x}) \to (X, x)$ by partitioning $\bar{X}(H)$ into the path components of $q_H^{-1}(M)$, $q_H^{-1}(N)$ and $q_H^{-1}(T)$.

---

[10] The subgroup A is *malnormal* in H if whenever $c \in H - A$, $c^{-1} A \subset \cap A = \{1\}$.

Poincaré Duality in Manifolds and Groups

Poincaré Duality on an orientable n-manifold gives a canonical isomorphism between homology and cohomology. This isomorphism links dimension k with dimension $n - k$. Ordinary homology is Poincaré dual to cohomology based on finite chains, and ordinary cohomology is Poincaré dual to homology based on infinite chains. The geometric treatment given here exhibits these duality isomorphisms at the level of chains in an intuitively satisfying way. Historically, it is how things were first done. A more sophisticated treatment in which Poincaré Duality is presented as "cap product with a fundamental class" can be found in many modern books on algebraic topology.

We end the chapter with a discussion of Poincaré Duality groups and duality groups.

15.1 CW manifolds and dual cells

Let e be a cell of a regular CW complex X. The *dual cone* of e, denoted by e^{dual}, is the smallest subcomplex of the abstract first derived sd X containing all simplexes of the form $\{e, e_1, \cdots, e_k\}$. The *base* of e^{dual}, denoted by $b(e^{\text{dual}})$, is the subcomplex of e^{dual} consisting of all simplexes $\{e_1, \cdots, e_k\}$ such that the cell e is a proper face of the cell e_1. See Fig. 15.1. Clearly, e^{dual} is the cone $e * b(e^{\text{dual}})$ with vertex $e \in V_{\text{sd} X}$ and base $b(e^{\text{dual}})$.

One may think of e^{dual} as "orthogonal" to the cell e so that, when $|\text{sd} X|$ is identified with X (see Sect. 5.3), the cell e and the subspace $|e^{\text{dual}}|$ together "span" a compact neighborhood of the barycenter $\hat{e}$. To explain this, recall from 5.3.2 that as sets $C(e) = e$. So there is a subcomplex sd $C(e)$ of sd X with $|\text{sd} C(e)| = e$. The inclusions sd $C(e) \hookrightarrow$ sd X and $b(e^{\text{dual}}) \hookrightarrow$ sd X together define a simplicial map $\alpha : (\text{sd} C(e)) * b(e^{\text{dual}}) \to$ sd X which is clearly a simplicial isomorphism onto a full subcomplex of sd X. Its topological significance is:

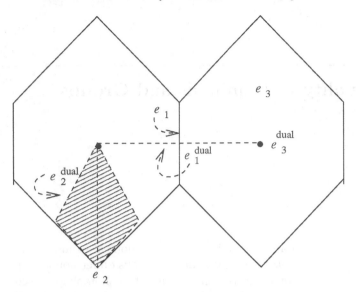

Fig. 15.1.

Proposition 15.1.1. *The image of the closed embedding*

$$|\alpha| : |(\text{sd } C(e)) * b(e^{\text{dual}})| \rightarrow |\text{sd } X| = X$$

is a neighborhood of every point of $\overset{\circ}{e}$.

Proof. We use the notation and results of Sect. 1.5. Clearly, the image of $|\alpha|$ is $N_{|\text{sd } X|}(\hat{e})$. Let $Y = |\text{sd } X| \overset{c}{-} \{\hat{e}\}$. Then Y, being a subcomplex, is closed in $|\text{sd } X|$, and $\overset{\circ}{e} \cap Y = \emptyset$. By 1.5.5, $|\text{sd } X| = (\text{image } |\alpha|) \cup Y$. So the open set $|\text{sd } X| - Y$ contains $\overset{\circ}{e}$ and lies in the image of $|\alpha|$. $\qquad\square$

We call the regular CW complex X a CW *n-manifold* if for each k, and each k-cell e of X, $|b(e^{\text{dual}})|$ is homeomorphic either to S^{n-k-1} or to B^{n-k-1}. If X has this property and e is a k-cell, then, by 5.2.7, $|e^{\text{dual}}|$ is an $(n-k)$-ball which we call the *dual cell* of e. Thus X is n-dimensional and locally finite.

Proposition 15.1.2. *The underlying space of a CW n-manifold X is a topological n-manifold. Moreover, the boundary of this manifold, ∂X, is the subcomplex consisting of all cells e of X for which $|b(e^{dual})|$ is homeomorphic to a ball rather than a sphere.*

Proof. Let $x \in X$. By 10.1.25 X is metrizable. Let e be the unique cell of X such that $x \in \overset{\circ}{e}$. By 15.1.1 and 5.2.7, x has a neighborhood homeomorphic to $\mathbb{R}^n$ or to $\mathbb{R}^n_+$, and the latter iff $|b(e^{\text{dual}})|$ is homeomorphic to a ball. It follows that ∂X is a subcomplex, for ∂X is closed in X and therefore if $\overset{\circ}{e}$ lies in ∂X so does e. By 5.3.2, every face of e also lies in ∂X. $\qquad\square$

Combining 15.1.2 with 10.1.8 and 10.1.12 we get:

Corollary 15.1.3. *If X is a CW manifold, the regular CW complex X is strongly locally finite.* $\square$

The term CW n-manifold is not standard but it is precisely what is needed for our treatment of duality in the following sections. Of course, the reader will wish to know how CW n-manifolds arise. A sufficient condition is:

Proposition 15.1.4. *If K is a combinatorial n-manifold then $|K|$ is a CW n-manifold.* $\square$

If X is a CW n-manifold and e is a k-cell of the subcomplex ∂X, e has a dual $(n-k)$-cell $|e^{\text{dual}}|$ in X, and e has a dual $(n-k-1)$-cell in ∂X which we will denote by $|e^{\text{dual}}|_{\partial}$.

Theorem 15.1.5. *Let X be a CW n-manifold. There is a regular CW complex structure X^* on X whose set of $(n-k)$-cells is*

$$\{|e^{\text{dual}}| \mid e \text{ is a } k\text{-cell of } X\} \cup \{|e^{\text{dual}}|_{\partial} \mid e \text{ is a } (k-1)\text{-cell of } \partial X\}$$

Proof. We show that X^* satisfies (i)–(v) of Proposition 1.2.14. First we note that, by 5.2.7, for every cell e of X which is not in ∂X we have $|e^{\text{dual}}|^{\circ} \subset \overset{\circ}{X}$, and for every cell e of ∂X we have $(|e^{\text{dual}}|_{\partial})^{\circ} \subset \partial X$. If $x \in X$, there is exactly one simplex $\sigma = \{e_0, e_1, \cdots, e_k\}$ of sd X such that $x \in |\sigma|^{\circ}$, where, as usual, $e_0 \subsetneqq e_1 \subsetneqq \cdots \subsetneqq e_k$. If $x \in \overset{\circ}{X}$, this implies that $x \in |e^{\text{dual}}|^{\circ}$ iff $e = e_0$. If $x \in \partial X$, then $|\sigma| \subset \partial X$ and $x \in (|e^{\text{dual}}|_{\partial})^{\circ}$ iff $e = e_0$. This proves (i). The next two parts, (ii) and (iii), are clear; (iv) follows from 15.1.3. For (v), let $F \subset X$ be such that $F \cap |e^{\text{dual}}|$ is closed in the ball $|e^{\text{dual}}|$ for every cell e of X, and $F \cap |e^{\text{dual}}|_{\partial}$ is closed in $|e^{\text{dual}}|_{\partial}$ for every cell e of ∂X. Then all these intersections are compact, hence they are closed subsets of the Hausdorff space X. Since they form a locally finite collection of closed sets in the locally compact space X, their union, namely F, is closed in X. $\square$

The CW complex X^* is called the *dual complex* of X. Clearly we have:

Proposition 15.1.6. *In the CW n-manifold X, let e_{β} be a face of e_{α}. Then $|e_{\alpha}^{\text{dual}}|$ is a face of $|e_{\beta}^{\text{dual}}|$.* $\square$

By 15.1.5, we have two CW complex structures X and X^* on the same underlying n-manifold. If these CW complexes X and X^* are oriented, we get an isomorphism of graded R-modules $\phi_k : C_k(X; R) \to C_{n-k}(X^*, \partial X^*; R)$ defined by $\phi_k(e) = |e^{\text{dual}}|$, and a similar isomorphism $\phi_k^{\infty} : C_k^{\infty}(X; R) \to C_{n-k}^{\infty}(X^*, \partial X^*; R)$. In Sect. 15.2 we investigate when the orientations on X and X^* can be chosen so that $\delta \circ \phi_k = (-1)^k \phi_{k-1} \circ \partial$ and $\delta \circ \phi_k^{\infty} =$

$(-1)^k \phi_{k-1}^\infty \circ \partial$. Whenever this can be done, one has important duality isomorphisms $H_k(X; R) \to H_f^{n-k}(X^*, \partial X^*; R)$, $H_f^k(X; R) \to H_{n-k}(X^*, \partial X^*; R)$, $H_k^\infty(X; R) \to H^{n-k}(X^*, \partial X^*; R)$ and $H^k(X; R) \to H_{n-k}^\infty(X^*, \partial X^*; R)$. Note that when R has characteristic 2 it follows from 5.3.10 and 15.1.5 that such relations hold for any choice of orientations on X and on X^*, since in that case $1 = -1$. Thus, we have in fact proved the Poincaré and Lefschetz Duality theorems for these coefficients. Rather than state the details, we simply point out that 15.2.3 and 15.2.4 have already been proved here when $2 = 0$ in R. One special case is needed immediately:

Proposition 15.1.7. *If X is a path connected CW n-manifold, then we have $H_n^\infty(X, \partial X; \mathbb{Z}_2) \cong \mathbb{Z}_2$.*

Proof. By the last paragraph, $H_n^\infty(X^*, \partial X^*; \mathbb{Z}_2)$ is isomorphic to $H^0(X; \mathbb{Z}_2)$. Apply 12.1.2 and 11.1.10. $\qquad\square$

This proposition yields a useful feature of CW manifolds:

Corollary 15.1.8. *If $e_\alpha^n \neq e_\beta^n$ are cells of a path connected CW n-manifold X, there is a finite sequence $e_\alpha^n = e_{\alpha_0}^n, e_{\alpha_1}^n, \cdots, e_{\alpha_k}^n = e_\beta^n$ of n-cells of X such that each $e_{\alpha_i}^n \cap e_{\alpha_{i+1}}^n$ is an $(n-1)$-cell of X. Hence X is a CW n-pseudomanifold.*

Proof. Let $\mathcal{A}$ index the n-cells of X. Say $\alpha, \beta \in \mathcal{A}$ are "equivalent" iff $\alpha = \beta$ or e_α^n and e_β^n are related in the manner described. This is an equivalence relation on $\mathcal{A}$. For each equivalence class $\mathcal{B} \subset \mathcal{A}$, $\sum_{\beta \in \mathcal{B}} e_\beta^n \in Z_n^\infty(X, \partial X; \mathbb{Z}_2)$ by 5.3.6. Since $B_n^\infty(X, \partial X; \mathbb{Z}_2) = 0$, it follows from 15.1.7 that there is only one equivalence class: i.e., $\mathcal{B} = \mathcal{A}$. The last sentence of the Corollary follows by 5.3.6. $\qquad\square$

Another special case of $\mathbb{Z}_2$-Lefschetz Duality was used in Sect. 12.3:

Proposition 15.1.9. *If X is a CW n-manifold, $H_{n-1}^\infty(X^*, \partial X^*; \mathbb{Z}_2)$ is isomorphic to $H^1(X; \mathbb{Z}_2)$.* $\qquad\square$

Exercises

1. Let K be an abstract simplicial complex, let σ be a simplex of K and let $x \in |\overset{\circ}{\sigma}| \subset |K|$. Prove that $|\text{st}_K \sigma|$ is a neighborhood of x in $|K|$. (*Hint:* use 15.1.1.)
2. Prove 15.1.4.

15.2 Poincaré and Lefschetz Duality

Let X be an orientable CW n-manifold. Orient the cells of X, in the sense of Sect. 2.5. Choose a fundamental cycle $\sum \epsilon_\delta e_\delta^n$ for X, thereby specifying an orientation on (the pseudomanifold) X. We wish to orient the cells of the dual complex X^* in a desirable way, and thereby prove the duality theorems 15.2.3 and 15.2.4. We will also discuss an equivariant version, and a version involving twisted coefficients.

Let e_α^k be a k-cell of X. Then $|e_\alpha^{\text{dual}}|$ is an $(n-k)$-cell, and is the union of certain $(n-k)$-simplexes of $|\text{sd } X|$. More precisely, if $e^0 \subset e^1 \subset \cdots \subset e^{k-1} \subset e^k = e_\alpha^k \subset e^{k+1} \subset \cdots \subset e^n = e_\gamma^n$ are cells of X, consider the following simplexes[1] of sd X : $\sigma^n = \{\hat{e}^0, \cdots, \hat{e}^n\}$, $\sigma^k = \{\hat{e}^0, \cdots, \hat{e}^k\}$, and $\sigma^{n-k} = \{\hat{e}^k, \cdots, \hat{e}^n\}$. Note that $|\sigma^k| \subset e_\alpha^k$ and $|\sigma^{n-k}| \subset |e_\alpha^{\text{dual}}|$.

As usual, we use the natural ordering for each simplex of sd X. By 5.4.3, this ordering determines an orientation on each simplex of sd X. IN THIS SECTION ONLY, we introduce the notation that $(|\sigma^n|, 1)$ denotes $|\sigma^n|$ with this orientation, and $(|\sigma^n|, -1)$ denotes $|\sigma^n|$ with the opposite orientation. By 5.4.3, we may think of $(|\sigma^n|, -1)$ as being obtained from $(|\sigma^n|, 1)$ by applying an odd permutation to the vertices. Similar notation will be used for σ^k, σ^{n-k}, etc.

Here is the rule for orienting $|e_\alpha^{\text{dual}}|$. We pick $\xi, \eta \in \{1, -1\}$ in the following way. Let $e^0, \cdots, e^k = e_\alpha^k, \cdots, e^n = e_\gamma^n$ be as above. Let $(|\sigma^k|, \xi)$ represent the orientation which $|\sigma^k|$ inherits from the chosen orientation on e_α^k, and let $(|\sigma^n|, \eta\epsilon_\gamma)$ represent the orientation which $|\sigma^n|$ inherits from the chosen orientation on e_γ^n. Orient $|e_\alpha^{\text{dual}}|$ so that $|\sigma^{n-k}|$ inherits the orientation $(|\sigma^{n-k}|, \xi\eta)$ from $|e_\alpha^{\text{dual}}|$.

Proposition 15.2.1. *This orientation on $|e_\alpha^{\text{dual}}|$ is well defined.*

Proof. If, instead, we look at $d^0 \subset d^1 \subset \cdots \subset d^{k-1} \subset e^k = e_\alpha^k \subset \cdots \subset e^n = e_\gamma^n$, and let $\tau^n = \{\hat{d}^0, \cdots, \hat{d}^{k-1}, \hat{e}^k, \cdots, \hat{e}^n\}$ and $\tau^k = \{\hat{d}^0, \cdots, \hat{d}^{k-1}, \hat{e}^k\}$, we are led to $(|\tau^k|, \xi')$ and $(|\tau^n|, \eta'\epsilon_\gamma)$ as the inherited orientations from e_α^k and e_γ^n. By 5.4.4, $\xi'\eta' = \xi\eta$, so we get the same orientation for $|e_\alpha^{\text{dual}}|$, namely, that which yields $(|\sigma^{n-k}|, \xi'\eta')$.

Next, consider $e^0 \subset e^1 \subset \cdots \subset e^k = e_\alpha^k \subset \cdots \subset e^{i-1} \subset d^i \subset e^{i+1} \subset \cdots \subset e^n = e_\gamma^n$, where $i < n$. Let $\mu^n = \{\hat{e}^0, \cdots, \hat{e}^{i-1}, \hat{d}^i, \hat{e}^{i+1}, \cdots, \hat{e}^n\}$ and let $\mu^{n-k} = \{\hat{e}^k, \cdots, \hat{e}^{i-1}, \hat{d}^i, \hat{e}^{i+1}, \cdots, \hat{e}^n\}$. The 0th through kth vertices of μ^n are those of σ^k; and the n-simplexes σ^n and μ^n differ only in one vertex. Hence $(|\mu^n|, -\eta\epsilon_\gamma)$ represents the orientation which $|\mu^n|$ inherits from e_γ^n. So $(|\mu^{n-k}|, -\xi\eta)$ is to be the inherited orientation from $|e_\alpha^{\text{dual}}|$. This is the previously defined orientation, since σ^{n-k} and μ^{n-k} also differ by exactly one vertex.

Now consider $e^0 \subset \cdots \subset e^k = e_\alpha^k \subset \cdots \subset e^{n-1} \subset e_{\gamma'}^n$, where $\gamma' \neq \gamma$. Let $\mu^n = \{\hat{e}^0, \cdots, \hat{e}^{n-1}, \hat{e}_{\gamma'}^n\}$ and let $\mu^{n-k} = \{\hat{e}^k, \cdots, \hat{e}^{n-1}, \hat{e}_{\gamma'}^n\}$. A similar argument shows that $(|\mu^n|, -\eta\epsilon_{\gamma'})$ represents the orientation which $|\mu^n|$ inherits from $e_{\gamma'}^n$; here we use the fact that $\epsilon_\gamma e_\gamma^n + \epsilon_{\gamma'} e_{\gamma'}^n$ is part of a fundamental

---

[1] Note the identification introduced after 5.3.9.

cycle in X. So, as before, $(|\mu^{n-k}|, -\xi\eta)$ is to be the inherited orientation from $|e_\alpha^{\text{dual}}|$, and again this is compatible with that of $(|\sigma^{n-k}|, \xi\eta)$.

Finally, the subcomplex e_α^{dual} of sd X is a pseudomanifold (exercise) so the cases dealt with in the last two paragraphs are sufficient to complete the proof – just pass from one $(n-k)$-simplex to another using the pseudomanifold property, altering one vertex at a time. $\qquad\square$

This orientation on $|e_\alpha^{\text{dual}}|$ is the *dual orientation*. The fundamental fact is:

Proposition 15.2.2. *Let $e_\beta^{k-1} \subset e_\alpha^k$. With these orientations we have $[e_\alpha^k : e_\beta^{k-1}] = (-1)^k [|e_\beta^{dual}| : |e_\alpha^{dual}|]$.*

Proof. We saw in 15.1.6 that $|e_\alpha^{\text{dual}}|$ is a face of $|e_\beta^{\text{dual}}|$. Consider $e^0 \subset e^1 \subset \cdots \subset e^{k-1} = e_\beta^{k-1} \subset e^k = e_\alpha^k \subset \cdots \subset e^n = e_\gamma^n$. Let $\sigma^n = \{\hat{e}^0, \cdots, \hat{e}^n\}$ as before; similarly σ^k and σ^{n-k}. Let $\tau^{k-1} = \{\hat{e}^0, \cdots, \hat{e}^{k-1}\}$ and $\tau^{n-k+1} = \{\hat{e}^{k-1}, \cdots, \hat{e}^n\}$. Let $\xi, \eta \in \{-1, 1\}$ be as above, and let $\xi' = \pm 1$ be such that $(|\tau^{k-1}|, \xi')$ is the orientation inherited from e_β^{k-1}. Then

$$[e_\alpha^k : e_\beta^{k-1}] = \xi\xi'[|\sigma^k| : |\tau^{k-1}|]$$
$$= (-1)^k \xi\xi', \quad \text{by 5.4.5,}$$
$$[|e_\beta^{\text{dual}}| : |e_\alpha^{\text{dual}}|] = (\eta\xi)(\eta\xi')[|\tau^{n-k+1}| : |\tau^{n-k}|]$$
$$= \xi\xi', \quad \text{by 5.4.5.}$$

$\qquad\square$

An immediate consequence of 15.2.2 is:

Theorem 15.2.3. *Let X be an oriented CW complex which is also an oriented CW n-manifold. Give X^* the dual orientation. Let $\phi_k : C_k(X; R) \to C_{n-k}(X^*, \partial X^*; R)$ and $\phi_k^\infty : C_k^\infty(X; R) \to C_{n-k}^\infty(X^*, \partial X^*; R)$ be the isomorphisms described by $e \mapsto |e^{dual}|$ in Sect. 15.1. Then $\delta \circ \phi_k = (-1)^k \phi_{k-1} \circ \partial$ and $\delta \circ \phi_k^\infty = (-1)^k \phi_{k-1}^\infty \circ \partial$.* $\qquad\square$

When X^* has the dual orientation with respect to X, we call ϕ_k and ϕ_k^∞ *duality isomorphisms*.

Corollary 15.2.4. *The isomorphisms ϕ_k and ϕ_k^∞ induce isomorphisms*

$$H_k(X; R) \to H_f^{n-k}(X^*, \partial X^*; R),$$
$$H_f^k(X; R) \to H_{n-k}(X^*, \partial X^*; R),$$
$$H_k^\infty(X; R) \to H^{n-k}(X^*, \partial X^*; R),$$
$$H^k(X; R) \to H_{n-k}^\infty(X^*, \partial X^*; R).$$

$\qquad\square$

Remark 15.2.5. We saw in Sect. 15.1 that if $2 = 0$ in R, then 15.2.3 and 15.2.4 hold for non-orientable CW manifolds too.

The statements in 15.2.3 and 15.2.4 are known as *Lefschetz Duality*. The special case in which $\partial X = \emptyset$ is known as *Poincaré Duality*. Note that $|\mathrm{sd}\ X|$ is a common subdivision of X and X^* and the identity maps $|\mathrm{sd}\ X| \to X$ and $|\mathrm{sd}\ X| \to X^*$ are cellular maps which are proper homotopy equivalences rel boundary, so one can restate 15.2.4 in terms of the single CW complex $|\mathrm{sd}\ X|$ to give isomorphisms $H_k(|\mathrm{sd}\ X|; R) \to H_f^{n-k}(|\mathrm{sd}\ X|, \partial|\mathrm{sd}\ X|; R)$, etc.

We now apply Theorem 15.2.3 to covering spaces. Let X be a path connected (not necessarily orientable) CW n-manifold whose cells are oriented. Pick a base vertex v and write $G = \pi_1(X, v)$. Orient the cells of the universal cover $(\tilde{X}, \tilde{v})$ as usual so that each covering transformation preserves the orientations of the cells of X. By 12.3.12, $\tilde{X}$ is orientable. Let $H \leq G$ be the subgroup of all orientation preserving covering transformations. We saw in 12.3.15 that $G = H$ or $[G: H] = 2$ depending on whether X is orientable or non-orientable. The dual orientation for $|e_\alpha^{\mathrm{dual}}|$ (in $\tilde{X}^*$) depends on the given orientation for the cell e_α and on the pseudomanifold orientation of $\tilde{X}$. Thus if $g \in H$, g preserves the dual orientations on the cells of $\tilde{X}^*$. In other words:

Proposition 15.2.6. *When we regard* $C_*(\tilde{X}; R)$, $C_*(\tilde{X}^*, \partial\tilde{X}^*; R)$, $C_*^\infty(\tilde{X}; R)$ *and* $C_*^\infty(\tilde{X}^*, \partial\tilde{X}^*; R)$ *as RH-chain complexes, then the duality isomorphisms* ϕ_* *and* ϕ_*^∞ *are RH-module isomorphisms (RG-module isomorphisms if X is orientable).* $\qquad\square$

In the aspherical case we have:

Corollary 15.2.7. *Let X be orientable and aspherical with $\partial X = \emptyset$. Then* $H_f^k(\tilde{X}^*; R) = 0$ *when $k \neq n$ and (since $H_0(\tilde{X}; R)$ is canonically identifiable with the trivial RG-module R)* $\phi_{0*} : R \to H_f^n(\tilde{X}^*; R)$ *is an isomorphism of RG-modules.* $\qquad\square$

With this corollary we have reached the essence of Poincaré Duality for aspherical CW complexes. Up to now it has appeared to hold because it holds locally (cells and dual cells); only the hypothesis of orientability was global. But Corollary 15.2.7 has a "converse" (again in the aspherical case) which involves no local hypotheses at all. Let Y be an n-dimensional locally finite aspherical CW complex. Write $G = \pi_1(Y, w)$ and orient the cells of $(\tilde{Y}, \tilde{w})$ as usual. Assume that $H_f^k(\tilde{Y}; R) = 0$ when $k \neq n$. Then

$$\cdots \longrightarrow C_{n-1}(\tilde{Y}; R) \overset{\delta}{\longrightarrow} C_n(\tilde{Y}; R) \longrightarrow H_f^n(\tilde{Y}; R) \longrightarrow 0$$

is a free RG-resolution of the RG-module $H_f^n(\tilde{Y}; R)$. If we further assume that $H_f^n(\tilde{Y}; R)$ is isomorphic to the trivial RG-module R, and if we call a choice of isomorphism a *formal R-orientation* of Y, we get from 8.1.1 a Poincaré Duality property:

Proposition 15.2.8. *A formal R-orientation on Y induces a canonical chain homotopy equivalence $(C_k(\tilde{Y}; R), \partial) \to (C_{n-k}(\tilde{Y}; R), \delta)$.*   □

Corollary 15.2.9. *A formal R-orientation on Y induces canonical isomorphisms $H_f^{n-k}(Y; R) \to H_k(Y; R)$ and $H^{n-k}(Y; R) \to H_k^\infty(Y; R)$.*

Proof. By 13.2.1, the chain complexes $(R \otimes_G C_k(\tilde{Y}; R), \mathrm{id} \otimes \partial)$ and $(C_k(Y; R), \partial)$ are isomorphic; similarly the cochain complexes $(R \otimes_G C_{n-k}(\tilde{Y}; R), \mathrm{id} \otimes \delta)$ and $(C_{n-k}(Y; R), \delta)$ are isomorphic. By 13.2.8, the cochain complexes $(\mathrm{Hom}_G(C_{n-k}(\tilde{Y}; R), R), \partial^*)$ and $(C_{n-k}^\infty(Y; R), \delta)$ are isomorphic; similarly the chain complexes $(\mathrm{Hom}_G(C_k(\tilde{Y}; R), R), \delta^*)$ and $(C_k^\infty(Y; R), \partial)$ are isomorphic. These isomorphisms together with 15.2.8 give the required conclusion.   □

Our focus is on groups and we will see in the next section that this study of Y leads us naturally to Poincaré Duality groups. Here are some remarks comparing 15.2.4 with 15.2.9.

(i) In 15.2.4 X is an oriented manifold, not necessarily aspherical. In 15.2.9 Y has a formal orientation and is aspherical. If X is in fact aspherical and if we take $R = \mathbb{Z}$ and $X = Y$, the notions of orientation and formal $\mathbb{Z}$-orientation coincide in an obvious sense; and, once we identify the cohomology of X^* with that of $X = Y$, the only question to ask is: are the isomorphisms in 15.2.4 and 15.2.9 the same? Up to sign the answer is yes, as can be seen using a more conventional modern treatment of Poincaré Duality in terms of cap product with a fundamental class, as is found in many books, e.g., [146], [50], [74].

(ii) 15.2.9 shows that Poincaré Duality is a global property which happens to be visualizable locally in the manifold case (provided the manifold is orientable – a global property!).

(iii) The duality asserted in 15.2.4 and in 15.2.9 is invariant under proper homotopy. It is a deep issue in topology to decide whether the space Y (with properties as above) is always proper homotopy equivalent to a manifold. Even when Y is assumed to be compact, this is an open question at time of writing. Indeed the more general non-aspherical version of this issue involves defining "Poincaré Duality spaces" (of which our Y when compact is an example) and using surgery theory to provide L-theoretic obstructions to a positive answer. At present, non-vanishing obstructions are known only in non-aspherical cases.

We end with a brief discussion of duality for non-orientable CW n-manifolds. As before, let H be the group of orientation preserving covering transformations. Let ^tR denote the right RG-module whose underlying R-module is R, where the G-action is given by $r.g = r$ [resp. $-r$] if $g \in H$ [resp. $g \notin H$]. Here "t" stands for *twisted*. Assume X is non-orientable (otherwise ^tR is the trivial RG-module R). Then we have seen that the duality isomorphisms $\tilde{\phi}_*$ and $\tilde{\phi}^\infty$ are not RG-module isomorphisms. Nevertheless, $\tilde{\phi}_k : C_k(\tilde{X}; R) \to C_{n-k}(\tilde{X}^*, \partial \tilde{X}^*; R)$ satisfies: $\tilde{\phi}_k(g\tilde{e}_\alpha^k) = g\,|\tilde{e}_\alpha^{\mathrm{dual}}|$ [resp. $-g\,|\tilde{e}_\alpha^{\mathrm{dual}}|$] if $g \in H$ [resp. if $g \notin H$]. Hence we have isomorphisms of R-modules

$$^tR \otimes_G C_k(\tilde{X};R) \to R \otimes_G C_{n-k}(\tilde{X}^*, \partial \tilde{X}^*; R)$$

and

$$R \otimes_G C_k(\tilde{X};R) \to {}^tR \otimes_G C_{n-k}(\tilde{X}^*, \partial \tilde{X}^*; R)$$

given by $r \otimes c \mapsto r \otimes \tilde{\phi}_k(c)$. The homology R-modules of $\{^tR \otimes_G C_k(\tilde{X};R), \mathrm{id} \otimes \partial\}$ are denoted by $H_*(X; {}^tR)$, with similar definitions for $H_f^*(X^*, \partial X^*; {}^tR)$, etc. In particular we have *twisted (Poincaré and) Lefschetz Duality*:

Proposition 15.2.10. *Let X be a non-orientable CW n-manifold. The above isomorphisms induce isomorphisms* $H_k(X; {}^tR) \to H_f^{n-k}(X^*, \partial X^*; R)$ *and* $H_k(X;R) \to H_f^{n-k}(X^*, \partial X^*; {}^tR)$. $\qquad\qquad\square$

Remark 15.2.11. It should be repeated that there is a version of Poincaré and Lefschetz Duality, using singular homology/cohomology and cap product, which works on topological manifolds, with no reference to a CW complex structure. See [146], [50], [74]. For a rather different treatment see [110]. From one point of view this is the "right" approach for experts to use. One of our purposes has been to bring out the geometrical content of an older approach which has tended to be ignored in modern books, but which many topologists understand in a "folk" sense.

Exercises

1. Show that the following diagrams commute where the horizontal lines come from Sect. 11.4 and Sect. 12.2, and the vertical isomorphisms are induced by the duality isomorphisms.

$$
\begin{array}{ccccccc}
H_k(X) & \longrightarrow & H_k^\infty(X) & \longrightarrow & H_{k-1}^e(X) & \longrightarrow & H_{k-1}(X) \\
\downarrow \cong & & \downarrow \cong & & \downarrow \cong & & \downarrow \cong \\
H_f^{n-k}(X^*, \partial X^*) & \longrightarrow & H^{n-k}(X^*, \partial X^*) & \longrightarrow & H_e^{n-k}(X^*, \partial X^*) & \longrightarrow & H_e^{n-k+1}(X^*, \partial X^*)
\end{array}
$$

$$
\begin{array}{ccccccc}
H_f^k(X) & \longrightarrow & H^k(X) & \longrightarrow & H_e^k(X) & \longrightarrow & H_f^{k+1}(X) \\
\downarrow \cong & & \downarrow \cong & & \downarrow \cong & & \downarrow \cong \\
H_{n-k}(X^*, \partial X^*) & \longrightarrow & H_{n-k}^\infty(X^*, \partial X^*) & \longrightarrow & H_{n-k-1}^e(X^*, \partial X^*) & \longrightarrow & H_{n-k-1}(X^*, \partial X^*)
\end{array}
$$

   (Note that this gives Lefschetz/Poincaré Duality "at the end.")

2. Let X be an open CW n-manifold such that $H_*(X;R)$ is finitely generated. Prove that $H_e^*(X;R)$ is finitely generated.

3. By definition a CW n-manifold is a regular CW complex. Give an example where $\tilde{X}$ is a CW n-manifold but X is not regular. Extend the proofs of 15.2.3, 15.2.4 and 15.2.6 to cover this case.

4. Suppose that in 15.2.4 $\partial X = \partial_0 X \cup \partial_1 X$ where each $\partial_i X$ is the union of some path components of ∂X, and $\partial_0 X \cap \partial_1 X = \emptyset$. Find a duality theorem giving an isomorphism $H_k(X, \partial_0 X; R) \to H_f^{n-k}(X^*, \partial_1 X^*; R)$. Find similar isomorphisms for the other cases treated in 15.2.4.
5. Establish the duality theorems for filtered homology and cohomology indicated in Remark 14.2.4.
6. Show that if X is a contractible open n-manifold and $n \geq 2$ then[2] $H_k^e(X; R)$ is trivial when $k \neq n-1$, while $H_{n-1}^e(X; R)$ and $H_{n-1}^e(X; R)$ are isomorphic to R.

15.3 Poincaré Duality groups and duality groups

We defined the homological finiteness property "G has type FP over R" in Sect. 8.2. Such a group G is an *n-dimensional orientable Poincaré Duality group over R* if $H^n(G, RG)$ is isomorphic (as a right RG-module – see 13.2.16) to the trivial RG-module R, and $H^k(G, RG) = 0$ when $k \neq n$. A choice of such an isomorphism is an *R-orientation* on G. Topological motivation for this comes from Sect. 15.2, where we saw that fundamental groups of closed orientable aspherical CW n-manifolds are examples. Motivated by 15.2.7 and 15.2.9 we establish "Poincaré Duality" for these groups:

Theorem 15.3.1. *The following are equivalent:*

(i) *G is an n-dimensional orientable Poincaré Duality group over R;*
(ii) (a) *G has type[3] FP, and*
  (b) *For each k and each right RG-module M there is an isomorphism $H^{n-k}(G, M) \to H_k(G, M)$ which is natural when $H^{n-k}(G, -)$ and $H_k(G, -)$ are regarded as covariant functors: Right RG-modules $\to R$-modules.*

The statement (ii) in 15.3.1 is normally given as the definition of an n-dimensional orientable Poincaré Duality group over R. It is convenient for our exposition to adopt our equivalent definition instead.

Before proving 15.3.1, we discuss the meaning of (ii)(b). If $\beta : M \to M'$ is a homomorphism of right RG-modules and

$$0 \longrightarrow F_n \xrightarrow{\partial} \cdots \xrightarrow{\partial} F_0 \xrightarrow{\epsilon} R \longrightarrow 0$$

is a projective resolution of R, then we have a chain map $\beta \otimes \mathrm{id} : M \otimes_G F_* \to M' \otimes_G F_*$ inducing a homomorphism $H_*(G, M) \to H_*(G, M')$. This is the sense in which $H_*(G, -)$ is a covariant functor. The discussion for $H^*(G, -)$ is similar: when M and M' are turned into left RG-modules in the usual way, β is also a homomorphism of left RG-modules, so $\beta_\# : \mathrm{Hom}_G(F_*, M) \to \mathrm{Hom}_G(F_*, M')$ induces a homomorphism $H^*(G, M) \to H^*(G, M')$, etc.

---

[2] Thus, from a homological point of view, X "looks like" S^{n-1} at infinity. We will see in Sect. 16.6 that this is not necessarily true from a homotopical point of view.
[3] It can be shown that (ii)(b) implies (ii)(a); see pages 140-141 of [14].

Proof (of 15.3.1). Let $0 \longrightarrow F_n \stackrel{\partial}{\longrightarrow} \cdots \stackrel{\partial}{\longrightarrow} F_1 \stackrel{\epsilon}{\longrightarrow} R \longrightarrow 0$ be a projective RG-resolution of the trivial RG-module R. Some statements in this proof are more obvious when the modules F_k are free; we will assume they are free, leaving the projective case to the reader. Each F_k is a finitely generated projective left module, and, as in Sect. 13.2, RG is, in the first instance, a right RG-module, so $F_k^* := \mathrm{Hom}_G(F_k, RG)$ becomes a projective left RG-module via $(g.f)(x) = f(x)g^{-1}$.

(i) $\Rightarrow$ (ii): By (i) we have an exact sequence

$$0 \longleftarrow R \longleftarrow F_n^* \stackrel{\partial^*}{\longleftarrow} \cdots \stackrel{\partial^*}{\longleftarrow} F_0^* \longleftarrow 0$$

which is another finitely generated projective resolution of R. For each right RG-module M there are R-isomorphisms $\alpha_k : M \otimes_G F_k^* \to \mathrm{Hom}_G(F_k, M)$ given by $\alpha_k(m \otimes f)(x) = m\overline{f(x)}$, where $\overline{\sum_\alpha r_i g_i} := \sum_i r_i g_i^{-1}$, and they fit together to give an isomorphism of chain complexes (over R) $\{M \otimes_G F_*^*, \mathrm{id} \otimes \partial^*\} \to \{\mathrm{Hom}_G(F_*, M), \partial^*\}$. Thus $H_k(G, M) \cong H^{n-k}(G, M)$, and naturality is clear.

(ii) $\Rightarrow$ (i): From the resolution we see that $H_k(G, RG) = 0$ when $k > 0$ and $H_0(G, RG)$ is isomorphic to the trivial RG-module R. By (ii), $H^k(G, RG) = 0$ when $k \neq n$ and $H^n(G, RG)$ is isomorphic to R at least as an R-module. Left multiplication by $g \in G$, $\lambda_g : RG \to RG$, is a homomorphism of right RG-modules and thus induces $\lambda_{g*} : H^n(G, RG) \to H^n(G, RG)$, which is the identity by the naturality hypothesis. Thus by 13.2.16, $\lambda_g : C_*(\tilde{X}; R) \to C_*(\tilde{X}; R)$ induces this, so right multiplication by g^{-1} in $C_*(\tilde{X}; R)$ induces the identity on $H^*(G, RG)$. Hence $H^n(G, RG)$ is isomorphic to the trivial RG-module R. $\qquad\square$

Recall that a right R-module M is *flat* if the exactness of $A \to B \to C$ always implies the exactness of $M \otimes_R A \to M \otimes_R B \to M \otimes_R C$. A group G of type FP is an *n-dimensional duality group over* R if $H^k(G, RG) = 0$ when $k \neq n$ while $H^n(G, RG)$ is non-trivial[4] and flat as an R-module. We write $D = H^n(G, RG)$ and call it the *dualizing module*.

In parallel with 15.3.1 we have:

Theorem 15.3.2. *The following are equivalent:*

(i) *G is an n-dimensional duality group over R;*

(ii) (a); *G has type[5] FP*

   (b) *For each k and each right RG-module M there is an isomorphism $H^{n-k}(G, M) \to H_k(G, M \otimes_R D)$ which is natural when $H^{n-k}(G, -)$ and $H_k(G, - \otimes_R D)$ are regarded as covariant functors: Right RG-modules $\to$ R-modules. Here G acts on $M \otimes_R D$ "diagonally" via $(m \otimes d).g = m.g \otimes d.g$.*

---

[4] Here, "non-trivial" is redundant by 13.10.1.

[5] As with 15.3.1, it can be shown that (ii)(b) implies (ii)(a); see [14, pp. 140–141].

The proof of 15.3.2 runs parallel to the proof we have given for 15.3.1, but involves some more general homological algebra. It can be found in [14, pp. 140–141] and [29, Chap. 8, Sect. 10].

The statement (ii) in 15.3.2 is usually given as the definition of an n-dimensional duality group. As with the Poincaré Duality case, our equivalent definition is more convenient here. Note that when D is the trivial RG-module R, 15.3.2 reduces to 15.3.1.

For the rest of this section we assume that R is a PID. Then "flat" is equivalent to "torsion free" (see for example III.9.1.3 of [76]). Let G have type FD (equivalently G is finitely presented and has type FP). By 7.2.13, there is a $K(G,1)$-complex X of finite type. Let $\{K_i\}$ be a finite type filtration of $\tilde{X}$. Then 13.3.2 and 13.3.3 give:

Proposition 15.3.3. *Under these hypotheses G is an n-dimensional duality group iff for all $k \neq n - 1$ $\{\tilde{H}_k(\tilde{X} \stackrel{c}{-} K_i; R)\}$ is pro-trivial and $\{H_{n-1}(\tilde{X} \stackrel{c}{-} K_i; R)\}$ is pro-torsion free and not pro-trivial.* ☐

The more general notion of n-dimensional Poincaré Duality group, G, over R is defined as in the orientable case, above, except that $H^n(G, RG)$ is only required to be isomorphic to R as an R-module. Then G is a duality group, so 15.3.2 applies. In this case we get a sharpening of 15.3.3:

Proposition 15.3.4. *Under the hypotheses of 15.3.3, G is an n-dimensional Poincaré Duality group iff for all $k \neq n - 1$ $\{\tilde{H}_k(\tilde{X} \stackrel{c}{-} K_i; R)\}$ is pro-trivial and $\{H_{n-1}(\tilde{X} \stackrel{c}{-} K_i; R)\}$ is stable with free inverse limit of rank 1 (i.e., stably R).*

We turn to examples. The primary examples[6] of n-dimensional [orientable] Poincaré Duality groups are the fundamental groups of closed [orientable] aspherical n-manifolds (CW n-manifolds in our treatment) – see 15.2.7. By 15.3.4, the "homology at infinity" of an n-dimensional [orientable] Poincaré Duality group is that of an $(n-1)$-sphere [on which G acts trivially]. In Sect. 16.6 we will see examples which are not simply connected at infinity.

By 15.3.3, finitely generated non-trivial free groups are 1-dimensional duality groups over R and those of rank ≥ 2 are not Poincaré Duality groups. A group G for which there exists a finite 2-dimensional $K(G,1)$-complex is a 2-dimensional duality group over R iff G has one end, by 13.5.7. For example, the simple groups G mentioned in 9.4.4 have one end (by Theorem 6.3 of [14]) and are therefore 2-dimensional duality groups. It is proved in [11] that if X is the "spine of Outer Space" K_n of Sect. 9.5, then the condition in 15.3.4 is satisfied, hence every torsion free subgroup of finite index in $\mathrm{Out}(F_n)$ is a $(2n - 3)$-dimensional duality group.

Many other examples of duality groups are known, some discussed in [29].

Source Notes: 15.3.1 and 15.3.2 are found in [14].

---

[6] Indeed, the only known examples at time of writing.

Exercises

1. Prove 15.3.3.
2. What can be deduced from 15.3.3 about $H_*^e(\tilde{X}; R)$?

PART V: HOMOTOPICAL GROUP THEORY

In Part IV, we presented topics from homological group theory as "homology at the end" of universal covers of suitable $K(G, 1)$-complexes. Here we do the same with homotopy. The difference is that while there is a well established, and totally algebraic, subject called Homology of Groups based on the idea of a resolution, there is no corresponding established algebraic theory called Homotopy of Groups. We attempt to organize the beginnings of Homotopy Theory of Groups, making full use of topological methods. At each step we develop the necessary topology at the level of proper homotopy theory and then apply it to universal covers of suitable CW complexes.**

In Chap. 16 we deal with the lowest dimensional cases: connectedness at infinity and various inequivalent definitions of fundamental group at infinity. We prove Wright's Theorem which places severe restrictions on when a locally finite CW complex can be a non-trivial covering space. We also treat some important examples in detail: Whitehead's Contractible 3-manifold, and Davis' examples of closed manifolds whose universal covers are not simply connected at infinity.

And in Chap. 17 we discuss basic topics in higher-dimensional homotopical group theory.

** To keep the two ideas apart, we tend (in Part V) to use Y for a general strongly locally finite CW complex and $\tilde{X}$ for the universal cover of a finite CW complex X.

The Fundamental Group At Infinity

16.1 Connectedness at infinity

Let Y be a strongly locally finite path connected CW complex. In this section we discuss various meanings of the vague sentence "Y is connected at infinity". One possible meaning is that Y has one end. As we saw in Sect. 13.4, this means that for any two proper rays ω and τ in Y, $\omega \mid \mathbb{N}$ and $\tau \mid \mathbb{N}$ are properly homotopic. Another possible meaning is that Y is *strongly connected at infinity* by which we mean that any such ω and τ are themselves properly homotopic. A third possible meaning is that the infinite 1-chains over the ring R defined by any such (cellular) ω and τ are properly homologous, in which case we will say that Y is strongly R-*homology connected at infinity*. Then the distinctions multiply: if Y has more than one end we can ask: is Y strongly connected or strongly R-homology connected at a particular end? To deal with all these matters we need a vocabulary. So we begin again.

We define a *strong end* of Y to be a proper homotopy class of proper rays in Y. We denoted the set of ends of Y by $\mathcal{E}(Y)$. Now we denote the set of strong ends of Y by $\mathcal{SE}(Y)$. The function $\gamma : \mathcal{SE}(Y) \to \mathcal{E}(Y)$, sending a proper homotopy class of proper rays to the end it determines, is surjective; its failure to be injective is related to fundamental group questions, as we now explain.

In order to discuss inverse sequences of fundamental groups, we need a replacement for the base point. A *base ray* in Y is a chosen proper ray ω : $[0, \infty) \to Y$. We write (Y, ω) for the space Y equipped with the base ray ω. This ω is *well parametrized* with respect to the finite filtration $\{L_i\}$ of Y if $\omega([i, \infty)) \subset Y \stackrel{c}{-} L_i$ for all i. Any proper ray can be reparametrized by a proper homotopy to achieve this with respect to a given $\{L_i\}$. Given $(Y, \{L_i\}, \omega)$ with ω well parametrized, consider the inverse sequence of groups

$$\{\pi_1(Y \stackrel{c}{-} L_i, \omega(i))\},$$

where the suppressed bond:

$$\pi_1(Y \stackrel{c}{-} L_{i+1}, \omega(i+1)) \to \pi_1(Y \stackrel{c}{-} L_i, \omega(i))$$

is $[\tau] \to [\omega_{[i,i+1]}.\tau.\omega_{[i,i+1]}^{-1}]$. Here, "." denotes path-multiplication, and $\omega_{[a,b]}$: $I \to Y$ denotes the path $t \mapsto \omega((1-t)a+tb)$, i.e., the $[a,b]$-segment of the ray ω. For fixed ω, let $\mathcal{SE}(Y,\omega) \subset \mathcal{SE}(Y)$ be the set of strong ends of Y which define the same end as ω. We define a canonical bijection:

$$\eta : \mathcal{SE}(Y,\omega) \to \varprojlim{}^1\{\pi_1(Y \xrightarrow{c} L_i, \omega(i))\}$$

as follows. Take a proper ray τ in Y defining the same end as ω. Then there is a proper homotopy $H : \tau \mid \mathbb{N} \cong \omega \mid \mathbb{N}$. Let σ_i be the path $H \mid \{i\} \times I$ from $\tau(i)$ to $\omega(i)$. There is a sequence $n_1 < n_2 < \cdots$ such that, for all i, σ_{n_i} lies in $Y \xrightarrow{c} L_i$ and $n_i \geq i$. Let μ_i be the loop $\omega_{[i,n_i]}.\sigma_{n_i}^{-1}.\tau_{[n_i,n_{i+1}]}.\sigma_{n_{i+1}}.\omega_{[i,n_{i+1}]}^{-1}$, which is based at $\omega(i)$. See Fig. 16.1. Then $([\mu_i])$ defines an element of the set $\varprojlim{}^1\{\pi_1(Y \xrightarrow{c} L_i, \omega(i))\}$.

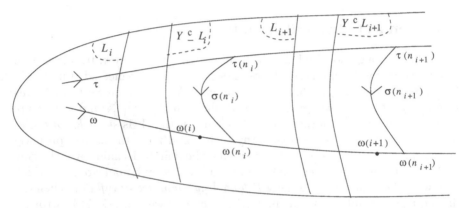

Fig. 16.1.

Proposition 16.1.1. *The correspondence* $\tau \mapsto ([\mu_i])$ *induces a bijection* η : $\mathcal{SE}(Y,\omega) \to \varprojlim{}^1\{\pi_1(Y \xrightarrow{c} L_i, \omega(i))\}$.

Proof. We indicate the definition of η^{-1}, leaving it to the reader to check that η and η^{-1} are well defined and are mutually inverse. The omitted details are tedious but instructive.

For each $i \geq 0$, let ν_i be an edge loop in $Y \xrightarrow{c} L_i$ based at $\omega(i)$. Consider the sequence of paths $\nu_0, \omega_{[0,1]}, \nu_1, \omega_{[1,2]}, \ldots$. There is a proper ray $\zeta : [0,\infty) \to Y$ whose restriction to each closed interval $[n,n+1]$ agrees (up to parametrization) with the nth path in that sequence ($n \geq 0$). Then η^{-1} is well defined: it takes $([\nu_i])$ to the strong end defined by ζ. See Fig. 16.2. $\square$

Recall from Sect. 11.3 that an inverse sequence $\{G_n\}$ of groups is *semistable* if for each m there exists $\phi(m) \geq m$ such that for all $k \geq \phi(m)$ image $f_m^{\phi(m)}$ = image f_m^k, and that, by 11.3.2, when each G_n is countable this is equivalent to the triviality of $\varprojlim{}_n^1\{G_n\}$. Summarizing:

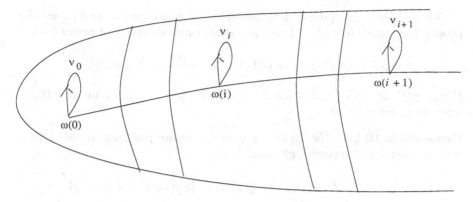

Fig. 16.2.

Proposition 16.1.2. *The following are equivalent:*

(i) *Every proper ray defining the same end as ω is properly homotopic to ω;*
(ii) $\{\pi_1(Y \stackrel{c}{-} L_i, \omega(i))\}$ *is semistable;*
(iii) $\varprojlim^1 \{\pi_1(Y \stackrel{c}{-} L_i, \omega(i))\}$ *is trivial.*　　　　　□

It follows from this and the discussion in Sect. 11.3 that whenever $\mathcal{SE}(Y, \omega)$ contains more than one strong end, it contains uncountably many strong ends. Example 11.4.15 provides an inverse mapping telescope with one end but uncountably many strong ends.

Remark 16.1.3. Using 16.1.1 we can easily prove 11.3.6 in the case of an inverse sequence $\{G_n\}$ of finitely presented groups. For each n let (X_n, v_n) be a pointed finite connected CW complex with $\pi_1(X_n, v_n) \cong G_n$. Choose a cellular map $(X_{n+1}, v_{n+1}) \to (X_n, v_n)$ inducing the bond $G_{n+1} \to G_n$. Let Y be the inverse mapping telescope and ω the obvious base ray with $\omega(i)$ equal to (the equivalence class of) v_i. The following diagram commutes, and by 16.1.1, η and η' are bijections.

$$\varprojlim^1 \{\pi_1(Y \stackrel{c}{-} L_i, \omega(i))\}$$

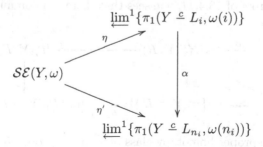

$$\varprojlim^1 \{\pi_1(Y \stackrel{c}{-} L_{n_i}, \omega(n_i))\}$$

Hence α is a bijection as claimed in 11.3.6. In the case of arbitrary groups G_n, a similar argument works using filtered ends, filtered strong ends, etc., in the spirit of Sect. 14.3.

We write $e(\omega)$ for the end determined by the proper ray ω, and $[\omega]$ for the proper homotopy class of ω. There is a short exact sequence of pointed sets

$$(\mathcal{SE}(Y,\omega),[\omega]) \longleftrightarrow (\mathcal{SE}(Y),[\omega]) \xrightarrow{\gamma} (\mathcal{E}(Y), e(\omega)).$$

Define $\pi_0^e(Y,\omega)$ to be the pointed set $(\mathcal{SE}(Y),[\omega])$. Then 13.4.6 and 16.1.1 allow us to rewrite this:

Proposition 16.1.4. *The functions η and γ define functions a and b in a natural short exact sequence of pointed sets:*

$$\varprojlim{}^1\{\pi_1(Y \xrightarrow{c} L_i, \omega(i))\} \xrightarrow{a} \pi_0^e(Y,\omega) \xrightarrow{b} \varprojlim\{\pi_0(Y \xrightarrow{c} L_i, \omega(i))\}.$$

$\square$

This[1] should be compared with the $n = 0$ case of 11.4.8: indeed, one can also describe a and b directly in the spirit of Remark 11.4.9.

Turning to homology, let Y be oriented and let R be a ring. By 10.1.14, the proper ray ω can be taken to be cellular. With the usual CW complex structure and orientation on $[0,\infty)$, an oriented 1-cell of $[0,\infty)$ has the form $[n, n+1]$. Let $\bar{\omega} = \sum_{n=0}^{\infty} \omega_\#([n, n+1])$. We call $\bar{\omega}$ the *chain defined by* ω. This $\bar{\omega}$ is a 0-cycle of the end of Y in the sense of Sect. 11.4. If τ is another cellular proper ray in Y, we say ω and τ are *properly R-homologous* if $\bar{\omega} - \bar{\tau} \in B_0^e(Y; R)$; i.e., if $\bar{\omega} - \bar{\tau}$ differs from the boundary of an infinite 2-chain by a finite 1-chain. A *strong R-homology end* of Y is a proper R-homology class of chains defined by a proper ray in Y. We denote the set of strong R-homology ends of Y by $\mathcal{SHE}(Y; R)$. Thus $\mathcal{SHE}(Y; R) \subset H_0^e(Y; R)$. From 11.4.8, we have an exact sequence of R-modules

$$0 \to \varprojlim{}^1\{H_1(Y \xrightarrow{c} L_i; R)\} \xrightarrow{a} H_0^e(Y; R) \xrightarrow{b} \varprojlim\{H_0(Y \xrightarrow{c} L_i; R)\} \to 0.$$

Using the geometric explanation of b given in Remark 11.4.9, together with 13.4.7 and the proof of 13.4.11, one sees that there is a commutative diagram in Sets:

$$
\begin{array}{ccc}
\mathcal{SE}(Y) \xrightarrow{\alpha} & \mathcal{SHE}(Y; R) \lhook\joinrel\longrightarrow & H_0^e(Y; R) \\
\gamma \downarrow \quad {\scriptstyle\beta} \swarrow & & \downarrow {\scriptstyle -b} \\
\mathcal{E}(Y) \rightarrowtail \varprojlim\{\pi_0(Y \xrightarrow{c} L_i)\} \xrightarrow[h_{\mathcal{E}(Y)}]{} & \varprojlim\{H_0(Y \xrightarrow{c} L_i; R)\}. &
\end{array}
$$

Here α takes the proper homotopy class of ω to the proper R-homology class $\{\bar{\omega}\}$ of $\bar{\omega}$, and β takes $\{\bar{\omega}\}$ to $e(\omega)$. The bottom line was defined at the

---

[1] We caution the reader against trying to deduce too much from exactness in the category Pointed Sets: see the footnote following the proof of 11.3.2.

end of Sect. 13.4. The minus sign in "$-b$" occurs because the proper ray is parametrized away from its initial point and towards infinity.

Let $\mathcal{SHE}(Y, \omega; R) \subset H_0^e(Y; R)$ be the set of strong R-homology ends which define the same end as ω. Then $\mathcal{SHE}(Y, \omega; R)$ is a subset of the inverse image under $-b$ of an element of $\varprojlim\{H_0(Y \overset{c}{-} L_i; R)\}$. By subtracting $\{\bar{\omega}\} \in H_0^e(Y; R)$ we translate $\mathcal{SHE}(Y, \omega; R)$ into $\ker b$. Indeed, identifying the $\varprojlim^1$ expression with $\ker b$ via a, this gives us an injection $\bar{\eta}$ defined by $\bar{\eta}(\{\bar{\tau}\}) = \{\bar{\tau}\} - \{\bar{\omega}\}$, making the following diagram commute:

$$
\begin{array}{ccc}
\mathcal{SE}(Y, \omega) & \overset{\eta}{\rightarrowtail} & \varprojlim^1\{\pi_1(Y \overset{c}{-} L_i, \omega(i))\} \\
\big\downarrow{\scriptstyle \alpha|} & & \big\downarrow{\scriptstyle h_R} \\
\mathcal{SHE}(Y, \omega; R) & \overset{\bar{\eta}}{\rightarrowtail} & \varprojlim^1\{H_1(Y \overset{c}{-} L_i; R)\}
\end{array}
$$

where h_R is induced by the composition

$$
\pi_1(Y \overset{c}{-} L_i, \omega(i)) \overset{h}{\twoheadrightarrow} H_1(Y \overset{c}{-} L_i; \mathbb{Z}) \overset{1 \otimes \cdot}{\longrightarrow} R \otimes_{\mathbb{Z}} H_1(Y \overset{c}{-} L_i; \mathbb{Z}) \overset{\cong}{\longrightarrow} H_1(Y \overset{c}{-} L_i; R)
$$

(see 3.1.20 and 12.4.2).

The function $\bar{\eta}$ can also be defined by analogy with the definition of η, above.

Proposition 16.1.5. *If Y has one end, $\bar{\eta}$ is a bijection.*

Proof. The proof is analogous to that of 16.1.1. Alternatively, when $R = \mathbb{Z}$ or more generally when the homomorphism $1 \otimes \cdot$ is surjective, the function h_R is surjective by 3.1.19 and 11.3.4, and so the injection $\bar{\eta}$ is a bijection. $\qquad \square$

Proposition 16.1.6. *$\bar{\eta}$ is a homomorphism of R-modules, hence an isomorphism when Y has one end.*

Proof. Let the proper ray τ define the same end as ω. It is implicit in the proof of 16.1.1 that τ is properly homotopic to a proper ray which has the form of an "infinite product" $\nu_0.\omega_{[0,1]}.\nu_1.\omega_{[1,2]}.\ldots$, where ν_i is a loop at $\omega(i)$. If τ' is another such, represented by $\nu'_0.\omega_{[0,1]}.\nu'_1.\omega_{[1,2]}.\ldots$, then the proper ray ζ represented by $\nu_0.\nu'_0.\omega_{[0,1]}\nu_1.\nu'_1.\omega_{[1,2]}.\ldots$ has the property that $\bar{\eta}(\{\bar{\zeta}\}) = \bar{\eta}(\{\bar{\tau}\}) + \bar{\eta}(\{\bar{\tau}'\})$. The rest is clear. $\qquad \square$

We say Y is *strongly connected* at the end $e(\omega)$ if any two proper rays in Y defining $e(\omega)$ are properly homotopic (see 16.1.2). We say Y is *strongly homology connected* at the end $e(\omega)$ if the chains defined by any two proper cellular rays in Y defining $e(\omega)$ are properly R-homologous. When Y has one end we say Y is *strongly connected at infinity* or *strongly R-homology connected at infinity* when it has the appropriate property at its end.

When Y has one end, Y is strongly R-homology connected at infinity iff $\{H_1(Y \overset{c}{-} L_i; R)\}$ is semistable iff (by 11.3.2) $\varprojlim^1\{H_1(Y \overset{c}{-} L_i; R)\}$ is trivial. But when Y has more than one end, care is needed in making the corresponding statement (16.1.8 below). Note first that, by 13.4.7, an end e of Y can be regarded as an element (Z_i) of $\varprojlim\{\pi_0(Y \overset{c}{-} L_i)\}$; here Z_i denotes a path component of $Y \overset{c}{-} L_i$, and $Z_i \supset Z_{i+1}$ for all i:

Proposition 16.1.7. *Y is strongly R-homology connected at the end e iff $\varprojlim^1\{H_1(Z_i; R)\}$ is trivial.*

Proof. One defines a bijection $\eta' : \mathcal{SHE}(Y, \omega; R) \to \varprojlim^1\{H_1(Z_i; R)\}$ by analogy with the definition of η. The inverse of η' is defined as in the proof of 16.1.1. The details are an exercise. $\qquad\square$

Proposition 16.1.8. *Let $H_1(Y; R)$ be finitely generated. The following are equivalent when R is a PID:*

(i) *Y is strongly R-homology connected at every end;*
(ii) *$H_e^1(Y; R)$ is a free R-module;*
(iii) *$\{H_1(Y \overset{c}{-} L_i; R)\}$ is semistable;*
(iv) *$\varprojlim^1\{H_1(Y \overset{c}{-} L_i; R)\}$ is trivial.*

Proof. The equivalence of (ii), (iii) and (iv) comes[2] from 12.5.10 and 11.3.2. The equivalence of (i) and (iii) follows from 16.1.22 in the Appendix as explained in 16.1.21. $\qquad\square$

Example 16.1.9. Let T be the dyadic solenoid inverse mapping telescope of 11.4.15. T has one end, and $H_1(T; R)$ is a finitely generated R-module. T is not strongly $\mathbb{Z}$-homology connected at infinity though (like all one-ended Y; see 3.5.9) T is strongly $\mathbb{Q}$-homology connected at infinity.

Example 16.1.10. Let $\phi : \langle a, b \rangle \to \langle a, b \rangle$ be the endomorphism of a free group on two generators given by $\phi(a) = aba^{-1}b^{-1}$, $\phi(b) = a^2b^2a^{-2}b^{-2}$. Let $f : S^1 \vee S^1 \to S^1 \vee S^1$ be the obvious map inducing ϕ on fundamental group. The *Case-Chamberlin inverse mapping telescope* C is obtained from the inverse sequence

$$S^1 \vee S^1 \xleftarrow{\;f\;} S^1 \vee S^1 \xleftarrow{\;f\;} \cdots.$$

C is strongly $\mathbb{Z}$-homology connected at infinity, but is not strongly connected at infinity.

We leave as an exercise:

Proposition 16.1.11. *If Y is strongly connected at the end e then Y is strongly R-homology connected at e.* $\qquad\square$

---

[2] The hypotheses that $H_1(Y; R)$ be finitely generated and that R be a PID are only needed for the equivalence of (ii) and (iii).

Combining 16.1.8 and 16.1.11 we get:

Proposition 16.1.12. *Let Y be strongly connected at each end, let R be a PID, and let $H_1(Y;R)$ be finitely generated. Then $H_e^1(Y;R)$ is a free R-module.* □

Dependence on skeleta is given by:

Proposition 16.1.13. *Let $i^{(n)} : Y^n \hookrightarrow Y$ be the inclusion map. Let ω be a cellular base ray (in $Y^1 \subset Y$) which is well parametrized with respect to $\{L_i\}$.*

(i) $i_\#^{(1)} : \mathcal{E}(Y^1) \to \mathcal{E}(Y)$ *is a bijection;*

(ii) $i_\#^{(2)} : \mathcal{SHE}(Y^2;R) \to \mathcal{SHE}(Y;R)$ *is a bijection;*

(iii) $i_\#^{(2)} : \mathcal{SE}(Y^2,\omega) \to \mathcal{SE}(Y,\omega)$ *is a bijection;*

(iv) $i_\#^{(2)} : \mathcal{SHE}(Y^2,\omega;R) \to \mathcal{SHE}(Y,\omega;R)$ *is a bijection.*

Proof. These follow from the Cellular Approximation Theorems 1.4.3 and 10.1.14. Indeed, (i) is essentially 13.4.1. □

Thus the invariance question reduces to low dimensional skeleta. More precisely, we leave as an exercise:

Proposition 16.1.14. *Let Y and Z be path connected locally finite 2-dimensional CW complexes, let $f : Y \to Z$ be a proper 2-equivalence, and let ω be a base ray in Y. Then f induces isomorphisms (in the appropriate categories) $\mathcal{E}(Y) \to \mathcal{E}(Z)$, $\mathcal{SE}(Y) \to \mathcal{SE}(Z)$, $\mathcal{SHE}(Y;R) \to \mathcal{SHE}(Z;R)$, $\mathcal{SE}(Y,\omega) \to \mathcal{SE}(Z,f \circ \omega)$, and $\mathcal{SHE}(Y,\omega;R) \to \mathcal{SHE}(Z,f \circ \omega;R)$.* □

The exact sequences in this section are natural. We remarked in Sect. 11.3 that it is possible to express $\lim^1$ as a functor on towers-Groups or on towers-(R-modules). However, here we have identified our $\lim^1$ terms with topologically simpler objects via the canonical bijections η and $\bar{\eta}$, so we may avoid this functor. By 10.1.14, we have:

Proposition 16.1.15. *Let $f : Y \to Z$ be a proper map between strongly locally finite, infinite, path connected CW complexes. Then the function on proper rays $\tau \mapsto f \circ \tau$ induces functions $f_\# : \mathcal{E}(Y) \to \mathcal{E}(Z)$, $f_\# : \mathcal{SE}(Y) \to \mathcal{SE}(Z)$, and $f_\# : \mathcal{SHE}(Y;R) \to \mathcal{SHE}(Z;R)$. Indeed, $\mathcal{E}(\cdot)$, $\mathcal{SE}(\cdot)$ and $\mathcal{SHE}(\cdot;R)$ can be regarded as covariant functors from the proper homotopy category to the category Sets.* □

A proper map $f : Y \to Z$ together with a base ray ω in Y define a *base ray preserving proper map* $f : (Y,\omega) \to (Z,\omega')$ where $\omega' = f \circ \omega$. A *base ray preserving proper homotopy* is a proper base ray preserving map $H : (Y \times I, \omega(\cdot,0)) \to (Z,\omega')$, i.e., $H_t(\omega(s,0)) = \omega'(s)$ for all $t \in I$ and $s \in [0,\infty)$. Thus one defines *the base ray preserving proper homotopy category*. There are similar definitions for pairs.

Addendum 16.1.16. *The function on rays $\tau \mapsto f \circ \tau$ induces functions of pointed sets $f_{\#} : \mathcal{SE}(Y, \omega) \to \mathcal{SE}(Z, f \circ \omega)$ and $f_{\#} : \mathcal{HSE}(Y, \omega; R) \to \mathcal{HSE}(Z, f \circ \omega; R)$, where the base points of these sets are defined by ω and $f \circ \omega$. In fact, $\mathcal{SE}(\cdot, \omega)$ and $\mathcal{HSE}(\cdot, \omega; R)$ define covariant functors from the base ray preserving proper homotopy category to the category* Pointed Sets. $\square$

Corollary 16.1.17. *The short exact sequence in 16.1.4 is natural with respect to base ray preserving proper homotopy classes of proper maps.*[3] $\square$

Recall that proper edge rays are defined in Sect. 11.1. Sometimes it is convenient to describe ends combinatorially. A *proper parametrization* of the proper edge ray $\tau := (\tau_1, \tau_2, \ldots)$ is a map $h_\tau : [0, \infty) \to Y$ such that for each integer $k \geq 0$, $h_\tau \mid [k, k + 1]$ is a characteristic map for the edge τ_k (after reparametrizing to $[-1, 1]$) in the non-degenerate case, and is constant at the point τ_k in the degenerate case. This h_τ is a proper ray because no edge appears infinitely often in $(\tau_1, \tau_2, \ldots)$. Using 10.1.14 and the proof of 3.1.1, it is easy to show:

Proposition 16.1.18. *Every proper ray in Y is properly homotopic to the proper parametrization of some proper edge ray in Y. All proper parametrizations of a proper edge ray are properly homotopic.* $\square$

Proper edge rays $\sigma := (\sigma_1, \sigma_2, \ldots)$ and $\tau := (\tau_1, \tau_2, \ldots)$ in Y *define the same end* of Y if there are edge paths λ_n in Y where: λ_n and σ_n have the same initial point, the final point of λ_n is the initial point of τ_n, and for any finite subcomplex K of Y only finitely many of the edge paths λ_n meet K. This is equivalent to saying that proper parametrizations of these proper edge rays define the same end.

By 3.4.1 we have:

Proposition 16.1.19. *Proper parametrizations of $(\sigma_1, \sigma_2, \ldots)$ and $(\tau_1, \tau_2, \ldots)$ are properly homotopic iff there exist edge paths λ_n as above such that for any finite subcomplex K of Y all but finitely many of the edge loops $\sigma_n.\lambda_{n+1}.\tau_n^{-1}.\lambda_n^{-1}$ are equivalent*[4] *in $Y \stackrel{c}{=} K$ to trivial edge loops.* $\square$

Appendix. Semistability and trees of modules

The combinatorial definition of proper rays in 16.1.18 and 16.1.19 is useful in the case of a locally finite tree T. Choose a base vertex w for T. For each $e \in \mathcal{E}(T)$ there is a unique proper edge ray τ_e with initial point w such that every initial segment of τ_e is a reduced edge path and the proper ray h_{τ_e} determines the end e. The space $\mathcal{E}(T)$ is compact, and we clearly have:

---

[3] If one is prepared to consider $\lim^1$ as a functor from towers-Groups to Pointed Sets, or from towers-(R-modules) to R-modules (see Sect. 11.3), one can check that the bijections η and $\bar{\eta}$ are natural.

[4] In the sense of Sect. 3.1; see Fig. 2.1.

Proposition 16.1.20. *If a sequence $\langle e_n \rangle$ converges to $e \in \mathcal{E}(T)$ then as $n \to \infty$ more and more of the edges of τ_{e_n} agree with the edges of τ_e.* $\qquad\square$

For each vertex v of T let τ_v denote the unique reduced edge path with initial vertex w and final vertex v. The set of vertices T^0 becomes a poset under the relation $u \leq v$ iff u is a vertex of an edge in τ_v. Each $v \neq w$ has a unique *immediate predecessor* $\pi(v)$, the last vertex before v on τ_v. We write $d(v, w)$ for the distance from v to w (as defined in Sect. 11.1).

Let $\{M_v\}_{v \in T^0}$ be a family of R-modules. The poset T^0 is not usually a directed set, but it contains many directed subsets. In fact, for each $e \in \mathcal{E}(T)$ the set $\{\tau_e(n)\}_{n \in \mathbb{N}}$ is such, where $\tau_e(n)$ denotes the nth vertex of τ_e (i.e., the final point of the nth initial segment). We will assume given a homomorphism $M_v \to M_{\pi(v)}$ for all vertices $v \neq w$. Then $\{M_{\tau_e(n)}\}$ is an inverse sequence of R-modules for each $e \in \mathcal{E}(T)$.

Associated with this is the *total inverse sequence* $\{N_n\}$ where $N_n := \oplus\{M_v \mid d(v, w) = n\}$ and the bond $N_n \to N_{n-1}$ is defined by the bonds $M_v \to M_{\pi(v)}$ for all v with $d(v, w) = n$.

Recall that a vertex of the tree T is a *leaf* if it is a face of only one 1-cell. We will assume that T has no leaves.

Example 16.1.21. To prove 16.1.8 we take T to be "the tree of path components of complements." With notation as above, a vertex of T is a path component of some $Y \stackrel{c}{-} L_i$, and there is an edge joining the path component C of $Y \stackrel{c}{-} L_i$ to the path component D of $Y \stackrel{c}{-} L_{i+1}$ iff $D \subset C$. Then T has no leaves iff every path component of every $Y \stackrel{c}{-} L_i$ is unbounded, as we may always assume, by 13.4.9. In applying the following Theorem 16.1.22 to this, the module M_C corresponding to the "vertex" C will be $H_1(C; R)$, and the bond $H_1(D; R) \to H_1(C; R)$ will be induced by inclusion. The module N_n will be $H_1(Y \stackrel{c}{-} L_n; R)$. Then 16.1.8 follows from:

Theorem 16.1.22. *Let the locally finite pointed tree (T, w) have no leaves and let $\{M_v\}_{v \in T^0}$ be a family of R-modules with given homomorphisms $M_v \to M_{\pi(v)}$ for all $v \neq w$. The total inverse sequence $\{N_n\}$ is semistable [resp. protrivial] iff for every $c \subset \mathcal{E}(T)$ the inverse sequence $\{M_{\tau_e(n)}\}$ is semistable [resp. pro-trivial].*

The proof involves the compactness of the space $\mathcal{E}(T)$. Semistability will be analyzed by means of "eventual images." If $\{U_n\}$ is an inverse sequence of R-modules the *eventual image* in U_m is $EI(U_m) := \bigcap_{n \geq m} \text{image}(U_n \to U_m)$.

Thus $\{U_n\}$ is semistable iff for each m there exists $k \geq m$ such that $EI(U_m) = \text{image}(U_k \to U_m)$.

Proof (of 16.1.22). "Only if" is clear. We prove "if" for the semistable case, leaving the (similar) proof of the pro-trivial case as an exercise.

We may assume $M_w = 0$, hence $N_0 = 0$. For each n let $d(n)$ be the least integer $\leq n$ such that $EI(N_{d(n)}) = \text{image}(N_n \to N_{d(n)})$. Then $d(n) \geq 0$ and

$EI(N_k) \subsetneqq \text{image}(N_n \to N_k)$ when $n \geq k > d(n)$. Thus $\{N_n\}$ is semistable iff $d(n) \to \infty$ as $n \to \infty$.

Suppose $d(n) \not\to \infty$ as $n \to \infty$. Then there exists m_0 and a cofinal subsequence $\{N_{n_k}\}$ such that $d(n_k) < m_0$ for all k. We may assume $n_1 > m_0$. Then for all $k \geq 1$ $EI(N_{m_0}) \subsetneqq \text{image}(N_{n_k} \to N_{m_0})$. It follows easily that $EI(N_{m_0}) \subsetneqq \text{image}(N_n \to N_{m_0})$ for all $n \geq m_0$. Letting V_n denote the set of vertices of T whose distance from w is n, the last sentence can be rewritten: for each $n \geq m_0$

$$\bigoplus_{v \in V_{m_0}} \bigcap_{i \geq n} \text{image}\left(\left(\bigoplus_{\substack{u \in V_i \\ v \leq u}} M_u\right) \to M_v\right) \subsetneqq \bigoplus_{v \in V_{m_0}} \text{image}\left(\left(\bigoplus_{\substack{u \in V_n \\ v \leq u}} M_u\right) \to M_v\right).$$

So for some $v_0 \in V_{m_0}$,

$$E := \bigcap_{i \geq n} \text{image}\left(\left(\bigoplus_{\substack{u \in V_i \\ v_0 \leq u}} M_u\right) \to M_{v_0}\right) \subsetneqq \text{image}\left(\left(\bigoplus_{\substack{u \in V_n \\ v_0 \leq u}} M_u\right) \to M_{v_0}\right).$$

Since V_{m_0} is finite we may assume (passing to a subsequence if necessary) that v_0 is independent of n. This inequality of modules shows that for each $n \geq m_0$ there is $u_n \in V_n$ such that $\text{image}(M_{u_n} \to M_{v_0})$ does not lie in E (which is independent of n).

The tree T has no leaves, so for each $n \geq m_0$ there exists $e_n \in \mathcal{E}(T)$ such that $\tau_{e_n}(n) = u_n$. Since $\mathcal{E}(T)$ is compact we may assume (again passing to a subsequence) that the sequence of e_n's converges to some $e \in \mathcal{E}(T)$. By 16.1.20, for each m there exists $r(m)$ such that, when $n > r(m)$, $\tau_e(m) = \tau_{e_n}(m) < \tau_{e_n}(n) = u_n$. Thus if $m \geq m_0$, $\text{image}(M_{\tau_e(m)} \to M_{v_0})$ does not lie in E. But (referring to the inverse sequence $\{M_{\tau_e(m)}\}$) $EI(M_{\tau_e(m_0)}) \subset E$. So $EI(M_{\tau_e(m_0)}) \subsetneqq \text{image}(M_{\tau_e(m)} \to M_{v_0})$ for all $m \geq m_0$, contradicting the fact that $\{M_{\tau_e(m)}\}$ is semistable. $\square$

Exercises

1. Fill in the omitted details in 16.1.1.
2. Give an elementary proof (i.e., not using 16.1.1) that Y is strongly connected at $e(\omega)$ iff $\{\pi_1(Y \xrightarrow{c} L_i, \omega(i))\}$ is semistable.
3. Check the details of 16.1.4.
4. Give an example where $\mathcal{SHE}(Y; R)$ is a proper subset of $H_0^e(Y; R)$.
5. Prove that a CW-proper 1-equivalence induces a bijection of ends.
6. Draw a picture of a proper ray representing $\bar{\eta}(\{\bar{\tau}\}) - \bar{\eta}(\{\bar{\tau}'\})$.
7. What is meant by saying that the exact sequence in 16.1.4 is natural? Prove naturality.
8. Develop a theory of "strong ends" and "homology strong ends" in filtered homotopy.

16.2 Analogs of the fundamental group

Let Y be a strongly locally finite path connected CW complex, let e be an end of Y, and let the proper ray ω represent e. When K is a finite subcomplex of Y we denote by $K(e)$ the path component of $Y \xrightarrow{c} K$ containing all but a compact subset of $\omega([0, \infty))$. We say that Y is *simply connected at* e if for every finite subcomplex K of Y there is a finite subcomplex $L \supset K$ such that every map $f : S^1 \to L(e)$ extends to a map $F : B^2 \to K(e)$. If Y has only one end and satisfies this condition we say that Y is *simply connected at infinity*. For example, $\mathbb{R}^n$ is simply connected at infinity when $n \geq 3$, but not when $n = 2$. The line, $\mathbb{R}$, is simply connected at both ends, but $\mathbb{R} \times S^1$ is not simply connected at either end.

In this section we discuss various ways of associating a "fundamental group" with e. The three definitions we discuss are not, in general, equivalent. Indeed, for the second and third definitions it is not true (without further hypotheses) that triviality implies simple connectivity at e. Nevertheless all three deserve to be known and the relationships among them should be understood.

Let $\{L_i\}$ be a finite filtration of Y, and let ω be well parametrized with respect to $\{L_i\}$. The *fundamental pro-group of* Y *based at* ω (with respect to $\{L_i\}$) is the inverse sequence $\{\pi_1(Y \xrightarrow{c} L_i, \omega(i))\}$ described in detail in Sect. 16.1. If $\{M_i\}$ is another finite filtration of Y with respect to which ω is also well parametrized, then there is an obvious pro-isomorphism from this sequence to $\{\pi_1(Y \xrightarrow{c} M_i, \omega(i))\}$. Thus we may ignore dependence on $\{L_i\}$. We will discuss dependence on ω in a moment.

The triviality of the fundamental pro-group is equivalent to simple connectivity at $e(\omega)$; more precisely:

Proposition 16.2.1. *The following are equivalent:*

(i) Y *is simply connected at* e;
(ii) $\{\pi_1(Y \xrightarrow{c} L_i, \omega(i))\}$ *is pro-trivial;*
(iii) $\varprojlim\{\pi_1(Y \xrightarrow{c} L_i, \omega(i))\}$ *and* $\varprojlim^1\{\pi_1(Y \xrightarrow{c} L_i, \omega(i))\}$ *are trivial.*

Proof. For (iii) $\Leftrightarrow$ (ii), use Exercise 10 of Sect. 11.3; (i) $\Leftrightarrow$ (ii) is clear. ⊔

By 16.1.2, it follows that if Y is simply connected at e then Y is strongly connected at e.

Dependence on skeleta (compare 16.1.13) is given by:

Proposition 16.2.2. $i^{(2)}_{\#} : \{\pi_1(Y^2 \xrightarrow{c} L_i^2, \omega(i))\} \to \{\pi_1(Y \xrightarrow{c} L_i, \omega(i))\}$ *is an isomorphism in* pro-Groups. □

So the invariance question reduces to the 2-skeleton:

Proposition 16.2.3. *Let* $f : Y \to Z$ *be as in 16.1.14, let* $\{L_i\}$ *and* $\{M_j\}$ *be finite filtrations of* Y *and* Z *respectively. Then* ω *can be reparametrized by a proper homotopy so that* $f_{\#} : \{\pi_1(Y \xrightarrow{c} L_i, \omega(i))\} \to \{\pi_1(Z \xrightarrow{c} M_j, f \circ \omega(j))\}$ *is an isomorphism in* pro-Groups.

Proof. Let $H : g \circ f \mid\simeq$ inclusion and $\bar{H} : f \circ g \mid\simeq$ inclusion be proper cellular homotopies. Give Y the cellular base ray ω. By reparametrizing ω if necessary we can assume that $H_t \circ \omega$ [resp. $\bar{H}_t \circ f \circ \omega$] is well parametrized with respect to $\{L_i\}$ [resp. $\{M_j\}$] for all t. Let α_i be the path in Y given by $\alpha_i(t) = H_t \omega(i)$, and let β_j be the path in Z given by $\beta_j(t) = \bar{H}_t f \omega(j)$. The isomorphisms $\pi_1(Y \xrightarrow{c} L_i, g f \omega(i)) \xrightarrow{\alpha_{i\#}} \pi_1(Y \xrightarrow{c} L_i, \omega(i))$ given by $[\sigma_i] \mapsto [\alpha_i^{-1}.\sigma_i.\alpha_i]$ fit together to give an isomorphism in pro-Groups

$$\alpha_\# : \{\pi_1(Y \xrightarrow{c} L_i, g f \omega(i))\} \to \{\pi_1(Y \xrightarrow{c} L_i, \omega(i))\}.$$

There is a similar definition for $\beta_\#$. The proper homotopy H is needed for the proof that $\alpha_\#$ is an isomorphism. We have morphisms in pro-Groups

$$f_\# : \{\pi_1(Y \xrightarrow{c} L_i, \omega(i))\} \to \{\pi_1(Z \xrightarrow{c} M_j, f \omega(j))\}$$

and

$$g_\# : \{\pi_1(Z \xrightarrow{c} M_j, f \omega(j)\} \to \{\pi_1(Y \xrightarrow{c} L_i, g f \omega(i)\}$$

where, again, some details of definition are left to the reader. The upshot is that $\alpha_\# \circ g_\# \circ f_\# = \mathrm{id}$ and $\beta_\# \circ f_\# \circ g_\# = \mathrm{id}$. Together with 16.2.2 this proves what was claimed. $\qquad\square$

The proper ray ω plays the role of the base point. If $F : \omega \simeq \tau$ is a cellular proper homotopy between cellular base rays (which is well parametrized[5] with respect to $\{L_i\}$), F induces a "change of base ray" isomorphism

$$\{\pi_1(Y \xrightarrow{c} L_i, \omega(i))\} \to \{\pi_1(Y \xrightarrow{c} L_i, \tau(i))\}$$

entirely analogous to what is described in Sect. 3.3; the details are omitted. This means that up to pro-isomorphism the fundamental pro-group only depends on the strong end defined by ω. However, if Y is not strongly connected at e, the fundamental pro-groups with respect to different base rays defining e may not be pro-isomorphic, indeed may not even have isomorphic inverse limits. Here is a one-ended example[6] where with one base ray the inverse limit is infinite cyclic, while with another base ray the inverse limit is trivial:

Example 16.2.4. Let $W = S_1^1 \vee S_2^1$, a wedge of two circles with wedge point v. Let $f : W \to W$ be a map taking each circle to itself (hence $f(v) = v$), agreeing with the map $f_{1,2}$ (of degree 2 – see Sect. 2.4) on S_1^1 and with the identity map on S_2^1. Writing a and b for generators of the fundamental groups of the two circles based at v, f induces the homomorphism ϕ defined by $a \mapsto a^2$, $b \mapsto b$. The one-ended space in question is the inverse mapping telescope of

---

[5] The references to "well parametrized" can always be dispensed with by passing to a subsequence $\{L_{n_i}\}$.

[6] The connection between examples of this kind and strong shape theory is explained in Sect. 17.7.

$\{W \xleftarrow{\ f\ } W \xleftarrow{\ f\ } \cdots\}$, which we denote by Y. The sequence $(v, v, \ldots)$ defines a base ray ω in Y using straight lines in the mapping cylinders whose union is Y. By restricting the map f to S_1^1 we find a copy of T, the dyadic solenoid inverse mapping telescope (see 11.4.15) as a subset of Y, and the image of ω actually lies in T. Letting L_n denote the union of the first n mapping cylinders in Y, we have a finite filtration $\{L_i\}$ of Y. Clearly, $\varprojlim\{\pi_1(Y \xrightarrow{c} L_i, \omega(i))\}$ is isomorphic to $\mathbb{Z}$ (generated by S_2^1), since $\varprojlim\{\pi_1(T \xrightarrow{c} L_i, \omega(i))\}$ is trivial. Now $\varprojlim^1\{\pi_1(T \xrightarrow{c} L_i, \omega(i))\}$ is non-trivial by 11.4.15 and thus, by 14.1.1, there is a proper ray τ in T which is not properly homotopic to ω in T. Since T is a retract of Y, such a ray τ is not properly homotopic to ω in Y. We claim $\varprojlim\{\pi_1(Y \xrightarrow{c} L_i), \tau(i)\}$ is trivial. The proof of the contrapositive, that if $\varprojlim\{\pi_1(Y \xrightarrow{c} L_i), \tau(i)\}$ is non-trivial then τ is properly homotopic to ω in Y, is set out in Exercise 4.

Now for our second definition of "fundamental group associated with e:" with notation as before, the *Čech fundamental group of Y based at ω* is $\check{\pi}_1(Y, \omega) := \varprojlim\{\pi_1(Y \xrightarrow{c} L_i, \omega(i))\}$. In view of Example 16.2.4, its isomorphism class depends on ω. We regard $\check{\pi}_1(Y, \omega)$ as a topological group: each $\pi_1(Y \xrightarrow{c} L_i, \omega(i))$ is regarded as a discrete topological group and the inverse limit is interpreted in the category Topological Groups. When thus topologized, $\check{\pi}_1(Y, \omega)$ carries the same information as the fundamental pro-group if (and only if) Y is strongly connected at e. More precisely, in that case the inverse sequence $\{\pi_1(Y \xrightarrow{c} L_i, \omega(i))\}$ can be recovered, up to pro-isomorphism, from the topological group $\check{\pi}_1(Y, w)$. Indeed, let $\{G_i\}$ be a semistable inverse sequence of countable groups with inverse limit G. Topologized as above, G is a complete first countable zero-dimensional[7] topological group. Choose a basis $U_1 \supset U_2 \supset \cdots$ for the neighborhoods of $1 \in G$ which are closed and open subgroups of G. The following proposition is explained more fully in Sect. 16.7:

Proposition 16.2.5. *Under these hypotheses the inverse sequence of (discrete) groups $\{G/U_1 \leftarrow G/U_2 \leftarrow \cdots\}$, where each bond is the obvious quotient epimorphism, is pro-isomorphic to $\{G_i\}$.* □

Note that $\check{\pi}_1(Y, \omega)$ has already appeared in 16.2.1. Its invariance properties (up to isomorphism of topological groups) are covered implicitly by 16.2.2 and 16.2.3. Notice the similarity between 16.2.5 and the fact (following from 13.4.15) that the space of ends $\mathcal{E}(Y)$ determines $\{\pi_0(Y \xrightarrow{c} L_i)\}$ up to pro-isomorphism.

Our third definition is by analogy with $H_1^e(Y; R)$. We define the *strong (or Steenrod) fundamental group* $\pi_1^e(Y, \omega)$: its elements are the base ray preserving proper homotopy classes of proper maps $(S^1 \times [0, \infty), \{v\} \times [0, \infty)) \to (Y, \omega)$, where v is a base point of S^1, and $\{v\} \times [0, \infty)$ denotes the base ray $t \mapsto (v, t)$ in $S^1 \times [0, \infty)$. Multiplication of two such proper maps restricts to ordinary

---

[7] These terms are defined in Sect. 16.7.

loop multiplication on $S^1 \times \{t\}$ for each $t \in [0, \infty)$; similarly for inversion. One obtains a group structure on $\pi_1^e(Y, \omega)$ just as in Sect. 3.3. The trivial element is represented by the map $(e^{2\pi i \theta}, t) \mapsto \omega(t)$.

As in Sect. 3.3, π_1^e is a covariant functor from the base ray preserving proper homotopy category to the category Groups. By analogy with 11.4.8, we define $b : \pi_1^e(Y, \omega) \to \varprojlim \{\pi_1(Y \overset{c}{-} L_i, \omega(i))\}$ as follows. Let

$$\alpha : (S^1 \times [0, \infty), \{v\} \times [0, \infty)) \to (Y, \omega)$$

be a proper map. We can alter α by a proper homotopy, rel$\{v\} \times [0, \infty)$, so that $\alpha(S^1 \times [i, \infty)) \subset Y \overset{c}{-} L_i$ for all i. Then $\alpha \mid S^1 \times \{i\} =: \alpha_i$ is a loop in $Y \overset{c}{-} L_i$ at $\omega(i)$. The required b maps $[\alpha]$ to $([\alpha_i])$ where $[\cdot]$ denotes a base ray preserving proper homotopy class or a pointed homotopy class, as appropriate. It is a routine exercise (compare 11.4.8) to prove:

Proposition 16.2.6. *The function b is well defined and is an epimorphism of groups. It fits into a natural short exact sequence of groups:*

$$\varprojlim{}^1 \{\pi_2(Y \overset{c}{-} L_i, \omega(i))\} \overset{a}{\rightarrowtail} \pi_1^e(Y, \omega) \overset{b}{\twoheadrightarrow} \varprojlim \{\pi_1(Y \overset{c}{-} L_i, \omega(i))\}$$

$\square$

It is clear from 16.2.6 and 16.2.1 that the vanishing of $\pi_1^e(Y, \omega)$ need not imply that Y is simply connected at e, but it does imply that $\check{\pi}_1(Y, \omega)$ is trivial.

Let $e \in \mathcal{E}(Y)$, and for each i let Z_i be the corresponding path component of $Y \overset{c}{-} L_i$ (compare 16.1.7). We say Y is 1-*acyclic at e* (with respect to R) if for each i there is j such that every cellular 1-cycle (over R) in Z_j bounds a cellular 2-chain in Z_i. If Y has only one end this is abbreviated to 1-*acyclic at infinity*. By analogy with 16.2.1 we have:

Proposition 16.2.7. *Let R be a PID and let $H_1(Y; R)$ be finitely generated. The following are equivalent:*

(i) *Y is 1-acyclic at every end with respect to R;*
(ii) *$H_e^1(Y; R) = 0$;*
(iii) *$\{H_1(Y \overset{c}{-} L_i; R)\}$ is pro-trivial;*
(iv) *$\varprojlim \{H_1(Y \overset{c}{-} L_i; R)\}$ and $\varprojlim{}^1 \{H_1(Y \overset{c}{-} L_i; R)\}$ are trivial.*

Proof. Similar to that of 16.1.8 (including its footnote). Again (i) $\Rightarrow$ (iii) uses 16.1.22. $\square$

By analogy with 16.1.11 and 16.1.12 we have:

Proposition 16.2.8. *If Y is simply connected at $e \in \mathcal{E}(Y)$ then Y is 1-acyclic at e (with respect to R).* $\square$

Proposition 16.2.9. *Let Y be simply connected at each end and let $H_1(Y; R)$ be finitely generated, where R is a PID. Then $H_e^1(Y; R) = 0$.* $\square$

Source Notes. Čech homotopy groups first appeared in [40]. For proper homotopy they are developed in [60]. Example 16.2.4 is adapted from a shape theoretic example of Borsuk [20]. Some delicate issues involving the various fundamental groups at infinity are discussed in [26].

Exercises

1. Prove that the inverse sequence $\{H_n\}$ in Example 16.2.4 has trivial inverse limit.
2. Prove that $\check{\pi}_1(Y, \omega)$ is either discrete and countable or non-discrete and uncountable.
3. What is meant by saying that the exact sequence in 16.2.6 is natural? Prove naturality.
4. Fill in the missing proof in Example 16.2.4 by proving that if $\varprojlim\{\pi_1(Y \xrightarrow{c} L_i, \tau(i))\}$ is non-trivial then τ is properly homotopic to the ray ω described in that example. Here are the steps:
   (a) By performing a proper homotopy in T if necessary, it may be assumed that $f(\tau(i+1)) = \tau(i)$, for each i, and that the ray τ consists of straight lines in the mapping cylinders whose union is Y. *Hint:* The bonding maps in the dyadic solenoid inverse sequence (restrictions of f to the copies of S_1^1) are covering projections.
   (b) If (λ_i) is a non-trivial element of $\varprojlim\{\pi_1(Y \xrightarrow{c} L_i, \tau(i))\}$, where we think of λ_i as a loop in W based at $\tau(i) \in S_1^1$, then all but finitely many of the loops λ_i cannot be homotoped, rel base point, off S_2^1. *Hint:* The dyadic solenoid (see Sect. 17.7) is a topological group, hence $\varprojlim\{\pi_1(T \xrightarrow{c} L_i, \tau(i))\}$ is trivial for any τ.
   (c) The loop λ_i is a product $\alpha_i.\mu_i.\beta_i^{-1}$ where α_i and β_i are paths in S_1^1 from $\tau(i)$ to v, and μ_i is a loop in W based at v, such that, in the free groups $\pi_1(W, v)$, the element $[\mu_i]$ is a word in the alphabet defined by a and b which begins and ends with a non-trivial power of b.
   (d) There is a homotopy $H_i : \alpha_i \simeq f \circ \alpha_{i+1}$ rel$\{-1, 1\}$ and these homotopies can be pieced together to give the desired proper homotopy $\tau \underset{p}{\simeq} \omega$.

16.3 Necessary conditions for a free $\mathbb{Z}$-action

Continuing with the set up and notation of Sect. 16.2, we discuss further properties of the fundamental pro-group $\{\pi_1(Y \xrightarrow{c} L_i, \omega(i))\}$. We saw that it is pro-trivial iff Y is simply connected at the end $e(\omega)$; and that it is semistable iff Y is strongly connected at $e(\omega)$. "Semistable" means "pro-isomorphic to a sequence of epimorphisms" (11.3.1). Here we discuss the dual notion "pro-isomorphic to a sequence of monomorphisms" and the combination of the two, which is called "stable." The main result is Wright's Theorem 16.3.4.

First, some terminology. An inverse sequence $\{G_n\}$ of groups is (i) *pro-monomorphic*, (ii) *stable*, (iii) *stably H* (for a given group H), (iv) *pro-trivial*, (v) *pro-free*, (vi) *pro-finitely generated* if it is pro-isomorphic to an inverse sequence (i) whose bonds are monomorphisms, (ii) whose bonds are isomorphisms, (iii) whose groups are isomorphic to H and whose bonds are isomorphisms, (iv) whose bonds are trivial, (v) whose groups are free, (vi) whose groups are finitely generated. This list and the following proposition should be compared with Proposition 12.5.3 and the list preceding it.

Proposition 16.3.1. *The above definitions* (i)–(vi) *are equivalent to the following intrinsic definitions:*

(i) *there is a cofinal subsequence* $\{G_{n_i}\}$ *such that image* $(G_{n_{i+2}} \to G_{n_{i+1}})$ *is mapped by the bond monomorphically into image* $(G_{n_{i+1}} \to G_{n_i})$;

(ii) *there is a cofinal subsequence* $\{G_{n_i}\}$ *such that image* $(G_{n_{i+2}} \to G_{n_{i+1}})$ *is mapped by the bond isomorphically onto image* $(G_{n_{i+1}} \to G_{n_i})$;

(iii) *same as (ii) with each image* $(G_{n_{i+1}} \to G_{n_i})$ *isomorphic to* H;

(iv) $\forall m \; \exists n$ *such that the bond* $G_n \to G_m$ *is trivial;*

(v) $\forall m \; \exists n$ *such that the bond* $G_n \to G_m$ *factors through a free group;*

(vi) $\forall m \; \exists n$ *such that image* $(G_n \to G_m)$ *lies in a finitely generated subgroup of* G_m. □

More useful, also an exercise, is:

Proposition 16.3.2. *The inverse sequence of groups* $\{G_n\}$ *is pro-monomorphic iff* $\exists \; m_0$ *such that* $\forall \; n \geq m_0 \; \exists \; k \geq n$ *such that* $\ker(G_k \to G_{m_0}) \subset \ker(G_k \to G_n)$. □

Let e be an end of Y and let the proper ray ω represent e. We say that Y has a *pro-monomorphic fundamental pro-group at* e if $\{\pi_1(Y \stackrel{c}{-} L_i, \omega(i))\}$ is pro-monomorphic. This depends only on e rather than on the proper ray ω because of the following consequence[8] of 16.3.2.

Proposition 16.3.3. $\{\pi_1(L_i(e), \omega(i))\}$ *is pro-monomorphic iff* $\exists \; m_0$ *such that* $\forall \; n \geq m_0 \; \exists \; k \geq n$ *so that any loop in* $L_k(e)$ *which bounds[9] in* $L_{m_0}(e)$ *bounds in* $L_n(e)$. □

We are interested in the question: for Y simply connected, what prevents Y from being the universal cover of a finite CW complex? The next theorem addresses a more basic question: when can $\mathbb{Z}$ act freely on Y?

Theorem 16.3.4. (Wright's Theorem) *Let* Y *be a strongly locally finite and simply connected free* $\mathbb{Z}$-*CW complex with one end. If* Y *has pro-monomorphic fundamental pro-group at infinity then the fundamental pro-group at infinity is pro-free and pro-finitely generated.*

---

[8] As in Sect. 16.2, $L_i(e)$ denotes the path component of $Y \stackrel{c}{-} L_i$ containing all but a compact subset of $\omega([0, \infty))$.

[9] For spaces $A \subset B$, we say a loop $f : S^1 \to A$ *bounds in* B if (inclusion $\circ f$) : $S^1 \to B$ is homotopically trivial.

Proof. We start with the main ideas. With reindexing, we may rewrite the condition in 16.3.3 as: $\forall i \geq 2$ any loop in $Y \xrightarrow{c} L_i$ which bounds in $Y \xrightarrow{c} L_1$ bounds in $Y \xrightarrow{c} L_{i-1}$. We may assume each $Y \xrightarrow{c} L_i$ is path connected. At the end of this proof we will define two subcomplexes A and B of Y such that $Y \xrightarrow{c} (A \cup B)$ is finite and contains L_1, and every loop in A or in B bounds in $Y \xrightarrow{c} L_1$. We choose $q \geq 2$ so that $N(Y \xrightarrow{c} (A \cup B)) \subset L_q$. For $i \geq q$, we write $A_i := A \cap (Y \xrightarrow{c} L_i)$ and $B_i := B \cap (Y \xrightarrow{c} L_i)$. Then $Y \xrightarrow{c} L_i = A_i \cup B_i$; it is the union of subcomplexes C_j where each C_j is a path component of A_i or of B_i. No point of $Y \xrightarrow{c} L_i$ lies in more than two of the sets C_j, so by 6.2.11, the fundamental group of $Y \xrightarrow{c} L_i$ is isomorphic to $\pi_1(\mathcal{G}_i, \Gamma_i; T_i)$ for some generalized graph of groups $(\mathcal{G}_i, \Gamma_i)$ and maximal tree T_i in Γ_i, where the vertex groups become trivial in $Y \xrightarrow{c} L_1$ and hence also in $Y \xrightarrow{c} L_{i-1}$. By 6.2.12, there is an epimorphism $\pi_1(\mathcal{G}_i, \Gamma_i; T_i) \twoheadrightarrow \pi_1(\Gamma_i; T_i)$ whose kernel is generated by the vertex groups. So there is a commutative diagram:

$$
\begin{array}{ccccc}
\pi_1(Y \xrightarrow{c} L_i, \omega(i)) & \longrightarrow\!\!\!\!\!\rightarrow & \pi_1(\Gamma_i; T_i) & & = F(E_i) \\
\downarrow & \swarrow & \downarrow {\scriptstyle \phi_i} & & \downarrow \\
\pi_1(Y \xrightarrow{c} L_{i-1}, \omega(i-1)) & \longrightarrow\!\!\!\!\!\rightarrow & \pi_1(\Gamma_{i-1}; T_{i-1}) & & = F(E_{i-1})
\end{array}
$$

Here, the diagonal arrow has been explained, and ϕ_i is chosen to make the lower triangle commute. Letting E_i denote the set of edges of Γ_i which are not in T_i, we can identify $\pi_1(\Gamma_i; T_i)$ with $F(E_i)$, the free group generated by E_i. Abelianizing, and denoting the free abelian group functor by FA, we get a commutative diagram:

$$
\begin{array}{ccccc}
H_1(Y \xrightarrow{c} L_i; \mathbb{Z}) & \longrightarrow\!\!\!\!\!\rightarrow & H_1(\Gamma_i; \mathbb{Z}) & & = FA(E_i) \\
\downarrow & \swarrow & \downarrow {\scriptstyle \phi'_i} & & \downarrow \\
H_1(Y \xrightarrow{c} L_{i-1}; \mathbb{Z}) & \longrightarrow\!\!\!\!\!\rightarrow & H_1(\Gamma_{i-1}; \mathbb{Z}) & & = FA(E_{i-1}).
\end{array}
$$

By 12.5.9, each $H_1(Y \xrightarrow{c} L_i; \mathbb{Z})$ is finitely generated, so each $FA(E_i)$ is free abelian of finite rank equal to the rank of $F(E_i)$ as a free group. Hence $\{F(E_i)\}$ is pro-finitely generated and pro-free.

It only remains to define A and B with the stated properties.

Let the automorphism $j : Y \to Y$ generate an infinite cyclic group acting freely on Y. The proof involves carefully choosing some positive integers m, n, p and s.

Choose $m \geq 2$ so that $j^{-1}(L_1) \cup L_1 \cup j(L_1) \subset L_m$. Let

$$
A = Y \xrightarrow{c} \bigcup_{i \in \mathbb{Z}} j^i(L_{m-1}).
$$

Claim 1: Every loop in A bounds in $Y \xrightarrow{c} L_1$.

Proof (of Claim 1). Let $f : S^1 \to A$ be a loop; f bounds in Y, so f bounds in $Y \overset{c}{-} j^k(L_1)$ for some k; so f is a loop in $Y \overset{c}{-} j^k(L_m)$ which bounds in $Y \overset{c}{-} j^k(L_1)$; so f bounds in $Y \overset{c}{-} j^k(L_{m-1})$, hence also in $Y \overset{c}{-} (j^{k-1}(L_1) \cup j^k(L_1) \cup j^{k+1}(L_1))$. Thus when f bounds in $Y \overset{c}{-} j^k(L_1)$ it also bounds in $Y \overset{c}{-} j^{k\pm1}(L_1)$. By induction f bounds in $Y \overset{c}{-} L_1$, and the claim is proved.

Choose n so that, for all i with $|i| \geq n$, $j^i(L_{m+1}) \cap L_{m+1} = \emptyset$. Let $K = \bigcup_{i=0}^{n-1} j^i(L_{m+1})$ and let $h = j^n$. Then for any k,

$$h^m(K) \cap h^k(K) = j^{nm}(K) \cap j^{nk}(K);$$

so $h^m(K) \cap h^k(K) = \emptyset$ when $|m - k| \geq 2$. Let P be a finite path connected subcomplex of Y containing $h(K) \cap K$. Choose p so that

$$h^{-1}(P) \cup K \cup P \subset L_p$$

and so that every loop in $h^{-1}(P) \cup K \cup P$ bounds in L_p.

Claim 2: For every q and r, every loop in $\bigcup_{i=q}^{q+r} h^i(K)$ bounds in $\bigcup_{i=q}^{q+r} h^i(L_p)$.

Proof (of Claim 2). For any q and r, any loop in $\bigcup_{i=q}^{q+r} h^i(K)$ is a loop in $h^{q-1}(P) \cup h^q(K) \cup \cdots \cup h^{q+r}(K) \cup h^{q+r}(P)$. Since P is path connected, this can be written as a product of loops each of which is in $h^{i-1}(P) \cup h^i(K) \cup h^i(P)$ for some i and which therefore bounds in $h^i(L_p)$. The Claim follows.

Choose $s > 0$ so that $h^k(L_p) \subset Y \overset{c}{-} L_1$ for all k such that $|k| \geq s$. Then $\bigcup_{|i| \geq ns} j^i(L_{m+1}) = \bigcup_{|i| \geq s} h^i(K)$. By Claim 2 any loop in $\bigcup_{|i| \geq ns} j^i(L_{m+1})$ bounds in $\bigcup_{|i| \geq s} h^i(L_p)$, and hence bounds in $Y \overset{c}{-} L_1$, by definition of s. Let $B = \bigcup_{|i| \geq ns} j^i(L_{m+1})$. Then we have proved:

Claim 3: Every loop in B bounds in $Y \overset{c}{-} L_1$.

Now, $Y \overset{c}{-} (A \cup B) \subset \left(\bigcup_{i=-ns+1}^{ns-1} j^i(L_{m+1}) \right)$ so $A \cup B$ covers all but a finite subcomplex of Y. Thus A and B have the required properties. $\square$

Theorem 16.3.4 imposes severe restrictions on the kind of space which can be an infinite cyclic covering space:

Remark 16.3.5. In the next section we describe Whitehead's contractible open 3-manifold, W, which has one end but is not simply connected at infinity (and therefore is not proper homotopy equivalent to, much less homeomorphic to, $\mathbb{R}^3$). We will see that W has pro-monomorphic fundamental group at infinity, but that $\{\pi_1(W \xrightarrow{c} L_i, \omega(i))\}$ is not pro-finitely generated. So Theorem 16.3.4 implies that no non-trivial group G acts freely on W; indeed, G could have no element of infinite order by 16.3.4, and no non-trivial element of finite order by 7.2.12.

Remark 16.3.6. Theorem 16.3.4 would not be changed if Y was assumed only to have finitely many ends; by Exercises 6 and 7 of Sect. 13.5, Y would have to have one end or two ends, and when Y has two ends G would have to be a two-ended group, in which case the fundamental group at each end would be pro-trivial.

Example 16.3.7. Let M be a compact path connected n-manifold with path connected (hence non-empty) boundary. Pick a ray ω in $\overset{\circ}{M}$ approaching $x \in \partial M$. Then the fundamental pro-group of $\overset{\circ}{M}$ based at ω is stably $\pi_1(\partial M, x)$. If M is simply connected and if $\mathbb{Z}$ acts as covering transformations on $\overset{\circ}{M}$, then 16.3.4 implies that $\pi_1(\partial M, x)$ must be a free (and finitely generated) group.

Example 16.3.8. If, in the last example, M is contractible then (see Exercise 6 of Sect. 15.2) the homology of ∂M with $\mathbb{Z}$-coefficients is the same as that of S^{n-1}; one says that ∂M is a "homology $(n-1)$-sphere." If $\overset{\circ}{M}$ is a non-trivial covering space then $\mathbb{Z}$ must act as covering transformations, by 7.2.12. Hence, by 16.3.4, $\pi_1(\partial M, x)$ must be a free group. Assume $n \geq 3$. Then $H_1(M; \mathbb{Z}) = 0$, so ∂M must be simply connected.[10] Thus, if a non-simply connected homology sphere bounds a compact contractible manifold M, then $\overset{\circ}{M}$ is not a non-trivial covering space.

Source Notes: Wright's Theorem appeared in [155].

Exercises

1. Prove 16.3.2.
2. Give a counterexample to 16.3.4 when Y has infinitely many ends. *Hint*: Let P be a finite CW complex and let $Q = P \times [0, \infty)/P \times \{0\}$ be the "open cone" on P. Let $X = S^1 \vee Q$ and consider $Y = \tilde{X}$.
3. What change in the hypotheses of 16.3.4 would make it true for CW complexes Y with infinitely many ends?

16.4 Example: Whitehead's contractible 3-manifold

Every contractible open 1-manifold is homeomorphic to $\mathbb{R}$ and every contractible open 2-manifold is homeomorphic to $\mathbb{R}^2$. But there are uncountably

---

[10] See Remark 16.4.14, which applies here too.

many contractible open 3-manifolds no two of which are homeomorphic. For group theory the interesting question is: can any of these, other than $\mathbb{R}^3$, be universal covers of closed 3-manifolds? If there were such a closed 3-manifold it would be an object of intense interest in 3-dimensional topology, and its fundamental group G would be a remarkable Poincaré duality group of dimension 3 over $\mathbb{Z}$.

Here, we describe the original example in [153] of a contractible open subset W of $\mathbb{R}^3$ which is not simply connected at infinity. The properties of W ensure that no non-trivial group acts freely on W; in other words, the only covering projections $W \to V$ are homeomorphisms. By varying the details of the construction of W one can obtain uncountably many examples, no two of which are homeomorphic; we will not pursue that; see [113].

A background discussion is useful here. A *fake 3-sphere* is a closed simply connected 3-manifold not homeomorphic to S^3; the *Poincaré Conjecture* says that there are no fake 3-spheres. A *fake 3-ball* is obtained from a fake 3-sphere by removing the interior of an unknotted 3-ball. Thus, the boundary of a fake 3-ball is a 2-sphere. If the Poincaré Conjecture were false, then, given any 3-manifold M, one could obtain a new 3-manifold M' by deleting the interiors of a locally finite pairwise disjoint collection of unknotted 3-balls in M and gluing[11] in fake 3-balls instead. Then M and M' would have the same proper homotopy type but would not be homeomorphic, and we would say that M and M' are *the same modulo the Poincaré Conjecture*. It has been known for many years that every contractible open 3-manifold which is strongly connected at infinity is the same as $\mathbb{R}^3$ modulo the Poincaré Conjecture (see [80] or [27]) and hence is simply connected at infinity. So our W and all the uncountably many other exotic examples must have non-semistable fundamental pro-groups.

To begin the construction of W we note:

Proposition 16.4.1. *For $n \geq 2$ every contractible open n-manifold has one end.*

Proof. This follows from Poincaré Duality together with 13.4.11 and the exact sequence in Sec. 12.2. □

For ease of exposition we will not make explicit the CW complex structures on the spaces to be discussed. Once W has been constructed it will not be difficult to see that W admits the structure of a simplicial complex. Indeed, while we draw our pictures smoothly, they can also be realized as subcomplexes of suitable simplicial complex structures on S^3.

We begin with the trefoil knot $K \subset S^3$ illustrated in Fig. 16.3. Let $N(K)$ be a compact neighborhood of K such that $(N(K), K)$ is homeomorphic to $(S^1 \times B^2, S^1 \times \{0\})$, i.e., a solid torus "fattening" of K. We may choose $N(K)$ so that $C_K := \mathrm{cl}(S^3 - N(K))$ is a compact manifold with boundary; for

---

[11] See the definition of "connected sum" in Sect. 5.1.

example, make $N(K)$ a regular neighborhood in the sense of piecewise linear topology, or a tubular neighborhood in the sense of differential topology.

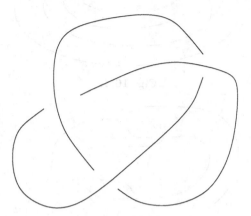

Fig. 16.3.

We will assume two standard facts from elementary knot theory; proofs can be found in [43] or [135]. Apart from these two propositions, our treatment of W will be self-contained, if informal.

Proposition 16.4.2. *For $v \in \partial(N(K))$ the inclusion map induces a monomorphism $\pi_1(\partial(N(K)), v) \to \pi_1(C_K, v)$.* $\square$

Proposition 16.4.3. *$\pi_1(C_K, v)$ is not abelian.* $\square$

Consider Fig. 16.4. It consists of a solid torus T_1 (i.e., a copy of $S^1 \times B^2$) in $\mathbb{R}^3 \subset S^3$ and a solid torus[12] $T_2 \subset \overset{\circ}{T}_1$. Let $h : T_1 \to T_2$ be a homeomorphism, and let $T_3 = h(T_2)$; then T_3 sits in T_2 as T_2 sits in T_1. Iterating this, we get a nested sequence of solid tori $T_1 \supset T_2 \supset \cdots$ whose intersection is a compact non-empty set $Z \subset S^3$. Define $W := S^3 - Z$. Note that if we remove a point of Z we may also regard[13] W as an open subset of $\mathbb{R}^3$. This W is *Whitehead's Contractible 3-manifold* which we now study.

Consider the two solid tori T and L illustrated in Fig. 16.5.

Proposition 16.4.4. *There is a homeomorphism $f : \mathbb{R}^3 \to \mathbb{R}^3$ such that $f(T) = L$ and $f(L) = T$; f extends to a homeomorphism of S^3.*

---

[12] T_2 is drawn as a circle, but should be viewed as "fat," i.e., as a solid torus. Similarly in Fig. 16.5.

[13] We will frequently identify $\mathbb{R}^3$ with $S^3 - \{\text{point}\}$.

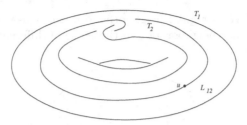

Fig. 16.4.

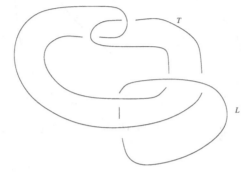

Fig. 16.5.

Proof. Take two circles of string embedded in $\mathbb{R}^3$ as illustrated in Fig. 16.5. Stretch and reshape the string L to occupy the space originally occupied by T. Then the string T can easily be moved to occupy the space originally occupied by L. Since faraway points of $\mathbb{R}^3$ need not move during this process, the homeomorphism extends to S^3.   □

Proposition 16.4.5. *Let $T_1' = \mathrm{cl}(S^3 - T_1)$. Then T_1' is also a solid torus, and $T_1' \cap T_1 = \partial T_1$.*

Proof. $S^3 = \partial B^4$ is homeomorphic to $\partial(B^2 \times B^2) = (B^2 \times S^1) \cup (S^1 \times B^2)$, and $(B^2 \times S^1) \cap (S^1 \times B^2) = S^1 \times S^1$.   □

Proposition 16.4.6. *Let $L_{1,2} = \mathrm{cl}(T_1 - T_2)$ so that $\partial L_{1,2} = \partial T_1 \cup \partial T_2$. There is a homeomorphism of S^3 mapping $L_{1,2}$ to itself, mapping ∂T_1 to ∂T_2, and mapping ∂T_2 to ∂T_1.*

Proof. Interpret Fig. 16.5 as consisting of two solid tori T and L. By 16.4.5, there is a homeomorphism $k : S^3 \to S^3$ taking T to T_2 and L to $\mathrm{cl}(S^3 - T_1)$; in other words, k maps the copy of S^3 in Fig. 16.5 to the copy of S^3 in Fig. 16.4 as indicated. By 16.4.4, there is a homeomorphism of S^3 mapping T_2 to $\mathrm{cl}(S^3 - T_1)$ and mapping $\mathrm{cl}(S^3 - T_1)$ to T_2. This homeomorphism must then map $L_{1,2}$ to itself, interchanging the boundary components.   □

Fig. 16.6.

Now consider Fig. 16.6. It is obtained from Fig. 16.4 by cutting the solid torus T_1 along the disk D, twisting through one full rotation, and then regluing. This is a homeomorphism of T_1 which maps the solid torus T_2 onto the solid torus R. Thus $L'_{1,2} := \mathrm{cl}(T_1 - R)$ is homeomorphic to $L_{1,2}$ by a homeomorphism j taking ∂R to ∂T_2 and ∂T_1 to itself. Now if we ignore T_1 in Fig. 16.6, we see that R is $N(K)$, the "fattening" of a trefoil knot[14] in S^3; of course, the homeomorphism j does not extend to S^3, since T_2 is unknotted in S^3. The following diagram commutes:

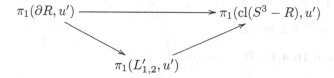

where all arrows are induced by inclusion and u' is a base point. Thus 16.4.2 and 13.8.1 imply:

Proposition 16.4.7. *If $u \in \partial T_2$ and $v \in \partial T_1$, then the homomorphisms $\pi_1(\partial T_1, v) \to \pi_1(L_{1,2}, v)$ and $\pi_1(\partial T_2, u) \to \pi_1(L_{1,2}, u)$ induced by the inclusion maps are monomorphisms.* □

Proposition 16.4.8. *The monomorphisms in 16.4.7 are not epimorphisms.*

Proof. $\mathrm{cl}(S^3 - R) = L'_{1,2} \cup \mathrm{cl}(S^3 - T_1)$ and $L'_{1,2} \cap \mathrm{cl}(S^3 - T_1) = \partial T_1$. If $\pi_1(\partial T_1, v) \to \pi_1(L'_{1,2}, v)$ were onto then, by 3.1.19, $\pi_1(\mathrm{cl}(S^3 - R), v)$ would be abelian; but, by 16.4.3, this is false. Thus, by 16.4.6, neither homomorphism in 16.4.7 is onto. □

Now we are ready to study W. Let $L_{0,1} = \mathrm{cl}(S^3 - T_1)$ and for $i \geq 1$ let $L_{i,i+1} = \mathrm{cl}(T_i - T_{i+1})$. For $k \geq 1$ we write $L_k = \bigcup_{i=0}^{k-1} L_{i,i+1}$. Then $\{L_k\}$ is a finite filtration of W.

Proposition 16.4.9. *The inclusion $L_1 \hookrightarrow L_2$ is homotopically trivial.*

---

[14] Provided the twist is done in the correct direction.

Proof. Fig. 16.5 makes it obvious that the inclusion $T \hookrightarrow \mathrm{cl}(S^3 - L)$ is homotopically trivial. By 16.4.4, it follows that the inclusion $L \hookrightarrow \mathrm{cl}(S^3 - T)$ is homotopically trivial. So, by means of the homeomorphism k, the inclusion $\mathrm{cl}(S^3 - T_1) \hookrightarrow \mathrm{cl}(S^3 - T_2)$ is homotopically trivial.   □

By 16.4.5, the homeomorphism $h : T_1 \to T_2$ used in the definition of W extends to a homeomorphism of pairs $(S^3, T_1) \to (S^3, T_2)$ and therefore maps L_1 homeomorphically onto L_2. Now $h(T_2) = T_3$ so $h(L_1) = L_2$ and $h(L_2) = L_3$. Thus the following diagram commutes:

$$
\begin{array}{ccc}
L_1 & \!\!\!\hookrightarrow\!\!\! & L_2 \\
{\scriptstyle h|}\downarrow & & \downarrow{\scriptstyle h|} \\
L_2 & \!\!\!\hookrightarrow\!\!\! & L_3
\end{array}
$$

and the vertical arrows are homeomorphisms. So $L_2 \hookrightarrow L_3$ is homotopically trivial. Proceeding by induction we get:

Corollary 16.4.10. *The inclusion $L_k \hookrightarrow L_{k+1}$ is homotopically trivial for every $k \geq 1$.*   □

Hence, by 7.1.2, we have

Proposition 16.4.11. W *is contractible.*   □

Choose a base ray ω in W well parametrized with respect to $\{L_k\}$. Then for any k we have $\mathrm{cl}_W(W - L_k) = L_{k,k+1} \cup L_{k+1,k+2} \cup \cdots$. We may assume $\omega(i) \in L_{i,i+1} \cap L_{i+1,i+2}$; then 16.4.7 and 16.4.8 tell us that for $i \geq 1$ the homomorphisms induced by inclusion $\pi_1(L_{i,i+1} \cap L_{i+1,i+2}, \omega(i)) \to \pi_1(L_{i,i+1}, \omega(i))$ and $\pi_1(L_{i,i+1} \cap L_{i+1,i+2}, \omega(i)) \to \pi_1(L_{i+1,i+2}, \omega(i))$ are monomorphisms but not epimorphisms. We form a graph of groups as follows: the oriented graph Γ_1 has underlying space $[1, \infty)$ with vertices at the integer points and edges oriented in the positive direction. The vertex group at the vertex i is $\pi_1(L_{i,i+1}, \omega(i))$, the edge group corresponding to the edge $[i, i+1]$ is $\pi_1(L_{i,i+1} \cap L_{i+1,i+2}, \omega(i))$. Using change of base point via ω as usual, the monomorphisms from edge groups to vertex groups are those non-epimorphisms just specified. We take the maximal tree $T_1 = \Gamma_1$ and form $\pi_1(\mathcal{G}_1, \Gamma_1; T_1)$. For each integer $k \geq 1$ let Γ_k denote the subgraph of Γ_1 corresponding to $[k, \infty)$. Then there is an obvious graph of groups $(\mathcal{G}_k, \Gamma_k)$, and $\pi_1(\mathcal{G}_k, \Gamma_k; T_k)$ can be identified with a proper subgroup of $\pi_1(\mathcal{G}_{k-1}, \Gamma_{k-1}; T_{k-1})$ for all $k \geq 2$. The inverse sequences $\{\pi_1(\mathrm{cl}_W(W - L_k), \omega(k))\}$ and $\{\pi_1(\mathcal{G}_k, \Gamma_k; T_k)\}$ can be identified.

Proposition 16.4.12. *These inverse sequences are not semistable and are not pro-finitely generated.*

Proof. The first statement follows from Exercise 2 and the second from Exercise 3.   □

Summarizing, and applying Remark 16.3.5:

Theorem 16.4.13. *W is a contractible open subset of $\mathbb{R}^3$ which has pro-monomorphic fundamental pro-group at infinity but is not strongly connected at infinity. The inverse sequence of groups $\{\pi_1(W - L_k, \omega(k))\}$ is not pro-finitely generated. Hence no non-trivial group acts as a group of covering transformations on W.* □

Remark 16.4.14. In this book almost all spaces have been CW complexes. Thus what we have actually proved is that when W is given a CW complex structure, no non-trivial group acts freely on W by automorphisms in the sense of Sect. 3.2. However, the reader familiar with the theory of absolute neighborhood retracts and topological manifolds can check that our proof of 16.4.13 (and the supporting material such as 16.3.4) goes through, with appropriate adaptation, to prove 16.4.13 as stated.[15]

Remark 16.4.15. For deeper examples of this type, including examples of contractible open 3-manifolds on which $\mathbb{Z}$ acts as a group of covering transformations but which do not cover closed 3-manifolds, see [124].

Exercises

1. Why is the first sentence in the proof of 16.4.9 true?
2. If $G \cong A \underset{C}{*} B$ is a free product with amalgamation[16] and neither $C \rightarrowtail A$ nor $C \rightarrowtail B$ is onto, show that $A \rightarrowtail G$ and $B \rightarrowtail G$ are not onto.
3. Let $G \cong A \underset{C}{*} B$ where A is finitely generated. Prove that if B lies in a finitely generated subgroup of G then G is finitely generated.
4. Show that $W \times \mathbb{R}$ is homeomorphic to $\mathbb{R}^4$.

16.5 Group invariants: simple connectivity, stability, and semistability

We can apply the previous sections to define properties of finitely presented groups. If (X_1, v_1) and (X_2, v_2) are finite path connected pointed 2-dimensional CW complexes whose fundamental groups are isomorphic to G and if $\phi : \pi_1(X_1, v_1) \to \pi_1(X_2, v_2)$ is an isomorphism then it follows easily from 7.1.7 (the details were set as an exercise in Sect. 7.1) that ϕ is induced by a 2-equivalence $f : (X_1, x_1) \to (X_2, x_2)$. By 10.1.23 (the details are an exercise in Sect. 11.1) such a 2-equivalence lifts to a proper 2-equivalence $\tilde{f} : \tilde{X}_1 \to \tilde{X}_2$.

---

[15] Alternatively, a deep classical result of 3-dimensional topology says that every 3-manifold is triangulable, but it is overkill to use this.

[16] Exercises 3 and 4 follow from Britton's Lemma 6.2.1. See also the proof of 18.3.19 for normal forms in $A \underset{C}{*} B$.

By 16.1.14 , $\tilde{f}$ induces bijections on the set of ends, the set of strong ends, and the set of strong homological ends. Thus the cardinal numbers of these sets are invariants[17] of G. The map $\tilde{f}$ also induces isomorphisms on $H_1^e(\cdot; R)$ and $H_e^1(\cdot; R)$ so that their isomorphism classes as R-modules (indeed, as RG-modules) are also invariants of G, but we already knew this via the discussion of resolutions in Chap. 8. When ω is a proper base ray in $\tilde{X}_1$, we discussed in Sect. 16.2 how the fundamental pro-group, the Čech fundamental group, and the strong fundamental group based at ω depend on the choice of ω. Thus (see 16.2.4) one must be careful in asserting that these, or even their isomorphism classes, are invariants of G. However, no such caution is needed for certain related invariants, as we now explain.

Let G be a finitely presented group. We saw in Example 1.2.17 that there is a finite path connected pointed CW complex (X, v) with $\pi_1(X, v) \cong G$. We say that G is *semistable at each end* if $\tilde{X}$ is strongly connected at each end. When G has one end and $\tilde{X}$ is strongly connected at its end, we say[18] that G is *semistable at infinity*.

The connection with homological invariants is:

Theorem 16.5.1. *If G is semistable at each end then $H^2(G, \mathbb{Z}G)$ is a free abelian group.*

Proof. By 16.1.8 and 16.1.11, $H_e^1(\tilde{X}; \mathbb{Z})$ is free abelian. By 13.2.9 and 13.2.13, $H_e^1(\tilde{X}; \mathbb{Z}) \cong H^1(G, \mathbb{Z}G^e) \cong H^2(G, \mathbb{Z}G)$.   □

Similarly, we say that G is *simply connected at each end* if $\tilde{X}$ is simply connected at each end. When G has one end and $\tilde{X}$ is simply connected at its end, we say that G is *simply connected at infinity*.[19]

Theorem 16.5.2. *If G is simply connected at each end then $H^2(G, \mathbb{Z}G) = 0$.*

Proof. By 16.2.7, $H_e^1(\tilde{X}; \mathbb{Z}) = 0$. But once again $H_e^1(\tilde{X}; \mathbb{Z}) \cong H^2(G, \mathbb{Z}G)$.   □

We saw in 13.5.2 that the number of ends of a finitely presented group does not change on passing to a subgroup of finite index (which is also finitely presented by 3.2.13). Similarly, we have:

Proposition 16.5.3. *Let H be a subgroup of finite index in the finitely presented group G. Then H is semistable at each end [resp. simply connected at each end] iff G is semistable at each end [resp. simply connected at each end].*   □

---

[17] Indeed, the homeomorphism types of the spaces of ends are invariants of G; see 13.5.8.

[18] At the time of writing it is unknown if there is a finitely presented group which is not semistable at each end.

[19] It follows that a finite group is simply connected at each end but is not simply connected at infinity. This is analogous to saying that each path component of the empty space is simply connected, which is true because there are none.

Corollary 16.5.4. *Every two-ended group is simply connected at each end.*
□

Remark 16.5.5. If G is semistable at infinity then the fundamental pro-group, the Čech fundamental group, and the strong fundamental group are well-defined by G (i.e., independent of X and ω) up to isomorphism in the appropriate category; see Sect. 16.2. Thus, just as we can speak of "the number of ends" of G and "the homeomorphism type of the ends" of G, so in the semistable case we can speak of "the pro-isomorphism type of the fundamental pro-group" of G, "the isomorphism type of the Čech fundamental group" of G, etc.

We say G is *stable at each end* if $\tilde{X}$ has stable fundamental pro-group at each end. Since "stable" is equivalent to "semistable and pro-monomorphic," this is well-defined. For one-ended groups we use the term *stable at infinity*. Recall that if the stable fundamental pro-group is isomorphic to $\mathbb{Z}$, as in the next theorem, we say G is *stably $\mathbb{Z}$ at infinity*. Here is a remarkable consequence of Wright's Theorem 16.3.4:

Theorem 16.5.6. *Let the one-ended finitely presented group G be stable at infinity and assume G contains an element of infinite order. Then G is either simply connected at infinity or stably $\mathbb{Z}$ at infinity.*

Proof. By 16.3.4, the stable inverse sequence $\{\pi_1(\tilde{X} \overset{c}{-} K_i, \omega(i))\}$ must be pro-isomorphic to a finitely generated free group, so $\{H_1(\tilde{X} \overset{c}{-} K_i; \mathbb{Z})\}$ is pro-isomorphic to a finitely generated free abelian group of the same rank. By 13.3.2(iv) $H^2(G, \mathbb{Z}G)$ is free abelian of that same rank, and, by 13.7.12, that rank is 0 or 1. □

Example 16.5.7. We will see in 16.9.7 that Thompson's group F is simply connected at infinity. The same is true of Thompson's group T; see [30]. Certain right angled Coxeter groups are shown to be simply connected or semistable at infinity in 16.6.1; for Coxeter groups in general, see [117]. The group $\mathrm{Out}(F_n)$ is simply connected at infinity when $n \geq 3$ (see [11]), and is semistable at each end when $n = 2$ (being virtually free).

Source Notes: The question of whether every one-ended finitely presented group is semistable at infinity was first addressed in [70] and [114], though the issue had been remarked on in passing by Houghton, earlier, in [86]. There is now a substantial literature showing that the answer is positive for many classes of groups; see, for example, the bibliography of [119]. "Simply connected at infinity" has a much longer history in geometric topology. See, for example, [144], [93] and [95].

Exercises

1. Give a counterexample to the converse of 16.5.2.
2. For the pseudomanifold W in Sect. 13.8, prove that $G = \pi_1(W, w)$ is semistable at infinity. (It follows that $H^2(G, \mathbb{Z}G)$ is free abelian.)

16.6 Example: Coxeter groups and Davis manifolds

All finitely generated Coxeter groups are (finitely presented[20] and) semistable at each end; the latter is proved in [117]. Here we will discuss this only for the right angled Coxeter groups described in Theorem 13.9.5. Recall that these are defined as follows: Starting with an integer $d \geq 2$ and a non-empty finite connected closed combinatorial $(d-1)$-manifold L which is also a flag complex, we let (G, S) be the corresponding right angled Coxeter system (whose nerve K is sd L). In this section G is understood to be such a group. Here, we show that G is semistable and that there is a torsion free subgroup H of finite index in G for which there is a closed 4-manifold $K(H, 1)$-complex whose universal cover is contractible but is not simply connected at infinity, and hence is not homeomorphic to $\mathbb{R}^4$.

Theorem 16.6.1. *The group G is semistable at infinity; and G is simply connected at infinity iff $|L|$ is simply connected.*

Proof. By 16.5.3, these properties are the same for G and for any torsion free subgroup of finite index; hence they can be checked in $|D|$. The homotopy commutative diagram in the paragraph preceding Theorem 13.9.5 shows that the inclusion map of the $(n + 1)^{\text{th}}$ neighborhood of the end into the nth neighborhood of the end induces an epimorphism on fundamental group. In fact, if $d = 2$, the inverse sequence of fundamental groups is stably $\mathbb{Z}$. If $d \geq 4$, the inclusion induces $G_n * G_0 \to G_n$ where G_n denotes the n-fold free product of copies of $G_0 = \pi_1(|L|, v)$; the epimorphism is the identity on G_n, and kills the final free summand G_0. This is because $\pi_1(M_1 \# M_2, x)$ is isomorphic to $\pi_1(M_1, x) * \pi_1(M_2, x)$ when the manifolds have dimension ≥ 3. The case $d = 3$ is an exercise. The last sentence of the proposition is the special case when G_0 is trivial. □

Now let L be the boundary of a finite contractible combinatorial 4-manifold J. It follows from Lefschetz duality (see 15.2.4) that the homology of the 3-manifold $|L|$ is the same as that of S^3. A closed manifold having the same homology groups as S^n is a *homology n-sphere*, so this $|L|$ is a homology 3-sphere which bounds the compact contractible 4-manifold $|J|$. Examples exist in which $|L|$ is not simply connected, a well-known example being the "Mazur sphere" – see [135, p. 356]. We will assume $|L|$ is not simply connected.[21]

We built D out of copies of F, the cone on K where $K =$ sd L. Now for every n we replace $g_n F$ in D by a copy of J, called $g_n J$, identifying $g_n(\partial J) = g_n K$ with $g_n K$ in D. There results an abstract simplicial G-complex D_0, finite mod G, whose vertex-stabilizers are finite, (where (G, S) corresponds to L as above); D_0 is a combinatorial 4-manifold. From the proof of 9.1.3 we get:

---

[20] See Theorem 9.1.7.

[21] Compare Example 16.3.8.

Theorem 16.6.2. $|D_0|$ *is a contractible CW 4-manifold on which G acts rigidly with finite cell-stabilizers so that $G\backslash|D_0|$ is finite.*

By 9.1.10, G has a torsion free subgroup H of finite index. By 16.6.1 and 16.6.2 we have:

Corollary 16.6.3. $H\backslash|D_0|$ *is a closed aspherical 4-manifold whose universal cover is not homeomorphic to $\mathbb{R}^4$.*

One can build such examples in any dimension ≥ 4 since non-simply connected homology spheres bounding compact contractible n-manifolds exist for all $n \geq 4$. We call such closed aspherical manifolds *Davis manifolds*. They first appeared in [46], and until then it was unknown if a contractible open n-manifold not homeomorphic to $\mathbb{R}^n$ could be the universal cover of a compact manifold. Compare this with Whitehead's manifold in dimension 3: it is contractible but, by 16.4.13, it is not the universal cover of a closed manifold.

Exercises

1. Fill in the missing details of 16.2.4 for the cases $d = 2$ and $d = 3$.
2. Prove that if $|L|$ is a $(d-1)$-dimensional homology sphere, then G is a Poincaré duality group of dimension d (whether or not $|L|$ bounds a compact contractible manifold); see Sec. 15.3.

16.7 Free topological groups

The Čech fundamental group is a topological group. To study it in the next section we review some basics here.

A group G equipped with a topology is a *topological group* if multiplication $G \times G \to G$ and inversion $G \to G$ are continuous. A sequence $\{g_n\}$ in G is a *Cauchy sequence* if given a neighborhood U of $1 \in G$ there exists $N \in \mathbb{N}$ such that for all $i, j \geq N$ $g_i^{-1}g_j \in U$. A topological group is *complete* if every Cauchy sequence converges. Discrete groups are complete, and the countable product of complete groups is a complete group. Closed subgroups of complete groups are complete. Thus if $\mathcal{G} := \{G_1 \leftarrow G_2 \leftarrow \cdots\}$ is an inverse sequence of discrete groups, its inverse limit is a complete group.

A topological group G is *zero-dimensional* if there is a basis for the neighborhoods of $1 \in G$ consisting of closed-and-open subgroups. The complete group $\varprojlim \mathcal{G}$, above, is zero-dimensional and first countable (being a metrizable space).

If G is a complete first countable zero-dimensional group and $U_1 \supset U_2 \supset \cdots$ is a basis for the neighborhoods of 1 with each U_i a closed-and-open subgroup, then $\mathcal{H}(G) := \{G/U_1 \leftarrow G/U_2 \leftarrow \cdots\}$ is a semistable inverse sequence of discrete groups. Indeed $\mathcal{H}$ is a sort of inverse for $\varprojlim$ in the following sense:

Proposition 16.7.1. *If $\mathcal{G}$ is semistable then $\mathcal{H}(\varprojlim \mathcal{G})$ and $\mathcal{G}$ are pro-isomorphic. If G is complete, first countable, and zero-dimensional, then the topological groups $\varprojlim \mathcal{H}(G)$ and G are isomorphic.* $\qquad\square$

So although $\mathcal{H}$ depends on $\{U_n\}$, up to isomorphism the choice of $\{U_n\}$ does not matter.

There is a forgetful functor Groups $\rightarrow$ Pointed Sets and hence the notion of *the free group generated by the pointed set* (S,s). This consists of a group $F(S,s)$ and a pointed function $i : (S,s) \rightarrow (F(S,s),1)$ satisfying the usual universal property that for any group H and pointed function $f : (S,s) \rightarrow (H,1)$ there is a unique homomorphism $\bar{f} : F(S,s) \rightarrow H$ such that $\bar{f} \circ i = f$. Note the difference from the usual notion of "free group generated by a set." In the pointed case, the free group $F(S,s)$ has rank (cardinality of S)-1 rather than cardinality of S.

We need a topological analog of this. Let $\mathcal{C}$ be the category of metrizable 0-dimensional pointed spaces and pointed maps. Let $\mathcal{D}$ be the category of complete first countable zero-dimensional topological groups and continuous homomorphisms. There is a forgetful functor $\mathcal{D} \rightarrow \mathcal{C}$. As usual, we define the free object in $\mathcal{D}$ generated by an object (Z,z) of $\mathcal{C}$ to consist of a group $F(Z,z)$ in $\mathcal{D}$ and a map $i : (Z,z) \rightarrow (F(Z,z),1)$ satisfying the appropriate universal property. Uniqueness of $F(Z,z)$ up to isomorphism follows at once. For existence, we will be satisfied with a special case:

Proposition 16.7.2. *The free object in $\mathcal{D}$ generated by a compact object (Z,z) in $\mathcal{C}$ exists.*

Proof. First note that the "free group generated by a pointed set" is a functor $F :$ Pointed Sets $\rightarrow$ Groups. By 13.4.13, $(Z,z) = \varprojlim_n (Z_n, z_n)$ where each (Z_n, z_n) is a finite pointed set. The required group $F(Z,z)$ is $\varprojlim_n \{F(Z_n, z_n)\}$ where the functor F is applied to the whole inverse sequence $\{(Z_n, z_n)\}$. The universal property to be checked is summarized in the diagram

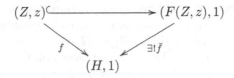

As explained, $H = \varprojlim_n \{H_n\}$ where the groups H_n are discrete and the bonds are epimorphisms. Since each Z_n is finite there is a morphism of pro-Pointed Sets $\{(Z_n, z_n)\} \rightarrow \{(H_n, 1)\}$ whose inverse limit is f; it can be represented by a sequence $f_{m(n)} : (Z_{m(n)}, z_{m(n)}) \rightarrow (H_n, 1)$ of pointed functions commuting with the bonds. The required $\bar{f}$ is $\varprojlim_n \{\bar{f}_{m(n)} : F(Z_{m(n)}, z_{m(n)}) \rightarrow H_n\}$. $\qquad\square$

Remark 16.7.3. The proof of 16.7.2 also shows that the free object of towers-Groups generated by an object of towers-(Finite Pointed Sets) exists and is semistable.

Source Notes: Proposition 16.7.1 appears in [4].

Exercises

1. Prove 16.7.1.
2. Assuming $\mathcal{G}$ semistable, prove $\mathcal{G}$ is stable iff $\varprojlim \mathcal{G}$ is discrete iff $\varprojlim \mathcal{G}$ is countable.
3. Give an example where the underlying group of $F(Z, z)$ is not isomorphic to the free group generated by the pointed set underlying (Z, z).
4. Suppose we are given for each n a commutative diagram of countable abelian groups whose horizontal rows are exact:

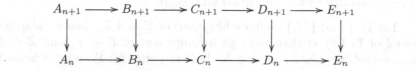

If $\{A_n\}$, $\{B_n\}$, $\{D_n\}$ and $\{E_n\}$ are stable (with verticals as bonds), prove that $\{C_n\}$ is stable.

16.8 Products and group extensions

We first make a general computation of the Čech fundamental group of a product (16.8.1). Then we apply that in 16.8.5 to compute the Čech fundamental group of any group G which fits into a short exact sequence $N \rightarrowtail G \twoheadrightarrow Q$ of infinite finitely presented groups.

If (A, a) and (B, b) are pointed spaces their *smash product* is the pointed space $(A \wedge B, p)$ where $A \wedge B := A \times B/\{a\} \times B \cup A \times \{b\}$ and p is the image of $\{a\} \times B \cup A \times \{b\}$ in $A \wedge B$. Compactness and 0-dimensionality are preserved under smash product. If A and B are Cantor sets, so is $A \wedge B$, by 13.4.16.

The smash product occurs naturally in computing the Čech fundamental group of $Y \times Z$ where Y and Z are infinite strongly locally finite path connected CW complexes. It is clear (see Exercise 2 in Sect. 13.4) that $Y \times Z$ has one end. Pick base points e_1 and e_2 for the spaces of ends $\mathcal{E}(Y)$ and $\mathcal{E}(Z)$, and pick a base ray ω in $Y \times Z$.

Theorem 16.8.1. *If Y and Z are simply connected, then $Y \times Z$ is strongly connected at infinity and $\check{\pi}_1(Y \times Z, \omega)$ is isomorphic to the free complete first countable zero-dimensional topological group generated by the smash product $(\mathcal{E}(Y), e_1) \wedge (\mathcal{E}(Z), e_2)$ of the two spaces of ends.*

The proof is given after 16.8.3.

If Y and Z have r and s ends respectively, where r and s are finite, then 16.8.1 implies that $\check{\pi}_1(Y \times Z, \omega)$ is discrete and is free of rank $(r-1)(s-1)$. If Y has one end, 16.8.1 says that $Y \times Z$ is simply connected at infinity (by 16.2.1 and 16.1.2).

When A and B are discrete disjoint non-empty spaces, their *topological join* is the graph[22] $A * B$ whose set of vertices is $A \coprod B$ and which has an edge joining each point of A to each point of B. If a and b are base points for A and B, then the subgraph of $A * B$ consisting of all edges having a or b as a vertex is a maximal tree in $A * B$ which we call the *canonical maximal tree*. If we orient the graph $A * B$, then 3.1.16 allows us to regard the edges outside the maximal tree as free generators of the fundamental group. A better way of saying this is:

Proposition 16.8.2. *The inclusion* $(A \wedge B, p) \hookrightarrow (\pi_1(A * B, a), 1)$ *is the free group generated by the pointed set* $(A \wedge B, p)$. □

Let $\{L_i\}$ and $\{M_i\}$ be finite filtrations of Y and Z, chosen (adapting the proof of 13.4.9) so that every path component of $Y \overset{c}{-} L_i$ and $Z \overset{c}{-} M_i$ is unbounded. Let $y \in L_1$ and $z \in M_1$ be base points for Y and Z. Choose finite sets $A_i = \{a_{ij}\}$ and $B_i = \{b_{ik}\}$ where a_{ij} [resp. b_{ik}] lies in the path component U_{ij} of $Y \overset{c}{-} L_i$ [resp. V_{ik} of $Z \overset{c}{-} M_i$] (one for each path component). Choose paths σ_{ij} in Y [resp. τ_{ik} in Z] from y to $a_{ij} \in A_i$ [resp. from z to $b_{ik} \in B_i$] for each point $a_{ij} \in A_i$ and $b_{ik} \in B_i$. Define a map $f_i : A_i * B_i \to Y \times Z \overset{c}{-} L_i \times M_i$ as follows: $f_i(a_{ij}) = (a_{ij}, z)$, $f_i(b_{ik}) = (y, b_{ik})$, f_i maps the mid-point of the edge joining a_{ij} to b_{ik} to (a_{ij}, b_{ik}), and f_i follows the paths σ_{ij} and τ_{ik} as appropriate. This should be done so that f_i embeds $A_i * B_i$ as a subcomplex.

Proposition 16.8.3. *The space* $Y \times Z \overset{c}{-} L_i \times M_i$ *is path connected and the embedding* f_i *induces an isomorphism of fundamental groups.*

Proof. The paths σ_{ij} and τ_{ik} give path connectedness. The cover of

$$Y \times Z \overset{c}{-} L_i \times M_i$$

by the sets $Y \times V_{ik}$ and $U_{ij} \times Z$ has the property that no point lies in more than two of those sets. By 6.2.11, the fundamental group of $Y \times Z \overset{c}{-} L_i \times M_i$ is isomorphic to $\pi_1(\mathcal{G}_i, A_i * B_i; T_i)$ where $\mathcal{G}_i$ is a generalized graph of groups for the graph $A_i * B_i$, T_i being the canonical maximal tree in that graph. The diagram of vertex and edge groups for a typical edge is (suppressing base points):

$$\pi_1(Y) \times \pi_1(V_{ik}) \leftarrow \pi_1(U_{ij}) \times \pi_1(V_{ik}) \to \pi_1(U_{ij}) \times \pi_1(Z).$$

Since Y and Z are simply connected, the fundamental group collapses to that of $A_i * B_i$ and the proposition follows. □

---

[22] Here, as always in graphs, edges intersect only in vertices. For more on joins, see Sect. 5.2.

Proof (of 16.8.1). Write $W_i := Y \times Z \xrightarrow{c} L_i \times M_i$. Choose the points a_{ij} so that $f_i(a_{i1})$ lies on ω for all i. The following diagram commutes:

$$
\begin{array}{ccccc}
(A_i \wedge B_i, p_i) & \hookrightarrow & \pi_1(A_i * B_i, a_{i1}) & \xrightarrow{\sim_=} & \pi_1(W_i, f_i(a_{i1})) \\
\uparrow & & \uparrow & & \uparrow \\
(A_{i+1} \wedge B_{i+1}, p_{i+1}) & \hookrightarrow & \pi_1(A_{i+1} * B_{i+1}, a_{i+1,1}) & \xrightarrow{\cong} & \pi_1(W_{i+1}, f_{i+1}(a_{i+1,1}))
\end{array}
$$

where the vertical arrows have obvious meanings. The result now follows from 16.8.2 and 16.8.3 on taking inverse limits. □

We can apply 16.8.1 to group extensions. Let $N \rightarrowtail G \twoheadrightarrow Q$ be a short exact sequence of infinite finitely presented groups. Let Y and Z be finite path connected CW complexes whose fundamental groups are isomorphic to N and Q respectively.

Theorem 16.8.4. *There is a finite CW complex X whose fundamental group is isomorphic to G such that $\tilde{X}$ is proper 2-equivalent to $\tilde{Y} \times \tilde{Z}$.*

To avoid repetition we postpone the proof until Sect. 17.3, where we prove the more general version 17.3.4. From 16.8.4 and 16.8.1 we deduce:

Corollary 16.8.5. *The group G has one end, is semistable at infinity, and (the isomorphism type of) its Čech fundamental group is freely generated by the smash product of (the homeomorphism types of) the ends of N and Q in the sense of 16.8.1. In particular, if N or Q has one end, then G is simply connected at infinity.* □

Source Notes: For another use of the free topological group generated by a pointed compact metric space – a "1, 2 or infinity" theorem for the fundamental group of a group, see [72].

Exercises

1. What does 16.8.1 become if Y or Z is finite?
2. What survives of 16.8.1 if Y or Z is not simply connected?
3. Prove that if there is an exact sequence $N \rightarrowtail G \twoheadrightarrow Q$ of infinite finitely generated groups, then G has one end.

16.9 Sample theorems on simple connectivity and semistability

In this section we give some group theoretic conditions (Theorems 16.9.1 and 16.9.5) which imply that a finitely presented group is semistable at infinity or is simply connected at infinity. The methods are elementary but illustrate how such questions are often dealt with in the literature. We begin with a strengthening of the first part of Corollary 16.8.5.

Theorem 16.9.1. *Let $N \rightarrowtail G \overset{\pi}{\twoheadrightarrow} Q$ be a short exact sequence of infinite finitely generated groups. If G is finitely presented then G is semistable at infinity. Hence $H^2(G, \mathbb{Z}G)$ is free abelian.*

Before proving this we set things up. To begin, we choose a presentation $\langle h_i, q_j \mid r_k, q_j w_{ij} q_j^{-1} h_i^{-1} \rangle$ of G where i, j and k range over finite sets of indices, and the set of generators $\{h_i, q_j\}$ is closed under inversion. In detail: the generators h_i and q_j are "N-generators" and "Q-generators" respectively, chosen so that the set $\{\pi(q_j)\}$ generates Q and the set $\{h_i\}$ generates N. Since N is normal, each $q_j^{-1} h_i q_j$ is a word in the N-generators, denoted by w_{ij}. Starting with any finite presentation of G, one can get to this special presentation by Tietze transformations.

Let (X, v) be the corresponding presentation complex. Orient the cells of X. The (oriented) 1-cells are $\{e^1(h_i), e^1(q_j)\}$. The (oriented) 2-cell corresponding to the relation $q_j w_{ij} q_j^{-1} h_i^{-1}$ is e_{ij}^2. There are finitely many other 2-cells corresponding to the relations r_k but they will play no role in the proof. The universal cover is $(\tilde{X}, \tilde{v})$. We choose lifts $\tilde{e}^1(h_i), \tilde{e}^1(q_i)$, oriented compatibly with their images in X, having initial point $\tilde{v}$. All other 1-cells of $\tilde{X}$ are translates of these and carry translated orientations. We choose lifts $\tilde{e}_{ij}^2$, oriented compatibly with e_{ij}^2, so that $\Delta(\tilde{e}_{ij}^2)$ is represented by the edge loop $(\tilde{e}^1(q_j), q_j \tilde{w}_{ij}, h_i \tilde{e}^1(q_j)^{-1}, \tilde{e}^1(h_i)^{-1})$. See Fig. 16.7.

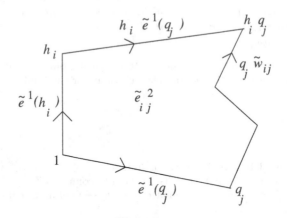

Fig. 16.7.

An edge in $\tilde{X}^1$ is either an "N-edge" $g.\tilde{e}^1(h_i)$ or a "Q-edge" $g.\tilde{e}^1(q_j)$; to simplify notation we will usually use the letter σ to denote an N-edge and the letter τ to denote a Q-edge (with the preferred orientation). Since the set of generators is closed under inversion, all edge rays can be chosen so that the exponent of every σ or τ in every such ray is 1 rather than -1.

For each vertex u of $\tilde{X}$ we select a proper Q-edge ray with initial point u as follows. We have the covering projection[23] $p_N : \tilde{X} \to \bar{X}(N)$; for each vertex $\bar{u}$ of $\bar{X}(N)$ we select a proper edge ray, containing no edges which are loops, with initial point $\bar{u}$; we do this so that whenever C is a finite subcomplex of $\bar{X}(N)$ only finitely many of these edge rays involve an edge of C. The required Q-edge rays in $\tilde{X}$ are the lifts of these rays in $\bar{X}(N)$. We call them "selected" edge rays: they are proper; there is exactly one for each vertex w; and they are Q-edge rays because p_N maps N-edges in $\tilde{X}$ to edges in $\bar{X}(N)$ which are loops. If $\{L_r\}$ is a finite filtration of $\bar{X}(N)$, then by 14.1.6 $(\tilde{X}, \{p_N^{-1}(L_r)\})$ is a well filtered CW complex and the selected proper edge rays are filtered rays in this sense. The set of selected edge rays is N-invariant. Initial segments of chosen or selected edge rays are "selected edge paths."

We now describe some proper homotopies:

(i) *Sliding τ along $(\sigma_1, \sigma_2, \ldots)$*: Here τ is a Q-edge and $(\sigma_1, \sigma_2, \ldots)$ is a proper N-edge ray with the same initial vertex as τ. We describe a homotopy $[0, \infty) \times [0, 1] \to \tilde{X}$ whose domain subdivision is illustrated in Fig. 16.8(a). The homotopy itself, illustrated in Fig. 16.9(a), is obtained by patching together characteristic maps for 2-cells. Its restriction to $[0, \infty) \times \{0\}$ is a proper parametrization of $(\sigma_1, \sigma_2, \ldots)$, and each $[k, k+1] \times [0, 1]$ is mapped into the CW neighborhood of the edge σ_k, so this is a proper homotopy by 10.1.15.

(ii) *Sliding σ along $(\tau_1, \tau_2, \ldots)$*: Here σ is an N-edge and $(\tau_1, \tau_2, \ldots)$ is a selected Q-edge ray with the same initial vertex as σ. We describe a homotopy $[0, \infty) \times [0, 1] \to \tilde{X}$ whose domain subdivision is illustrated in Fig. 16.8(b). The homotopy itself is illustrated in Fig. 16.9(b). The restriction to $[0, \infty) \times \{0\}$ is a proper parametrization of $(\tau_1, \tau_2, \ldots)$ and its restriction to $[0, \infty) \times \{1\}$ is an N-translate of this, and is therefore a proper parametrization of another selected Q-edge ray. The homotopy $[0, \infty) \times [0, 1] \to \bar{X}(N)$ obtained by composing with p_N is proper, by 10.1.15, so the homotopy (into $\tilde{X}$) is proper, by 10.1.17.

(iii) *Sliding $(\sigma_1, \sigma_2, \ldots)$ along $(\tau_1, \tau_2, \ldots)$*: Here $(\tau_1, \tau_2, \ldots)$ is a selected Q-edge ray and $(\sigma_1, \sigma_2, \ldots)$ is an N-edge ray having the same initial vertex. The desired homotopy $[0, \infty) \times [0, \infty) \to \tilde{X}$ is obtained by writing the domain as $\bigcup_k [0, \infty) \times [k, k+1]$ and stacking the homotopies described for sliding each σ_k along $(\tau_1, \tau_2, \ldots)$ in (ii). If $\{L_r\}$ is a finite filtration of $\bar{X}(N)$, this homotopy into $\tilde{X}$ is filtered with respect to the filtration $\{p_N^{-1}(L_r)\}$ of $\tilde{X}$ and the filtration $\{[0, r] \times [0, \infty)\}$. And for each r its restriction to $[0, r] \times [0, \infty)$ is certainly proper (compare (i)). So this is a proper homotopy.

We will also restrict these to "slides of edges along edge paths" $(\sigma_1, \ldots, \sigma_m)$ in (i) or $(\tau_1, \ldots, \tau_m)$ in (ii).

---

[23] As always, $\bar{X}(N) = N \backslash \tilde{X}$.

(a)

(b)

Fig. 16.8.

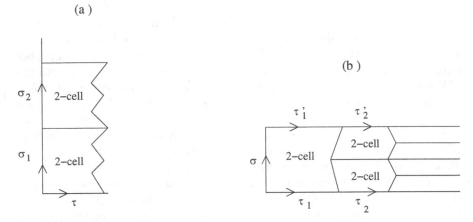

Fig. 16.9.

Proposition 16.9.2. *Given a finite subcomplex K of $\tilde{X}$ there is a finite subcomplex M such that if the given σ's and τ's in (i)–(iii) all lie in $\tilde{X} \overset{c}{-} M$, then the proper homotopies constructed in (i)–(iii) all take place in $\tilde{X} \overset{c}{-} K$.*

Proof. Choose k so that $p_N(K) \subset L_k$. Let $m > 0$ be such that whenever $(\bar{\tau}_1, \bar{\tau}_2, \dots)$ is the image under p_N of a selected proper edge ray in $\tilde{X}$, then the edges $\bar{\tau}_m, \bar{\tau}_{m+1}, \dots$ all lie in $\bar{X}(N) \overset{c}{-} N(L_k)$; here we use the fact that L_k meets only finitely many of the chosen edge rays. There are only finitely many pairs $(\tilde{e}^1(h_i), \alpha)$ where α is a Q-edge path in $\tilde{X}$ of length $\leq m$ having initial vertex $\tilde{v}$; choose $n \geq 1$ so large that for every such pair the slide of $\tilde{e}^1(h_i)$ along α takes place in $N^n(\tilde{v})$. Let[24] $M = N^n(K)$; M is finite by 11.4.4. $\square$

We pause for a useful fact about CW neighborhoods:

---

[24] Recall from Sect. 11.4 that the nth CW neighborhood of K is $N^n(K) = N(N^{n-1}(K))$ where $N^0(K) = K$.

Lemma 16.9.3. *Let A and B be subcomplexes of a CW complex Y, and let s be a positive integer. If $B \subset Y \overset{c}{-} N^s(A)$, then $N(B) \subset Y \overset{c}{-} N^{s-1}(A)$. Hence $N^s(B) \subset Y \overset{c}{-} A$, so $A \subset Y \overset{c}{-} N^s(B)$.*

Proof. Suppose $N(B)$ is not a subset of $Y \overset{c}{-} N^{s-1}(A)$. Then $N(B)$ and $N^{s-1}(A)$ share a vertex u. So there is a cell e of Y whose carrier $C(e)$ contains both u and a vertex w of B. Since $C(e) \subset N^s(A)$, $N^s(A) \cap B \neq \emptyset$, a contradiction. The last part of the Lemma now follows by induction. $\square$

Proof (of 16.9.2 (concluded)). For homotopies of type (i) the assertion follows from 16.9.3, for $(\sigma_1, \sigma_2, \ldots)$ lies in $\tilde{X} \overset{c}{-} M$, the homotopy takes place in the union of the CW neighborhoods of these edges and $n \geq 1$. A homotopy of type (ii) is a slide of σ along $(\tau_1, \tau_2, \ldots)$. Let $g.\tilde{v}$ be the initial vertex. The slide of σ along the selected edge path $(\tau_1, \tau_2, \ldots, \tau_m)$ takes place in $N^n(g\tilde{v})$, which lies in $\tilde{X} \overset{c}{-} K$, by 16.9.3. The rest of the slide also takes place in $\tilde{X} \overset{c}{-} K$ because its image under p_N takes place in $\bar{X}(N) \overset{c}{-} L_k \subset \bar{X}(N) - p_N(K)$. The claim for homotopies of type (iii) follows immediately. $\square$

Proof (of 16.9.1). We will show that 16.1.2(ii) holds, i.e., that we can "push loops to infinity." Initially we will ignore base ray issues to keep things simple.

Let K be a finite subcomplex of $\tilde{X}$ and let $M = N^n(K)$ as in 16.9.2. Only finitely many selected Q-edge rays contain an edge of M. Let M' be a finite subcomplex containing $N(M)$ such that whenever a selected Q-edge ray has initial point in $\tilde{X} \overset{c}{-} M'$ then it contains no edges of M.

The graph $\tilde{X}^1$ is the Cayley graph of G with respect to $\{h_i, q_j\}$. The vertex $\tilde{v}$ lies in a path connected subgraph $\Gamma(N)$ which is the Cayley graph of N with respect to $\{h_i\}$. Enlarging M' if necessary, we may assume by 13.4.9 that for each $g \in G$ the graph $g\Gamma(N) \overset{c}{-} (M' \cap \Gamma(N))$ has only infinite path components; this is justified because $g\Gamma(N) \cap M' = \emptyset$ for all but finitely many of the graphs $g\Gamma(N)$.

Let $\alpha := (\alpha_1, \ldots, \alpha_r)$ be an edge loop in $\tilde{X} \overset{c}{-} M'$. Let $h_\alpha : S^1 \to \tilde{X} \overset{c}{-} M'$ be a parametrization of α (see Sect. 3.4). We construct in pieces a proper homotopy $[0, \infty) \times S^1 \to \tilde{X} \overset{c}{-} K$ extending h_α as follows. If α_s is a Q-edge, let $F^{(s)} : [0, \infty) \times [0, 1] \to \tilde{X}$ be obtained as in (i) by sliding α_s along an infinite N-edge ray in $\tilde{X} \overset{c}{-} M'$ with the same initial vertex as α_s. If α_s is an N-edge, let $F^{(s)} : [0, \infty) \times [0, 1] \to \tilde{X}$ be obtained as in (ii) by sliding α_s along the selected infinite Q-edge ray with the same initial vertex as α_s. By 16.9.2, these homotopies $F^{(s)}$ can be chosen to take place in $\tilde{X} \overset{c}{-} K$. Moreover the "top," $F_1^{(s)}$, of the homotopy, $[0, \infty) \times \{1\} \to \tilde{X}$ has its image in $\tilde{X} \overset{c}{-} M$ because in one case it parametrizes a proper N-edge ray in the CW neighborhood of the "bottom" $F_0^{(s)}$, and $N(M) \subset M'$, so it misses M, by 16.9.3; and in the other case it parametrizes a selected Q-edge ray with initial point in $\tilde{X} \overset{c}{-} M'$. The homotopies $F^{(s)}$ are illustrated in Fig. 16.10.

We consider three situations. (a) If α_s and α_{s+1} are N-edges then $F^{(s)}$ and $F^{(s+1)}$ can be properly fitted together, since $F_1^{(s)}$ and $F_0^{(s+1)}$ are proper parametrizations of the same selected Q-edge ray. (b) If α_s and α_{s+1} are Q-edges then $F_1^{(s)}$ and $F_0^{(s+1)}$ parametrize proper N-edge rays in $\tilde{X} \stackrel{c}{=} M$ with the same initial point. (c) If one of α_s and α_{s+1} is an N-edge and the other is a Q-edge, then one of the proper rays at the final point of α_s parametrizes a proper N-edge ray and the other a selected proper Q-edge ray. See Fig. 16.10. In (c) we can link up these two proper rays by a proper homotopy in $\tilde{X} \stackrel{c}{=} K$ of type (iii). In (b) we can do this twice, linking both of the infinite N-edges by proper homotopies of type (iii) to the selected Q-edge which begins at that initial point. The fitting together of all these homotopies gives the required proper homotopy $[0, \infty) \times S^1 \to \tilde{X} - K$ extending h_α.

We have ignored base rays. In choosing the selected Q-edge rays it is easy to arrange that if $(\tau_1, \tau_2, \ldots)$ is selected, then so is $(\tau_s, \tau_{s+1}, \ldots)$ for every $s \geq 1$; this simply requires care in the choice of rays in $\bar{X}(N)$. Assume this is done. Then if we are given a selected Q-edge ray as base ray and if the base point of α is on that base ray, our proper homotopy extending h_α "moves" the base point to infinity along a parametrization of the given base ray. Thus (ii) of 16.1.2 is satisfied and $\tilde{X}$ is strongly connected at each end.

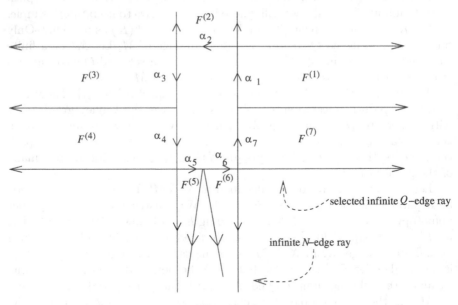

Fig. 16.10.

It only remains to prove that $\tilde{X}$ has one end. Indeed, this was an exercise in Sect. 16.8. To see it directly, join two "far out" vertices of $\tilde{X}$ by an edge path α in $\tilde{X}$, choose proper N-edge rays starting at those two points, and properly deform α, sending its end points along those N-edge rays in the manner described above for edge loops. Thus the points can be joined "far out," and $\tilde{X}$ has one end and is therefore strongly connected at infinity, i.e., G is semistable at infinity.

The final sentence of the theorem now follows, by 16.5.1.   □

In view of 16.9.1 one might expect that the second part of 16.8.5 could also be strengthened by weakening the hypothesis on N from "finitely presented" to "finitely generated." This is not so:

Example 16.9.4. Let F_2 be the free group of rank 2 and let $G = F_2 \times F_2$. Then G has one end by 16.8.1, and $H^2(G, \mathbb{Z}G) \neq 0$ by 13.2.9 and 12.6.1; hence G is not simply connected at infinity. But there is a short exact sequence $N \rightarrowtail G \twoheadrightarrow \mathbb{Z}$ with N a finitely generated group having one end. To see this we will now construct G in another way. Let F_2 be generated by a and b, let $x_i = a^i b a^{-i}$ and let $B \triangleleft F_2$ be the subgroup (freely) generated by $\{x_i \mid i \in \mathbb{Z}\}$. The inner automorphism $\phi : F_2 \to F_2$, $g \mapsto aga^{-1}$ maps B onto B. Let $N = F_2 *_B F_2$ and let $\psi : N \to N$ be the automorphism determined by ϕ. Let $\bar{G} = N*_\psi = \langle N, t \mid t^{-1}xt = \psi(x) \ \forall \ x \in N \rangle$. Then $N \hookrightarrow \bar{G} \twoheadrightarrow Q$ is exact where $Q \cong \mathbb{Z}$ is generated by the image of t. By Tietze transformations one sees (exercise) that $\bar{G}$ is isomorphic to G. Clearly N is finitely generated. That N has one end is left as an exercise.

Other examples to illustrate the sharpness of 16.8.5 can be found in [116]. In particular one can have $N \rightarrowtail G \twoheadrightarrow Q$ with N finitely generated and one-ended but not finitely presented, $Q \cong \mathbb{Z}^n$ and $G \cong (\mathbb{Z}^n * \mathbb{Z}) \times (\mathbb{Z}^n * \mathbb{Z})$ for any $n \geq 1$: this G is not simply connected at infinity.

We turn to ascending HNN extensions.

Theorem 16.9.5. *Let H be finitely presented, let $\phi : H \rightarrowtail H$ be a monomorphism, and let $G = H*_\phi$ be the resulting ascending HNN extension. If H is infinite then G has one end and is semistable at infinity. If H has one end then G is simply connected at infinity.*

Proof. Let $\{h_i\}$ be a finite set of generators for H. The group G has the finite presentation $\langle H, t \mid t^{-1}h_i t\phi(h_i)^{-1}, \ \forall i \rangle$ (where a finite presentation of H with generators $\{h_i\}$ is understood – see Sect. 3.1 Appendix). Let (X, v) be a corresponding presentation complex. Orient the cells of X and of $\tilde{X}$ compatibly. We use terminology similar to that in the proof of 16.9.1. In particular, $\tilde{X}$ has H-edges and t-edges.

There is a finite subcomplex Z of X so that $(Z, v) \hookrightarrow (X, v)$ induces the inclusion $H \hookrightarrow G$, and each vertex $g.\tilde{v}$ of $\tilde{X}$ lies in a translate $g\tilde{Z}$ of the universal cover, $\tilde{Z}$, of Z (where $\tilde{Z}$ is a subcomplex of $\tilde{X}$). There is a canonical homomorphism $\psi : G \to \mathbb{Z}$ defined by sending H to $\{0\}$ and t to $1 \in \mathbb{Z}$. The *level* of a vertex $g.\tilde{v}$ of $\tilde{X}$ is the integer $\psi(g)$; indeed, $\psi(g)$ is the t-exponent sum of any word in the generators representing g. Giving $\mathbb{R}$ the usual CW complex structure, ψ extends to a cellular map $f : \tilde{X} \to \mathbb{R}$, taking each H-edge to a vertex, each t-edge to an edge of $\mathbb{R}$, and each 2-cell into the convex hull of the f-image of its boundary. Any subset of $f^{-1}(-\infty, k)$ [resp. $f^{-1}(k, \infty)$] will be said to be *below level k* [resp. *above level k*].

Let K be a finite subcomplex of $\tilde{X}$. There are integers m_- and m_+ such that K is below level m_+ and above level m_-. Let $\alpha := (\alpha_1, \ldots, \alpha_r)$ be an edge loop.

Lemma 16.9.6. *If every α_s is above level m_+ then α is equivalent in $\tilde{X} \stackrel{c}{-} K$ to a trivial edge loop.*

Proof. If every α_s is an H-edge then the whole loop lies in a simply connected subcomplex $g\tilde{Z}$, and the f-image of such a subcomplex is an integer $> m_+$, so $g\tilde{Z} \subset \tilde{X} \stackrel{c}{-} K$. If some α_s is a t-edge, then the edge loop contains an edge path $(t^{-1}, \alpha_p, \ldots, \alpha_q, t)$ (indices written modulo r); this is because the t-exponent sum around any loop is zero (as many t-edges oriented one way as the other). That subpath can be changed to $(t^{-1}, \alpha_p, t, t^{-1}, \alpha_{p+1}, t, \cdots, \alpha_q, t)$ and slid "upwards" rel its end points (compare the proof of 16.9.1) to a path of H-edges. Proceeding thus, the edge loop α is shown to be equivalent in $\tilde{X} \stackrel{c}{-} K$ to an H-edge loop, and, as we have seen, that is enough. $\square$

Proof (of 16.9.5 (concluded)). As in the proof of 16.9.1, for each vertex $g.\tilde{v}$ of $\tilde{X}$ we select the proper edge ray with that initial vertex whose vertices are $g.\tilde{v}, gt.\tilde{v}, gt^2.\tilde{v}, \ldots$. And, just as in that proof, we use the conjugation relations to slide an H-edge along the selected ray (or finite segments thereof) at its initial vertex. Let n be such that the slide of any edge $\tilde{e}(h_i)$ (with initial vertex $\tilde{v}$) along the first $m_+ - m_- + 2$ edges of the selected infinite edge ray beginning at $\tilde{v}$ takes place in $N^n(\tilde{v})$. By 16.9.3, the slide of any H-edge in $\tilde{X} \stackrel{c}{-} N(K)$ along the first $m_+ - m_- + 2$ edges of the selected edge ray beginning at its initial point takes place in $\tilde{X} \stackrel{c}{-} K$, and if that H-edge is above $m_- - 2$ then it is moved to an H-path above m_+.

Choose M to be a finite subcomplex of $\tilde{X}$ containing $N^n(K)$ such that for each $g \in G$ every path component of $g\tilde{Z} \stackrel{c}{-} (g\tilde{Z} \cap M)$ is infinite. Assuming (for now) that H has one end, we are to show that any edge loop α in $\tilde{X} \stackrel{c}{-} M$ is equivalent in $\tilde{X} \stackrel{c}{-} K$ to a trivial edge loop.

If α is above level $m_- - 2$ then we can slide it in $\tilde{X} \stackrel{c}{-} K$ above level m_+, and then we are done, by 16.9.6. If α is below level m_- then we can slide it up to a loop at level $m_- - 1$, where it lies in a simply connected subcomplex $g\tilde{Z}$ at that level, so again we are done. Thus, after suitable adjustment, the only case left is when α contains an edge path of the form $(t^{-1}, \alpha_p, \ldots, \alpha_q, t)$ where $(\alpha_p, \ldots, \alpha_q)$ is an H-edge path at level $m_- - 1$ with end points in $\tilde{X} \stackrel{c}{-} M$. Then $(\alpha_p, \ldots, \alpha_q)$ lies in some $g\tilde{Z}$ at level $m_- - 1$. The fact that this $g\tilde{Z}$ is one-ended and simply connected allows us to replace $(\alpha_p, \ldots, \alpha_q)$ by an edge path in $g\tilde{Z}$ lying in $g\tilde{Z} \stackrel{c}{-} (g\tilde{Z} \cap M)$. In this way, our edge loop is homotopic in $\tilde{X} \stackrel{c}{-} K$ to an edge loop which can be slid forward to level $m_+ + 1$ in $\tilde{X} \stackrel{c}{-} K$. By 16.9.6, that is enough.

These ideas can easily be modified to show that G has one end and, assuming H infinite rather than one-ended, that G is semistable at infinity. □

If H in 16.9.5 is finite then G has two ends and is simply connected at both ends. By 16.5.1 and 16.5.2, it follows from 16.9.5 that $H^2(G, \mathbb{Z}G)$ is free abelian or trivial as appropriate.

Example 16.9.7. By 9.2.5 and 16.9.5, Thompson's group F is simply connected at infinity.

Remark 16.9.8. In this section we have considered decomposition properties of a finitely presented group G which imply that G is semistable at each end (and hence that $H^2(G, \mathbb{Z}G)$ is free abelian). There are other theorems of this type in the literature, stronger than the ones given here and more difficult to prove. For some of these, see [119], [118] and the references therein. For a homological result see [102]. In connection with 16.9.5, we point out that (at time of writing) it is unknown whether semistability holds if H is assumed only to be finitely generated, G being finitely presented.

Source Notes: 16.9.1 appeared in [114]; 16.9.5 appeared in [115]; Example 16.9.4 is taken from [129]. I am indebted to Michael Mihalik for help with the writing of this section.

Exercises

1. Prove that the group N in Example 16.9.4 has one end.
2. Find the Tietze transformations mentioned in Example 16.9.4.
3. Fill in the details at the end of the proof of 16.9.5.
4. In the proof of 16.9.1 the relations r_k played no role, so a more general theorem was proved. State that theorem.

Higher homotopy theory of groups

In Chap. 16 we concentrated on π_0 and π_1 issues at the end of a CW complex. Here we continue with a discussion of higher homotopy invariants.

17.1 Higher proper homotopy

Let Y be a countable strongly locally finite path connected CW complex, let $\{L_i\}$ be a finite filtration of Y, and let $n \geq -1$ be an integer. We say Y is *n-connected at infinity* if for each $-1 \leq k \leq n$ and each i there exists $j \geq i$ such that every map $S^k \to Y \overset{c}{-} L_j$ extends to a map $B^{k+1} \to Y \overset{c}{-} L_i$. Thus Y is (-1)-connected at infinity iff Y is infinite; Y is 0-connected at infinity iff Y has one end, and Y is 1-connected at infinity iff Y is simply connected at infinity. We say Y is *n-acyclic at infinity* with respect to the ring R if $\{\tilde{H}_k(Y \overset{c}{-} L_i; R)\}$ is pro-trivial for each $-1 \leq k \leq n$. By 1.4.3 and 2.7.6 these properties only depend on the $(n+1)$-skeleton of Y.

When A is a subcomplex of Y and $n \geq 0$, we say (Y, A) is *properly n-connected* if (Y, A) is n-connected and for every $0 \leq k \leq n$ and every i there exists $j \geq i$ such that every map $(B^k, S^{k-1}) \to (Y \overset{c}{-} L_j, A \overset{c}{-} (L_j \cap A))$ is homotopic rel S^{k-1} in $Y \overset{c}{-} L_i$ to a map whose image lies in A. For example, (Y, A) is properly 0-connected iff A is non-empty and the function $\mathcal{E}(A) \to \mathcal{E}(Y)$ induced by inclusion is surjective.

Proposition 17.1.1. (Proper Whitehead Theorem) *Let Y be finite-dimensional. The following are equivalent:*

(i) *A is a proper strong deformation retract of Y;*

(ii) *$A \overset{i}{\hookrightarrow} Y$ is a proper homotopy equivalence;*

(iii) *(Y, A) is properly n-connected for all n such that $Y - A$ contains an n-cell.*

Proof. The proof is similar to the proof of the Whitehead Theorem 4.1.4. In an exercise, the reader is asked to consider why one needs Y to be finite-dimensional. $\square$

This last proposition is used in a proper version of Theorem 4.1.5:

Theorem 17.1.2. *If in Theorem 4.1.5 X and X' are finite-dimensional and locally finite, $f : X \to X'$ is proper, and $f \mid A$, $f \mid B$ and $f \mid A \cap B$ are proper homotopy equivalences, then f is a proper homotopy equivalence. Moreover, the map g in 4.1.5 can be chosen to be a proper homotopy inverse for f so that all the indicated homotopies are proper.*   □

Just as 4.1.5 leads to proofs of 4.1.7 and 4.1.8, the last theorem implies:

Theorem 17.1.3. *If in Theorems 4.1.7 and 4.1.8 all the spaces are finite-dimensional locally finite CW complexes and all the given maps are proper, then the homotopy equivalences in the conclusions of those theorems are proper homotopy equivalences.*   □

Corollary 17.1.4. *If Y is obtained from A by properly attaching n-cells then the proper homotopy type of Y only depends on the proper homotopy class of the simultaneous attaching map.*   □

For $n \geq -1$, we say Y is *properly n-connected* if Y is n-connected and n-connected at infinity. A proper analog of 7.1.2 is:

Proposition 17.1.5. *Let Y be infinite, let $\omega : [0, \infty) \to Y$ be a cellular proper ray and an embedding, and let $n \geq 0$. Y is properly n-connected iff the inclusion $Y^n \hookrightarrow Y$ is properly homotopic to a map into $\omega([0, \infty))$.*

Proof. The case $n = 0$ is clear. Let Y be properly n-connected. By induction, assume $Y^{n-1} \hookrightarrow Y$ is properly homotopic to a map into the ray $\omega([0, \infty))$. Then 17.1.4 implies that Y^n has the proper homotopy type of the locally finite CW complex $(\{p\} \times [0, \infty)) \cup (S^n \times \mathbb{N}) \subset S^n \times [0, \infty)$, where $p \in S^n$ is a base point; here, S^n and $[0, \infty)$ have the usual CW complex structures. Thus, since Y is properly n-connected, the inclusion $Y^n \hookrightarrow Y$ has the desired property. The converse is clear.   □

We say Y is *properly n-acyclic* with respect to R if $H_k(Y; R) = 0$ when $k \leq n$, and Y is n-acyclic at infinity with respect to R. This is related to "properly n-connected" by the following analog of 4.5.1:

Theorem 17.1.6. (Proper Hurewicz Theorem) *Let $n \geq 2$ and let Y be properly 1-connected. Then Y is properly n-connected iff Y is properly n-acyclic with respect to $\mathbb{Z}$.*

Proof. Simply observe that the proof of 4.5.1 gives this; use 17.1.3 and 17.1.5 in place of 7.1.2 and 4.1.8 in that proof.   □

Source Notes: The material here is developed in analogy with the corresponding shape theory – see the bibliography in [109] for details.

Exercises

1. Let $\omega : [0, \infty) \to Y$ be a cellular proper ray in Y which is also an embedding. Prove that Y is properly n-connected iff the pair $(Y, \omega([0, \infty)))$ is properly n-connected.
2. Rewrite this section in the CW-proper context (Sect. 10.2).
3. Rewrite this section in the filtered homotopy context (Sect. 14.1).
4. Why is "finite-dimensional" needed in 16.5.1?
5. State and prove the Proper Rebuilding Lemma, i.e., the proper version of 6.1.4.
6. State and prove the Proper Relative Hurewicz Theorem. *Hint*: Compare 4.5.1.
7. Here we gave proper analogs of 4.1.3 through 4.1.8. Are there proper analogs of the other propositions in Sect. 4.1? What might their statements be?

17.2 Higher connectivity invariants of groups

Let G be a group of type F_n where $n \geq 0$. Let (X, x) be a $K(G, 1)$-complex with finite n-skeleton. We say that G is $(n - 1)$-*connected at infinity* [resp. $(n - 1)$-*acyclic at infinity* with respect to R] if $\tilde{X}^n$ is $(n - 1)$-connected at infinity [resp. $(n - 1)$-acyclic at infinity with respect to R]. As in Sect. 16.5, one sees that these are properties of G rather than of X.

A group G is (-1)-connected at infinity iff G is (-1)-acyclic with respect to some (equivalently, any) ring iff G is infinite. A finitely generated group G is 0-connected at infinity iff G is 0-acyclic at infinity with respect to some (equivalently, any) ring iff G has one end. If a finitely presented group G is simply connected at infinity then G is 1-acyclic at infinity with respect to any ring. By 13.3.3 (ii), G is $(n - 1)$-acyclic at infinity with respect to R iff $H^k(G, RG) = 0$ when $k \leq n$ and the R-module $H^{n+1}(G, RG)$ is torsion free.

By 17.1.6 we have:

Theorem 17.2.1. *Let the group G have type F_n and assume G is simply connected at infinity. Then G is $(n-1)$-connected at infinity iff G is $(n-1)$-acyclic at infinity with respect to $\mathbb{Z}$.* $\square$

Example 17.2.2. Hence Thompson's group F is n-connected at infinity for all n. This follows from 9.3.19, 13.10.1, 13.3.3 and 16.9.7.

Now let G have type F_∞. We have seen that connectivity properties of G such as n-connectedness, n-acyclicity and semistability at infinity can be read off from a contractible free G-CW complex of locally finite type which is of finite type mod G. In real situations the naturally occurring contractible G-CW complexes are sometimes not free, but rather are rigid with finite cell-stabilizers. The connectivity properties of G can often be read off in that situation. One useful case is:

Theorem 17.2.3. *Let Y be a contractible rigid G-CW complex which is finite mod G, and let the stabilizer of each cell be finite. Then G is semistable* *at each end or n-connected at infinity or n-acyclic at infinity iff Y has the corresponding property. Moreover, for any ring R, $H^*(G, RG)$ is isomorphic to $H_f^*(Y; R)$.*

Proof. The CW complex Y is strongly locally finite by 10.1.12. Apply the Borel Construction using a $K(G, 1)$-complex X of finite type to get the usual commutative diagram

$$\tilde{X} \times Y \xrightarrow{\text{projection}} Y$$
$$\downarrow \qquad\qquad \downarrow$$
$$Z \xrightarrow{q} \Gamma := G\backslash Y.$$

Here, the diagonal action of G on the contractible CW complex $\tilde{X} \times Y$ is free, so Z is a $K(G, 1)$-complex. Then $q : Z \to \Gamma$ is a stack of CW complexes which, by 6.1.4, can be rebuilt to give a commutative diagram

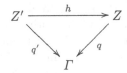

in which $q' : Z' \to \Gamma$ is a stack of CW complexes and h is a homotopy equivalence. Here, the fiber F_e of q over the cell e of Γ is a $K(G_e, 1)$-complex where G_e is finite, and the fiber F_e' of q' over e is a $K(G_e, 1)$-complex of finite type (see 7.2.5). From Sect. 6.1 and Sect. 7.3 one sees that the map $\tilde{Z}' \xrightarrow{\tilde{h}} \tilde{X} \times Y \xrightarrow{\text{projection}} Y$ is a stack of CW complexes in which the fiber over a cell $\tilde{e}$ of Y is homeomorphic to the universal cover of F_e' (where $\tilde{e}$ lies over the cell e of Γ). Thus that fiber is a contractible CW complex of finite type. It follows that the indicated map $\tilde{Z}' \to Y$ is a CW-proper homotopy equivalence.[1] Indeed this holds hereditarily just as in that exercise, and that is crucial: the connectivity properties under discussion are invariants of CW-proper homotopy, and they hold for G iff they hold for $\tilde{Z}'$, hence iff they hold for Y. The claim about $H^*(G, RG)$ holds for the same reason.   □

A similar proof gives the following useful variation on 17.2.3:

Theorem 17.2.4. *Let Y be an $(m-1)$-connected m-dimensional G-CW complex which is finite mod G, where the stabilizer of each cell is finite and $m \geq 2$. Then G satisfies the conclusions of Theorem 17.2.3 for all $n \leq m - 1$.*   □

There is an analog of 7.3.1 in the present context; for a proof see [34]:

---

[1] To see this, compare with the exercise in Sect. 6.1: the hint given there applies here too.

Theorem 17.2.5. *Let Y be an $(m-1)$-connected rigid G-CW complex having finite m-skeleton mod G. If the stabilizer of each i-cell has type F_{m-i} and is $(m-i-1)$-connected at infinity, then G is $(m-1)$-connected at infinity.* $\square$

A homological version of 17.2.5 can also be found in [34].

Remark 17.2.6. 17.2.5 is not a generalization of 17.2.4 but rather covers a different kind of hypothesis. In 17.2.4 the stabilizers are finite and hence are not (-1)-connected at infinity.

Exercises

1. Prove that the connectivity properties discussed in this section do not change on passing to a subgroup of finite index or to a quotient by a finite normal subgroup.
2. State and prove a version of 17.2.4 for the case $m = 1$; compare 13.5.12.

17.3 Higher invariants of group extensions

Let $N \rightarrowtail G \twoheadrightarrow Q$ be a short exact sequence of infinite groups. One expects G to have better connectivity properties at infinity than N and Q. In this section we establish theorems of that type.

Let (X, x) [resp. (Y, y), (Z, z)] be a $K(G, 1)$-complex [resp. $K(N, 1)$-complex, $K(Q, 1)$-complex]. We saw at the end of Sect. 7.1 how to build a commutative diagram

$$
\begin{array}{ccc}
\tilde{X} \times \tilde{Z} & \xrightarrow{\text{projection}} & \tilde{Z} \\
\downarrow & & \downarrow \\
W & \xrightarrow{\quad q \quad} & Z
\end{array}
$$

in which q is a stack of CW complexes all of whose fibers are homeomorphic to the quotient CW complex $N \backslash \tilde{X}$, and W is a $K(G, 1)$-complex.[2] We can use this to get information about $H^*(G, \mathbb{Z}G)$ in terms of $H^*(N, \mathbb{Z}N)$ and $H^*(Q, \mathbb{Z}Q)$ under suitable finiteness hypotheses.

We choose a cellular homotopy equivalence $g : N \backslash \tilde{X} \to Y$. By 6.1.4 (see also the last part of Sect. 7.1 on group extensions) there is a commutative diagram

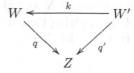

with k a cellular homotopy equivalence and q' a stack of CW complexes in which every fiber is homotopy equivalent to Y.

---

[2] Indeed, q is a fiber bundle.

We first consider the case in which N and Q have type F. Then we may take Y and Z to be finite CW complexes and in that case W' is a finite $K(G, 1)$-complex. We have a commutative diagram in which the unmarked arrows are projections:

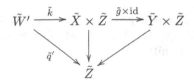

Proposition 17.3.1. *The map* $(\tilde{g} \times \text{id}) \circ \tilde{k}$ *is a proper homotopy equivalence.*

Proof. The maps $\tilde{q}' : \tilde{W}' \to \tilde{Z}$ and projection: $\tilde{Y} \times \tilde{Z} \to \tilde{Z}$ are stacks of CW complexes with every fiber $\tilde{Y}$, and the projection map: $\tilde{X} \times \tilde{Z} \to \tilde{Z}$ is a stack of CW complexes with every fiber $\tilde{X}$. Let $f : Y \to N\backslash\tilde{X}$ be a homotopy inverse for g, lifting to $\tilde{f} : \tilde{Y} \to \tilde{X}$. The construction of k using f (see 6.1.4) shows that, before identifications are made, $(\tilde{g} \times \text{id}) \circ \tilde{k}$ is built from compositions of the form

$$\tilde{Y} \times B^n \xrightarrow{\tilde{f} \times \text{id}} \tilde{X} \times B^n \xrightarrow{\tilde{g} \times \text{id}} \tilde{Y} \times B^n$$

and hence $(\tilde{g} \times \text{id}) \circ \tilde{k}$ is a proper homotopy equivalence by the Proper Rebuilding Lemma (Exercise 5 of Sect. 17.1). $\qquad\square$

From 17.3.1, 12.6.1 and 13.2.9 we obtain:

Theorem 17.3.2. *Let R be a PID, and let N and Q have type F. Then G has type F and for all p there are split short exact sequences*

$$0 \to \bigoplus_{i+j=p} H^i(N, RN) \otimes_R H^j(Q, RQ) \to H^p(G, RG) \to \bigoplus_{i+j=p+1} \text{Tor}_R(H^i(N, RN), H^j(Q, RQ)) \to 0.$$

$\qquad\square$

Applying this with 13.3.3(ii) we get:

Corollary 17.3.3. *With respect to the PID R, if N is s-acyclic at infinity and Q is t-acyclic at infinity, then G is $(s + t + 2)$-acyclic at infinity.* $\quad\square$

More generally, if the infinite groups N and Q have type F_n we may take Y and Z to have finite n-skeleta, implying that W' is a $K(G, 1)$-complex with finite n-skeleton, so G has type F_n. In that case the appropriate analog of 17.3.1 is:

Proposition 17.3.4. *Under these hypotheses the map $(\tilde{g} \times \text{id}) \circ \tilde{k} : \tilde{W}' \to \tilde{Y} \times \tilde{Z}$ is a CW-proper n-equivalence.*

Proof. The proof is similar to that of 17.3.1. That proof depends on 17.1.3, which in turn is merely an observation about the proof of 4.1.7 (in which 17.1.1 and 17.1.2 play the roles of 4.1.4 and 4.1.5). Here we are asserting that a parallel observation, using appropriately modified versions of 17.1.1 and 17.1.2, proves what is claimed.   □

As above, this gives:

Theorem 17.3.5. *Let R be a PID, and let N and Q be infinite and have type F_n. Then G has type F_n and for all $p \leq n - 1$ there are split short exact sequences*

$$0 \to \bigoplus_{i+j=p} H^i(N, RN) \otimes_R H^j(Q, RQ) \to H^p(G, RG) \to \bigoplus_{i+j=p+1} \mathrm{Tor}_R(H^i(N, RN), H^j(Q, RQ)) \to 0.$$

Moreover, there is a free R-module F such that

$$H^n(G, RG) \oplus F \cong \left(\bigoplus_{i+j=n} H^i(N, RN) \otimes_R H^j(Q, RQ) \right) \oplus \left(\bigoplus_{i+j=n+1} \mathrm{Tor}_R(H^i(N, RN), H^j(Q, RQ)) \right).$$

Proof. The CW complexes $\tilde{Y}^n$ and $\tilde{Z}^n$ are locally finite, and $(\tilde{Y}^n \times \tilde{Z}^n)^n = (\tilde{Y} \times \tilde{Z})^n$. By 12.6.1, for every p we have a split short exact sequence

$$0 \to \bigoplus_{i+j=p} H_f^i(\tilde{Y}^n; R) \otimes_R H_f^j(\tilde{Z}^n; R) \to H_f^p(\tilde{Y}^n \times \tilde{Z}^n; R) \to \bigoplus_{i+j=p+1} \mathrm{Tor}_R(H_f^i(\tilde{Y}^n; R), H_f^j(\tilde{Z}^n; R)) \to 0.$$

By a straightforward use of chain homotopies, 17.3.4 implies that for $p \leq n-1$ $H_f^p((\tilde{W}')^n; R) \cong H_f^p((\tilde{Y} \times \tilde{Z})^n; R) = H_f^p(\tilde{Y}^n \times \tilde{Z}^n; R)$. By 13.2.9, $H_f^i(\tilde{Y}^n; R) \cong H^i(N, RN)$ when $i \leq n - 1$, $H_f^j(\tilde{Z}^n; R) \cong H^j(Q, RQ)$ when $j \leq n - 1$, and $H_f^p((\tilde{W}')^n; R) \cong H^p(G, RG)$ when $p \leq n - 1$. This establishes the theorem for $p \leq n - 1$. By 13.2.11 and 12.1.2, the extremes $i = n$ and $j = n$ cause no trouble in the Tor term.

Now let $p = n$. Again we wish to identify H_f^* of various CW complexes with $H^*(\Gamma, R\Gamma)$ for $\Gamma = G$, N or Q. The term $H^n(G, RG) \oplus F$ comes from 13.2.17. In the other cases where 13.2.9 does not give these identifications, namely, the extremes i or $j = n$ in the tensor product and i or $j = n$ or $n + 1$ in the Tor terms, the relevant modules are all trivial (by 12.1.2, 12.5.1, 12.5.10(ii), 13.2.11 and 13.3.2).   □

Theorem 17.3.6. *Let $N \rightarrowtail G \twoheadrightarrow Q$ be a short exact sequence of infinite groups of type F_n, and let R be a PID. Working with respect to R, let N be s-acyclic at infinity and let Q be t-acyclic at infinity where $s \leq n - 1$ and $t \leq n - 1$. Then G is u-acyclic at infinity where $u = \min\{s + t + 2, n - 1\}$. If $s \geq 0$ or $t \geq 0$ (i.e., if N or Q has one end) then G is u-connected at infinity.*

Proof. By 13.3.2, when $s \leq n-1$, N is s-acyclic at infinity iff $H^i(N, \mathbb{Z}N) = 0$ for $i \leq s+1$ and $H^{s+2}(N, \mathbb{Z}N)$ is torsion free; a similar statement holds for Q, replacing s by t. It follows from 16.8.5 that $H^p(G, \mathbb{Z}G) = 0$ for $p \leq \min\{s+t+3, n-1\}$ and $H^p(G, \mathbb{Z}G)$ is torsion free for $p \leq \min\{s+t+4, n-1\}$. So $H^p(G, \mathbb{Z}G) = 0$ for $p \leq u+1$, and $H^{u+2}(G, \mathbb{Z}G)$ is torsion free, as required for the first part. The second part then follows by 16.8.5 and 17.2.1. $\quad\square$

With 17.1.6 this gives:

Corollary 17.3.7. *If N is s-connected at infinity and Q is t-connected at infinity then G is u-connected at infinity.* $\quad\square$

Source Notes: A version of this material appeared in [70] and [71].

Exercises

1. What happens to the material in this section when N or Q is finite?
2. Together with 16.5.1 and 16.5.2, Corollary 16.8.5 implies that (for G as in that theorem) $H^2(G, \mathbb{Z}G)$ is free abelian, and is trivial when N or Q has one end. Deduce these conclusions directly from 17.3.5.
3. Give an example of $H \leq G$ where H and G/H are infinite while G has more than one end; compare 16.8.5.

17.4 The space of proper rays

Let Y be a strongly locally finite CW complex which is path connected. In this section and the next we discuss spaces which model the end of Y in the sense that their ordinary algebraic topology invariants are isomorphic to the invariants "at the end of Y" which we have been discussing.

To begin, we must define higher homotopy groups at the end of Y. The nth *strong* (or *Steenrod*) *homotopy group* of Y with respect to the proper base ray ω, denoted $\pi_n^e(Y, \omega)$, is defined by analogy with $\pi_0^e(Y, \omega)$ in Sect. 16.1 and $\pi_1^e(Y, \omega)$ in Sect. 16.2: its elements are the base ray preserving proper homotopy classes of proper maps $(S^n \times [0, \infty), \{v\} \times [0, \infty)) \rightarrow (Y, \omega)$ where v is a base point for S^n, and $\{v\} \times [0, \infty)$ is the base ray $t \mapsto (v, t)$ as in Sect. 16.2. Multiplication (compare Sect. 4.4) is as in the π_1^e-case.

Extending 16.1.4 and 16.2.6, we have:

Proposition 17.4.1. *If $\mathcal{L} := \{L_i\}$ is a finite filtration of Y and if ω is well parametrized with respect to $\mathcal{L}$, then there is a natural short exact sequence*

$$\varprojlim{}^1\{\pi_{n+1}(Y \stackrel{c}{-} L_i, \omega(i))\} \stackrel{a}{\rightarrowtail} \pi_n^e(Y, \omega) \stackrel{b}{\twoheadrightarrow} \varprojlim\{\pi_n(Y \stackrel{c}{-} L_i, \omega(i))\}.$$

Proof. Similar to the corresponding homological proof indicated in Remark 11.4.9. The details are an exercise. $\quad\square$

Let $PR(Y)$ denote the space of all proper rays in Y with the compact-open topology. The strong homotopy groups of Y are canonically isomorphic to the homotopy groups of $PR(Y)$. To see this, define $\sigma : \pi_n^e(Y, \omega) \to \pi_n(PR(Y), \omega)$ by $\sigma([\hat{f}]) = [f]$, where $\hat{f} : (S^n \times [0, \infty), \{v\} \times [0, \infty)) \to (Y, \omega)$ is a proper map and $f : (S^n, v) \to (PR(Y), \omega)$ is its adjoint.

Proposition 17.4.2. σ *is an isomorphism when $n \geq 1$ and a bijection when $n = 0$.*

The proof of 17.4.2 would be immediate if it were true that the adjoint of a map $K \to PR(Y)$ is always a proper map $K \times [0, \infty) \to Y$, but this can fail to be true even when K is a circle. To get around this, we consider a finite filtration $\mathcal{L}$ of Y as above and we assume ω is well parametrized. We say that a map $g : K \times [0, \infty) \to Y$ is $\mathcal{L}$-*proper* if, for each $i \geq 1$, $g(K \times [i, \infty)) \subset Y \stackrel{c}{-} L_i$. Let $PR_{\mathcal{L}}(Y)$ be the subspace of $PR(Y)$ consisting of $\mathcal{L}$-proper rays[3] in Y. There are corresponding homotopy groups $\pi_n(Y, \mathcal{L}, \omega)$ whose elements are the base ray preserving $\mathcal{L}$-proper homotopy classes of $\mathcal{L}$-proper maps $(S^n \times [0, \infty), \{v\} \times [0, \infty)) \to (Y, \omega)$.

Proof (of 17.4.2). There are obvious functions

$$\pi_n^e(Y, \omega) \xleftarrow{\alpha} \pi_n(Y, \mathcal{L}, \omega) \xrightarrow{\beta} \pi_n(PR_{\mathcal{L}}(Y), \omega) \xrightarrow{\gamma} \pi_n(PR(Y), \omega)$$

To see that β is an isomorphism observe that the adjoint of a map $K \to PR_{\mathcal{L}}(Y)$ is an $\mathcal{L}$-proper map $K \times [0, \infty) \to Y$ and vice versa.[4] Since $\mathcal{L}$-proper implies proper, α and γ are well-defined; they are isomorphisms because a proper map $g : K \times [0, \infty) \to Y$ can be reparametrized by a proper homotopy to be $\mathcal{L}$-proper, and if J is a closed subset of K such that g is $\mathcal{L}$-proper on $J \times [0, \infty)$ then this proper homotopy can be chosen to be rel J. Clearly $\sigma = \gamma \circ \beta \circ \alpha^{-1}$. □

In view of the similarity between H_*^e and π_*^e exhibited by 17.4.1 and 11.4.8, one might guess that $H_*^{\Delta}(PR(Y); R)$ is isomorphic[5] to $H_*^e(Y; R)$ by analogy with 17.4.2, but that is not the case in general:

Example 17.4.3. (This uses standard facts about the topology of surfaces – see 5.1.7 and 5.1.8.) The surface of genus i with two boundary circles is denoted by $T_{i,2}$. There is an obvious embedding of $T_{i,2}$ in $T_{i+1,2}$ illustrated in Fig. 17.1, and there is a CW complex structure L_i for $T_{i,2}$ such that L_i becomes a subcomplex of L_{i+1} under this embedding. Let $Y = \bigcup_{i=1}^{\infty} L_i$. Then $\{L_i\}$ is a finite filtration of the one-ended CW complex Y. By using the exact sequence

---

[3] The $\mathcal{L}$-proper rays are precisely the rays well parametrized with respect to $\mathcal{L}$.

[4] β is defined by $[f] \mapsto [\hat{f}]$ where f and $\hat{f}$ are adjoint.

[5] $PR(Y)$ is not a CW complex: recall that H_*^{Δ} denotes singular homology.

in 16.2.6, one shows that $\pi_1^e(Y, \omega)$ is trivial for every well parametrized proper ray ω. Hence, by 17.4.2, $\pi_1(PR(Y), \omega)$ is trivial for every base point ω. It follows[6] from 3.1.19 that every path component of $PR(Y)$ has trivial first homology with coefficients $\mathbb{Z}$. Hence $H_1^\Delta(PR(Y); \mathbb{Z}) = 0$. But, either from the definition of H_1^e or by 11.4.8, one sees that $H_1^e(Y; \mathbb{Z}) \cong \mathbb{Z}$.

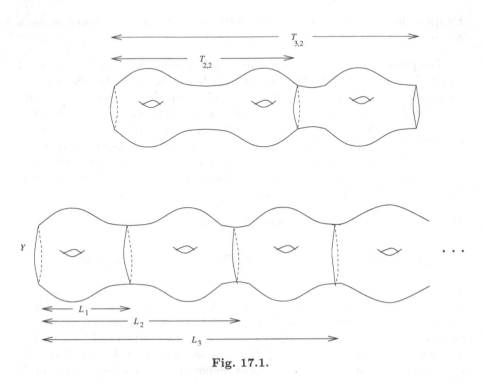

Fig. 17.1.

Remark 17.4.4. The space $PR(Y)$ is, up to homotopy equivalence, the "homotopy inverse limit" of the inverse sequence $\{Y \xleftarrow{c} L_i\}$ where $\mathcal{L} := \{L_i\}$ is a finite filtration of Y. See, for example, [21] or [60] for more on this subject.

Source Notes: The homotopy inverse limit was introduced in [21]. The material in this section is adapted from the shape theory literature, in the spirit of [131], [59], [60].

Exercises

1. Find a map $S^1 \times [0, \infty) \to Y$ which is not proper but whose restriction to every $\{x\} \times [0, \infty)$ is proper.

---

[6] Strictly, we only proved 3.1.19 for CW complexes, but it is well known to hold for all spaces with respect to singular homology.

2. For any finite pointed CW complex K, establish a short exact sequence of pointed sets

$$\{p\} \to \varprojlim^1[\Sigma K, L_n] \to [K, PR(Y)] \to \varprojlim[K, L_n] \to \{p\}$$

where $[X, Y]$ denotes the set of pointed homotopy classes of maps $X \to Y$.
3. Fill in the missing details in Example 17.4.3.

17.5 Z-set compactifications

Let Y be a locally compact metrizable space which is path connected; thus if Y is a CW complex it is locally finite. Here we discuss how Y may be compactified so that the compactifying space "models" the end of Y. By a *compactification* of Y we mean[7] a compact metrizable space W containing (a space homeomorphic to) Y – we write $Y \subset W$ – so that Y is a dense open subset of W. The nowhere dense compact set $C := W - Y$ is the *compactifying space*. In general, a closed subset $D \subset W$ is a Z-*set* in W if for every open set U in W the inclusion map $U - D \hookrightarrow U$ is a homotopy equivalence. If the compactifying space C is a Z-set in W, we say that W is a Z-*set compactification* of Y.

Example 17.5.1. If W is a compact manifold (see Sect. 5.1), every closed subset $C \subset \partial W$ is a Z-set in W. Thus the manifold W is a Z-set compactification of the manifold $W - C$. In particular, C can be ∂W, so W is a Z-set compactification of the open manifold $\overset{\circ}{W}$.

When Y is a (locally finite) CW complex, the existence of a Z-set compactification imposes restrictions on Y:

Proposition 17.5.2. *If such a Y admits a Z-set compactification W, then for any finite filtration $\mathcal{L} := \{L_i\}$ each $Y \overset{c}{-} L_i$ is finitely dominated and $\{Y \overset{c}{-} L_i\}$ is equivalent in pro-Homotopy to an inverse sequence of finite CW complexes.*

Proof. Write $C = W - Y$. The subspace $A_i := (Y \overset{c}{-} L_i) \cup C$ is a compact neighborhood of C in W since $W \overset{c}{-} N(L_i)$ is compact[8] (by 1.5.4). Thus $Y \overset{c}{-} L_i \hookrightarrow A_i$ is a homotopy equivalence. Letting $f : A_i \to Y \overset{c}{-} L_i$ be a homotopy inverse, there is a finite subcomplex J_i of $Y \overset{c}{-} L_i$ containing $f(A_i)$ (by 1.2.13). Since the composition

$$Y \overset{c}{-} L_i \hookrightarrow A_i \overset{f'}{\longrightarrow} J_i \hookrightarrow Y \overset{c}{-} L_i$$

---

[7] The only compactifications discussed in this book are metrizable compactifications which we abbreviate to "compactification." In other contexts, W would be compact Hausdorff; for example, the Stone-Čech compactification βY.

[8] Y is a CW complex but W, in general, is not. Even when W has the structure of a CW complex, the (dense open) subset Y is not a subcomplex. Nonetheless, $W \overset{c}{-} N(L_i)$ makes sense.

is homotopic to the identity map (where f' is the corestriction of f), J_i dominates $Y \stackrel{c}{\longrightarrow} L_i$. Thus each inclusion $Y \stackrel{c}{\longrightarrow} L_{i+1} \hookrightarrow Y \stackrel{c}{\longrightarrow} L_i$ factors up to homotopy through J_i, so 11.2.2 implies that $\{Y \stackrel{c}{\longrightarrow} L_i\}$ is pro-isomorphic to $\{J_i\}$ for suitable bonding maps $J_{i+1} \to J_i$. □

With W as in 17.5.2, let $p \in C = W - Y$ be a base point. Since C is a Z-set there is a proper base ray ω in Y which extends to a map $[0, \infty] \to W$, sending ∞ to p, and ω can be chosen to be well parametrized with respect to $\mathcal{L}$. Then a variant on the proof of 17.5.2 shows:

Proposition 17.5.3. *Under these conditions the fundamental pro-group* $\{\pi_1(Y \stackrel{c}{\longrightarrow} L_i, \omega(i))\}$ *is pro-finitely presented.*[9] □

We will see examples in Sect. 16.4 where the fundamental pro-group is not pro-finitely presented, and hence the space does not admit a Z-set compactification. But here we will discuss some of the principal sources (other than 17.5.1) of locally compact spaces Y which have Z-set compactifications. We begin with a rather general construction which yields interesting examples as special cases. Let C_∞ be a compact subset of the space of proper rays $PR(Y)$. For each $t \in [0, \infty)$ let[10] $C_t \subset C([0, \infty), Y)$ be defined by: $\omega_t \in C_t$ iff $\omega_t \mid [0, t] = \omega \mid [0, t]$ for some $\omega \in C_\infty$ and $\omega_t([t, \infty)) = \omega_t(t)$. Thus C_t is the set of restrictions of members of C to $[0, t]$ prolonged to be constant on $[t, \infty)$. Let $D_{t_0} = \bigcup\{C_t \mid 0 \le t \le t_0\}$ and $D_\infty = \bigcup\{C_t \mid 0 \le t < \infty\}$. Write $W = D_\infty \cup C_\infty \subset C([0, \infty), Y)$. All these spaces inherit the compact-open topology.

Proposition 17.5.4. *W is a Z-set compactification of D_∞, with C_∞ as compactifying space.*

Proof. There is an obvious retraction $r_i : D_{i+1} \to D_i$ under which, for $i \le t \le i + 1$, $\omega_t \mapsto \omega_i$. The inverse limit in Spaces of the inverse sequence $\{D_0 \xleftarrow{r_0} D_1 \xleftarrow{r_1} \cdots\}$ is homeomorphic to W. For all t there are continuous surjections $C_\infty \to C_t$ and $C_t \times I \to D_t$, so each D_i is compact. The space D_t is homeomorphic to a subset of $C(I, Y)$ and is therefore metrizable. So W is compact and metrizable, being a closed subset of the compact metrizable space $\prod_{i=0}^{\infty} D_i$. Clearly D_∞ is open in W. The map $\omega \mapsto \omega_t$ of C_∞ onto C_t is homotopic in W to the inclusion $C_\infty \hookrightarrow W$. Using this, it is straightforward to show that C_∞ is a Z-set in W. □

The interest of 17.5.4 is that in at least two important cases C_∞ can be chosen so that D_∞ is homeomorphic to Y, making W a Z-set compactification of Y:

---

[9] *Pro-finitely presented* means pro-isomorphic to an inverse sequence of finitely presented groups.

[10] Recall from Sect. 1.1 that $C(X, Y)$ denotes the space of all maps $X \to Y$, with the compact-open topology.

Example 17.5.5. Let Y admit a proper metric d so that the usual weak topology on Y is induced by d. Assume that d is a CAT(0) metric or, more generally, a unique-geodesic metric space. Choose a base point $v \in Y$ and let C_∞ be the space of geodesic rays starting at v. Then for $0 \le t < \infty$, D_t is homeomorphic to the space of geodesic segments starting at v and of length $\le t$, hence also to $B_t(v)$, the ball of radius t in Y centered at v. It follows that each D_t is compact and hence that $D_\infty \cup C_\infty$ is compact. The "terminal point map" $D_\infty \to Y$ is a homeomorphism. Thus, up to identification of Y with D_∞, the space of geodesic rays is a Z-set compactifying space for Y. In the language of CAT(0) geometry, C_∞ is the "boundary" of Y with the "cone topology".[11]

Example 17.5.6. Here we show that every compact metrizable space C can play the role of $W - Y$ for a Z-set compactification of a locally finite CW complex Y. Let C be homeomorphic to the inverse limit in Spaces of an inverse sequence $W_1 \leftarrow W_2 \leftarrow \cdots$ of finite CW complexes.[12] Let Y be the inverse mapping telescope of $W_0 := \{v\} \xleftarrow{f_0} W_1 \xleftarrow{f_1} W_2 \xleftarrow{f_2} \cdots$; we have added a one-point space $\{v\}$ to make Y contractible. The space Y is $\coprod_{i=0}^{\infty} M(f_i)/\sim$ (see Sect. 11.4). Now $M(f_i)$ is $W_i \cup_{f_i} (W_{i+1} \times I)$ as in Sect. 4.1, so for each $w_{i+1} \in W_{i+1}$, the path $t \mapsto (w_{i+1}, t)$ in $W_{i+1} \times I$ defines a path in $M(f_i)$ from $f_i(w_{i+1})$ to w_{i+1}. Call the corresponding path in Y a "canonical segment" from $f_i(w_{i+1})$ to w_{i+1}; we can form "canonical rays" in Y as infinite products of canonical segments just as in Sect. 3.3 (where we formed finite products only). All these canonical rays start at v and the ith restriction to $[i, i+1]$ is a canonical segment. Let $C_\infty \subset PR(Y)$ be the space of all canonical segments in Y. Then it is clear that the space D_∞ in 17.5.4 is homeomorphic to Y in this instance, while C_∞ is homeomorphic to $\varprojlim_i W_i$, hence to C.

Remark 17.5.7. In the last two examples the space of proper rays C_∞ is compact. In general, the Arzela-Ascoli Theorem [51] is a useful criterion for recognizing compact subsets[13] of $C([0, \infty), Y)$.

Remark 17.5.8. In 17.4.3 we described a space Y such that $H_1^\Delta(PR(Y); \mathbb{Z}) = 0$ and $H_1^e(Y; \mathbb{Z}) \ne 0$. That example Y does not admit a Z-set compactification.

---

[11] Strictly, this definition depends on the base point v. An equivalent definition independent of base point is given in [24].

[12] A well-known theorem [61, Chap. 10, Sect. 10], says that every compact metrizable space C has this property. Of course, many different inverse sequences $\{W_i\}$ have homeomorphic inverse limits. For example, if $C \subset \mathbb{R}^n$ then C is the intersection of compact polyhedral neighborhoods W_i (see Sect. 5.2), so $W_1 \hookleftarrow W_2 \hookleftarrow \cdots$ has inverse limit C. Or the W_i's can be nerves of ever finer finite open covers of C. See Sect. 9.3E for more on nerves.

[13] Note that Y is metrizable by 10.1.25, a necessary condition in the Arzela-Ascoli Theorem.

Examples of the opposite phenomenon, $H_k^{\triangle}(PR(Y); \mathbb{Z}) \neq 0$ and $H_k^e(Y; \mathbb{Z}) = 0$, which are inverse mapping telescopes as in 17.5.6 (and so admit Z-set compactifications) have been given in [97] and [98]. A particular case is as follows: Y is an inverse mapping telescope of $\{W_1 \leftarrow W_2 \leftarrow \cdots\}$ where $n \geq 2$ is fixed, $W_i := \bigvee_{j=1}^{i} S^n$ (the wedge of i copies of S^n) and $W_{i+1} \rightarrow W_i$ is the identity on $W_i \subset W_{i+1}$ while sending the remaining copy of S^n to the base point. Specifically: $H_{2n-1}^{\triangle}(PR(Y); \mathbb{Z}) \neq 0$ and $H_{2n-1}^e(Y; \mathbb{Z}) = 0$. In this case, C_{∞} is the n-dimensional analog of the Hawaiian Earring, a "compact countable wedge" of copies of S^n. This strange behavior arises from the fact that the singular homology of an n-dimensional space can fail to vanish in dimensions $> n$.

Remark 17.5.9. A well-known (metrizable) compactification of Y is the *one-point* (or *Alexandroff*) *compactification* $\hat{Y} = Y \cup \{p\}$ where $p \notin Y$ is a point. The topology of $\hat{Y}$ is the smallest topology containing the open sets of Y and the complements in $\hat{Y}$ of all compact subsets of Y. If $Y = [0, 1)$ then $\hat{Y}$ is homeomorphic to $[0, 1]$. If $Y = \mathbb{R}$ then $\hat{Y}$ is homeomorphic to S^1. A more delicate compactification of Y is the *end-point* (or *Freudenthal*) *compactification*[14] $\bar{Y} = Y \cup \mathcal{E}(Y)$. As in Sect. 16.1, we regard each end e as an element $(Z_i(e))$ of $\varprojlim\{\pi_0(Y \overset{c}{\leftarrow} L_i)\}$ where $\{L_i\}$ is a finite filtration of Y. The topology of $\bar{Y}$ is the smallest topology containing all open sets of Y and all sets of the form $Z_i(e) \cup \{e\}$. For fixed e, the sets $Z_i(e) \cup \{e\}$ then form a basis for the neighborhoods of e in $\bar{Y}$. If Y has one end, then $\hat{Y} = \bar{Y}$. If, for example, $Y = (0, 1)$ then $\bar{Y}$ is homeomorphic to $[0, 1]$.

Source Notes: The concept of Z-set comes from infinite dimensional topology – see, for example, [38] or [8]. The relevance to group theory was pointed out in [68]. Other relevant sources are [39], [12] and [9]. A weaker definition, possibly useful in group theory, is given in [143].

Exercises

1. Show that a non-empty compact subset of $\mathbb{R}^n$ is never a Z-set.
2. Show that the space Y in 17.4.3 does not admit a Z-set compactification.
3. Let W be a compact contractible manifold (see Sect. 5.1) with non-empty[15] path connected boundary ∂W. Show that if the fundamental group of ∂W is not $\mathbb{Z}$ or $\{1\}$, then $\overset{\circ}{W}$ is not the universal cover of a finite CW complex. *Hint*: use 16.5.5 and 7.2.12. *Remark*: There exist such manifolds W with non-trivial perfect fundamental groups, e.g., homology spheres.
4. Show that the one-point and end-point compactifications of Y really are (metrizable) compactifications.

---

[14] Recall from Sect. 13.4 that $\mathcal{E}(Y)$ denotes the set of ends of Y.
[15] By 12.3.7, if ∂W is empty W cannot be contractible.

5. Assume[16] that in Example 17.5.6 the bonds $W_{i+1} \to W_i$ are Serre fibrations. Prove that the canonical map $C \to PR(Y)$ is a weak homotopy equivalence. (Part of this exercise is to decide what the "canonical map" is.)
6. In Example 17.5.6, take each W_i to be $S^1 \times S^1$ with the product of two degree 2 covering projections as bonds. Prove that $H_1^{\Delta}(PR(Y); \mathbb{Z}) = 0$ and $H_1^e(Y; \mathbb{Z}) \neq 0$.

17.6 Compactifiability at infinity as a group invariant

Let G be a group of type F, let (X, x) be a finite $K(G, 1)$-complex (an identification of G with $\pi_1(X, x)$ is chosen), and let $(\tilde{X}, \tilde{x})$ be the (pointed) universal cover. Then $\tilde{X}$ is a contractible finite-dimensional locally finite G-CW complex which is finite mod G. We choose a proper base ray ω in $\tilde{X}$ starting at $\tilde{x}$. One might wish to say that the strong homotopy groups $\pi_*^e(\tilde{X}, \omega)$ are invariants of G, but the issue of dependence on ω makes this doubtful, as discussed in Sect. 16.3. However, if G is semistable at infinity, then all proper rays are properly homotopic and one can say that the isomorphism class of $\pi_k^e(\tilde{X}, \omega)$ is an invariant of G for each k. By 17.4.2, this is isomorphic to $\pi_k(PR(\tilde{X}), \omega)$.

The (isomorphism class of the) homology $H_k^{\Delta}(PR(\tilde{X}); R)$ is also an invariant of G which can be different from $H_k^e(\tilde{X}; R) \cong H_{k+1}(G, RG^e)$ (see 13.2.3).

In another direction we ask:

Question 17.6.1. Let $\{K_i\}$ be a finite filtration of $\tilde{X}$. Is $\{\tilde{X} \xrightarrow{c} K_i\}$ equivalent in pro-Homotopy to an inverse sequence of finite CW complexes?

Note that the condition in 17.6.1 only depends on G, not on the choice of X or $\{K_i\}$.

Indeed, 17.5.3 suggests a simpler question:

Question 17.6.2. If G is semistable at infinity, is the fundamental pro-group of $\tilde{X}$ (with respect to ω) pro-finitely presented?

There are interesting strengthenings of 17.6.1:

Question 17.6.3. Can X be chosen so that $\tilde{X}$ admits a Z-set compactification?

As posed, X is to be a finite CW complex but it is better to allow more generality in 17.6.3; namely, X can be a compact connected ANR (see Sec. 17.7) with fundamental group G and contractible universal cover. There is literature relevant to this which is beyond the scope of this book; e.g., [144], [39], [126], [75].

---

[16] This exercise involves the algebraic topology of spaces which are not CW complexes: for the definitions of "Serre fibration" and "weak homotopy equivalence" see [146].

Question 17.6.4. Can X be chosen so that $\tilde{X}$ admits a Z-set compactification W such that the free G-action on $\tilde{X}$ by covering transformations extends to a G-action on W by homeomorphisms?

The answer to 17.6.4 is yes, with important consequences, when $\tilde{X}$ is a CAT(0) space on which G acts freely with $X := G\backslash\tilde{X}$ compact. Then X is a compact ANR as above. A classical example is the compactification of the hyperbolic plane by the circle at infinity, with G the fundamental group of a surface of genus ≥ 2. The answer is also yes in the case of the Rips complex of a hyperbolic group [24].

Exercise

Prove that the right angled Coxeter groups discussed in Sec. 13.9 provide positive examples for Question 17.6.3; in fact, also for 17.6.4, since they are CAT(0) groups; see [122].

17.7 Strong shape theory

This book is influenced by strong shape theory and shape theory. Here we briefly explain, confining ourselves to versions of those theories for compact metrizable spaces (also called *compacta*).

A space is *locally contractible* if for any point p and any neighborhood U of p there is a neighborhood V of p, $V \subset U$, such that $V \hookrightarrow U$ is homotopically trivial. Roughly speaking, a compactum is considered to be "locally pathological" if it is not locally contractible. For example, the *dyadic solenoid* and the *Case-Chamberlin compactum* (which are the inverse limits corresponding to the mapping telescopes described in 11.4.15 and 16.1.10) are not locally contractible; nor is the compactum described in 17.5.8. We say "roughly speaking" because a concept somewhat more restrictive than local contractibility turns out to be better. A metrizable space X is an *absolute neighborhood retract* (abbrev. ANR) if whenever Y is a metrizable space, Z is a closed subset of Y, and $f : Z \to X$ is a map, then there exists a continuous extension $F : U \to X$ of f to some open subset U of Y containing Z. An ANR X is an *absolute retract* (abbrev. AR) if in this definition of ANR the neighborhood U can always be taken to be all of Y.

Examples and Properties 17.7.1. The metrizable space X is an AR iff X is a contractible ANR. Every open subset of an ANR is an ANR. Finite products of ANR's are ANR's and retracts of ANR's are ANR's. By a variation of the Tietze Extension Theorem ([51]) $\mathbb{R}$ is an AR. Thus every open subset of $\mathbb{R}^n$, as well as every retract of such an open set, is an ANR. If every point of X has an ANR neighborhood, then X is an ANR. Thus every manifold is an ANR. Every ANR is locally contractible. Every locally finite CW complex is an ANR (by 10.1.25 they are metrizable). A Z-set compactification of a locally compact AR is a (compact) AR. References for ANR's are [19] and [89]. A theorem of J.H.C. Whitehead says that every ANR has the homotopy type of a CW complex; a reference for this is [109, Appendix 1].

For compacta which are not ANR's, the strong shape category and the shape category are often better "places to do algebraic topology" than the homotopy category. All three categories (as considered here) have compacta as objects; the morphisms in Homotopy are homotopy classes of maps; the morphisms in the other two categories need not be described yet. There are canonical functors Homotopy $\xrightarrow{A}$ Strong Shape $\xrightarrow{B}$ Shape which are the identity on objects; if X and Y are compacta then, in general, the induced function $A_{\#}$: Homotopy$(X, Y) \to$ Strong Shape(X, Y) is neither injective nor surjective, while $B_{\#}$: Strong Shape$(X, Y) \to$ Shape(X, Y) is surjective but not, in general, injective (see Example 17.7.2). However, shape theory, strong shape theory and homotopy theory all agree on compact ANR's in the sense that $A_{\#}$ and $B_{\#}$ are bijections when X and Y are ANR's.

We can avoid giving the technically complicated definitions of morphisms in Shape and Strong Shape by the following trick. Let SS be the category whose objects are pairs (W, C) where W is a compact AR and C is a (closed) Z-set in W; a morphism $(W_1, C_1) \to (W_2, C_2)$ in SS is by definition a proper homotopy class of proper maps $Y_1 \to Y_2$ where $Y_i = W_i - C_i$. There is a canonical functor: SS $\to$ Strong Shape which takes (W, C) to C and is a bijection on each set of morphisms. Thus SS and Strong Shape are philosophically the same, even though in SS an object is an embedded compactum and in Strong Shape it is an abstract compactum.

There is also a category S having a similar relation to the category Shape. Its objects are the same as those of SS, namely pairs (W, C) as above. Define two proper maps $f, g : Y_1 \to Y_2$ to be *weakly properly homotopic* if for any compact subset L of Y_2 there is a compact subset K of Y_1 and a homotopy $H : f \simeq g$ such that $H((Y_1 - K) \times I) \subset Y_2 - L$. A morphism $(W_1, C_1) \to (W_2, C_2)$ in S is by definition a weak proper homotopy class of proper maps $Y_1 \to Y_2$ where $Y_i = W_i - C_i$.

When $Y = W - C$ can be given the structure of a strongly locally finite CW complex, then $H_n^e(Y; R)$ is the nth *Steenrod homology* of C with coefficients in R, denoted $H_n^S(C; R)$. Picking a finite filtration $\{L_i\}$ of Y, $\varprojlim_i H_n(Y \xrightarrow{c} L_i; R)$ is the nth *Čech homology* of C with coefficients in R, denoted $\check{H}_n(C; R)$. We remarked in Sect. 11.4 that $H_n^e(\cdot; R)$ is a proper homotopy invariant, hence $H_n^S(\cdot; R)$ is a strong shape invariant. Now we add that $\varprojlim_i H_n(Y \xrightarrow{c} L_i; R)$ is a weak proper homotopy invariant (though $H_n^e(Y; R)$ is not) and hence $\check{H}_n(\cdot; R)$ is a shape invariant (though $H_n^S(\cdot; R)$ is not). Clearly, some details concerning independence of Y and $\{L_i\}$ have been omitted here. The following commutative diagram of categories and functors summarizes the situation, where H denotes the category whose objects are pairs (W, C) as above, and whose morphisms from $(W_1, C_1) \to (W_2, C_2)$ are homotopy classes of maps $C_1 \to C_2$:

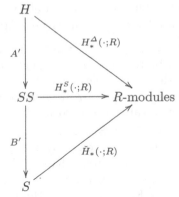

Here, the functors A' and B' are the identity on objects. A map $f : C_1 \to C_2$ defines (using the fact that W_2 is an ANR and C_2 is a Z-set) a (non-unique) proper map $\bar{f} : W_1 - C_1 \to W_2 - C_2$, so that if another map g is homotopic to f then $\bar{g}$ is properly homotopic to $\bar{f}$. This indicates how to define A' on morphisms of H. The definition of B' on morphisms is obvious since properly homotopic maps are certainly weakly properly homotopic.

Although the morphisms in SS and in S are different, it is a theorem that two objects are isomorphic in one category iff they are isomorphic in the other ("shape equivalent iff strong shape equivalent") – see [60].

Example 17.7.2. To illustrate the differences between H, SS and S we give two examples of the set of morphisms $(W_1, C_1) \to (W_2, C_2)$. The dependence on W_1 and W_2 is only a convenience, as we have explained; we are really considering morphisms $C_1 \to C_2$ in Homotopy, Strong Shape and Shape. Therefore we will sometimes omit reference to the W's. In both examples C_1 is a single point. First, let C_2 be the dyadic solenoid. Then there is only one shape morphism $C_1 \to C_2$; this is because the solenoid is connected. However, the solenoid has $\mathfrak{c}$ path components (where $\mathfrak{c}$ is the cardinality of $\mathbb{R}$) and there are indeed $\mathfrak{c}$ morphisms from $C_1 \to C_2$ both in Homotopy and in Strong Shape – in fact, $A_\#$ is a bijection in this case.[17] In the second example, C_2 is the Case-Chamberlin compactum, which is also connected. Again, there is only one shape morphism $C_1 \to C_2$. There are $\mathfrak{c}$ strong shape morphisms $C_1 \to C_2$; these can be partitioned into those which are in the image of $A_\#$ (there are $\mathfrak{c}$ of them), and those not in the image of $A_\#$ (there are $\mathfrak{c}$ of those too). A geometric translation of these strange statements is this: embed this (1-dimensional) compactum C_2 in S^3 so that it becomes a Z-set in $W_2 := B^4$. We are asserting that there are $\mathfrak{c}$ strong ends of $B^4 - C_2$ representable by proper rays $\omega : [0, \infty) \to B^4 - C_2$ which can be extended continuously to paths $[0, \infty] \to B^4$; and there are also $\mathfrak{c}$ strong ends whose representative

---

[17] The similarity occurs because the solenoid is the inverse limit of fibrations.

proper rays cannot be so extended ("empty strong shape components" of C_2, i.e., not representable by points of C_2). For details, see [69]. For another example of this kind of behavior at the π_1-level, see [157].

Remark 17.7.3. Here is a more practical, but equivalent, definition of morphisms in the category S. Given objects (W_1, C_1) and (W_2, C_2), let $\{U_\alpha\}$ and $\{V_\beta\}$ be the directed sets of all compact ANR neighborhoods of C_1 in W_1 and C_2 in W_2 respectively. A morphism in S from (W_1, C_1) to (W_2, C_2) is a morphism of pro-Homotopy $\{U_\alpha\} \to \{V_\beta\}$. Indeed, there are cofinal towers in these large inverse systems, and to discuss a morphism one would normally first choose such towers $\{U_{\alpha_n}\}$ and $\{V_{\beta_n}\}$ where $\bigcap_n U_{\alpha_n} = C_1$ and $\bigcap_n V_{\beta_n} = C_2$; but there are no "canonical" towers, so it is easier to define things functorially using the whole inverse systems. Note that since C_1 and C_2 are Z-sets, we could equally well use $\{U_{\alpha_n} - C_1\}$ and $\{V_{\beta_n} - C_2\}$ to describe a morphism.

Central to this book is the proper homotopy category of locally finite CW complexes, especially the full subcategory of those which are contractible universal covers of finite CW complexes (so that their fundamental groups have type F). When such a universal cover $\tilde{X}$, corresponding to the group G of type F, admits a Z-set compactification, then the "homotopical group theory" of G as described in this chapter is essentially the strong shape theory of the compactifying space. We have set things up here so that that statement is more or less a tautology.

Source Notes. Some sources on shape theory: [58], [109], [108] and [60].

Exercises

1. Write a version of this section for groups of type F_n.
2. State a version of 17.6.2 when G has more than one end and is semistable at each end.
3. Let (X, x) be a finite $K(G, 1)$-complex with universal cover $(\tilde{X}, \tilde{x})$. Assume that $\tilde{X}$ has a Z-set compactification $W = \tilde{X} \cup C$ where the compactifying space C is connected and locally path connected. Prove that $\tilde{X}$ is strongly connected at infinity. *Hint:* Use the properties of ANR's in 17.7.1.

PART VI: THREE ESSAYS

This final part of the book consists of three unrelated essays intended to point the reader towards some interesting topics which have natural connections with our main themes: l_2-Poincaré duality, quasi-isometry, and the Bieri-Neumann-Strebel "geometric" invariant of a finitely generated group.

Three Essays

18.1 l_2-Poincaré duality

Here we use $\mathbb{R}$-coefficients. A countable CW complex Y *has bounded geometry* if there is a number N such that every cell meets at most N other cells; Y has *bounded geometric type* if every skeleton has bounded geometry. These are stronger conditions than "locally finite" and "locally finite type"; in particular, once Y is oriented, $C_n^\infty(Y;\mathbb{R}) \overset{\partial_n}{\underset{\delta_{n-1}}{\rightleftarrows}} C_{n-1}^\infty(Y;\mathbb{R})$ make sense. Throughout this Appendix, Y is assumed to have bounded geometric type.

Let $C_n^{(2)}(Y)$ be the vector subspace of $C_n^\infty(Y;\mathbb{R})$ consisting of l_2-*chains*, i.e., chains $\sum_\alpha r_\alpha e_\alpha^n$ such that $\sum_\alpha r_\alpha^2 < \infty$. Since Y is countable, $\{e_\alpha^n\}$ is an orthonormal basis for the separable Hilbert space $C_n^{(2)}(Y)$, the inner product being $\left(\sum_\alpha r_\alpha e_\alpha^n, \sum_\alpha s_\alpha e_\alpha^n \right) = \sum_\alpha r_\alpha s_\alpha$. Because Y has bounded geometric type, ∂ and δ take l_2-chains to l_2-chains and are bounded linear operators. Write $Z_n^{(2)}(Y)$ and $Z_{(2)}^n(Y)$ for ker ∂_n and ker δ_n respectively; elements are the l_2-*cycles* and l_2-*cocycles*. Write $\bar{B}_n^{(2)}(Y)$ and $\bar{B}_{(2)}^n(Y)$ for the closures of image ∂_{n+1} and image δ_{n-1} respectively; elements are the l_2-*boundaries* and l_2-*coboundaries*. Clearly $C_n^{(2)}(Y) \overset{\partial_n}{\underset{\delta_{n-1}}{\rightleftarrows}} C_{n-1}^{(2)}(Y)$ are adjoint operators in the sense that (omitting subscripts on ∂ and δ from now on) $(c, \delta d) = (\partial c, d)$. As usual, if W is a closed linear subspace of a Hilbert space V, then $W^\perp$ denotes $\{v \in V \mid (v, w) = 0$ for all $w \in W\}$ and we write $V = W \perp W^\perp$ for the direct sum decomposition. It is easy to see that $Z_n^{(2)} = (\bar{B}_{(2)}^n)^\perp$, that $Z_{(2)}^n = (\bar{B}_n^{(2)})^\perp$, and that $C_n^{(2)}(Y) = \bar{B}_n^{(2)}(Y) \perp \bar{B}_{(2)}^n(Y) \perp (Z_n^{(2)}(Y) \cap Z_{(2)}^{(n)}(Y))$. We write $\mathcal{H}_n^{(2)}(Y) = Z_n^{(2)}(Y) \cap Z_{(2)}^n(Y)$ and we call its elements (cycles which are also cocycles) *harmonic chains*. Clearly, $Z_n^{(2)}(Y) = \bar{B}_n^{(2)}(Y) \perp \mathcal{H}_n^{(2)}(Y)$

and $Z_{(2)}^n(Y) = \bar{B}_{(2)}^n(Y) \perp \mathcal{H}_n^{(2)}(Y)$. The (reduced) l_2-homology and l_2-cohomology of Y are the vector spaces $H_n^{(2)}(Y) = Z_n^{(2)}(Y)/\bar{B}_n^{(2)}(Y)$ and $H_{(2)}^n(Y) = Z_2^n(Y)/\bar{B}_{(2)}^n(Y)$ respectively. The quotient maps give canonical isomorphisms $H_n^{(2)}(Y) \xleftarrow{\cong} \mathcal{H}_n^{(2)}(Y) \xrightarrow{\cong} H_{(2)}^n(Y)$ which give canonical Hilbert space structures to $H_n^{(2)}(Y)$ and $H_{(2)}^n(Y)$. By these isomorphisms we may think of harmonic chains as unique "best" representatives of homology classes and of cohomology classes.[1]

Define $\Delta : C_*^{(2)}(Y) \to C_*^{(2)}(Y)$ to be $\Delta := \delta \circ \partial + \partial \circ \delta$. This is the *combinatorial Laplacian operator*. The name "harmonic" is explained by:

Proposition 18.1.1. *Let c be an l_2-chain. Then c is harmonic iff $\Delta c = 0$.* $\square$

The equation $\Delta c = 0$ is a close mathematical relative of the classical Laplace Equation.

Note that when Y is finite all this reduces to ordinary cellular chains, giving for a finite CW complex a best representative (co)-cycle in each (co)-homology class when the coefficient ring is $\mathbb{R}$.

Now let X be an oriented CW n-manifold which has bounded geometry. By assuming X has only countably many path components, we ensure it is a countable CW complex (by 11.4.3). Orient the cells of X and give dual orientations to the cells of X^*. Assume $\partial X = \emptyset$. We saw that the Poincaré duality isomorphisms ϕ_k send cycles to cocycles and cocycles to cycles. So they induce a canonical Hilbert space isomorphism $\mathcal{H}_k^{(2)}(X) \to \mathcal{H}_{n-k}^{(2)}(X^*)$, hence also a canonical Hilbert space isomorphism $*_k : \mathcal{H}_k^{(2)}(|\text{sd } X|) \to \mathcal{H}_{n-k}^{(2)}(|\text{sd } X|)$ called the *combinatorial Hodge star operator*.

Remark 18.1.2. In Exercise 2 we claim that $H_*^{(2)}$ is a functor on the category of CW complexes having bounded geometry and CW Lipschitz maps. Universal covers of finite CW complexes have bounded geometry. It follows that when G is a group of type F_∞, then $H_*^{(2)}(G)$ is well defined.[2] In Sect. 18.2 we will see that l_2-homology is a quasi-isometry invariant of such groups.

A full discussion of this subject is beyond the scope of this book. For a thorough treatment of l_2-homology see [104].

Exercises

1. Prove that $*_{n-k} \circ *_k = (-1)^{k(n-k)}$ identity.
2. Prove that $H_n^{(2)}$ and $H_{(2)}^n$ are functors from the category of CW complexes having bounded geometry and CW Lipschitz maps to the category of real Hilbert spaces and bounded linear operators.
3. Prove that $H_*^{(2)}(\mathbb{R}) = 0$, while the 0th unreduced l_2-homology of $\mathbb{R}$ is non-zero.

---

[1] There is also *non-reduced* l_2-homology and l_2-cohomology, $Z_n^{(2)}(Y)/B_n^{(2)}(Y)$ and $Z_{(2)}^n(Y)/B_{(2)}^n(Y)$. We will not discuss this.

[2] Compare 8.1.1 and the convention which follows it to interpret this precisely.

18.2 Quasi-isometry invariants

Quasi-isometry is a geometrical relation among finitely generated groups which is weaker than isomorphism. Here we cover just enough of the subject to prove that many of the homological and homotopical properties of groups discussed in previous chapters are invariant under quasi-isometry.

Let G be a finitely generated group, and let $\{g_1, \ldots, g_s\}$ be a subset which generates G. Recall that the *length* of $g \in G$ with respect to this set of generators is the least integer m such that $g = h_1 \cdots h_m$ where each $h_i \in \{g_1^{\pm 1}, \cdots, g_s^{\pm 1}\}$. We write $l(g)$ for the length of g; by convention $l(1) = 0$. The *word metric* on the set G is defined by $d(g, h) = l(h^{-1}g)$. This is easily seen to make G into a metric space.

When (X, d) and (X', d') are metric spaces, a function $f : X \to X'$ is an *isometric embedding* if $d'(f(x), f(y)) = d(x, y)$ for all $x, y \in X$. This implies f is injective; if in addition f is surjective it is called an *isometry*. An isometric embedding is an embedding (in the topological sense) and an isometry is a homeomorphism. The isometries $X \to X$ form a subgroup of the group of all homeomorphisms $X \to X$. (The action of G on itself by left multiplication is an action by isometries when G carries the word metric.)

A (not necessarily continuous) function $f : X \to X'$ is a *quasi-isometric embedding* if there are positive constants λ and ϵ such that for all $x, y \in X$

$$\frac{1}{\lambda}d(x, y) - \epsilon \le d'(f(x), f(y)) \le \lambda d(x, y) + \epsilon.$$

If in addition there is a constant $C \ge 0$ such that for every $z \in X'$ there is some $x \in X$ with $d'(f(x), z) \le C$, the quasi-isometric embedding f is said to be *quasi-surjective*, and then f is called a *quasi-isometry*. We say X and X' are *quasi-isometric* if there exists a quasi-isometry $X \to X'$. The following is an exercise:

Proposition 18.2.1. *If d and d' are word metrics on a group G with respect to two finite sets of generators, then id_G is a quasi-isometry between (G, d) and (G, d').* □

Two finitely generated groups G and H are *quasi-isometric* if for some (equivalently, any) choice of finite generating sets the metric spaces G and H (with respect to the word metrics) are quasi-isometric.

In order to decide when two groups are not quasi-isometric it would be useful to know some quasi-isometry invariants. In particular, we will see that the finiteness properties of groups and many of the homological and homotopical properties of groups discussed in this book are indeed quasi-isometry invariants. Thus, for example, a finitely generated non-finitely presented group is not quasi-isometric to a finitely presented group, and a one-ended group is not quasi-isometric to a group which does not have one end. We show in 18.2.6 that commensurable groups are quasi-isometric.[3]

The spaces of choice in this book are CW complexes. They do not necessarily carry natural metrics, but they do carry natural pseudometrics. Recall that a *pseudometric* on a set Z is a function $\psi : Z \times Z \to [0, \infty]$ which satisfies: (i) $\psi(x, x) = 0$; (ii) $\psi(x, y) = \psi(y, x)$; and (iii) $\psi(x, z) \leq \psi(x, y) + \psi(y, z)$. What makes this different from a metric is that ∞ is permitted as a value, and $\psi(x, y)$ can be 0 when $x \neq y$. A *pseudometric space* (Z, ψ) consists of a set Z and a pseudometric ψ on Z. The example we have in mind is the *CW pseudometric* ρ on a CW complex X: $\rho(x, y)$ is defined to be the least integer k such that there is a path in X joining x and y which meets the interiors of $(k + 1)$ different cells of X. Thus $\rho(x, y) \in \mathbb{N} \cup \{\infty\}$; it is 0 iff $x \in \overset{\circ}{e}$ and $y \in \overset{\circ}{e}$ for some cell e of X. If x and y lie in different path components of X then $\rho(x, y)$ is defined to be ∞.

Our definitions of quasi-isometric embedding, quasi-isometry, etc. are valid for pseudometric spaces too – just replace the metric by the pseudometric – so we will use the language of quasi-isometry in connection with pseudometric spaces.

Proposition 18.2.2. *Let Y be a path connected rigid G-graph which is finite mod G. Assume that the stabilizer of each vertex is finite. Let ρ denote the CW pseudometric on Y and let d be the word metric on G with respect to some finite set of generators. Then (Y, ρ) is quasi-isometric to (G, d).*

Proof. Pick a base vertex $v \in Y$ and a compact fundamental domain $C \subset Y$ for the G-action on Y so that $v \in C$. For the pseudometric ρ, as for metrics, we write $B_r(y) = \{y' \in Y \mid \rho(y, y') < r\}$. Choose r so that $C \subset B_r(v)$. Let $A = \{g \in G \mid gB_r(v) \cap B_r(v) \neq \emptyset\}$. One easily sees that A is finite and generates G. By 18.2.1, we are permitted to use A as our given set of generators in terms of which d is defined. The required quasi-isometry $F : G \to Y$ is given by $f(g) = gv$, as we now prove.

Let $\delta_1 = \max\{\rho(v, av) \mid a \in A\}$. The G-action on Y clearly preserves the pseudo-distance ρ. It follows that, for any $g \in G$, $\rho(v, gv) \leq \delta_1 d(1, g)$; indeed, this would be clear with respect to any finite set of generators. For $g \in G$ choose an edge path $(\tau_1, \ldots, \tau_k)$ in Y from v to gv with k minimal. Thus the τ_i's are non-degenerate edges which are not edge loops, and $\rho(v, gv) = 2k + 1$. Working along this edge path, one finds $a_1, \ldots, a_s \in A$ such that the edge path is covered by $B_r(v), B_r(a_1v), B_r(a_1a_2v), \ldots, B_r(a_1a_2\ldots a_sv)$ with $s \leq k$. It follows that $d(v, gv) \leq s + \delta_1 \leq k + \delta_1$, giving $d(v, gv) - \delta_1 < \rho(v, gv)$. Thus f is a quasi-isometric embedding. By cocompactness, f is a quasi-isometry. $\square$

Proposition 18.2.3. *Let Y be a path connected rigid G-CW complex which is finite mod G. Let ρ and ρ_1 be the CW pseudometrics on Y and Y^1 respectively,*

---

[3] We remind the reader that all groups for which quasi-isometry is discussed must be finitely generated.

and let $i : Y^1 \hookrightarrow Y$ *denote the inclusion map. Then* i *is a quasi-isometry between* (Y^1, ρ_1) *and* (Y, ρ).

Proof. We first prove that (Y^1, ρ_1) and $(Y^1, \rho|)$ are quasi-isometric. There are only finitely many different isomorphism types of carriers $C(e)$ of cells e of Y. If ω is a path in Y joining two points of Y^1 and if $e_1, \cdots, e_r$ are the cells of Y whose interiors meet the image of ω, then, by 1.4.3, ω is homotopic rel end points to a path in $\bigcup_{i=1}^{r} C(e_i) \cap Y^1$ so that the homotopy is controlled as described in 1.4.4. The last two sentences imply the existence of $\lambda \geq 1$ such that for all $x, y \in Y^1$, $\rho(x,y) \leq \rho_1(x,y) \leq \lambda\rho(x,y)$. Thus id: $Y^1 \to Y^1$ is a quasi-isometry as claimed.

The inclusion $i : (Y^1, \rho|) \hookrightarrow (Y, \rho)$ is an isometric embedding and by 1.5.1 is quasi-surjective, hence is a quasi-isometry. $\qquad\square$

Combining the last two results we get:

Theorem 18.2.4. *If* Y *is a path connected rigid* G-CW *complex which is finite mod* G *and has finite vertex stabilizers, then* (Y, ρ) *is quasi-isometric to* (G, d). $\qquad\square$

Corollary 18.2.5. *If* H *is a subgroup of finite index in the finitely generated group* G, *then* G *and* H *are quasi-isometric.* $\qquad\square$

In Sect. 14.5 we defined commensurability for subgroups of a given group. More generally, two groups G and H are *commensurable* if there is a finite sequence of groups $G = G_0, G_1, \ldots, G_n = H$ such that for each $i < n$ either G_i is isomorphic to a subgroup of finite index in G_{i+1}, or G_{i+1} is isomorphic to a subgroup of finite index in G_i. This is compatible with the definition of commensurability given in Sect. 14.5 (Exercise 2).

Corollary 18.2.6. *Commensurable finitely generated groups are quasi-isometric.* $\qquad\square$

Two quasi-isometric embeddings $f_1, f_2 : (Z_1, \psi_1) \to (Z_2, \psi_2)$ are *quasi-homotopic* if there exists $C \geq 0$ such that for all $z \in Z_1$, $\psi_2(f_1(z), f_2(z)) \leq C$; i.e., f_1 and f_2 are pointwise "boundedly close." We define the category Quasi-homotopy to have pseudometric spaces as objects and quasi-homotopy classes of quasi-isometric embeddings as morphisms. It is tempting to use the term "quasi-homotopy equivalence" for a quasi-isometric embedding which becomes invertible in this category, but we already have a term for this by the following (whose proof is an exercise):

Proposition 18.2.7. *A quasi-isometric embedding becomes invertible in Quasi-homotopy iff it is a quasi-isometry.* $\qquad\square$

Now let X and Y be path connected CW complexes. Their CW pseudometrics restrict to metrics on X^0 and Y^0. Let $f_0 : X^0 \to Y^0$ be a quasi-isometric embedding with respect to these metrics.

Proposition 18.2.8. *If Y is an n-dimensional and $(n-1)$-connected rigid G-CW complex which is finite mod G, then f_0 extends to a CW-Lipschitz map $f : X^n \to Y$. The restriction of f to X^{n-1} is unique up to CW-Lipschitz homotopy.*

Proof. The map f_0 is CW-Lipschitz. One extends this map skeleton by skeleton to get the cellular map f. The connectivity assumptions on Y make the extensions possible, and the fact that some group G acts rigidly with finite quotient ensures that the number of cells of Y involved in each extension to a cell of X is bounded. The same remarks apply to the uniqueness claim. The details are a standard exercise in homotopy theory similar to proofs in Sect. 1.4. □

Addendum 18.2.9. *If $g_0 : X^0 \to Y^0$ is another quasi-isometric embedding which is quasi-homotopic to f_0, and if g is the extension of g_0 as in 18.2.8, then $f \mid X^{n-1}$ and $g \mid X^{n-1}$ are CW-Lipschitz homotopic.* □

Proposition 18.2.10. *The map $f : X^n \to Y$ in 18.2.8 is proper.*

Proof. $\{N(v) \mid v \in X^0\}$ is a cover[4] of X. Suppose there is a finite subcomplex L of Y such that $f^{-1}(L)$ is not compact. By 11.4.4 and 10.1.12 there is an infinite set $\{v_i\}$ of vertices of X such that $N(v_i) \cap f^{-1}(L) \neq \emptyset$ for all i. Thus $f(v_i) \cap N(L) \neq \emptyset$ for all i. But this contradicts the fact that $f_0 : X^0 \to Y^0$ is a quasi-isometric embedding. □

Theorem 18.2.11. *Let G and H be finitely generated groups of type F_n. Let X be a $K(G,1)$ with finite n-skeleton, and let Y be a $K(H,1)$ with finite n-skeleton. If G and H are quasi-isometric then there is a proper n-equivalence $\tilde{X}^n \to \tilde{Y}^n$.*

Proof. We may assume X and Y have one vertex each so we may identify $\tilde{X}^0$ and $\tilde{Y}^0$ with G and H. By 18.2.8–18.2.10, there is a proper map $\tilde{X}^n \to \tilde{Y}^n$ which is a CW-Lipschitz n-equivalence. By 14.1.11, this is a proper n-equivalence.

□

It follows from 18.2.11 that many group theoretic invariants discussed in previous chapters are quasi-isometry invariants. Before listing some of these we prove:

Theorem 18.2.12. *If G and H are finitely generated quasi-isometric groups and if G has type F_n, then H has type F_n.*

Proof. Let X and Y be $K(G,1)$ and $K(H,1)$ complexes respectively, each having one vertex. Since G and H are finitely generated they are countable, so, by exercises in Sections 4.5 and 11.4, we may assume X and Y are countable

---

[4] See Sect. 11.4 for the definition of $N(v)$.

and locally finite. By 11.4.6, we may choose finite filtrations $\{X_i\}$ and $\{Y_j\}$ for X and Y. Let $\bar{X}_i$ [resp. $\bar{Y}_j$] denote the pre-image of X_i [resp. Y_j] in the universal cover $\tilde{X}$ [resp. in $\tilde{Y}$]. Thus we have a G-filtration of $\tilde{X}$ and an H-filtration of $\tilde{Y}$. Each $\bar{X}_i$ is finite mod G and each $\bar{Y}_j$ is finite mod H. By 7.4.1, $\{\bar{X}_i\}$ is essentially $(n-1)$-connected.

The construction of the map f in the proof of 18.2.8 can be carried over to get a map $f : \tilde{X} \to \tilde{Y}$ such that for each i there exists j such that $f(\bar{X}_i) \subset \bar{Y}_j$ and the induced map $\bar{X}_i \to \bar{Y}_j$ is CW Lipschitz. Just as in 18.2.8, we may conclude that f defines an isomorphism in the category ind-Homotopy. Such maps preserve essential $(n-1)$-connectedness, so the filtration $\{\bar{Y}_j\}$ is essentially $(n-1)$-connected, implying (by 7.4.1) that H has type F_n. □

In view of 18.2.11 and 18.2.12, we can list some quasi-isometry invariants of a finitely generated group G:

1) G has type F_n or type F_∞;
2) the number of ends of G;
3) $H^n(G, \mathbb{Z}G)$ when G has type F_n;
4) G is semistable at each end when G is finitely presented;
5) G is stable at each end when G is finitely presented;
6) G is simply connected at each end when G is finitely presented;
7) G is $(n-1)$-connected at infinity when G has type F_n;
8) G is $(n-1)$-acyclic at infinity with respect to a ring R when G has type F_n.

We turn to dimension. Geometric dimension and cohomological dimension are not quasi-isometry invariants: the trivial group has dimension 0 while non-trivial finite groups have infinite dimension. We saw in Exercise 1 of Sect. 13.10, that if a group G is of type F_∞ and has finite cohomological dimension then the cohomological dimension of G is equal to $\sup\{n \mid H^n(G, \mathbb{Z}G) \neq 0\}$. And the cohomological dimension of G is finite iff its geometric dimension is finite (see Theorem VIII.7.1 of [29]). Together with 7.2.13, these comments lead to an additional quasi-isometry invariant:

9) G has type FD.

By Remark 18.1.2 and 18.2.8, we can add another quasi-isometry invariant:

10) $H_*^{(2)}(G)$ when G has type F_∞.

There is considerable interest in which properties of finitely generated groups are quasi-isometry invariants. The list goes far beyond what is given here.

Remark 18.2.13. It is proved in [22] that there are uncountably many quasi-isometry classes of finitely generated groups.

Remark 18.2.14. There are finitely generated quasi-isometric groups G_1 and G_2 such that for all $H_1 \leq G_1$ the number of ends of the pair (G_1, H_1) is 1 or 0, while there is a subgroup $H_2 \leq G_2$ such that (G_2, H_2) has two ends. We outline an example. A theorem in [6] asserts that if G is a connected simple Lie group which has Kazhdan's Property T and has infinite cyclic fundamental group, if Γ is a cocompact lattice in G and if $\tilde{\Gamma} := p^{-1}(\Gamma)$, where p is the universal covering projection, then $\tilde{\Gamma}$ has Property T and is quasi-isometric to $\Gamma \times \mathbb{Z}$. We set $G_2 := \Gamma \times \mathbb{Z}$ and $H_2 := \Gamma \times \{0\}$, a normal subgroup. Then (G_2, H_2) has two ends, by 13.5.11. However, a theorem of Niblo and Roller [125] asserts that whenever a group G_1 has Property T and H_1 is a subgroup, then the number of ends of (G_1, H_1) is 1 or 0, depending on whether H_1 has infinite or finite index in G_1. So we may set $G_1 := \tilde{\Gamma}$. Note that this implies that G_2 does not have Property T.

Appendix: Quasi-isometry and geometry

Staying within the spirit of this book, we compared the word metric of a group G with the CW pseudometrics of suitable CW complexes on which G acts nicely. The reader should be aware that it is more usual to compare the word metric on G with metric spaces on which G acts. We briefly outline this.

Let $\omega : I \to X$ be a path in the metric space (X, d). The *length* of ω is

$$l(\omega) = \sup_P \sum_{i=1}^{n} d(\omega(t_{i-1}), \omega(t_i))$$

where P ranges over all partitions $0 = t_0 < t_1 < \cdots < t_n = 1$ of I (for all n); thus $0 \leq l(\omega) \leq \infty$. The metric d is a *length metric* if for all $x, y \in X$ $d(x, y) = \inf\{l(\omega) \mid \omega$ is a path in X from x to $y\}$. For more on length metrics see Chap. 1, Sect. 3 of [24].

An action of a group G on a metric space X by isometries is *proper* if for each $x \in X$ there exists $r > 0$ such that $\{g \in G \mid gB_r(x) \cap B_r(x) \neq \emptyset\}$ is finite. The action is *cocompact* if there is a compact set $C \subset X$ such that $\cup\{gC \mid g \in G\} = X$. The set C is a *fundamental domain* for the action.

These definitions set us up for a basic quasi-isometry relation between G and metric spaces on which G acts; a proof can be found in [24, p. 140]:

Theorem 18.2.15. (Švarc-Milnor Lemma) *If G acts properly and cocompactly by isometries on a length metric space X, then G is finitely generated and for any $x \in X$ the function $f : G \to X$, $g \mapsto gx$, is a quasi-isometry.* $\square$

Exercises

1. What goes wrong if one tries to define quasi-isometry for groups which are not finitely generated?
2. Let H_1, H_2 be subgroups of a group G. Prove that they are commensurable in the sense of Sect. 14.5 iff they are commensurable in the sense of this section.

3. Prove 18.2.1 .
4. Prove 18.2.5.
5. Prove 18.2.7.

18.3 The Bieri-Neumann-Strebel invariant

This is an introductory essay on the Sigma Invariants of group actions. Our aim is to convey the flavor by proving some of the basic theorems in a simple case – an action ρ of a finitely generated group G by translations on a finite-dimensional Euclidean space V. We treat only the "0-connected case", where the invariant $\Sigma^1(\rho)$ (a certain open subset of the unit sphere in V) is closely related to what is often called the "Bieri-Neumann-Strebel invariant" of G. This subject, and its higher generalizations, use filtered homotopy theory in a natural way.

A. Statement of the Boundary Criterion:

We start by stating the Boundary Criterion, a theorem whose proof exhibits the most important ideas.

Let V be a finite-dimensional real vector space equipped with an inner product $\langle \cdot, \cdot \rangle$. Each translation $T_v : V \to V$, $u \mapsto u + v$, is then an isometry of V and we denote the group of all translations by $\mathrm{Transl}(V)$. Let G be a finitely generated group. We consider a cocompact action of G on V by translations, i.e., a homomorphism $\rho : G \to \mathrm{Transl}(V)$ with compact fundamental domain. We do not assume that the action ρ is proper or that its orbits are discrete.

Let Γ be the Cayley graph of G with respect to some finite generating set $\{g_i\}$ of G. Let $h : \Gamma \to V$ be a G-map. Since Γ is a free G-graph the existence of such a map h is clear. We call h a *control map*.

There are various ways of filtering V, and each gives rise, via h^{-1}, to a filtration of Γ.

The filtration $\mathcal{F}$: This is the filtration of V by closed balls $B_n(0)$ of radius n centered at 0.

The filtrations $\mathcal{F}_u$: Each unit vector $u \in V$ defines a filtration of V by means of closed half-spaces whose boundaries are perpendicular to u. More precisely, let $H_u(0) := \{v \in V \mid \langle u, v \rangle \leq 0\}$ and for $r \in \mathbb{R}$ let $H_u(r) = H_u(0) + ru$. (The vector u "points out of" these spaces.) We define $\mathcal{F}_u$ to be the filtration $\{H_u(n) \mid n \in \mathbb{Z}\}$ of V.

The connection between these is seen in:

Theorem 18.3.1. (Boundary Criterion) *The following are equivalent:*

(i) *For every unit vector u, $(\Gamma, h^{-1}\mathcal{F}_u)$ has one filtered end;*

(ii) $\exists \, \lambda \geq 0$ *and* $n_0 \geq 0$ *such that for every* $n \geq n_0$ *any two points of* $h^{-1}(B_n(0))$ *can be joined by a path in* $h^{-1}(B_{n+\lambda}(0))$.

B. The Sigma Invariants:

Before proving this theorem we make some remarks:

1. The set of directions (= unit vectors) u for which $(\Gamma, h^{-1}\mathcal{F}_u)$ has one filtered end is denoted by $\Sigma^1(\rho)$, the first *sigma invariant* of the action ρ.

2. The condition (ii) in 18.3.1 implies that the filtration $h^{-1}\mathcal{F}$ of Γ is essentially 0-connected. Comparing with Brown's Criterion 7.4.1, one might therefore expect a connection between this condition and $\ker(\rho)$ being finitely generated. This connection is given in 18.3.12 below.

3. We write $N = \ker(\rho)$. This is a normal subgroup of G containing the commutator subgroup G'. Thus there is a canonical bijection[5] between the unit vectors in V and the equivalence classes $[\chi]$ of non-zero characters χ : $G \to \mathbb{R}$ such that $\chi(N) = 0$, where two characters are equivalent if one is a positive multiple of the other. The bijection is $u \mapsto [\langle u, \cdot \rangle]$. In the literature, $\Sigma^1(\rho)$ is often defined in terms of characters, namely, $\Sigma^1(\rho) = \{[\chi] \mid \chi$ is a non-zero character on $G, \chi(N) = 0$, and Γ_χ is path connected$\}$, where Γ_χ denotes the subgraph of Γ generated by the vertices $g \in G$ such that $\chi(g) > 0$. It will follow from 18.3.5 that this definition of $\Sigma^1(\rho)$ is equivalent to the one given in 1. above.

4. In the special case where $V = V_0 := G/G' \otimes_{\mathbb{Z}} \mathbb{R}$ and $\rho(g)(\bar{g}G' \otimes 1) = g\bar{g}G' \otimes 1$, $\Sigma^1(\rho)$ is denoted by $\Sigma^1(G)$ and is called the *Bieri-Neumann-Strebel invariant* of G.

5. In the literature one finds the *sphere of G* defined as

$$S(G) := \{[\chi] \mid \chi : G \to \mathbb{R} \text{ is a non-zero character}\}$$

and $\Sigma^1(G)$ is then defined as

$$\{[\chi] \in S(G) \mid \Gamma_\chi \text{ is path connected}\}.$$

This is more functorial insofar as no choice of inner product is involved. That it is equivalent to the definition in 4. is easily seen: every character χ determines a linear map $\lambda_\chi : V_0 \to \mathbb{R}$ and there is a natural bijection between $\{[\chi]\}$ and the oriented codimension 1 subspaces of V_0 (sending $[\chi]$ to $(\ker(\lambda_\chi)$, "+ side")). The chosen inner product on V_0 then determines the corresponding unit vector in V_0.

6. It is shown in [16] that our definition of $\Sigma^1(\rho)$ admits a substantial generalization in which ρ is an action of G on a proper CAT(0) space M. An analog of the Boundary Criterion holds (as do analogs of all the theorems in Sect. F below); the role of half-spaces in V is played by horoballs in M. The proofs are more difficult because the induced action of G on the CAT(0) boundary of M (see 17.5.5) is non-trivial in general, whereas in the case of V all translation actions induce trivial actions on $S_\infty(V)$, the sphere at infinity[6] of V.

---

[5] It is a bijection because we are assuming the action ρ is cocompact. Compare this with the non-cocompact action of $\mathbb{Z}$ on $\mathbb{R}^2$, where n takes (x, y) to $(x + n, y)$.

[6] The *sphere at infinity* of V is the space of geodesic (= straight) rays in V which start at 0, with the compact-open topology. If $\dim V = n$, then $S_\infty(V) \cong S^{n-1}$.

7. There are analogs of $\Sigma^1(\rho)$, namely, $\Sigma^n(\rho)$ for all integers $n \geq 0$. In the case of Σ^n, filtered path connectedness is replaced by filtered $(n-1)$-connectedness. Analogs of the Boundary Criterion and of the theorems in Sect. F also hold. The entire theory is set out in [16].

8. **Example.** Take $G = \mathbb{Z}^n$. Then, obviously, $\Sigma^1(G) = S_\infty(\mathbb{R}^n)$.

9. **Example.** Take $G = \mathbb{Z} * \mathbb{Z}$ with generators a and b for the two copies of $\mathbb{Z}$. Then $\Sigma^1(G) \subset S_\infty(\mathbb{R}^2)$. We will now show that $\Sigma^1(G)$ is a proper subset of $S_\infty(\mathbb{R}^2)$, leaving it as an exercise to show that $\Sigma^1(G)$ is actually empty.[7]

Take Γ to be the Cayley graph of G with respect to the generators $\{a, b, a^{-1}, b^{-1}\}$ and define $h : \Gamma \rightarrow \mathbb{R}^2$ as above. If $n \geq 1$ it is not hard to see that the vertex 1 and the commutator $[a, b^{n+1}]$ cannot be joined by a path over $B_n(0)$. So Condition (ii) in 18.3.1 fails.

10. **Example.** With notation as in 6.2.10, let $G = BS(1, 2)$. Then we have $G/G' \cong \mathbb{Z}$ so $\Sigma^1(G) \subset S_\infty(\mathbb{R}) = \pm\infty$. Identifying U with $Y \times \mathbb{R}$ as in 6.2.10, the G-tree Y is "rooted" in the sense that one of its ends is fixed by the G-action. Sending that end to $-\infty$ gives a G-map $Y \rightarrow \mathbb{R}$, and hence the required control map $\Gamma \rightarrow \mathbb{R}$ is the restriction to the 1-skeleton of $U = Y \times \mathbb{R} \rightarrow Y \rightarrow \mathbb{R}$. Then $-\infty \in \Sigma^1(G)$ while $+\infty \notin \Sigma^1(G)$.

11. The sigma invariants have interesting openness properties; these are stated in 18.3.15–18.3.17.

C. Preliminaries for the proof of 18.3.1:

The properties (i) and (ii) in 18.3.1 are easily seen to be independent of our choice of generating set $\{g_i\}$ and G-map h. So we make convenient choices:

- $\{g_i\}$ is to be invariant under inversion and is not to contain $1 \in G$. This ensures that Γ, while not being a simplicial graph, is a regular CW complex.
- $h(1) = 0$, and h maps each edge of Γ homeomorphically onto a line segment in V.

The CW neighborhood[8] of 1 in Γ is generated by the edges joining 1 to the vertices g_i. The h-images of these edges form a finite set, S, of line segments in V joining 0 to points of V. While some of these may be degenerate (joining 0 to 0) there are enough non-degenerate line segments that for every u some member of S does not lie in $H_u(0)$.

Define $h_u : \Gamma \rightarrow \mathbb{R}$ by $h_u(x) = \langle h(x), u \rangle$.

Proposition 18.3.2. *For any unit vector u in V and any vertex $g \in \Gamma$ there is a ray in Γ starting at g on which the map h_u is strictly increasing.*

Or, one may drop the "which start at 0" condition, and instead take the quotient space which identifies parallel rays. This makes the action by translations clear, and clearly trivial.

[7] Compare 18.3.18.

[8] One might wish to speak of the "star" of 1, but, because of our choice of generating set, Γ is not a simplicial graph: there are two edges joining 1 to each g_i.

Proof. The discussion shows that this is true when $g = 1$; translation by g preserves this property. □

Corollary 18.3.3. *The filtered space* $(\Gamma, h^{-1}\mathcal{F}_u)$ *is regular in the sense of Sect. 14.3 and has at least one filtered end.*

Proof. The existence of a filtered end follows directly from 18.3.2. Regularity holds because, by 18.3.2, complements have no $h^{-1}\mathcal{F}_u$-bounded path components. □

Proposition 18.3.4. $(\Gamma, h^{-1}\mathcal{F}_u)$ *is CW compatible.*

Proof. This follows from the fact that S is finite. □

Proposition 18.3.5. *If* $(\Gamma, h^{-1}\mathcal{F}_u)$ *has one filtered end then for all* $r \in \mathbb{R}$ *the subspace* $\Gamma - h^{-1}(H_u(r))$ *is path connected.*

Remark 18.3.6. This convenient strengthening of "having one filtered end" is not an invariant of filtered homotopy; rather, it holds because of our choices.

Proof (of 18.3.5). By hypothesis, $\forall\, n \in \mathbb{Z}\; \exists\, m \geq n$ such that any two points in $\Gamma - h^{-1}(H_u(m))$ can be joined[9] by a path in $\Gamma - h^{-1}(H_u(n))$. Let x and y be points of $\Gamma - h^{-1}(H_u(n)) = h_u^{-1}((n, \infty))$. It follows from 18.3.2 that there are paths in this space from x and y into $h_u^{-1}((m, \infty)) = \Gamma - h^{-1}(H_u(m))$. The end points of these paths can then be joined in $\Gamma - h^{-1}(H_u(n))$. □

D. Proof of (i) $\Rightarrow$ (ii) in 18.3.1:

The proof requires some careful analysis, so we begin with a sketch to motivate the details. The control map h is at the center of the discussion; we will say that a subset A of Γ is *over* a subset $\bar{A}$ of V if $A \subset h^{-1}(\bar{A})$. For example, the filtration $h^{-1}\mathcal{F}$ of Γ is by sets over the balls $B_n(0)$, and the filtration $h^{-1}\mathcal{F}_u$ is by sets over the half-spaces $H_u(n)$. So, using 18.3.5, we are to prove that if Γ is path connected over the complement of every $H_u(r)$, then λ and n_0 exist as in (ii). Starting with x and y over some $B_n(0)$, there is certainly an edge path ω in Γ joining x to y. We consider the minimum R such that ω is over $B_R(0)$ and then we modify ω near places where it is over $\partial B_R(0)$ to get a new path from x to y which is over $B_{R-\delta}(0)$ for some $\delta > 0$. An important feature is that δ should be independent of ω, x and y. Then, by repeated modification, we get edge paths joining x to y over $B_{R-k\delta}(0)$ until the process breaks down. This happens at $B_{n+\lambda}(0)$ for some λ. Analysis of the method shows that λ is independent of x, y and ω; there might be problems if n were too small, but the method shows that there is n_0 such that all works

---

[9] Recall from Sect. 14.1 that we need not confine ourselves to subgraphs as distinct from subspaces, and to CW complements as distinct from complements, since our filtration is CW compatible.

well when $n \geq n_0$. The key point is: in selecting our modifications of paths we always pick them to be G-translates of a finite set of "template paths".

We begin the details with some constructions and propositions (18.3.7–18.3.10).

For each unit vector $u \in V$ define $T(u) := \{\gamma \in G \mid \gamma = g_j$ for some j, and $h_u(\gamma) > 0\}$. If $\gamma \in T(u)$ then the line segment in S defined by γ does not lie in $H_u(0)$. For each g_i and each $\gamma \in T(u)$ we choose a path $\tau(u, g_i, \gamma)$ from γ to $g_i\gamma$ which is mapped by h_u into the closed interval $[\min\{h_u(\gamma), h_u(g_i\gamma)\}, \infty)$. (Such a path exists by 18.3.5 applied to $H_{-u}(s)$ for a suitable $s \in \mathbb{R}$.) There is then a neighborhood W_u of u in the unit sphere of V such that for all $u' \in W_u$ the image of $\tau(u, g_i, \gamma)$ under $h_{u'}$ lies in the open interval $(\min\{0, h_{u'}(g_i)\}, \infty)$. Since the unit sphere is compact we have (see Fig. 18.1 for this and what follows):

Proposition 18.3.7. *There is a finite set $\{u_k\}$ of unit vectors such that for every u and g_i there is some u_k such that for every $\gamma \in T(u_k)$ the image under h_u of $\tau(u_k, g_i, \gamma)$ lies in the open interval $(\min\{0, h_u(g_i)\}, \infty)$.*   □

We note that in 18.3.7 finitely many paths suffice for all unit vectors u in the following sense:

Corollary 18.3.8. *There exists $\epsilon > 0$ such that for each unit vector u and each generator g_i there exists k such that for each $\gamma \in T(u_k)$ the image under h_u of the path $\tau(u_k, g_i, \gamma)$ lies in the closed interval $[\min\{0, h_u(g_i)\} + \epsilon, \infty)$.*
□

Let $C(0, u, \theta)$ denote the closed infinite cone (in V)

$$\{y \in V \mid \langle \frac{y}{|y|}, u \rangle \geq \cos\theta\} \cup \{0\}$$

having vertex at 0, axis pointing in the direction u, and angle of amplitude $0 \leq \theta < \frac{\pi}{2}$. Given $0 < \epsilon < K$, the corresponding frustum of this cone is denoted by $F(0, u, \theta, \epsilon, K) := \{y \in C(0, u, \theta) \mid \epsilon \leq \langle y, u \rangle \leq K\}$. The translation of $C(0, u, \theta)$ having vertex v is $C(v, u, \theta) := C(0, u, \theta) + v$, and the corresponding translated frustum is $F(v, u, \theta, \epsilon, K) := F(0, u, \theta, \epsilon, K) + v$.

Since there are only finitely many paths $\tau(u_k, g_i, \gamma)$, each with compact image, we can strengthen 18.3.8:

Proposition 18.3.9. *There exist $0 < \epsilon < K$ and $0 < \theta < \frac{\pi}{2}$ such that for each unit vector u and each generator g_i there exists k such that for each $\gamma \in T(u_k)$ the path $\tau(u_k, g_i, \gamma)$ is over $F(v, u, \theta, \epsilon, K)$, where $v = h(g_i)$ if $h_u(g_i) < 0$ and $v = 0$ if $h_u(g_i) \geq 0$.*

□

The point of all this work is illustrated in Fig. 18.2. A "two-edge path" from γ to g_i through 1 can be replaced by a "modified path," namely, the

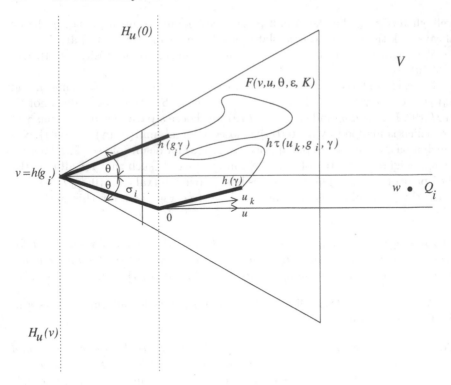

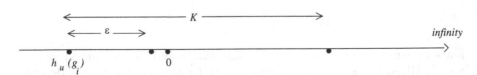

Fig. 18.1.

edge path $\tau(u_k, g_i, \gamma)$ from γ to $g_i\gamma$ followed by an edge joining $g_i\gamma$ to g_i. This "modified path" lies over $F(v, u, \theta, \epsilon, K)$.

Finally (before proving (i) $\Rightarrow$ (ii) in 18.3.1) we must say what it means for a point of V to be "far away" from the situation we have been studying. An edge of Γ joining 1 to g_i is over a (possibly degenerate) line segment σ_i in S which, together with u, defines a half-infinite strip (or half-line in the degenerate case) $Q_i \subset H_{-u}(v)$; see Fig. 18.1. Clearly we have:

Proposition 18.3.10. *If $w \in Q_i$ and the distance from w to σ_i is sufficiently large, then $F(v, u, \theta, \epsilon, K) \subset B_{|v-w|-\frac{\epsilon}{2}}(w)$.* $\square$

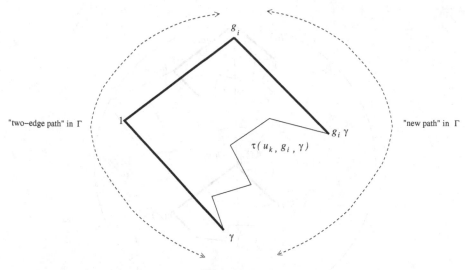

"two–edge path" in Γ "new path" in Γ

Fig. 18.2.

Proof (of (i) $\Rightarrow$ (ii) *in* 18.3.1). We start from the setup described in the proof-sketch above. We may assume $R > n$, so the piecewise linear path $h \circ \tau$ meets $\partial B_R(0)$ in finitely many points. We may consider ω as an edge path $(\omega_1, \dots, \omega_m)$ from x to y. The first vertex g on this path such that $h(g) \in \partial B_R(0)$ is the common vertex of two successive edges, say ω_{i-1} and ω_i. If g' is the other vertex of ω_{i-1}, then $h(g')$ lies in the interior of $B_R(0)$. Let u be the unit vector in V such that the line segment from $h(g)$ to 0 points in the direction u. We wish to replace the two-edge segment (ω_{i-1}, ω_i) from g' through g to g'' by a modified segment. To do this, we translate this two-edge path using g^{-1} to get a two-edge path from $\gamma := g^{-1}g'$ through 1 to $g_i := g^{-1}g''$. Applying the previous propositions to this latter two-edge path, and translating back the resulting modified path using g, we obtain a replacement of (ω_{i-1}, ω_i) by an edge path whose first contact with $\partial B_R(0)$ (if any) is further along the path ω; see Fig. 18.3. By 18.3.10, all of that modified path except its final edge is over $B_{R-\frac{\epsilon}{2}}(0)$. Proceeding thus, we eventually replace ω by a path over $B_{R-\frac{\epsilon}{2}}(0)$. This process can clearly be continued as described in the proof-sketch. The precise identification of n_0 and λ is left to the reader.   □

E. Proof of (ii) $\Rightarrow$ (i) in 18.3.1:

This is much easier and only an indication is needed. Given points x and y over $V - H_u(r)$, they lie over some ball $B_R(w) \subset V - H_u(r)$. By (ii) they can be joined by a path over $B_{R+\lambda}(w) \subset V - H_u(r - \lambda)$. So (i) holds.   □

F. Properties of Σ^1:

For a translation action ρ as above we have defined

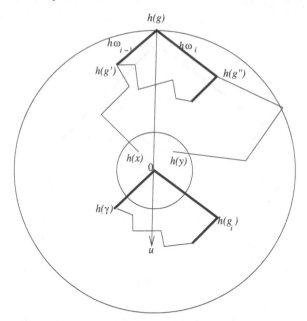

Fig. 18.3.

$$\Sigma^1(\rho) := \{u \mid (\Gamma, h^{-1}\mathcal{F}_u) \text{ has one filtered end}\}.$$

However, the definition of $\Sigma^1(\rho)$ usually given in the literature is slightly different. In the next proposition we prove the equivalence of the two definitions:

Proposition 18.3.11. $u \in \Sigma^1(\rho)$ *iff there exists* $\lambda \geq 0$ *such that for any n, any two points in* $\Gamma - h^{-1}(H_u(n))$ *can be joined by a path in the subspace* $\Gamma - h^{-1}(H_u(n - \lambda))$.

Proof. If $u \in \Sigma^1(\rho)$ there exists $\lambda \geq 0$ such that points in $\Gamma - h^{-1}(H_u(\lambda))$ can be joined by a path in $\Gamma - h^{-1}(H_u(0))$. There are elements of G which translate $\Gamma - h^{-1}(H_u(0))$ into $\Gamma - h^{-1}(H_u(n))$ for any n. Both directions of the "iff" follow easily from this. $\square$

The Boundary Criterion 18.3.1 says that (ii) holds iff $\Sigma^1(\rho) = S_\infty(V)$. For translation actions having discrete orbits, this leads us to a topological condition equivalent to finite generation of the kernel:

Theorem 18.3.12. *Let* $\rho(G)$ *have discrete orbits in* V. *The normal subgroup* $N := \ker(\rho)$ *is finitely generated iff* $\Sigma^1(\rho) = S_\infty(V)$.

Proof. In the commutative diagram

it is not hard to see that the map f is proper when the orbits are discrete. Thus the filtration $h^{-1}\mathcal{F}$ of Γ is by sets each of which is compact mod N. It is an exercise to show that this filtration is CW compatible. By Brown's Criterion and 18.3.1, the "if" part follows.

We now prove "only if" Since N is finitely generated we can choose the generators $\{g_i\}$ to include a generating set for N. The images of the g_i's in $Q := G/N$ form a generating set for Q. Let Γ_N and Γ_Q be the corresponding Cayley graphs for N and Q respectively. There is an obvious map $p : \Gamma \to \Gamma_Q$ and, for each vertex $q \in \Gamma_Q$, $p^{-1}(q)$ is a copy of Γ_N lying in Γ. There is a commutative diagram

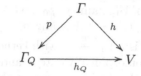

Given two points in $\Gamma - h^{-1}(H_u(r))$, they map to points in $\Gamma_Q - h_Q^{-1}(H_u(r))$ where they are clearly joinable by a path. That path lifts to a path in Γ joining one of the two points to a point in the fiber over Γ_Q containing the other point. Since Γ_N is path connected the two points can therefore be joined in $\Gamma - h^{-1}(H_u(r))$. Thus $\Sigma^1(\rho) = S_\infty(V)$. $\square$

Theorem 18.3.13. (Σ^1-Criterion) *A unit vector u in V lies in $\Sigma^1(\rho)$ iff there is an equivariant cellular map $\phi : \Gamma \to \Gamma$ such that, for all $x \in \Gamma$, $h_u \circ \phi(x) - h_u(x) > 0$.*

Proof. We use the notation of Part D. If $u \in \Sigma^1(\rho)$ choose $\gamma \in T(u)$. Define $\phi(g) = g\gamma$ for every $g \in G$. This is equivariant. Extend ϕ equivariantly to the edges of Γ so that the edge joining 1 to g_i is mapped to the path $\tau(u, g_i, \gamma)$. Then ϕ is as claimed. Conversely, if such an equivariant map ϕ exists, let x and y be two points of $\Gamma - h^{-1}(H_u(r))$, let ω be an edge path in Γ from x to y, and let τ be an edge path from 1 to $\phi(1)$. Then the product edge path $\tau.\phi \circ \tau. \ldots \phi^{n-1} \circ \tau$ runs from 1 to $\phi^n(1)$ and its translates run from x to $\phi^n(x)$, and from y to $\phi^n(y)$. The path $\phi^n \circ \omega$ joins $\phi^n(x)$ to $\phi^n(y)$. There exists $\lambda > 0$ such that τ is over $H_u(-\lambda)$. By equivariance of ϕ, the appropriate combination of the paths just mentioned joins x to y in $\Gamma - h^{-1}(H_u(r - \lambda))$. By 18.3.11, $u \in \Sigma^1(\rho)$. $\square$

Remark 18.3.14. Because of equivariance and compactness, the condition appearing in 18.3.13 can be equivalently stated as: there is an equivariant cellular map $\phi : \Gamma \to \Gamma$ and $\epsilon > 0$ such that for all $x \subset \Gamma$ $h_u \circ \phi(x) - h_u(x) > \epsilon$. This is useful in deducing the following corollaries.

Corollary 18.3.15. *$\Sigma^1(\rho)$ is an open subset of $S_\infty(V)$.* $\square$

Corollary 18.3.16. *In the space $\text{Hom}(G, \text{Transl}(V))$ of translation actions of G on V, with compact-open topology, the condition $\Sigma^1(\rho) = S_\infty(V)$ is an open condition.*

Proof. Combine 18.3.1 and 18.3.13.   □

Corollary 18.3.17. *In the subspace consisting of translation actions ρ whose orbits are closed discrete subsets of V, the property "$\ker(\rho)$ is finitely generated" is an open condition.*

Proof. Use 18.3.12 and 18.3.16.   □

G. Relation between $\Sigma^1(G)$ and $\Sigma^1(\rho)$:

We now coordinate the sets $\Sigma^1(\rho) \subset S_\infty(V)$ as ρ varies over all translation actions on V. Choose an inner product for $V_0 = G/G' \otimes_{\mathbb{Z}} \mathbb{R}$. There is a canonical G-action of G on V_0, namely, $\rho_0 : G \to \mathrm{Transl}(V_0)$ defined by $\rho_0(g)(\bar{g}(G/G')) \otimes 1 = g\bar{g}(G/G') \otimes 1$. For any $\rho : G \to \mathrm{Transl}(V)$ there is a canonical linear G-epimorphism $V_0 \overset{\pi}{\twoheadrightarrow} V$ with kernel V_π. This map takes the complementary space $V_\pi^\perp$ isomorphically onto V and thus imposes a preferred inner product on V. Using this inner product in the definition of $\Sigma^1(\rho)$ we obtain a canonical embedding $S_\infty(V) \hookrightarrow S_\infty(V_0)$ and $\Sigma^1(\rho) \hookrightarrow \Sigma^1(G)$. In these terms we see from 18.3.12, for example, that, when ρ has discrete orbits, $\ker(\rho)$ is finitely generated iff $S_\infty(V) \subset \Sigma^1(G)$; this is how the theorem is usually stated in the literature. It follows that if one understands $\Sigma^1(G)$, then one sees $\Sigma^1(\rho)$ as the intersection of $\Sigma^1(G)$ with the appropriate great sphere in $S_\infty(V_0)$.

H. An application to group theory:

Here is an elegant application of this theory:

Theorem 18.3.18. *Let G be finitely presented and let the rank of G/G' (as a $\mathbb{Z}$-module) be at least 2. If G has no non-abelian free subgroup, then there is a finitely generated normal subgroup $L \lhd G$ with G/L infinite cyclic.*

Proof. Let (X, x) be the presentation complex for some finite presentation of G. We may assume that the 1-skeleton of $\tilde{X}$ is a Cayley graph with the properties listed in Sect. C.

Claim. $S_\infty(V_0) = \Sigma^1(G) \cup (-\Sigma^1(G))$. To prove this Claim we recall that $V_0 = (G/G') \otimes_{\mathbb{Z}} \mathbb{R}$, ρ is the canonical translation action, and as usual (see Sect. B) we write $\Sigma^1(G)$ rather than $\Sigma^1(\rho)$ in this context. So for any non-zero character $\chi : G \to \mathbb{R}$ we are to show that either $[\chi]$ or $-[\chi] := [-\chi]$ lies in $\Sigma^1(G)$. Let u be the unit vector in V_0 corresponding to $[\chi]$. For some $r \geq 0$ we write $V_0 = H_u(r) \cup H_{-u}(-r)$. The control map on the 1-skeleton extends to a G-map $h : (\tilde{X}, \tilde{x}) \to (V_0, 0)$. Let $K := \ker(\chi)$ and $N := \ker(\rho)$. There is a commutative diagram

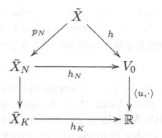

The spaces $Y^- := h_K^{-1}((-\infty, r])$ and $Y^+ := h_K^{-1}([-r, \infty))$ cover $\bar{X}_K$, and their intersection is $Y_0 := h_K^{-1}([-r, r])$. Choosing a base point $y_0 \in Y_0$, we apply the Seifert-Van Kampen Theorem 3.1.18. The spaces $Y^\pm$, Y_0 are path connected, for if r is the rank of the abelian group $\chi(G)$ then the 1-skeleton of $\bar{X}_K$ consists of the 1-skeleton of $\mathbb{R}^r$ with perhaps some other 1-cells attached; it is clear that for any linear map $\mathbb{R}^r \to \mathbb{R}$ the preimage of any interval has path connected intersection with the 1-skeleton of $\mathbb{R}^r$; hence our assertion that $Y^\pm$ and Y_0 are path connected.

Applying 3.1.18, there are three cases to consider:
Case 1: $\pi_1(Y^+, y_0) \to \pi_1(\bar{X}_K, y_0) = K$ is surjective. Then $h^{-1}(H_{-u}(-r))$ is path connected by 3.4.9, implying $[\chi] \in \Sigma^1(G)$. Similarly, if "plus" and "minus" are interchanged, $-[\chi] \in \Sigma^1(G)$.
Case 2: The image of $\pi_1(Y^+, y_0)$ in $\pi_1(\bar{X}_K, y_0)$ has finite index. Then by enlarging r we get back to Case 1. The same holds for the image of $\pi_1(Y^-, y_0)$.
Case 3: The images of $\pi_1(Y^\pm, y_0)$ in $\pi_1(\bar{X}_K, y_0)$ both have infinite index. But this cannot happen, since, by the discussion in Remark 6.2.13 together with 18.3.19, below, it would imply that K contains a non-abelian free group, something ruled out by the hypotheses.

Thus, the Claim is proved.

To complete the proof of the Theorem we note that $S_\infty(V_0)$ is a sphere of dimension ≥ 1, by our hypothesis on G/G'. By 18.3.15, the sets $\pm\Sigma^1(G)$ form an open cover of this connected sphere, so they have non-empty intersection. This intersection, being open, must contain some $[\chi_0]$ with χ_0 a rational character (i.e., one whose image is an infinite cyclic subgroup of $\mathbb{R}$). By 18.3.12, $L := \ker(\chi_0)$ is finitely generated. $\qquad\square$

The only missing piece in the last proof is a standard fact from combinatorial group theory:

Proposition 18.3.19. *If $G = G_1 *_{G_0} G_2$ is a free product with amalgamation, where G_0 has index ≥ 2 in G_1 and has index ≥ 3 in G_2, then G contains a free non-abelian subgroup.*

Proof (Sketch of Proof). The Normal Form Theorem for free products with amalgamation (see [106, p. 187]) says that a product $c_1 \dots c_n$ is non-trivial in G if (i) each c_i is in G_1 or G_2, (ii) c_i and c_{i+1} are not both in G_1 or both in G_2, (iii) when $n > 1$ no $c_i \in G_0$, and (iv) when $n = 1$, $c_1 \neq 1$. Choose $a \in G_1 - G_0$ and $b, c \in G_2 - G_0$ representing different cosets of G_0 in G_2. The Normal Form Theorem implies that ab and ac generate a free non-abelian subgroup of G. $\square$

Remark 18.3.20. It is known [25] that Thompson's group F has no non-abelian free subgroup. It follows then from 18.3.18 that there exists a short exact sequence $N \rightarrowtail F \twoheadrightarrow \mathbb{Z}$ with N finitely generated. By 16.9.1, we can conclude that F is semistable at infinity (something we already knew by 16.9.7). The existence of N implies, by 18.3.12, that $\Sigma^1(F)$ is non-empty, indeed, contains a diametrically opposite pair of points. In fact, $\Sigma^1(F)$ has been computed, see [17].

Source Notes:

The theorems in this section first appeared in [17]. The proofs here are adapted from [13]. The original definition of $\Sigma^1(G)$ used a finite generation property of the commutator subgroup (when regarded as an operator group over certain submomoids of the group). The definition given here is due to Bieri-Renz [BR88] and led to higher-dimensional versions of $\Sigma^n(G)$ for $n > 1$, both homotopical and homological; see [18] and the references therein. Later, the whole theory, for all n, was extended in [16] and [15] to the case where the translation action on the vector space V is replaced by an isometric action on a proper CAT(0) space.

Exercises

1. Prove that the group N in Remark 18.3.20 has one end.
2. Let M be a proper metric space and let L be a subgroup of Isom(M), the group of isometries of M with compact-open topology. Prove that the following are equivalent:
    (i) L is a discrete closed subset of Isom(M);
    (ii) Each orbit is a discrete closed subset of M and each point-stabilizer is a finite subgroup of L;
    (iii) The action of L on M is proper.
    (A subgroup L with these properties is called a *lattice*.)
3. Let V be a finite-dimensional real vector space equipped with an inner product. Show that Transl(V) is a closed subgroup of Isom(V).
4. Fill in the details of 18.3.4.
5. Show that if G is a finitely generated abelian group of rank n then $\Sigma^1(G) = S_\infty(\mathbb{R}^n)$.
6. Show that if G is a finitely generated free group then $\Sigma^1(G) = \emptyset$.
7. Extend the discussion of $\Sigma^1(BS(1,2))$ to the case of $BS(m,n)$. Is the result different?
8. Express the statement $u \in \Sigma^1(\rho)$ in terms of filtered locally finite homology, as in Sect. 14.2.

References

1. Herbert Abels and Kenneth S. Brown, *Finiteness properties of solvable S-arithmetic groups: an example*, J. Pure Appl. Algebra **44** (1987), no. 1-3, 77–83. MR MR885096 (88g:20104)
2. M. G. C. Andrade and E. L. C. Fanti, *A relative cohomological invariant for group pairs*, Manuscripta Math. **83** (1994), no. 1, 1–18. MR MR1265914 (94m:20101)
3. M. Artin and B. Mazur, *Etale homotopy*, Lecture Notes in Mathematics, No. 100, Springer-Verlag, Berlin, 1969. MR MR0245577 (39 #6883)
4. M. F. Atiyah and G. B. Segal, *Equivariant K-theory and completion*, J. Differential Geometry **3** (1969), 1–18. MR MR0259946 (41 #4575)
5. Werner Ballmann, *Lectures on spaces of nonpositive curvature*, DMV Seminar, vol. 25, Birkhäuser Verlag, Basel, 1995, With an appendix by Misha Brin. MR MR1377265 (97a:53053)
6. Bachir Bekka, Pierre de la Harpe, and Alain Valette, *Kazhdan's property T*, 2004.
7. P. H. Berridge and M. J. Dunwoody, *Nonfree projective modules for torsion-free groups*, J. London Math. Soc. (2) **19** (1979), no. 3, 433–436. MR MR540056 (80k:20041)
8. Czesław Bessaga and Aleksander Pełczyński, *Selected topics in infinite-dimensional topology*, PWN—Polish Scientific Publishers, Warsaw, 1975. MR MR0478168 (57 #17657)
9. Mladen Bestvina, *Local homology properties of boundaries of groups*, Michigan Math. J. **43** (1996), no. 1, 123–139. MR MR1381603 (97a:57022)
10. Mladen Bestvina and Noel Brady, *Morse theory and finiteness properties of groups*, Invent. Math. **129** (1997), no. 3, 445–470. MR MR1465330 (98i:20039)
11. Mladen Bestvina and Mark Feighn, *The topology at infinity of* $\text{Out}(F_n)$, Invent. Math. **140** (2000), no. 3, 651–692. MR MR1760754 (2001m:20041)
12. Mladen Bestvina and Geoffrey Mess, *The boundary of negatively curved groups*, J. Amer. Math. Soc. **4** (1991), no. 3, 469–481. MR MR1096169 (93j:20076)
13. R. Bieri and R. Strebel, *Geometric invariants for discrete groups - preprint of book*.
14. Robert Bieri, *Homological dimension of discrete groups*, Queen Mary College Mathematical Notes, Queen Mary College Department of Pure Mathematics, London, 1981. MR MR715779 (84h:20047)

15. Robert Bieri and Ross Geoghegan, *Controlled homological algebra over cat(0) spaces.*

16. _____, *Connectivity properties of group actions on non-positively curved spaces*, Mem. Amer. Math. Soc. **161** (2003), no. 765, xiv+83. MR MR1950396 (2004m:57001)

17. Robert Bieri, Walter D. Neumann, and Ralph Strebel, *A geometric invariant of discrete groups*, Invent. Math. **90** (1987), no. 3, 451–477. MR MR914846 (89b:20108)

18. Robert Bieri and Burkhardt Renz, *Valuations on free resolutions and higher geometric invariants of groups*, Comment. Math. Helv. **63** (1988), no. 3, 464–497. MR MR960770 (90a:20106)

19. Karol Borsuk, *Theory of retracts*, Monografie Matematyczne, Tom 44, Państwowe Wydawnictwo Naukowe, Warsaw, 1967. MR MR0216473 (35 #7306)

20. _____, *Some remarks concerning the shape of pointed compacta*, Fund. Math. **67** (1970), 221–240. MR MR0266168 (42 #1076)

21. A. K. Bousfield and D. M. Kan, *Homotopy limits, completions and localizations*, Springer-Verlag, Berlin, 1972. MR MR0365573 (51 #1825)

22. B. H. Bowditch, *Continuously many quasi-isometry classes of 2-generator groups*, Comment. Math. Helv. **73** (1998), no. 2, 232–236. MR MR1611695 (99f:20062)

23. Brian H. Bowditch, *Splittings of finitely generated groups over two-ended subgroups*, Trans. Amer. Math. Soc. **354** (2002), no. 3, 1049–1078 (electronic). MR MR1867372 (2002m:20065)

24. Martin R. Bridson and André Haefliger, *Metric spaces of non-positive curvature*, Grundlehren der Mathematischen Wissenschaften [Fundamental Principles of Mathematical Sciences], vol. 319, Springer-Verlag, Berlin, 1999. MR MR1744486 (2000k:53038)

25. Matthew G. Brin and Craig C. Squier, *Groups of piecewise linear homeomorphisms of the real line*, Invent. Math. **79** (1985), no. 3, 485–498. MR MR782231 (86h:57033)

26. Matthew G. Brin and T. L. Thickstun, *On the proper Steenrod homotopy groups, and proper embeddings of planes into 3-manifolds*, Trans. Amer. Math. Soc. **289** (1985), no. 2, 737–755. MR MR784012 (87g:57029)

27. _____, *3-manifolds which are end 1-movable*, Mem. Amer. Math. Soc. **81** (1989), no. 411, viii+73. MR MR992161 (90g:57015)

28. Edward M. Brown, *Proper homotopy theory in simplicial complexes*, Topology Conference (Virginia Polytech. Inst. and State Univ., Blacksburg, Va., 1973), Springer, Berlin, 1974, pp. 41–46. Lecture Notes in Math., Vol. 375. MR MR0356041 (50 #8513)

29. Kenneth S. Brown, *Cohomology of groups*, Graduate Texts in Mathematics, vol. 87, Springer-Verlag, New York, 1982. MR MR672956 (83k:20002)

30. _____, *Finiteness properties of groups*, J. Pure Appl. Algebra **44** (1987), no. 1-3, 45–75. MR MR885095 (88m:20110)

31. _____, *Buildings*, Springer-Verlag, New York, 1989. MR MR969123 (90e:20001)

32. Kenneth S. Brown and Ross Geoghegan, *An infinite-dimensional torsion-free* FP_∞ *group*, Invent. Math. **77** (1984), no. 2, 367–381. MR MR752825 (85m:20073)

33. _____, *Cohomology with free coefficients of the fundamental group of a graph of groups*, Comment. Math. Helv. **60** (1985), no. 1, 31–45. MR MR787660 (87b:20066)

34. Kenneth S. Brown and John Meier, *Improper actions and higher connectivity at infinity*, Comment. Math. Helv. **75** (2000), no. 1, 171–188. MR MR1760501 (2001h:57005)

35. Marc Burger and Shahar Mozes, *Lattices in product of trees*, Inst. Hautes Études Sci. Publ. Math. (2000), no. 92, 151–194 (2001). MR MR1839489 (2002i:20042)

36. Kai-Uwe Bux and Carlos Gonzalez, *The Bestvina-Brady construction revisited: geometric computation of Σ-invariants for right-angled Artin groups*, J. London Math. Soc. (2) **60** (1999), no. 3, 793–801. MR MR1753814 (2001f:20083)

37. J. W. Cannon, W. J. Floyd, and W. R. Parry, *Introductory notes on Richard Thompson's groups*, Enseign. Math. (2) **42** (1996), no. 3-4, 215–256. MR MR1426438 (98g:20058)

38. T. A. Chapman, *Lectures on Hilbert cube manifolds*, American Mathematical Society, Providence, R. I., 1976. MR MR0423357 (54 #11336)

39. T. A. Chapman and L. C. Siebenmann, *Finding a boundary for a Hilbert cube manifold*, Acta Math. **137** (1976), no. 3-4, 171–208. MR MR0425973 (54 #13922)

40. D. E. Christie, *Net homotopy for compacta*, Trans. Amer. Math. Soc. **56** (1944), 275–308. MR MR0010971 (6,97a)

41. Daniel E. Cohen, *Groups of cohomological dimension one*, Springer-Verlag, Berlin, 1972, Lecture Notes in Mathematics, Vol. 245. MR MR0344359 (49 #9098)

42. Marshall M. Cohen, *A course in simple-homotopy theory*, Springer-Verlag, New York, 1973. MR MR0362320 (50 #14762)

43. Richard H. Crowell and Ralph H. Fox, *Introduction to knot theory*, Springer-Verlag, New York, 1977. MR MR0445489 (56 #3829)

44. Marc Culler and Karen Vogtmann, *Moduli of graphs and automorphisms of free groups*, Invent. Math. **84** (1986), no. 1, 91–119. MR MR830040 (87f:20048)

45. Robert J. Daverman, *Decompositions of manifolds*, Pure and Applied Mathematics, vol. 124, Academic Press Inc., Orlando, FL, 1986. MR MR872468 (88a:57001)

46. Michael W. Davis, *Groups generated by reflections and aspherical manifolds not covered by Euclidean space*, Ann. of Math. (2) **117** (1983), no. 2, 293–324. MR MR690848 (86d:57025)

47. Michael W. Davis and Tadeusz Januszkiewicz, *Hyperbolization of polyhedra*, J. Differential Geom. **34** (1991), no. 2, 347–388. MR MR1131435 (92h:57036)

48. Pierre de la Harpe, *Topics in geometric group theory*, Chicago Lectures in Mathematics, University of Chicago Press, Chicago, IL, 2000. MR MR1786869 (2001i:20081)

49. Warren Dicks and M. J. Dunwoody, *Groups acting on graphs*, Cambridge Studies in Advanced Mathematics, vol. 17, Cambridge University Press, Cambridge, 1989. MR MR1001965 (91b:20001)

50. Albrecht Dold, *Lectures on algebraic topology*, Classics in Mathematics, Springer-Verlag, Berlin, 1995. MR MR1335915 (96c:55001)

51. James Dugundji, *Topology*, Allyn and Bacon Inc., Boston, Mass., 1978. MR MR0478089 (57 #17581)

456   References

52. M. J. Dunwoody, *The homotopy type of a two-dimensional complex*, Bull. London Math. Soc. **8** (1976), no. 3, 282–285. MR MR0425943 (54 #13893)

53. ———, *Accessibility and groups of cohomological dimension one*, Proc. London Math. Soc. (3) **38** (1979), no. 2, 193–215. MR MR531159 (80i:20024)

54. ———, *The accessibility of finitely presented groups*, Invent. Math. **81** (1985), no. 3, 449–457. MR MR807066 (87d:20037)

55. M. J. Dunwoody and M. E. Sageev, *JSJ-splittings for finitely presented groups over slender groups*, Invent. Math. **135** (1999), no. 1, 25–44. MR MR1664694 (2000b:20050)

56. Martin J. Dunwoody, *An inaccessible group*, Geometric group theory, Vol. 1 (Sussex, 1991), London Math. Soc. Lecture Note Ser., vol. 181, Cambridge Univ. Press, Cambridge, 1993, pp. 75–78. MR MR1238516 (94i:20067)

57. Jerzy Dydak, *A simple proof that pointed FANR-spaces are regular fundamental retracts of ANR's*, Bull. Acad. Polon. Sci. Sér. Sci. Math. Astronom. Phys. **25** (1977), no. 1, 55–62. MR MR0442918 (56 #1293)

58. Jerzy Dydak and Jack Segal, *Shape theory*, Lecture Notes in Mathematics, vol. 688, Springer, Berlin, 1978. MR MR520227 (80h:54020)

59. David A. Edwards and Ross Geoghegan, *Compacta weak shape equivalent to ANR's*, Fund. Math. **90** (1975/76), no. 2, 115–124. MR MR0394643 (52 #15444)

60. David A. Edwards and Harold M. Hastings, *Čech and Steenrod homotopy theories with applications to geometric topology*, Springer-Verlag, Berlin, 1976. MR MR0428322 (55 #1347)

61. Samuel Eilenberg and Norman Steenrod, *Foundations of algebraic topology*, Princeton University Press, Princeton, New Jersey, 1952. MR MR0050886 (14,398b)

62. Daniel S. Farley, *Finiteness and* CAT(0) *properties of diagram groups*, Topology **42** (2003), no. 5, 1065–1082. MR MR1978047 (2004b:20057)

63. F. Thomas Farrell, *The second cohomology group of G with Z_2G coefficients*, Topology **13** (1974), 313–326. MR MR0360864 (50 #13311)

64. ———, *Poincaré duality and groups of type* (FP), Comment. Math. Helv. **50** (1975), 187–195. MR MR0382479 (52 #3362)

65. Peter Freyd and Alex Heller, *Splitting homotopy idempotents. II*, J. Pure Appl. Algebra **89** (1993), no. 1-2, 93–106. MR MR1239554 (95h:55015)

66. K. Fujiwara and P. Papasoglu, *JSJ-decompositions of finitely presented groups and complexes of groups*, Geom. Funct. Anal. **16** (2006), no. 1, 70–125. MR MR2221253 (2007c:20100)

67. Ross Geoghegan, *A note on the vanishing of* lim^1, J. Pure Appl. Algebra **17** (1980), no. 1, 113–116. MR MR560787 (81i:18015a)

68. ———, *The shape of a group—connections between shape theory and the homology of groups*, Geometric and algebraic topology, Banach Center Publ., vol. 18, PWN, Warsaw, 1986, pp. 271–280. MR MR925870 (89a:20057)

69. Ross Geoghegan and Józef Krasinkiewicz, *Empty components in strong shape theory*, Topology Appl. **41** (1991), no. 3, 213–233. MR MR1135099 (92h:57034)

70. Ross Geoghegan and Michael L. Mihalik, *Free abelian cohomology of groups and ends of universal covers*, J. Pure Appl. Algebra **36** (1985), no. 2, 123–137. MR MR787167 (86h:20074)

71. ———, *A note on the vanishing of $H^n(G, \mathbf{Z}G)$*, J. Pure Appl. Algebra **39** (1986), no. 3, 301–304. MR MR821894 (87e:20094)

72. _____, *The fundamental group at infinity*, Topology **35** (1996), no. 3, 655–669. MR MR1396771 (97h:57002)

73. Brayton I. Gray, *Spaces of the same n-type, for all n*, Topology **5** (1966), 241–243. MR MR0196743 (33 #4929)

74. Marvin J. Greenberg and John R. Harper, *Algebraic topology*, Mathematics Lecture Note Series, vol. 58, Benjamin/Cummings Publishing Co. Inc. Advanced Book Program, Reading, Mass., 1981. MR MR643101 (83b:55001)

75. C. R. Guilbault, *A non-Z-compactifiable polyhedron whose product with the Hilbert cube is Z-compactifiable*, Fund. Math. **168** (2001), no. 2, 165–197. MR MR1852740 (2003a:57044)

76. Robin Hartshorne, *Algebraic geometry*, Springer-Verlag, New York, 1977. MR MR0463157 (57 #3116)

77. Allen Hatcher, *Algebraic topology*, Cambridge University Press, Cambridge, 2002. MR MR1867354 (2002k:55001)

78. Allen Hatcher and Karen Vogtmann, *Isoperimetric inequalities for automorphism groups of free groups*, Pacific J. Math. **173** (1996), no. 2, 425–441. MR MR1394399 (97f:20045)

79. Richard E. Heisey, *Manifolds modelled on the direct limit of lines*, Pacific J. Math. **102** (1982), no. 1, 47–54. MR MR682043 (84d:57009)

80. John Hempel, *3-manifolds*, AMS Chelsea Publishing, Providence, RI, 2004. MR MR2098385 (2005e:57053)

81. Graham Higman, *Finitely presented infinite simple groups*, Department of Pure Mathematics, Department of Mathematics, I.A.S. Australian National University, Canberra, 1974. MR MR0376874 (51 #13049)

82. P. J. Hilton, *An introduction to homotopy theory*, Cambridge Tracts in Mathematics and Mathematical Physics, no. 43, Cambridge, at the University Press, 1953. MR MR0056289 (15,52c)

83. P. J. Hilton and U. Stammbach, *A course in homological algebra*, Graduate Texts in Mathematics, vol. 4, Springer-Verlag, New York, 1997. MR MR1438546 (97k:18001)

84. P. J. Hilton and S. Wylie, *Homology theory: An introduction to algebraic topology*, Cambridge University Press, New York, 1960. MR MR0115161 (22 #5963)

85. Morris W. Hirsch, *Differential topology*, Graduate Texts in Mathematics, vol. 33, Springer-Verlag, New York, 1994. MR MR1336822 (96c:57001)

86. C. H. Houghton, *Ends of locally compact groups and their coset spaces*, J. Austral. Math. Soc. **17** (1974), 274–284. MR MR0357679 (50 #10147)

87. B. Hu, *Retractions of closed manifolds with nonpositive curvature*, Geometric group theory (Columbus, OH, 1992), Ohio State Univ. Math. Res. Inst. Publ., vol. 3, de Gruyter, Berlin, 1995, pp. 135–147. MR MR1355108 (96h:57023)

88. Sze-tsen Hu, *An exposition of the relative homotopy theory*, Duke Math. J. **14** (1947), 991–1033. MR MR0023052 (9,297h)

89. _____, *Theory of retracts*, Wayne State University Press, Detroit, 1965. MR MR0181977 (31 #6202)

90. J. F. P. Hudson, *Piecewise linear topology*, University of Chicago Lecture Notes prepared with the assistance of J. L. Shaneson and J. Lees, W. A. Benjamin, Inc., New York-Amsterdam, 1969. MR MR0248844 (40 #2094)

91. Bruce Hughes and Andrew Ranicki, *Ends of complexes*, Cambridge Tracts in Mathematics, vol. 123, Cambridge University Press, Cambridge, 1996. MR MR1410261 (98f:57039)

92. Witold Hurewicz and Henry Wallman, *Dimension Theory*, Princeton Mathematical Series, v. 4, Princeton University Press, Princeton, N. J., 1941. MR MR0006493 (3,312b)

93. Brad Jackson, *End invariants of group extensions*, Topology **21** (1982), no. 1, 71–81. MR MR630881 (83a:57002)

94. C. U. Jensen, *Les foncteurs dérivés de* $\varprojlim$ *et leurs applications en théorie des modules*, Springer-Verlag, Berlin, 1972. MR MR0407091 (53 #10874)

95. F. E. A. Johnson, *Manifolds of homotopy type $K(\pi, 1)$. I*, Proc. Cambridge Philos. Soc. **70** (1971), 387–393. MR MR0290358 (44 #7542)

96. Robion C. Kirby and Laurence C. Siebenmann, *Foundational essays on topological manifolds, smoothings, and triangulations*, Princeton University Press, Princeton, N.J., 1977. MR MR0645390 (58 #31082)

97. Akira Koyama, *Coherent singular complexes in strong shape theory*, Tsukuba J. Math. **8** (1984), no. 2, 261–295. MR MR767961 (86e:55012)

98. _____, *An example of coherent singular homology groups*, Proc. Japan Acad. Ser. A Math. Sci. **60** (1984), no. 9, 319–322. MR MR778517 (87g:55008)

99. Józef Krasinkiewicz and Piotr Minc, *Generalized paths and pointed 1-movability*, Fund. Math. **104** (1979), no. 2, 141–153. MR MR551664 (81b:55028)

100. P. H. Kropholler and M. A. Roller, *Relative ends and duality groups*, J. Pure Appl. Algebra **61** (1989), no. 2, 197–210. MR MR1025923 (91b:20069)

101. Serge Lang, *Algebra*, Addison-Wesley Publishing Co., Inc., Reading, Mass., 1965. MR MR0197234 (33 #5416)

102. Francisco F. Lasheras, *A note on fake surfaces and universal covers*, Topology Appl. **125** (2002), no. 3, 497–504. MR MR1935166 (2003j:57003)

103. Saul Lubkin, *A formula for* $\varprojlim^1$, Commutative algebra (Fairfax, Va., 1979), Lecture Notes in Pure and Appl. Math., vol. 68, Dekker, New York, 1982, pp. 257–273. MR MR655807 (83m:18015)

104. Wolfgang Lück, *L^2-invariants: theory and applications to geometry and K-theory*, Ergebnisse der Mathematik und ihrer Grenzgebiete. 3. Folge. A Series of Modern Surveys in Mathematics [Results in Mathematics and Related Areas. 3rd Series. A Series of Modern Surveys in Mathematics], vol. 44, Springer-Verlag, Berlin, 2002. MR MR1926649 (2003m:58033)

105. A. T. Lundell and S. Weingram, *The topology of CW complexes*, Van Nostrand Reinhold, New York, 1969.

106. Roger C. Lyndon and Paul E. Schupp, *Combinatorial group theory*, Classics in Mathematics, Springer-Verlag, Berlin, 2001. MR MR1812024 (2001i:20064)

107. Wilhelm Magnus, Abraham Karrass, and Donald Solitar, *Combinatorial group theory*, Dover Publications Inc., Mineola, NY, 2004. MR MR2109550 (2005h:20052)

108. Sibe Mardešić, *Shapes for topological spaces*, General Topology and Appl. **3** (1973), 265–282. MR MR0324638 (48 #2988)

109. Sibe Mardešić and Jack Segal, *Shape theory*, North-Holland Mathematical Library, vol. 26, North-Holland Publishing Co., Amsterdam, 1982. MR MR676973 (84b:55020)

110. William S. Massey, *Homology and cohomology theory*, Marcel Dekker Inc., New York, 1978. MR MR0488016 (58 #7594)

111. Michael Mather, *Counting homotopy types of manifolds*, Topology **3** (1965), 93–94. MR MR0176470 (31 #742)

112. Ralph McKenzie and Richard J. Thompson, *An elementary construction of unsolvable word problems in group theory*, Word problems: decision problems and the Burnside problem in group theory (Conf., Univ. California, Irvine, Calif. 1969; dedicated to Hanna Neumann), Studies in Logic and the Foundations of Math., vol. 71, North-Holland, Amsterdam, 1973, pp. 457–478. MR MR0396769 (53 #629)

113. D. R. McMillan, Jr., *Some contractible open 3-manifolds*, Trans. Amer. Math. Soc. **102** (1962), 373–382. MR MR0137105 (25 #561)

114. Michael L. Mihalik, *Semistability at the end of a group extension*, Trans. Amer. Math. Soc. **277** (1983), no. 1, 307–321. MR MR690054 (84d:57001)

115. _____, *Ends of groups with the integers as quotient*, J. Pure Appl. Algebra **35** (1985), no. 3, 305–320. MR MR777262 (86f:20060)

116. _____, *Semistability at infinity, simple connectivity at infinity and normal subgroups*, Topology Appl. **72** (1996), no. 3, 273–281. MR MR1406313 (97j:20035)

117. _____, *Semistability of Artin and Coxeter groups*, J. Pure Appl. Algebra **111** (1996), no. 1-3, 205–211. MR MR1394352 (97e:20060)

118. Michael L. Mihalik and Steven T. Tschantz, *One relator groups are semistable at infinity*, Topology **31** (1992), no. 4, 801–804. MR MR1191381 (95b:57005)

119. _____, *Semistability of amalgamated products and HNN-extensions*, Mem. Amer. Math. Soc. **98** (1992), no. 471, vi+86. MR MR1110521 (92k:57002a)

120. J. Milnor, *Morse theory*, Based on lecture notes by M. Spivak and R. Wells. Annals of Mathematics Studies, No. 51, Princeton University Press, Princeton, N.J., 1963. MR MR0163331 (29 #634)

121. Edwin E. Moise, *Geometric topology in dimensions 2 and 3*, Springer-Verlag, New York, 1977. MR MR0488059 (58 #7631)

122. Gabor Moussong, *Hyperbolic Coxeter groups*, University Microfilms, Ann Arbor, 1988.

123. James R. Munkres, *Topology: a first course*, Prentice-Hall Inc., Englewood Cliffs, N.J., 1975. MR MR0464128 (57 #4063)

124. Robert Myers, *Contractible open 3-manifolds which non-trivially cover only non-compact 3-manifolds*, Topology **38** (1999), no. 1, 85–94. MR MR1644087 (99g:57022)

125. Graham A. Niblo and Martin A. Roller, *Groups acting on cubes and Kazhdan's property (T)*, Proc. Amer. Math. Soc. **126** (1998), no. 3, 693–699. MR MR1459140 (98k:20058)

126. Gary O'Brien, *The missing boundary problem for smooth manifolds of dimension greater than or equal to six*, Topology Appl. **16** (1983), no. 3, 303–324. MR MR722123 (85m:57021)

127. Athanase Papadopoulos, *Metric spaces, convexity and nonpositive curvature*, IRMA Lectures in Mathematics and Theoretical Physics, vol. 6, European Mathematical Society (EMS), Zürich, 2005. MR MR2132506 (2005k:53042)

128. T. Porter, *Čech homotopy. I*, J. London Math. Soc. (2) **6** (1973), 429–436. MR MR0321080 (47 #9613)

129. Joseph S. Profio, *Using subnormality to show the simple connectivity at infinity of a finitely presented group*, Trans. Amer. Math. Soc. **320** (1990), no. 1, 281–292. MR MR961627 (90k:20057)

130. Charles Pugh and Michael Shub, *Axiom A actions*, Invent. Math. **29** (1975), no. 1, 7–38. MR MR0377989 (51 #14158)

131. J. Brendan Quigley, *Equivalence of fundamental and approaching groups of movable pointed compacta*, Fund. Math. **91** (1976), no. 2, 73–83. MR MR0413098 (54 #1219)

132. John G. Ratcliffe, *Foundations of hyperbolic manifolds*, Graduate Texts in Mathematics, vol. 149, Springer-Verlag, New York, 1994. MR MR1299730 (95j:57011)

133. E. Rips and Z. Sela, *Cyclic splittings of finitely presented groups and the canonical JSJ decomposition*, Ann. of Math. (2) **146** (1997), no. 1, 53–109. MR MR1469317 (98m:20044)

134. John Roe, *Lectures on coarse geometry*, University Lecture Series, vol. 31, American Mathematical Society, Providence, RI, 2003. MR MR2007488 (2004g:53050)

135. Dale Rolfsen, *Knots and links*, Mathematics Lecture Series, vol. 7, Publish or Perish Inc., Houston, TX, 1990. MR MR1277811 (95c:57018)

136. Colin Patrick Rourke and Brian Joseph Sanderson, *Introduction to piecewise-linear topology*, Springer Study Edition, Springer-Verlag, Berlin, 1982. MR MR665919 (83g:57009)

137. J. Hyam Rubinstein and Shicheng Wang, π_1-*injective surfaces in graph manifolds*, Comment. Math. Helv. **73** (1998), no. 4, 499–515. MR MR1639876 (99h:57039)

138. T. Benny Rushing, *Topological embeddings*, Academic Press, New York, 1973. MR MR0348752 (50 #1247)

139. Peter Scott, *Ends of pairs of groups*, J. Pure Appl. Algebra **11** (1977/78), no. 1-3, 179–198. MR MR487104 (81h:20047)

140. Peter Scott and Terry Wall, *Topological methods in group theory*, Homological group theory (Proc. Sympos., Durham, 1977), London Math. Soc. Lecture Note Ser., vol. 36, Cambridge Univ. Press, Cambridge, 1979, pp. 137–203. MR MR564422 (81m:57002)

141. Jean-Pierre Serre, *Arbres, amalgames,* SL$_2$, Société Mathématique de France, Paris, 1977. MR MR0476875 (57 #16426)

142. _____, *Trees*, Springer Monographs in Mathematics, Springer-Verlag, Berlin, 2003. MR MR1954121 (2003m:20032)

143. R. B. Sher, *Complement theorems in shape theory*, Shape theory and geometric topology (Dubrovnik, 1981), Lecture Notes in Math., vol. 870, Springer, Berlin, 1981, pp. 150–168. MR MR643529 (83a:57018)

144. L. C. Siebenmann, *The obstruction to finding a boundary for an open manifold of dimension greater than five - thesis submitted to Princeton university*, University Microfilms, Ann Arbor, 1966.

145. _____, *Infinite simple homotopy types*, Nederl. Akad. Wetensch. Proc. Ser. A 73 = Indag. Math. **32** (1970), 479–495. MR MR0287542 (44 #4746)

146. Edwin H. Spanier, *Algebraic topology*, Springer-Verlag, New York, 1981. MR MR666554 (83i:55001)

147. John Stallings, *Group theory and three-dimensional manifolds*, Yale University Press, New Haven, Conn., 1971. MR MR0415622 (54 #3705)

148. N. E. Steenrod, *A convenient category of topological spaces*, Michigan Math. J. **14** (1967), 133–152. MR MR0210075 (35 #970)

149. Norman Steenrod, *The topology of fibre bundles*, Princeton Landmarks in Mathematics, Princeton University Press, Princeton, NJ, 1999, Reprint of the 1957 edition, Princeton Paperbacks. MR MR1688579 (2000a:55001)

150. Melanie Stein, *Groups of piecewise linear homeomorphisms*, Trans. Amer. Math. Soc. **332** (1992), no. 2, 477–514. MR MR1094555 (92k:20075)

151. John C. Stillwell, *Classical topology and combinatorial group theory*, Graduate Texts in Mathematics, vol. 72, Springer-Verlag, New York, 1980. MR MR602149 (82h:57001)

152. G. A. Swarup, *On the ends of pairs of groups*, J. Pure Appl. Algebra **87** (1993), no. 1, 93–96. MR MR1222179 (94d:20037)

153. J. H. C. Whitehead, *A certain open manifold whose group is unity*, Quart. J. Math. **6** (1935), 268–279.

154. _____, *Combinatorial homotopy. I*, Bull. Amer. Math. Soc. **55** (1949), 213–245. MR MR0030759 (11,48b)

155. David G. Wright, *Contractible open manifolds which are not covering spaces*, Topology **31** (1992), no. 2, 281–291. MR MR1167170 (93f:57004)

156. Oscar Zariski and Pierre Samuel, *Commutative algebra. Vol. 1*, Springer-Verlag, New York, 1975. MR MR0384768 (52 #5641)

157. Smilka Zdravkovska, *An example in shape theory*, Proc. Amer. Math. Soc. **83** (1981), no. 3, 594–596. MR MR627700 (83b:55008)

Index

Graduate Texts in Mathematics

(continued from page ii)